T0213188

Image Processing

Tensor Transform and Discrete Tomography with MATLAB®

Image Processing

Tensor Transform and Discrete Tomography with MATLAB®

Artyom M. Grigoryan ▪ Merughan M. Grigoryan

CRC Press
Taylor & Francis Group
Boca Raton   London   New York

CRC Press is an imprint of the
Taylor & Francis Group, an **informa** business

CRC Press
Taylor & Francis Group
6000 Broken Sound Parkway NW, Suite 300
Boca Raton, FL 33487-2742

First issued in paperback 2017

© 2013 by Taylor & Francis Group, LLC
CRC Press is an imprint of Taylor & Francis Group, an Informa business

No claim to original U.S. Government works

ISBN-13: 978-1-4665-0994-8 (hbk)
ISBN-13: 978-1-138-07617-4 (pbk)

Library of Congress Cataloging-in-Publication Data

Grigoryan, Artyom.
    Image processing : tensor transform and discrete tomography with MATLAB /
Artyom Grigoryan, Merughan M. Grigoryan.
        p. cm.
    Includes bibliographical references and index.
    ISBN 978-1-4665-0994-8 (hardback)
    1. Image processing--Mathematics. 2. Image processing--Data processing. 3. Tensor algebra. 4. Geometric tomography. 5. MATLAB. I. Grigoryan, Merughan M. II. Title.

TA1637.G746 2012
006.4'2--dc23   2012025217

Visit the Taylor & Francis Web site at
http://www.taylorandfrancis.com

and the CRC Press Web site at
http://www.crcpress.com

TO STUDENTS IN ELECTRICAL ENGINEERING
COMPUTER SCIENCE
AND
RESEARCHERS WORKING IN COMPUTED TOMOGRAPHY

♣

Contents

Author Bios

Artyom M. Grigoryan received MS degrees in mathematics from Yerevan State University (YSU), Armenia, USSR, in 1978, in imaging science from Moscow Institute of Physics and Technology, USSR, in 1980, and in electrical engineering from Texas A&M University, USA, in 1999, and a Ph.D. degree in mathematics and physics from YUS, in 1990. In 1990-1996, he was a senior researcher with the Department of Signal and Image Processing at the Institute for Problems of Informatics and Automation, and Yerevan State University, National Academy Science of Armenia. In 1996-2000 he was a Research Engineer with the Department of Electrical Engineering, Texas A&M University. In December 2000, he joined the Department of Electrical Engineering, University of Texas at San Antonio, where he is currently an Associate Professor. He is author of the book entitled: *Multidimensional Discrete Unitary Transforms: Representation, Partitioning and Algorithms*, Marcel Dekker, Inc., 2003, and the textbook *Brief Notes in Advanced DSP: Fourier Analysis with MATLAB®*, CRC Press Taylor and Francis Group, 2009. He is the author of two book chapters and many journal papers and specializes in the theory and application of fast one- and multi-dimensional Fourier transforms, elliptic Fourier transforms, tensor and paired transforms, integer unitary heap transforms, design of robust linear and nonlinear filters, image cryptography, computerized 2-D and 3-D tomography, and processing of biomedical images.

Merughan M. Grigoryan received the MS degree in physics from Yerevan State University, Armenia, USSR, in 1979, and worked as a Postdoctoral Research Associate from 1979-1981 on the dispersion of ultrashort impulses in the Department of Radio-Physics and Electronics, YSU. From 1982-1995 he worked as a Senior Research Engineer in different science institutes as All-Union Scientific Associations "Astro," "Neitron," Scientific Research Institute of Non-Ferrous Metals (USSR), on topics which include electronics, signal and image processing, and acoustic emission. He is currently conducting a private research on the following topics: theory and application of quantum mechanics in signal processing, differential equations, Fourier analysis, elliptic Fourier transforms, fast integer unitary transformations, theory and methods of the fast unitary transforms generated by signals, and methods of encoding in cryptography. He is the coauthor of the book *Brief Notes in Advanced DSP: Fourier Analysis with MATLAB*, CRC Press Taylor and Francis Group, 2009.

Preface

This book is devoted to one of the most interesting applications of mathematical methods in digital image processing. It is computed tomography (CT) or computerized X-tomography, wherein the projection data of the reconstructing image are obtained by means of the roentgen radiation interaction with tissue. The result of the CT is a two-dimensional (2-D) or three-dimensional (3-D) image, which represents, with some degree of accuracy, the image through which the rays pass. Our task is to find the solution of the problem of image reconstruction by a finite number of projections. The fundamental principles and main methods of image reconstruction, which include the Fourier slice theorem, methods of summation and filtered backprojection, and the method of finite series expansion are well known and can be found in many books written on CT. The problem of reconstruction of the 2-D function from its ray-sums, or line-integrals, which was solved by Radon in 1917, faced its main obstacle in CT. The number of possible projections is finite. All known reconstructions in CT result in approximations of the image. The images in CT are not bounded, and Kotelnikov's theorem of sampling is not valid for them. We still do not know if the exact solution of the problem exists, and if it exists, then for what kind of images and models? Therefore, we will not follow the beaten path, but instead open slightly the door of the box where the solution of the problem can be found.

This book is a brief collection of some notes and results of our research in digital image reconstruction. We present the case of the 2-D image and the parallel projections. The model describing the process of projection data collection is simple. On both sides of the analyzed object, or image which is assumed to be immovable, the X-ray set and detectors are disposed and revolved around the object at different angles. Then, the measurement data of radiation and detection of X-rays are collected along other directions. This set of measurements, or projections, is used by specific mathematical methods to reconstruct the image of the observed object or tissue. Methods of image reconstruction must be fast and accurate because of the desired high-quality images for diagnostic purposes. More importantly, we need methods that reconstruct the image on the discrete Cartesian lattice with a minimal number of projections, in order to not overradiate the body in the CT.

Many concepts, ideas, and methods described in this book have not been presented or published anywhere else. This is written as a textbook with many examples described in detail and programs that are given in short form, to demonstrate the presented concepts, their properties, and methods of image reconstruction. New concepts include the methods of transferring the geometry of rays from the plane to the Cartesian lattice, the point map of projections, the particle and its field function, the statistical model of averaging, and others. Our goal is also to give graduate students and other readers solid material for the presented theory of image reconstruction, to benefit those interested in continuing this research and obtaining new results in image reconstruction.

The following describes the organization of the book. In Chapter 1 the concept of the 2-D tensor transform and splitting-signals of the image are described. The splitting of the 2-D discrete Fourier transform (DFT) by the tensor is analyzed. Chapter 2 is devoted to the paired representation of the image. This is the 2-D frequency and 1-D time transform of the image. The transform is not separable but unitary and leads to the effective calculation of the 2-D DFT. Both tensor and paired transforms are described by basic functions with binary coefficients located along the parallel rays; therefore, the image can be decomposed by direction images and used in the reconstruction. In Chapter 3, the method of transferring the geometry for calculating the paired transform of the image is presented. The examples for reconstructing images of sizes 4×4 and 8×8 are described in detail. The general algorithm for reconstructing the image on the Cartesian lattice of size $2^r \times 2^r$ and MATLAB®-based programs are given in Chapter 4. The case of images on a lattice of size $N \times N$, when N is prime, is described separately in Chapter 5. The tensor transform is calculated from the line-integrals and the reconstruction is obtained through the inverse tensor transform. In Chapter 6, the concept of particles and field functions and their application in reconstructing the image are described. The method of backprojection and new methods of summation are given in Chapter 7. The simple model with statistics is presented and applied for reconstructing the image on a Cartesian lattice of arbitrary size. Each chapter contains a list of problems that we suggest readers work on and solve. Difficult problems are marked with an asterisk; they require computations by hand and with a computer. In the Appendix, we include two translated papers published in the 1980s in Russian, wherein the concepts of the tensor and paired representation for images were presented.

We appreciate all who assisted in the preparation of the book. We are grateful to the reviewers for their suggestions and recommendations. Many thanks to Taylor & Francis/CRC Press, especially Ms. Nora Konopka, for giving us this great opportunity to prepare the main results of our research for publication. Finally, we express our gratitude and love of our families for their understanding, support, and extraordinary patience during the preparation of this book.

Artyom and Merughan M. Grigoryan
February 22, 2012.

Symbol Description

$f(x, y)$ — Image on square $[0, 1] \times [0, 1]$

$f_{n,m}$ — Discrete image

$F_{p,s}$ — 2-D DFT of the image $f_{n,m}$

F_p — 1-D DFT of the signal f_n

rect — Rectangle function on $[-0.5, 0.5]$

$X_{N,N}$ — Cartesian lattice $N \times N$

$\sigma = \sigma_J$ — Irreducible covering of the lattice

$\sigma' = \sigma_{J'}$ — Partition of the lattice

$J_{N,N}$ — Set of generators

$J'_{N,N}$ — Set of generators

$V_{p,s,t}$ — Set of pixels (n, m) of the lattice $X_{N,N}$, which satisfy the Diaphanous equation $np + ms = t \mod N$

$T_{p,s}$ — Cyclic group of frequencies generated by (p, s)

$T'_{p,s}$ — Subset (orbit) of frequencies generated by (p, s)

$f_{p,s,t}$ — Component of the splitting-signal in tensor representation

$f'_{p,s,t}$ — Component of the splitting-signal in paired representation

$\chi_{N,N}$ — Tensor transformation

$\chi'_{N,N}$ — 2-D paired transformation

$\chi_{p,s,t}$ — Basis function of the tensor transform

$\chi'_{p,s,t}$ — Basis paired function of the 2-D paired transform

$\chi'_{p,s,t;L}$ — L-paired function

$\chi'_{p,t}$ — Basis paired function of the 1-D paired transform

χ'_N — 1-D paired transformation

$\mathcal{A}_N$ — 1-D N-point DHdT

$\mathcal{F}_N$ — 1-D N-point DFT

$\mathcal{F}_{N,N}$ — 2-D $N \times N$-point DFT

$l = l_{p,s}$ — Line $np + ms = t$

$\Delta l_{p,s}$ — Length of intersection of the ray with the (n, m)-IE

$r_{p,s,t}$ — Wide ray

$\tilde{l} = \tilde{l}_{p,s}$ — Line $xp + my = t$

w_l — line-integral of the image along the l-ray

v_l — line-sum of the discrete image along the l-ray

b_t — Solution of the linear equation $\mathbf{b} = \mathbf{A}^{-1}\mathbf{w}$

$\mathbf{A}$ — Toeplitz matrix

A-ray — Arithmetical rays on the lattice

G-ray — Geometrical rays on the square $[0, 1] \times [0, 1]$

$CP(p, s)$ — Control points of the (p, s)-projection

$UH(K)$ — UH-square of the number K

$d^{p,s}_{n_1,n_2}$ — Direction image generated by (p, s)

$U_{N,N}$ — Set of triplets (p, s, t) of the 2-D paired transform

$U(p, s)$ — Set of triplets $2^k(p, s, t)$

$A(n, m)$ — Map of projections for point (n, m) of the lattice

$C(n, m)$ — Map of projections for G-particle $[n, m]$ on the square

$\psi_{n,m}$ — Field function of the A-particle (n, m)

$\phi_{n,m}$ — Field function of the G-particle $[n, m]$

Ψ — Base matrix of particles

$\Psi_{p,s}$ — 2-D DFT of the base matrix

R — Matrix of the field functions

$p^{(k)}_{p,s,t}$ — Probability density function

$ta(n, m)$ — Number $t(n, m; n_0, m_0)$ of A-rays passing trough the points (n, m) and (n_0, m_0) at the same time

$c(n, m)$ — Coefficient of intersection of rays with image element (n, m)

1

Discrete 2-D Fourier Transform

The theory of the continuous-time and discrete-time Fourier transformations has been well developed. The discrete Fourier transform has become a powerful technique in signal processing, and in particular in image processing. Effective methods, or fast algorithms, [1]-[6], of the discrete Fourier transforms (DFT) are used for solving many problems in image processing in the frequency domain, such as image filtration, restoration, enhancement, compression, and image reconstruction by projections [7]-[10]. Other unitary transformations are also used in signal and image processing. Considerable interest is given to many applications of the discrete Hartley transformation (DHT), since it relates closely to the DFT and has been created as an alternative form of the complex DFT to eliminate the necessity of using complex operations [11]-[15]. The discrete cosine transformation (DCT) is used in speech and image processing, especially in image compression and transform coding in telecommunication [16]-[22]. Another unitary transformation is the discrete Hadamard transformation (DHdT), whose basic functions take value ± 1 at each point [23]-[48]. The Hadamard transform has found useful applications in signal and image processing, communication systems, image coding, image enhancement, pattern recognition, and general two-dimensional filtering [2, 7].

The application of the two-dimensional transform involves the calculation of the transform, manipulation with transform coefficients, then calculation of the inverse transform. For images of large sizes, this process requires a great number of operations of multiplication and addition when performing the transforms. Different methods of effective calculation of two-dimensional unitary transforms have been developed to reduce the number of operations needed. In most cases, the calculation of the 2-D transform is reduced by partitioning the entire image by 1-D or 2-D blocks and calculating the transforms of these blocks. In the traditional "row-column" algorithm, 1-D transforms over all rows and then columns are calculated. For images of large sizes, the transposition slows down the process of calculating the transform. Other methods of fast calculation of the transform have been developed to avoid the transposition. We mention here the idea of generalization of the 2-D "butterfly" operation from the 1-D Cooley-Tukey algorithm [1] to the 4-D operation, when dividing the transforms by four parts of size $(N/2 \times N/2)$ each. This algorithm reduces the number of multiplications by almost 25%, when compared with the row-column method. The number of multiplications can be reduced approximately twice in the method of the polynomial transformations developed by Nussbaumer [38].

Our subject of study is the partitioning of two-dimensional discrete Fourier transform based on the concepts of the tensor and paired representations of images. In these representations, the images are described by sets of 1-D signals which carry the spectral information of images in different subsets of frequency-points. The processing of two-dimensional images is thus reduced to processing 1-D signals, called image-signals or splitting-signals, since they represent the images and split the transforms of these images. The splitting-signals are described here for the two-dimensional discrete Fourier transform, but they can be defined to other transforms, such as the Hartley, Hadamard, and cosine transforms [46, 48, 51, 55].

1.1   Separable 2-D transforms

Many of the two-dimensional transformations are separable, meaning that these transforms over 2-D images can be performed by calculating one-dimensional transforms consequently along both dimensions of the signal. For instance, for a separable two-dimensional (2-D) transformation T, the transform of a 2-D image $f = \{f_{n,m}\}$ of size $(N \times N)$, $N > 1$, can be obtained by first calculating the 1-D transforms over all columns of the image, and then calculating the 1-D transforms by the rows of the obtained 2-D data, as shown in Figure 1.1.

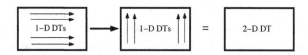

FIGURE 1.1
Block-diagram of calculation of the 2-D discrete transform (DT) (separable).

In matrix form, the transform of the image f can be written as

$$[\text{2-D } T][f] = [\text{1-D } T][f][\text{1-D } T]^t$$

where t denotes the matrix operation of transposition, and square brackets $[\cdot]$ are used to denote the matrices of the transformations T and image f.

As an example, we consider the 2-D DFT of the image $f_{n,m}$, which is defined by

$$F_{p,s} = (\mathcal{F}_{N,N} \circ f)_{p,s} = \sum_{n=0}^{N-1}\sum_{m=0}^{N-1} f_{n,m}W^{np+ms}, \quad p,s = 0 : (N-1), \quad (1.1)$$

where $W = W_N = \exp(-2\pi j/N)$ is the kernel of the transformation, and $j^2 = -1$. The designation $p = 0 : (N-1)$ denotes p as an integer that runs from 0 to $(N-1)$. (p, s) is the frequency-point. The kernel is separable, $W^{np+ms} = W^{np}W^{ms}$, and the transform can thus be written as

$$F_{p,s} = \sum_{n=0}^{N-1}\left[F_n(s) = \sum_{m=0}^{N-1} f_{n,m}W^{ms}\right]W^{np}, \quad p, s = 0 : (N-1), \qquad (1.2)$$

where $F_n(s)$ is the value of the 1-D DFT of row number n at point s. To calculate the 2-D DFT, $2N$ 1-D DFTs are used in the row-column algorithm. This algorithm is simple, but requires many operations of multiplication and addition. All twiddle coefficients, W^t, $t = 0 : (N-1)$, lie on N equidistant points of the unit circle, and many of them are irrational numbers.

We now consider the transformation whose kernel lies only on two points ± 1 on the unit circle. The 2-D separable discrete Hadamard transform of order $N \times N$, where $N = 2^r, r > 1$, is defined as

$$\begin{aligned}
A_{p,s} = (\mathcal{A}_{N,N} \circ f)_{p,s} &= \sum_{n=0}^{N-1}\sum_{m=0}^{N-1} f_{n,m}a(p;n)a(s;m) \\
&= \sum_{n=0}^{N-1}\left[\sum_{m=0}^{N-1} f_{n,m}a(s;m)\right]a(p;n).
\end{aligned} \qquad (1.3)$$

The kernel of the transformation is defined by the binary function

$$a(p;n) = (-1)^{n_0 p_0 + n_1 p_1 + \ldots + n_r p_r} \qquad (1.4)$$

where $(n_0, n_1, ..., n_r)$ and $(p_0, p_1, ..., p_r)$ are the binary representation of numbers n and p, respectively.

As an example, Figure 1.2 shows an image (2048×2048) in part a, along with the 2-D discrete Fourier and Hadamard transforms of the image in b and c, respectively. The realization of the Hadamard transformation requires only operations of addition (and subtraction). From the computational point of view, the 1-D Hadamard transform is faster than the complex Fourier transform. These two different transforms can share the same fast algorithm. For instance, the fast Fourier transform by the paired transforms can also be used for the fast Hadamard transform, when considering all twiddle coefficients W^t equal 1 [45].

We now present the tensor approach and its improvement, for dividing the calculation of the 2-D DFT into the minimal number of short 1-D transforms. The approach is universal because it can be implemented to calculate other discrete unitary transforms, such as the Hadamard, cosine, and Hartley transforms [39, 45, 48], and transforms of high dimensions.

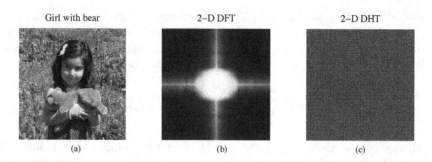

Girl with bear   2–D DFT   2–D DHT

(a)   (b)   (c)

FIGURE 1.2
(a) Original image of size 2048 × 2048, (b) 2-D DFT (in absolute mode and shifted to the center), and (c) 2-D DHdT of the image.

1.2   Vector forms of representation

When processing the two-dimensional image $f_{n,m}$, in the frequency domain by a specific unitary transformation, let's say the 2-D discrete Fourier transformation, the image can be represented in a form that splits the structure of the transformation in a way that yields an effective method of calculation of the transform with the following image processing. Such forms are not necessarily of the matrix form, but other figures. The presented work here does not rely on traditional methods of processing 2-D transforms and signals, but more effective methods which are based on the discovery that can be formulated briefly as follows. Two-dimensional spectra are split by appointed trajectories (such as orbits) and the movement of a spectral point along each such trajectory is of great interest in the process of formation of the spectra, as well as in processing the spectra. For many orders of the transform, trajectories do not intersect, and it is possible to extract the spectral information from such trajectories or change and put desired information into trajectories.

We discuss the theory of fast 2-D discrete Fourier transformation, which is based on the concept of partitions that reveal the two-dimensional and multi-dimensional transformations. The use of new forms of image and transform representation simplifies the calculation of 2-D transforms and leads to effective solutions of different problems in image processing, such as image enhancement, computerized tomography, image filtration, compression, and encryption. These new forms are the tensor and paired forms of representation. Their main task is to represent uniquely the image in the form of a set of 1-D signals which can be processed separately and then transferred back to the image, as shown in the diagram of Figure 1.3 (with or without block 2). The calculation of the 2-D DFT is reduced to calculation of 1-D DFTs and processing of the 2-D image to processing all or a few 1-D signals. The pro-

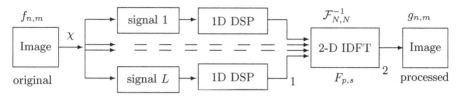

FIGURE 1.3
Block diagram of image processing by 1-D signals. (DSP = digital signal processing)

cessed image can be calculated by inverting the 2-D DFT, or by calculating directly from the processed 1-D signals. The mathematical structure of the 2-D DFT and other unitary transforms possess such representations [46, 48].

As an example, we consider the image enhancement by one such signal, instead of calculating the 2-D DFT of the image and manipulating all of its coefficients. Figure 1.4 shows the original image of size 256×256 in part a, along with a 1-D signal derived from the image in b, the 1-D DFT of the signal (in absolute scale and shifted to the center) in c, the coefficients to be multiplied pointwise by the 1-D DFT in d, a new 1-D signal in e, 256 frequency-points at which the spectral information of the new signal will be renewed in the 2-D DFT of the image in f, and the inverse 2-D DFT, or the enhanced image in g. The enhanced image can be obtained by the inverse 2-D DFT, as well as directly from the new signal (in e) [50, 49, 55]. Thus the problem of the 2-D image enhancement is reduced to processing one 1-D signal (or a few such signals), by passing the calculation of the 2-D DFT of the original image, as well as the inverse 2-D DFT (2-D IDFT) for the enhanced image. We now describe methods of deriving such 1-D signals, which lead to effective calculating and processing of the 2-D DFT.

1.3   Partitioning of 2-D transforms

Let a sequence $g = \{g_0, g_1, \ldots, g_{N-1}\}$ of length $N > 1$ be linearly and uniquely expressed by a sequence $f = \{f_0, f_1, \ldots, f_{N-1}\}$,

$$g_p = \sum_{n=0}^{N-1} \varphi_p(n) f_n, \quad p = 0 : (N-1). \tag{1.5}$$

The transformation of f into g, by using this equation is called a *linear transformation*, which we denote by $\mathcal{A}$. The set of coefficients $a_{p,n} = \varphi_p(n)$ forms the square $(N \times N)$ matrix $A = \| a_{p,n} \|$ of the transformation. The linear

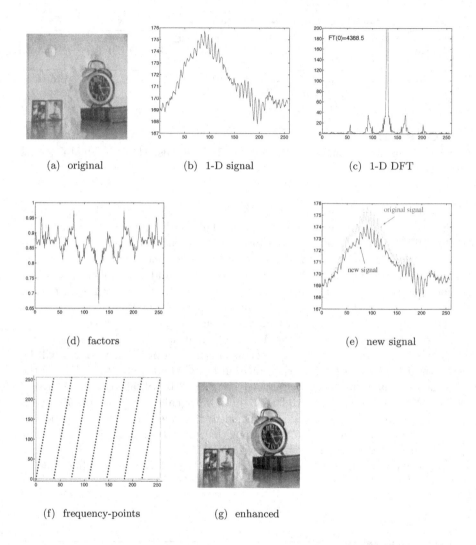

(a) original   (b) 1-D signal   (c) 1-D DFT

(d) factors   (e) new signal

(f) frequency-points   (g) enhanced

FIGURE 1.4 (See color insert)
Fast transform-based method of image enhancement. (a) The original image,
(b) the image-signal, (c) the magnitude spectrum of the signal, (d) factors, (e)
the processed signal, (f) marked locations of 256 frequency-points, at which
the 2-D DFT was changed, and (g) the image enhanced by one signal.

transformation can be written in matrix representation $[g] = A[f]$, where $[g]$
and $[f]$ denote the vector-columns of sequences g and f, respectively. Every
linear 1-D transformation uniquely determines a 2-D matrix A, and vice versa,
every 2-D matrix A determines the linear 1-D transformation. Similarly, each
four-dimensional matrix determines a certain linear transformation $\mathcal{A}$ of two-

dimensional sequences, or images

$$\mathcal{A} : f = \{f_{n_1, n_2}\} \rightarrow g = \{g_{p_1, p_2}\} \tag{1.6}$$

where $n_k, p_k = 0 : (N_k - 1)$ and $N_k > 1$, $k = 1 : 2$. The numbers N_1 and N_2 are called *orders* of the two-dimensional transformation $\mathcal{A} = \mathcal{A}_{N_1, N_2}$. $g = \mathcal{A} \circ f$ is called the two-dimensional transform of f.

The transform of the image f is described by the following relation:

$$g_{p_1, p_2} = \sum_{n_1=0}^{N_1-1} \sum_{n_2=0}^{N_2-1} \varphi_{p_1, p_2}(n_1, n_2) f_{n_1, n_2} \tag{1.7}$$

where $a_{p_1, p_2, n_1, n_2} = \varphi_{p_1, p_2}(n_1, n_2)$ are coefficients of the matrix of the transformation $\mathcal{A}$. Elements $(p_1, p_2) \in X$ are referred to as frequency-points. We assume that the image f and the transform $\mathcal{A} \circ f$ are defined on the same two-dimensional rectangular integer lattice of size $N_1 \times N_2$,

$$\begin{aligned} X = X_{N_1, N_2} &= \{(n_1, n_2); \; n_k = 0 : (N_k - 1), \; k = 1 : 2\} \\ &= \{(p_1, p_2); \; p_k = 0 : (N_k - 1), \; k = 1 : 2\}. \end{aligned} \tag{1.8}$$

This set X is called *the fundamental period* of the transformation $\mathcal{A}$.

The matrix of the $N \times N$-point discrete Fourier transformation $\mathcal{F}_{N,N}$ is the four-dimensional matrix

$$\left[\mathcal{F}_{N,N}\right] = \| a_{p_1, p_2, n_1, n_2} \| = \| W^{n_1 p_1 + n_2 p_2} \| .$$

The Fourier transformation is unitary. A discrete transformation $\mathcal{A}$ is called unitary if the matrix of the transformation is unitary, i.e., $AA^* = I$, where I is the identity matrix, and $A^* = \| \bar{a}_{n_1, n_2, p_1, p_2} \|$ is the complex conjugate to A, where the sign $^-$ denotes the transition to the complex conjugate value.

For a fixed frequency-point (p_1, p_2), the function $\varphi_{p_1, p_2}(n_1, n_2)$ is *the basis function* of the two-dimensional transformation $\mathcal{A}$, and the collection of such functions $\{\varphi\} = \{\varphi_{p_1, p_2}(n_1, n_2)\}$ is the *basis*, or *kernel* of $\mathcal{A}$. The unitary property of the transformation can be expressed as

$$\sum_{(n_1, n_2) \in X} \varphi_{p_1, p_2}(n_1, n_2) \bar{\varphi}_{s_1, s_2}(n_1, n_2) = \delta(p_1, s_1) \delta(p_2, s_2) \tag{1.9}$$

$$(p_1, p_2), \; (s_1, s_2) \in X,$$

where δ denotes the delta symbol, $\delta(p, s) = 1$, if $p = s$, and $\delta(p, s) = 0$, otherwise. If the collection of functions $\{\varphi\}$ satisfies this condition, then $\{\varphi\}$ is said to be a *complete* and *orthonormal* set of functions, and it is *orthogonal* if there is a factor different from 1 at $\delta(p_1, s_1) \delta(p_2, s_2)$ in (1.9). Thus, for the unitary transformation $\mathcal{A}$, this collection of functions is a complete and orthonormal set of basis functions.

Example 1.1 (1-D Fourier transformation) Let f be a 1-D sequence $f = \{f_0, f_1, ..., f_{N-1}\}$. The N-point Fourier transform of the sequence f is defined by

$$F_p = (\mathcal{F}_N \circ f)_p = \frac{1}{\sqrt{N}} \sum_{n=0}^{N-1} W^{np} f_n, \quad p = 0 : (N-1). \tag{1.10}$$

The basis functions $\varphi_p(n) = W^{np}$ are the pairs $(1/\sqrt{N} \cos(\omega_p n), 1/\sqrt{N} \sin(\omega_p n))$ of discrete-time cosine and sine waves with frequencies $\omega_p = (2\pi/N)p$. The waves are orthonormal, since

$$\sum_{n=0}^{N-1} \varphi_p(n) \bar{\varphi}_s(n) = \frac{1}{N} \sum_{n=0}^{N-1} W^{n(p-s)}$$

$$= \frac{1}{N} \sum_{n=0}^{N-1} \left[\cos \frac{2\pi(p-s)}{N} n - j \sin \frac{2\pi(p-s)}{N} n \right] = \begin{cases} 1, p = s \\ 0, p \neq s \end{cases}$$

and the transformation is unitary. The matrices of the transformation and its conjugate are symmetric,

$$[\mathcal{F}_N] = \frac{1}{\sqrt{N}} \parallel e^{-\frac{j2\pi}{N} np} \parallel_{n,p=0:(N-1)}, \quad [\mathcal{F}_N^*] = \frac{1}{\sqrt{N}} \parallel e^{\frac{j2\pi}{N} np} \parallel_{n,p=0:(N-1)},$$

and $[\mathcal{F}_N][\mathcal{F}_N^*] = I$. The conjugate matrix is thus the matrix of the inverse 1-D DFT.

The orthogonality of the basis functions of the 2-D DFT follows directly from the orthogonality of the basis functions of the 1-D DFT, because of separability $\varphi_{p,s}(n, m) = \varphi_p(n) \varphi_s(m)$. Indeed, the following calculations hold:

$$\sum_{n=0}^{N-1} \sum_{m=0}^{N-1} \varphi_{p_1,s_1}(n, m) \bar{\varphi}_{p_2,s_2}(n, m) = \sum_{n=0}^{N-1} \sum_{m=0}^{N-1} \varphi_{p_1}(n) \varphi_{s_1}(m) \bar{\varphi}_{p_2}(n) \bar{\varphi}_{s_2}(m)$$

$$= \sum_{n=0}^{N-1} \varphi_{p_1}(n) \bar{\varphi}_{p_2}(n) \sum_{m=0}^{N-1} \varphi_{s_1}(m) \bar{\varphi}_{s_2}(m) = \delta(p_1, p_2) \delta(s_1, s_2).$$

1.3.1   Tensor representation

In this section, we describe a concept of covering that reveals the mathematical structure of many multidimensional transforms [39]-[48]. This concept will be applied for the 2-D discrete Fourier transformation. The covering is considered to be composed of cyclic groups of frequency-points. Such covering leads to the tensor, or vector representation of images.

Suppose $\sigma = (T)$ is an irreducible covering of an 2-D lattice $X = X_{N_1,N_2}$. It means the set-theoretic union of all subsets T coincides with X and any smaller family of subsets of T from σ does not cover X. We use *card* to denote the cardinality of a set. If a discrete 2-D unitary transformation with the fundamental period X can be split into a set of $\operatorname{card} \sigma$ one-dimensional unitary transformations $\mathcal{A}$, then we say that the considered 2-D transformation *is revealed* by the covering σ, or, the covering σ *reveals* the transformation. Let f be an $N_1 \times N_2$ sequence, or image.

Definition 1.1 2-D $N_1 \times N_2$-point discrete transformation $\mathcal{P}$ is said to be *revealed* by a covering σ of X if, for each set $T \in \sigma$, there exists a 1-D orthogonal transformation $\mathcal{A} = \mathcal{A}(T)$ and a sequence f_T such that the restriction of the transform $\mathcal{P} \circ f$ on the set of frequency-points T equals the transform $\mathcal{A} \circ f_T$, i.e.,

$$\left(\mathcal{P} \circ f\right)_{|_T} = \mathcal{A} \circ f_T. \tag{1.11}$$

Each 1-D transformation $\mathcal{A}$ is determined by the corresponding subset T, not f. The set of the 1-D transforms $\{\mathcal{A}(T); T \in \sigma\}$ is called *the splitting* of the 2-D transformation $\mathcal{P}$ by the covering σ and denoted by $\mathcal{R}(\mathcal{P}; \sigma)$. The set of 1-D sequences $\{f_T; T \in \sigma\}$ is the σ-*representation* of f with respect to the transformation $\mathcal{P}$.

It should be noted that the covering σ not only splits the 2-D transformation by 1-D transformations, but also determines the corresponding representation of the 2-D image f as the set of 1-D sequences f_T. In other words, two representations are defined, one for the given image and another for its transform,

$$f \to \{f_T; T \in \sigma\}, \quad \text{and} \quad \mathcal{P} \circ f \to \{\mathcal{A} \circ f_T; T \in \sigma\}. \tag{1.12}$$

1.3.2 Covering with cyclic groups

We consider an irreducible covering σ composed only from additive cyclic groups

$$\sigma = \sigma_J = \left(T_{p_1, p_2}\right)_{(p_1, p_2) \in J} \tag{1.13}$$

with generators (p_1, p_2) from a certain subset $J \subset X = X_{N_1, N_2}$. The cyclic group T with a generator (p_1, p_2) is defined as a set of frequency-points which are integer multiples to the generator,

$$T = T_{p_1, p_2} = \{(\overline{kp_1}, \overline{kp_2}); \; k = 0 : (\operatorname{card} T - 1)\} \tag{1.14}$$

We use the short notation $\overline{kp_i} = (kp_i) \bmod N_i$ for $i = 1 : 2$.

Example 1.2 ((*Lattice* 3×3)) The covering of the lattice 3×3 can be defined by $\sigma = (T_{1,0}, T_{1,1}, T_{1,2}, T_{0,1})$ as shown

$$
\underbrace{\begin{bmatrix} \bullet & \bullet & \bullet \\ \bullet & \bullet & \bullet \\ \bullet & \bullet & \bullet \end{bmatrix}}_{X_{3,3}} = \underbrace{\begin{bmatrix} \bullet & \circ & \circ \\ \bullet & \circ & \circ \\ \bullet & \circ & \circ \end{bmatrix}}_{T_{1,0}}, \underbrace{\begin{bmatrix} \bullet & \circ & \circ \\ \circ & \bullet & \circ \\ \circ & \circ & \bullet \end{bmatrix}}_{T_{1,1}}, \underbrace{\begin{bmatrix} \bullet & \circ & \circ \\ \circ & \circ & \bullet \\ \circ & \bullet & \circ \end{bmatrix}}_{T_{1,2}}, \underbrace{\begin{bmatrix} \bullet & \bullet & \bullet \\ \circ & \circ & \circ \\ \circ & \circ & \circ \end{bmatrix}}_{T_{0,1}}.
$$

When the generator is $(1,1)$, all elements of the group $T_{1,1}$ are located along the diagonal of the lattice. All elements of the group $T_{1,0}$ are located on the first row of the lattice, and the elements of group $T_{0,1}$ are located on the first column of the lattice.

All other groups $T_{p,s}$ coincide with the groups of the covering σ. For instance, $T_{0,2} = T_{0,3} = T_{0,1}$, $T_{2,2} = T_{3,3} = T_{1,1}$, and $T_{2,1} = T_{1,2}$.

Example 1.3 (*Lattice* 4×4) The following covering of the lattice 4×4 holds:

$$
\begin{bmatrix} \bullet & \circ & \circ & \circ \\ \bullet & \circ & \circ & \circ \\ \bullet & \circ & \circ & \circ \\ \bullet & \circ & \circ & \circ \end{bmatrix}, \begin{bmatrix} \bullet & \circ & \circ & \circ \\ \circ & \bullet & \circ & \circ \\ \circ & \circ & \bullet & \circ \\ \circ & \circ & \circ & \bullet \end{bmatrix}, \begin{bmatrix} \bullet & \circ & \circ & \circ \\ \circ & \circ & \bullet & \circ \\ \bullet & \circ & \circ & \circ \\ \circ & \circ & \bullet & \circ \end{bmatrix}, \begin{bmatrix} \bullet & \circ & \circ & \circ \\ \circ & \circ & \circ & \bullet \\ \circ & \circ & \bullet & \circ \\ \circ & \bullet & \circ & \circ \end{bmatrix}, \begin{bmatrix} \bullet & \bullet & \bullet & \bullet \\ \circ & \circ & \circ & \circ \\ \circ & \circ & \circ & \circ \\ \circ & \circ & \circ & \circ \end{bmatrix}, \begin{bmatrix} \bullet & \circ & \bullet & \circ \\ \circ & \circ & \circ & \circ \\ \circ & \bullet & \circ & \bullet \\ \circ & \circ & \circ & \circ \end{bmatrix}.
$$
$$\;\;\; T_{1,0} \qquad\qquad T_{1,1} \qquad\qquad T_{1,2} \qquad\qquad T_{1,3} \qquad\qquad T_{0,1} \qquad\qquad T_{2,1}$$

Example 1.4 (*Lattice* 5×5) Let X be the lattice 5×5, that corresponds to the $N_1 = N_2 = 5$ case. The group T_{p_1,p_2} with a generator $(p_1,p_2) \in X = X_{5,5}$ is

$$T_{p_1,p_2} = \{(0,0), (p_1,p_2), (\overline{2p_1, 2p_2}), (\overline{3p_1, 3p_2}), (\overline{4p_1, 4p_2})\}.$$

There are six groups T that compose an irreducible covering $\sigma = \sigma_J$ of X, namely

$$
\begin{aligned}
T_{0,1} &= \{(0,0), (0,1), (0,2), (0,3), (0,4)\} \\
T_{1,1} &= \{(0,0), (1,1), (2,2), (3,3), (4,4)\} \\
T_{2,1} &= \{(0,0), (2,1), (4,2), (1,3), (3,4)\} \\
T_{3,1} &= \{(0,0), (3,1), (1,2), (4,3), (2,4)\} \\
T_{4,1} &= \{(0,0), (4,1), (3,2), (2,3), (1,4)\} \\
T_{1,0} &= \{(0,0), (1,0), (2,0), (3,0), (4,0)\}
\end{aligned}
\tag{1.15}
$$

and the set of generators

$$J = J_{5,5} = \{(0,1), (1,1), (2,1), (3,1), (4,1), (1,0)\}. \tag{1.16}$$

Figure 1.5 shows the location of all frequency-points of these six groups. These groups intersect only at point $(0,0)$. If the covering composed of these groups reveals the 5×5-point 2-D transformation, then this transformation can be split by six five-point 1-D transformations. This transformation is the Fourier transformation. To calculate, for instance, the 5×5-point 2-D DFT, only six 1-D five-point transforms are required, instead of ten transforms in the "row-column" method.

In the general $N \times N$ case, the elements of the group T_{p_1,p_2} lie on parallel lines at angle $\theta = \tan^{-1}(p_2/p_1)$ to the horizontal axis. The number l of such lines is determined as follows. If $p_1 = 0$ or $p_2 = 0$, then $l = 1$. For other cases, let k_1 and k_2 be the smallest integers satisfying the relations $\overline{k_1 p_1} = \overline{k_2 p_2} = N - 1$. Then, $l = k_1/k_2$ with $k_1 \geq k_2$, and $l = k_2/k_1$ with $k_1 < k_2$.

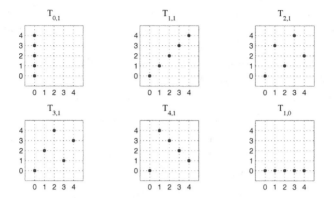

FIGURE 1.5
Arrangement of frequency-points of groups T_{p_1,p_2} covering the lattice 5×5.

For instance, when $N = 5$ and $(p_1, p_2) = (2, 1)$, we obtain $\overline{2p_1} = \overline{4p_2} = 4$ and $l = 4/2 = 2$. The frequency-points of the group $T_{2,1}$ lie on two parallel lines at angle $\theta = \tan^{-1}(1/2) = 26.5651°$ to the horizontal axis (see Figure 1.5). The points of this group can also be considered as lying on three parallel lines at angle $\theta_1 = \theta - 90° = -63.4349°$ to the horizontal axis.

It is important to note, that if we splice the opposite sides of the lattice bounds, then the lattice will be represented as a net traced on the surface of a three-dimensional torus and the mentioned l lines will compose a closed spiral on the torus, which will pass through those points on the net, which correspond to the points $(0, 0)$ and (p_1, p_2). All elements of the cyclic group will be the points of intersection of the spiral with the net. As an example,

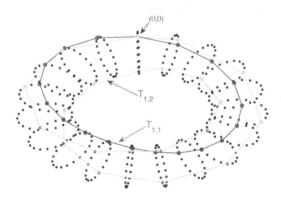

FIGURE 1.6 (See color insert)
Torus of the lattice 20×20 with two spirals corresponding to the groups $T_{1,1}$ and $T_{1,2}$.

Figure 1.6 shows the points of the lattice $X_{20,20}$ on the torus and two spirals with frequency-points of the groups $T_{1,1}$ and $T_{1,2}$ which intersect at the knot $(0,0)$.

The irreducible covering σ of the domain X composed from groups (1.14) is unique. To illustrate this property, we consider the square lattice $X_{5,5}$. The irreducible covering σ of $X_{5,5}$ is the family of six groups given in (1.15),

$$\sigma = (T_{0,1}, T_{1,1}, T_{2,1}, T_{3,1}, T_{4,1}, T_{1,0}).$$

The cyclic group T_{p_1,p_2} with any other generatrix $(p_1, p_2) \neq (0, 0)$, different from generators $(0,1), (1,1), (2,1), (3,1), (4,1)$ and $(1,0)$, coincides with one of the groups of the covering σ. For instance,

$$
\begin{aligned}
T_{1,2} &= \{(0,0),(1,2),(2,4),(3,1),(4,3)\} = T_{3,1}, \\
T_{3,2} &= \{(0,0),(3,2),(1,4),(4,1),(2,3)\} = T_{4,1}, \\
T_{2,2} &= \{(0,0),(2,2),(4,4),(1,1),(3,3)\} = T_{1,1}.
\end{aligned}
$$

1.4   Tensor representation of the 2-D DFT

In this section, we discuss in detail the construction and properties of the tensor representation of the image with respect to the two-dimensional discrete Fourier transformation.

Let $f = \{f_{n_1,n_2}\}$ be the image of size $N_1 \times N_2$, and let $N_0 = g.c.d.(N_1, N_2) > 1$, that is, $N_1 = N_0 N_1'$, $N_2 = N_0 N_2'$. Let σ be the irreducible covering of the rectangular lattice $X = X_{N_1,N_2}$, as defined in (1.13). We denote by $\mathcal{F} = \mathcal{F}_{N_1,N_2}$ the $N_1 \times N_2$-point 2-D discrete Fourier transformation. The 2-D DFT of the image f, accurate to the normalizing factor $1/N_1 N_2$, is defined by the following relation:

$$F_{p_1,p_2} = (\mathcal{F} \circ f)_{p_1,p_2} = \sum_{n_1=0}^{N_1-1} \sum_{n_2=0}^{N_2-1} f_{n_1,n_2} W_{N_1}^{n_1 p_1} W_{N_2}^{n_2 p_2}, \quad (p_1, p_2) \in X, \quad (1.17)$$

where $W_{N_k} = \exp(-2\pi j/N_k)$, $k = 1, 2$.

For an arbitrary frequency-point (p_1, p_2), we determine in the period X the sets of points

$$V_{p_1,p_2,t} = \{(n_1, n_2); \ N_2' n_1 p_1 + N_1' n_2 p_2 = t \bmod N\}, \quad t = 0 : (N-1), \quad (1.18)$$

where $N = N_1 N_2/N_0$. On these sets of points, we consider the sums of the sequence f, i.e., the following N quantities:

$$f_{p_1,p_2,t} = \sum \{f_{n_1,n_2}; \ (n_2, n_2) \in V_{p_1,p_2,t}\}, \quad t = 0 : (N-1). \quad (1.19)$$

For the spectral component F_{p_1,p_2}, the following calculations hold:

$$F_{p_1,p_2} = \sum_{n_1=0}^{N_1-1} \sum_{n_2=0}^{N_2-1} f_{n_1,n_2} W_N^{N_2' n_1 p_1 + N_1' n_2 p_2} = \sum_{t=0}^{N-1} f_{p_1,p_2,t} W^t, \quad (1.20)$$

where $W = W_N = e^{-j\frac{2\pi}{N}}$. The general equation is also valid,

$$F_{\overline{kp_1},\overline{kp_2}} = \sum_{t=0}^{N-1} f_{p_1,p_2,t} W^{kt}, \quad k = 0 : (N-1). \tag{1.21}$$

In other words, N components $F_{\overline{kp_1},\overline{kp_2}}$ of the 2-D DFT of the image f can be represented by the 1-D sequence of length N,

$$f_{T_{p_1,p_2}} = \{f_{p_1,p_2,0}, f_{p_1,p_2,1}, \ldots, f_{p_1,p_2,N-1}\}. \tag{1.22}$$

The sequence $f_{T_{p_1,p_2}}$ determines the spectral information of the image at frequency-points of the set T_{p_1,p_2}. Such a sequence is called *the splitting-signal*, or *the image-signal*. The components of the splitting-signal are numbered by the set of three, or triplets (p_1, p_2, t), where two components represent the frequency (p_1, p_2) and t is referred to as the time. Thus the splitting-signal is the (2-D frequency)-(1-D time) representation of the 2-D image f, which defines the full 2-D DFT of f at the frequency-points of the set T_{p_1,p_2}.

The set of splitting-signals

$$\{f_{T_{p_1,p_2}}; \; T_{p_1,p_2} \in \sigma\} \tag{1.23}$$

is called *the tensor representation*, or *tensor transform* of the image.

1.4.0.1 Code: Splitting-signal calculation

Below is the function with script ft_pst.m that accomplishes the calculation of the splitting-signal generated by (p, s). Here A stands for the image $f_{n,m}$ of size $N \times N$, and B for the splitting-signal.

```
% call: ft_pst        /   A.M. Grigoryan, 1996
function B=ft_pst(A,p,s)
   [M N]=size(A);
   B=zeros(1,N);
   ks=0;
   for m=1:M
      t=ks;
      for n=1:N
         if t>=N t=t-N; end
         t1=t+1;
         B(t1)=B(t1)+A(m,n); t=t+p;
      end
      ks=ks+s; if ks>=M ks=ks-M; end
   end
```

1.4.1 Tensor algorithm of the 2-D DFT

The algorithm of calculation of the 2-D discrete Fourier transform $\mathcal{F}_{N_1,N_2} \circ f$ of the image $f = f_{n_1,n_2}$ is performed by the following two steps.

Step 1. Calculate the 1-D splitting-signals f_T of the tensor representation of the image, i.e., calculate the tensor transform of the image,

$$\chi_\sigma : f \to \{f_T; T \in \sigma\}. \tag{1.24}$$

Step 2. Calculate the 1-D DFTs of the obtained splitting-signals, $\mathcal{F}_N(T)\circ$ $f_T, T \in \sigma$.

Step 3. Allocate the 1-D DFTs in the 2-D data by cyclic groups $T \in \sigma$,

$$F_{\overline{kp_1},\overline{kp_2}} = (\mathcal{F}_N \circ f_T)_k , \quad k = 0 : (N - 1),$$

where the number $N = N_1 N_2/g.c.d.(N_1, N_2)$.

The number of 1-D transforms required for calculating the 2-D DFT, coincides with the cardinality, $card\,\sigma$, of the covering σ, or the cardinality of the set J of generators of these groups,

$$\sigma = \sigma_J = \left(T_{p_1,p_2}\right)_{(p_1,p_2)\in J}. \tag{1.25}$$

We here separately consider the set of generators for the most interesting cases, when $N_1 = N_2 = N$, and N is a general prime, the product of two prime numbers, and then we describe the case when N is a power of prime.

1.4.2 N is prime

Let $N > 1$ be the prime. The irreducible covering σ_J of the set $X_{N,N}$ has the cardinality $N + 1$, i.e.,

$$card\,\sigma_J = N + 1. \tag{1.26}$$

Thus, the least number of cyclic groups $T_{p,s}$ that together cover the period X equals $(N + 1)$. Indeed, the irreducible covering σ_J is determined by the following set of generators:

$$J = J_{N,N} = \{(0,1), (1,1), (2,1), ..., (N - 1, 1), (1,0)\}. \tag{1.27}$$

Other sets of $(N + 1)$ generators can also be taken, for instance,

$$J = \{(1,0), (1,1), (1,2), ..., (1, N - 1), (0,1)\}.$$

Therefore, to calculate the $N \times N$-point 2-D DFT, it is sufficient to fulfill $(N + 1)$ one-dimensional N-point DFTs.

Example 1.5 Consider the 3×3-point DFT of the following two-dimensional sequence, or image:

$$f = \{f_{n_1,n_2}\} = \begin{bmatrix} 1 & 2 & 1 \\ 2 & 4 & 2 \\ 1 & 2 & 1 \end{bmatrix}.$$

The square lattice $X_{3,3} = \{(n_1, n_2); n_1, n_2 = 0, 1, 2\}$ is covered by the totality of sets $\sigma = (T_{0,1}, T_{1,1}, T_{2,1}, T_{1,0})$. The tensor representation of the image consists of four splitting-signals,

$$\chi_\sigma : \{f_{n_1,n_2}\} \to \{\{f_{0,1,t}\}, \{f_{1,1,t}\}, \{f_{2,1,t}\}, \{f_{1,0,t}\}\}, \quad t = 0, 1, 2. \quad (1.28)$$

Step 1: We denote by $\mathbf{f}$ the vector-column composed from rows of the sequence f_{n_1,n_2}, i.e., $\mathbf{f} = (1, 2, 1, 2, 4, 2, 1, 2, 1)'$. The first splitting-signal $\{f_{0,1,t}\}$ is calculated by

$$\begin{bmatrix} f_{0,1,0} \\ f_{0,1,1} \\ f_{0,1,2} \end{bmatrix} = \begin{bmatrix} 100100100 \\ 010010010 \\ 001001001 \end{bmatrix} \mathbf{f} = [4, 8, 4], \quad (1.29)$$

and the next three splitting-signals are calculated as follows:

$$\begin{bmatrix} f_{1,1,0} \\ f_{1,1,1} \\ f_{1,1,2} \end{bmatrix} = \begin{bmatrix} 100001010 \\ 010100001 \\ 001010100 \end{bmatrix} \mathbf{f} = [5, 5, 6], \quad (1.30)$$

$$\begin{bmatrix} f_{2,1,0} \\ f_{2,1,1} \\ f_{2,1,2} \end{bmatrix} = \begin{bmatrix} 100010001 \\ 010001100 \\ 001100010 \end{bmatrix} \mathbf{f} = [6, 5, 5], \quad (1.31)$$

$$\begin{bmatrix} f_{1,0,0} \\ f_{1,0,1} \\ f_{1,0,2} \end{bmatrix} = \begin{bmatrix} 111000000 \\ 000111000 \\ 000000111 \end{bmatrix} \mathbf{f} = [4, 8, 4]. \quad (1.32)$$

Step 2: The three-point DFT of the first signal $\{f_{0,1,0}, f_{0,1,1}, f_{0,1,2}\} = [4, 8, 4]$ is calculated by

$$\begin{bmatrix} A_0 \\ A_1 \\ A_2 \end{bmatrix} = \begin{bmatrix} 1 & 1 & 1 \\ 1 & W & W^2 \\ 1 & W^2 & W \end{bmatrix} \begin{bmatrix} 4 \\ 8 \\ 4 \end{bmatrix} = \begin{bmatrix} 16 \\ -2.0 - j3.4641 \\ -2.0 + j3.4641 \end{bmatrix},$$

where $W = \exp(-j2\pi/3)$. The sum of each splitting-signal equals 16, and there is no need to calculate the three-point DFTs of other splitting-signals at zero. Therefore, we can perform three incomplete three-point DFTs of these splitting-signals,

$$\begin{bmatrix} B_1 \\ B_2 \end{bmatrix} = \begin{bmatrix} 1 & W & W^2 \\ 1 & W^2 & W \end{bmatrix} \begin{bmatrix} 5 \\ 5 \\ 6 \end{bmatrix} = \begin{bmatrix} -0.5 + j0.8660 \\ -0.5 - j0.8660 \end{bmatrix},$$

$$\begin{bmatrix} C_1 \\ C_2 \end{bmatrix} = \begin{bmatrix} 1 & W & W^2 \\ 1 & W^2 & W \end{bmatrix} \begin{bmatrix} 6 \\ 5 \\ 5 \end{bmatrix} = \begin{bmatrix} 1 \\ 1 \end{bmatrix},$$

$$\begin{bmatrix} D_1 \\ D_2 \end{bmatrix} = \begin{bmatrix} 1 & W & W^2 \\ 1 & W^2 & W \end{bmatrix} \begin{bmatrix} 4 \\ 8 \\ 4 \end{bmatrix} = \begin{bmatrix} -2.0 - j3.4641 \\ -2.0 + j3.4641 \end{bmatrix}.$$

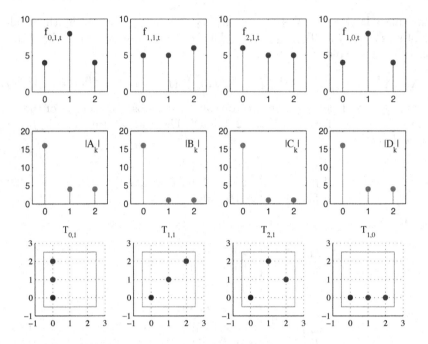

FIGURE 1.7
Four splitting-signals (the first row), the 3-point DFTs (in absolute scale) of the splitting-signals (the second rows), and the location of the frequency-points of the cyclic groups $T \in \sigma$ (the third row).

The splitting-signals and the magnitude spectrums of these signals are shown in the first and second rows of Figure 1.7, respectively.

Step 3: The location of frequency-points of four cyclic groups $T \in \sigma$ on the lattice 3×3, where the 1-D DFTs of the splitting-signals are placed, is shown in the last row of the figure. As a result, we obtain the following matrix expression for the 3×3-point DFT of the given sequence f:

$$\begin{bmatrix} F_{0,0} & F_{0,1} & F_{0,2} \\ F_{1,0} & F_{1,1} & F_{1,2} \\ F_{2,0} & F_{2,1} & F_{2,2} \end{bmatrix} = \begin{bmatrix} A_0 & A_1 & A_2 \\ D_1 & B_1 & C_2 \\ D_2 & C_1 & B_2 \end{bmatrix} = \begin{bmatrix} 16 & -2.0 - j3.4641 & -2 + j3.4641 \\ -2 - j3.4641 & -0.5 + j0.8660 & 1 \\ -2 + j3.4641 & 1 & -0.5 - j0.8660 \end{bmatrix}.$$

If we unite four binary matrices 9×3 in (1.29)-(1.32), we obtain the following matrix 12×9 of the tensor transformation in (1.28):

$$[\chi_\sigma] = \begin{bmatrix} 1 & 0 & 0 & 1 & 0 & 0 & 1 & 0 & 0 \\ 0 & 1 & 0 & 0 & 1 & 0 & 0 & 1 & 0 \\ 0 & 0 & 1 & 0 & 0 & 1 & 0 & 0 & 1 \\ 1 & 0 & 0 & 0 & 0 & 1 & 0 & 1 & 0 \\ 0 & 1 & 0 & 1 & 0 & 0 & 0 & 0 & 1 \\ 0 & 0 & 1 & 0 & 1 & 0 & 1 & 0 & 0 \\ 1 & 0 & 0 & 0 & 1 & 0 & 0 & 0 & 1 \\ 0 & 1 & 0 & 0 & 0 & 1 & 1 & 0 & 0 \\ 0 & 0 & 1 & 1 & 0 & 0 & 0 & 1 & 0 \\ 1 & 1 & 1 & 0 & 0 & 0 & 0 & 0 & 0 \\ 0 & 0 & 0 & 1 & 1 & 1 & 0 & 0 & 0 \\ 0 & 0 & 0 & 0 & 0 & 0 & 1 & 1 & 1 \end{bmatrix}. \tag{1.33}$$

In matrix form, the described tensor algorithm of the 3×3-point DFT can be written as

$$\begin{bmatrix} F_{0,0} \\ F_{0,1} \\ F_{0,2} \\ F_{1,1} \\ F_{2,2} \\ F_{2,1} \\ F_{1,2} \\ F_{1,0} \\ F_{2,0} \end{bmatrix} = \begin{bmatrix} 1 & 1 & 1 \\ 1 & W & W^2 \\ 1 & W^2 & W \\ & & & 1 & W & W^2 \\ & & & 1 & W^2 & W \\ & & & & & & 1 & W & W^2 \\ & & & & & & 1 & W^2 & W \\ & & & & & & & & & 1 & W & W^2 \\ & & & & & & & & & 1 & W^2 & W \end{bmatrix} [\chi_\sigma]\mathbf{f}.$$

The tensor algorithm of the 3×3-point DFT uses the following number of arithmetical operations: 4 real multiplications and 38 real additions, because the sequence f is real. Indeed, since $W^2 = -1 - W$, we have the following:

$$\begin{bmatrix} F_1 \\ F_2 \end{bmatrix} = \begin{bmatrix} 1 & W & W^2 \\ 1 & W^2 & W \end{bmatrix} \begin{bmatrix} x \\ y \\ z \end{bmatrix} = \begin{bmatrix} x + Wy - (z + Wz) \\ x - (y + Wy) + Wz \end{bmatrix}$$
$$= \begin{bmatrix} x - z + W(y - z) \\ x - y - W(y - z) \end{bmatrix}. \tag{1.34}$$

The incomplete 3-point Fourier transform can thus be calculated by one complex multiplication by $W^1 = (-\sqrt{3} - j)/2$. The three-point DFT of real data uses one operation of real multiplication, five additions, and one shifting. Three additions are required to calculate the incomplete three-point DFT, since $\bar{F}_2 = F_1$. Further, the direct calculation of the matrix $[\chi]$ of order 12×9 is fulfilled in the given example via 24 operations of real addition and subtraction. Therefore, the 3×3-point 2-D DFT requires $(5+3 \times 3)+24 = 38$ additions for the real sequence f. In the row-column algorithm, six 3-point DFTs are used, namely, three transforms with real inputs and three transforms with complex inputs. Therefore, the algorithm uses respectively $3 + 2(3) = 9$ real multiplications and $3 \times 5 + 3 \times 16 = 63$ additions, when data are real. Thus, in the tensor and polynomial algorithms, we get the advantage of the number of multiplications by 3 times, and 1.6 times for additions.

In the general case, for a prime number $N > 3$, in the traditional row-column algorithm, $2N$ one-dimensional N-point DFTs are used. Therefore, the tensor algorithm decreases the number of multiplications by $2N/(N+1)$ times, i.e., almost by 2 times, for large N. The tensor transform χ_σ requires $N^3 - N$ additions.

Example 1.6 (5×5-point DFT) Let $N = 5$ and f be the following image of size 5×5 :

$$f = \begin{vmatrix} \underline{1} & 2 & 1 & 3 & 1 \\ 2 & 0 & 1 & 1 & 2 \\ 1 & 3 & 2 & 2 & 1 \\ 4 & 1 & 0 & 1 & 3 \\ 2 & 4 & 1 & 2 & 1 \end{vmatrix}.$$

The underlined unit shows the location of the zero point. We consider the frequency-point $(p_1, p_2) = (2, 1)$. All values of t in the equations $n_1 p_1 + n_2 p_2 = t \bmod 5$ can be written in the following matrix:

$$\|t = (n_1 \cdot 2 + n_2 \cdot 1) \bmod 5\|_{n_2, n_1 = 0:4} = \begin{vmatrix} \underline{0} & 1 & 2 & 3 & 4 \\ 2 & 3 & 4 & 0 & 1 \\ 4 & 0 & 1 & 2 & 3 \\ 1 & 2 & 3 & 4 & 0 \\ 3 & 4 & 0 & 1 & 2 \end{vmatrix}.$$

Therefore, the components of the splitting-signal $f_{T_{2,1}}$ of f are defined as follows:

$$f_{T_{2,1}} = \begin{cases} f_{2,1,0} = f_{0,0} + f_{1,3} + f_{2,1} + f_{3,4} + f_{4,2} = 1 + 1 + 3 + 3 + 1 = 9 \\ f_{2,1,1} = f_{0,1} + f_{1,4} + f_{2,2} + f_{3,0} + f_{4,3} = 2 + 2 + 2 + 4 + 2 = 12 \\ f_{2,1,2} = f_{0,2} + f_{1,0} + f_{2,3} + f_{3,1} + f_{4,4} = 1 + 2 + 2 + 1 + 1 = 7 \\ f_{2,1,3} = f_{0,3} + f_{1,1} + f_{2,4} + f_{3,2} + f_{4,0} = 3 + 0 + 1 + 0 + 2 = 6 \\ f_{2,1,4} = f_{0,4} + f_{1,2} + f_{2,0} + f_{3,3} + f_{4,1} = 1 + 1 + 1 + 1 + 4 = 8 \end{cases}$$

and $f_{T_{2,1}} = \{9, 12, 7, 6, 8\}$. The sum of the this signal equals the sum of the image f, i.e.,

$$\sum_{t=0}^{4} f_{2,1,t} = 9 + 12 + 7 + 6 + 8 = 42 = \sum_{n_1=0}^{4} \sum_{n_2=0}^{4} f_{n_1, n_2}.$$

The five-point DFT of the splitting-signal equals

$$\{42, 4.6631 - 4.3920j, -3.1631 - 1.4001j, -3.1631 + 1.4001j, 4.6631 + 4.3920j\}.$$

This transform coincides with the 2-D DFT of the image f at frequency-points of the group

$$T_{2,1} = \{(0,0), (2,1), (4,2), (1,3), (3,4)\},$$

as shown in the following table:

$$\begin{bmatrix} 42 & 0 & 0 & 0 & 0 \\ 0 & 0 & 0 & -3.1631+1.4001j & 0 \\ 0 & 4.6631-4.3920j & 0 & 0 & 0 \\ 0 & 0 & 0 & 0 & 4.6631+4.3920j \\ 0 & 0 & -3.1631-1.4001j & 0 & 0 \end{bmatrix}.$$

We can fill the remaining values of the 2-D DFT, by calculating the 1-D DFTs of other splitting-signals. For instance, for the signal corresponding to the generator $(p_1, p_2) = (3, 1)$, we have the following table of time-points:

$$\|t = (n_1 \cdot 3 + n_1 \cdot 1) \bmod 5\|_{n_2, n_1 = 0:4} = \begin{vmatrix} 0 & 1 & 2 & 3 & 4 \\ 3 & 4 & 0 & 1 & 2 \\ 1 & 2 & 3 & 4 & 0 \\ 4 & 0 & 1 & 2 & 3 \\ 2 & 3 & 4 & 0 & 1 \end{vmatrix}.$$

The components of the splitting-signal $f_{T_{2,1}}$ are thus calculated as

$$f_{T_{3,1}} = \begin{cases} f_{3,1,0} = f_{0,0} + f_{1,2} + f_{2,4} + f_{3,1} + f_{4,3} = 1+1+1+1+2 = 6 \\ f_{3,1,1} = f_{0,1} + f_{1,3} + f_{2,0} + f_{3,2} + f_{4,4} = 2+1+1+0+1 = 5 \\ f_{3,1,2} = f_{0,2} + f_{1,4} + f_{2,1} + f_{3,3} + f_{4,0} = 1+2+3+1+2 = 9 \\ f_{3,1,3} = f_{0,3} + f_{1,0} + f_{2,2} + f_{3,4} + f_{4,1} = 3+2+2+3+4 = 14 \\ f_{3,1,4} = f_{0,4} + f_{1,1} + f_{2,3} + f_{3,0} + f_{4,2} = 1+0+2+4+1 = 8 \end{cases}$$

and $f_{T_{3,1}} = \{6, 5, 9, 14, 8\}$. The five-point DFT of this splitting-signal equals

$$\{42, -8.5902+5.7921j, 2.5902-2.9919j, 2.5902+2.9919j, -8.5902-5.7921j\}.$$

This transform defines the 2-D DFT of the image at frequency-points of the group $T_{3,1} = \{(0,0), (3,1), (1,2), (4,3), (2,4)\}$. In this stage, the 2-D DFT is filled as shown in the following table:

$$\begin{bmatrix} 42 & 0 & 0 & 0 & 0 \\ 0 & 0 & 2.5902-2.9919j & -3.1631+1.4001j & 0 \\ 0 & 4.6631-4.3920j & 0 & 0 & -8.5902-5.7921j \\ 0 & -8.5902+5.7921j & 0 & 0 & 4.6631+4.3920j \\ 0 & 0 & -3.1631-1.4001j & 2.5902+2.9919j & 0 \end{bmatrix}.$$

In a similar way, the 1-D DFTs of the splitting-signals $f_{T_{0,1}}, f_{T_{1,1}}, f_{T_{4,1}}$, and $f_{T_{1,0}}$ fill the rest of the table of the 2-D DFT of the image f. The tensor representation of the image is the following six splitting-signals:

$$f \rightarrow \begin{cases} f_{T_{0,1}} = \{10, 10, 5, 9, 8\} \\ f_{T_{1,1}} = \{9, 7, 7, 12, 7\} \\ f_{T_{2,1}} = \{9, 12, 7, 6, 8\} \\ f_{T_{3,1}} = \{6, 5, 9, 14, 8\} \\ f_{T_{4,1}} = \{5, 10, 11, 8, 8\} \\ f_{T_{1,0}} = \{8, 6, 9, 9, 10\} \end{cases}$$

1.4.2.1   Code: 2-D DFT by tensor transform

Below is the program with script TT2D_Nisprime.m, to accomplish the calculation of the $N \times N$-point 2-D DFT by the tensor transform. N is a prime.

```
% ==================================================================
% Call: TT2D_Nisprime.m      /    A.M. Grigoryan,   11/17/2001
% Demo code for calculating the NxN-point DFT by 2-D DTT.
% (N+1) N-point DFT are used, when N is a prime number.
% For each generator (p,s), two functions are called:
%    ft_pst(image,p,s)    - calculates the splitting-signal
%    cyclic_group(p,s,N) - calculates the cyclic group T(p,s)
% The "tree" image can be taken from the library of
% The Signal and Image Processing Institute, University of
% Southern California, by address http://sipi.usc.edu/database/
    fid=fopen('tree.img','rb');
    f=fread(fid,[256,256]); fclose(fid); clear fid; f=f';
% extend this image to the size 257x257 as
    f(:,257)=f(:,256);  f(257,:)=f(256,:);
% Set of generators J(257,257)
    N=257;   N1=N+1;
    Jps=ones(2,N1); Jps(1,1:N)=0:N-1; Jps(1:2,N1)=[1,0];
% 2-D DFT by the tensor transform
    FT=zeros(N);
    for k=1:N1
         s=Jps(1,k); p=Jps(2,k);
         Image_signal=zeros(1,N); Image_signal=ft_pst(f,p,s);
         TDFFT_signal=fft(Image_signal); T_ps=cyclic_group(p,s,N)+1;
         for k=1:N
              p1=T_ps(k,1); s1=T_ps(k,2); FT(s1,p1)=TDFFT_signal(k);
         end
    end
% Verify if the 2-D DFT has been calculated correctly
Inv_image=real(ifft2(FT));
Inv_image=Inv_image.*(Inv_image>=0); Inv_image=round(Inv_image);
% Display the original image and inverse 2-D DFT
figure; colormap(gray)
subplot(1,3,1); imagesc(f);
axis('image'); axis('off');  title('Original image');
```

```
subplot(1,3,2); image(abs(FT)/N);
axis('image'); axis('off');  title('2-D DFT by TT');
subplot(1,3,3); imagesc(Inv_image);
axis('image'); axis('off');  title('Inverse 2-D DFT');
% =============================================================
function T_ps=cyclic_group(p1,p2,N)
    T_ps=zeros(N,2);  p=0; s=0;
    for k=1:N
        if p>=N  p=p-N; end
        if s>=N  s=s-N; end
        T_ps(k,:)=[p,s]; p=p+p1; s=s+p2;
    end
end
```

This program also prints the original image, the 2-D DFT and its inverse, to verify the calculations.

1.4.3 N is a power of two

When $N = 2^r$, $r > 1$, the irreducible covering σ of the lattice $X = X_{2^r,2^r}$ can be taken as the following family of $3N/2$ cyclic groups:

$$\sigma_J = \left(\left(T_{p_1,1} \right)_{p_1=0:(2^r-1)}, \left(T_{1,2p_2} \right)_{p_2=0:(2^{r-1}-1)} \right). \tag{1.35}$$

Thus, to calculate the $2^r \times 2^r$-point 2-D DFT, $3 \cdot 2^{r-1}$ one-dimensional DFTs are used in the tensor algorithm, which is described by

$$\begin{aligned}
F_{\overline{kp},\overline{ks}} = (\mathcal{F}_{2^r,2^r} \circ f)_{\overline{kp},\overline{ks}} &= \sum_{n_1=0}^{2^r-1} \sum_{n_2=0}^{2^r-1} f_{n_1,n_2} W^{n_1 kp + n_2 ks} \\
&= (\mathcal{F}_{2^r} \circ f_T)_k = \sum_{t=0}^{2^r-1} f_{p,s,t} W^{kt}, \qquad k = 0 : (2^r - 1).
\end{aligned} \tag{1.36}$$

where the generators (p, s) are taken from the set

$$J = \{(0,1),(1,1),(2,1),...,(2^r-1,1)\} \cup \{(1,0),(1,2),(1,4),...,(1,2^r-2)\}. \tag{1.37}$$

The components of splitting-signals f_T are calculated by the characteristic functions of sets $V_{p,s,t}$,

$$f_{p,s,t} = \chi_{p,s,t} \circ f = \sum_{n_1=0}^{2^r-1} \sum_{n_2=0}^{2^r-1} \chi_{p,s,t}(n_1,n_2) f_{n_1,n_2} = \sum_{(n_1,n_2) \in V_{p,s,t}} f_{n_1,n_2}. \tag{1.38}$$

These binary functions determine the tensor transform χ_σ and are defined as

$$\chi_{p,s,t}(n_1,n_2) = \begin{cases} 1, & \text{if } n_1 p + n_2 s = t \bmod 2^r, \\ 0, & \text{otherwise,} \end{cases} \qquad (n_1,n_2) \in X. \tag{1.39}$$

All ones in the masks of the functions lie on parallel lines passing the knots of the corresponding sets $V_{p,s,t}$.

Example 1.7 (4×4-point DFT) Consider the following image 4×4 and its discrete Fourier transform:

$$\{f_{n,m}\} = \begin{vmatrix} 1 & 2 & 1 & 3 \\ 2 & 1 & 1 & 2 \\ 1 & 3 & 2 & 1 \\ 1 & 1 & 3 & 2 \end{vmatrix} \rightarrow \{F_{p,s}\} = \begin{vmatrix} 27 & j & 1 & -j \\ -2+j & 1 & -3j & 1+6j \\ -3 & -2+j & -5 & -2-j \\ -2-j & 1-6j & 3j & 1 \end{vmatrix}.$$

For $(p, s) = (1, 1)$, values of t in the equations $np + ms = t \bmod N$, $t = 0 : 3$, can be written in the form of the following matrix:

$$||t = (n \cdot 1 + m \cdot 1) \bmod 4||_{n,m=0:3} = \begin{vmatrix} 0 & 1 & 2 & 3 \\ 1 & 2 & 3 & 0 \\ 2 & 3 & 0 & 1 \\ 3 & 0 & 1 & 2 \end{vmatrix}.$$

Coefficients of this matrix show the points of sets $V_{1,1,t}$, $t = 0 : 3$. The image-signal $f_{T_{1,1}}$ is defined as

$$f_{T_{1,1}} = \begin{cases} f_{1,1,0} = f_{0,0} + f_{3,1} + f_{2,2} + f_{1,3} = 1 + 2 + 2 + 1 = 6 \\ f_{1,1,1} = f_{1,0} + f_{0,1} + f_{3,2} + f_{2,3} = 2 + 2 + 1 + 3 = 8 \\ f_{1,1,2} = f_{2,0} + f_{1,1} + f_{0,2} + f_{3,3} = 1 + 1 + 1 + 2 = 5 \\ f_{1,1,3} = f_{3,0} + f_{2,1} + f_{1,2} + f_{0,3} = 1 + 3 + 1 + 3 = 8. \end{cases} \quad (1.40)$$

One can note that

$$\sum_{t=0}^{3} f_{p,s,t} = \sum_{t=0}^{3} f_{1,1,t} = \sum_{n=0}^{3} \sum_{m=0}^{3} f_{n,m} = 27.$$

Therefore the last component $f_{1,1,3}$ can be calculated as $27 - (6 + 8 + 5) = 8$.

The 4-point DFT of the image-signal $f_{T_{1,1}}$ defines the 4×4-point DFT along the main diagonal as

$$f_{T_{1,1}} = \{6, 8, 5, 8\} \rightarrow F_k = \{27, 1, -5, 1\} \rightarrow \begin{vmatrix} 27 & \circ & \circ & \circ \\ \circ & 1 & \circ & \circ \\ \circ & \circ & -5 & \circ \\ \circ & \circ & \circ & 1 \end{vmatrix}.$$

For the generator $(p, s) = (1, 2)$, the corresponding matrix of equations $np + ms = t \bmod N$, $t = 0 : 3$, is calculated as follows:

$$||t = (n \cdot 1 + m \cdot 2) \bmod 4||_{n,m=0:3} = \begin{vmatrix} 0 & 2 & 0 & 2 \\ 1 & 3 & 1 & 3 \\ 2 & 0 & 2 & 0 \\ 3 & 1 & 3 & 1 \end{vmatrix}.$$

Therefore, the image-signal $f_{T_{1,2}}$ of $f_{n,m}$ is defined as

$$
f_{T_{1,2}} = \begin{cases} f_{1,2,0} = f_{0,0} + f_{2,1} + f_{0,2} + f_{2,3} = 1+1+1+3 = 6 \\ f_{1,2,1} = f_{1,0} + f_{3,1} + f_{1,2} + f_{3,3} = 2+2+3+2 = 9 \\ f_{1,2,2} = f_{2,0} + f_{0,1} + f_{2,2} + f_{0,3} = 1+2+2+1 = 6 \\ f_{1,2,3} = f_{3,0} + f_{1,1} + f_{3,2} + f_{1,3} = 3+1+1+1 = 6. \end{cases} \tag{1.41}
$$

The 4-point DFT of the image-signal $f_{T_{1,2}}$ defines the 4×4-point DFT for additional three samples of the set $T_{1,2} = \{(0,0),(1,2),(2,0),(3,2)\}$ as follows:

$$
f_{T_{1,2}} = \{6,9,6,6\} \rightarrow F_k = \{27,-3j,-3,3j\} \rightarrow \begin{vmatrix} 27 & \circ & \circ & \circ \\ \circ & \bullet & -3j & \circ \\ -3 & \circ & \bullet & \circ \\ \circ & \circ & 3j & \bullet \end{vmatrix}.
$$

To complete the 4×4-point DFT at frequency-points of groups $T_{1,0}, T_{1,3}, T_{0,1}$, and $T_{2,1}$, the four-point DFTs of the corresponding splitting-signals are used.

We also consider another example with different set of generators.

Example 1.8 (4×4-point DFT) Consider the following 2-D sequence f of size 4×4:

$$
f = \begin{vmatrix} f_{0,0} & f_{0,1} & f_{0,2} & f_{0,3} \\ f_{1,0} & f_{1,1} & f_{1,2} & f_{1,3} \\ f_{2,0} & f_{2,1} & f_{2,2} & f_{2,3} \\ f_{3,0} & f_{3,1} & f_{3,2} & f_{3,3} \end{vmatrix} = \begin{vmatrix} 1 & 2 & 1 & 3 \\ 2 & 0 & 1 & 1 \\ 1 & 3 & 2 & 2 \\ 2 & 4 & 1 & 2 \end{vmatrix}.
$$

The set of generators of the cyclic groups $T_{p,s}$ of the covering σ_J equals

$$
J = \{(0,1),(1,1),(2,1),(3,1)\} \cup \{(1,0),(1,2)\}. \tag{1.42}
$$

We first describe the splitting-signal corresponding to the frequency-point $(p,s) = (0,1)$. For that, we write all values of t in the equations $np + ms = t \bmod 4$ in the form of the following matrix:

$$
\|t = (n \cdot 0 + m \cdot 1) \bmod 4\|_{n,m=0:3} = \begin{vmatrix} 0 & 1 & 2 & 3 \\ 0 & 1 & 2 & 3 \\ 0 & 1 & 2 & 3 \\ 0 & 1 & 2 & 3 \end{vmatrix}.
$$

The components of this splitting-signal are calculated as follows:

$$
f_{0,1,0} = \chi_{0,1,0} \circ f = \begin{bmatrix} 1 & 0 & 0 & 0 \\ 1 & 0 & 0 & 0 \\ 1 & 0 & 0 & 0 \\ 1 & 0 & 0 & 0 \end{bmatrix} \circ \begin{bmatrix} 1 & 2 & 1 & 3 \\ 2 & 0 & 1 & 1 \\ 1 & 3 & 2 & 2 \\ 2 & 4 & 1 & 2 \end{bmatrix} = 1+2+1+2 = 6 \tag{1.43}
$$

$$
f_{0,1,1} = \chi_{0,1,1} \circ f = \begin{bmatrix} 0 & 1 & 0 & 0 \\ 0 & 1 & 0 & 0 \\ 0 & 1 & 0 & 0 \\ 0 & 1 & 0 & 0 \end{bmatrix} \circ \begin{bmatrix} 1 & 2 & 1 & 3 \\ 2 & 0 & 1 & 1 \\ 1 & 3 & 2 & 2 \\ 2 & 4 & 1 & 2 \end{bmatrix} = 2+0+3+4 = 9 \tag{1.44}
$$

$$f_{0,1,2} = \chi_{0,1,2} \circ f = \begin{bmatrix} \underline{0} & 0 & 1 & 0 \\ 0 & 0 & 1 & 0 \\ 0 & 0 & 1 & 0 \\ 0 & 0 & 1 & 0 \end{bmatrix} \circ \begin{bmatrix} \underline{1} & 2 & 1 & 3 \\ 2 & 0 & 1 & 1 \\ 1 & 3 & 2 & 2 \\ 2 & 4 & 1 & 2 \end{bmatrix} = 1 + 1 + 2 + 1 = 5 \quad (1.45)$$

$$f_{0,1,3} = \chi_{0,1,3} \circ f = \begin{bmatrix} \underline{0} & 0 & 0 & 1 \\ 0 & 0 & 0 & 1 \\ 0 & 0 & 0 & 1 \\ 0 & 0 & 0 & 1 \end{bmatrix} \circ \begin{bmatrix} \underline{1} & 2 & 1 & 3 \\ 2 & 0 & 1 & 1 \\ 1 & 3 & 2 & 2 \\ 2 & 4 & 1 & 2 \end{bmatrix} = 3 + 1 + 2 + 2 = 8. \quad (1.46)$$

The splitting-signal $f_{T_{0,1}} = \{6, 9, 5, 8\}$. The four-point DFT of this signal equals $(F_0, F_1, F_2, F_3) = (28, 1 - j, -6, 1 + j)$, which can be written in the table of the 2-D DFT of f at the frequency-points $(0,0), (0,1), (0,2)$, and $(0,3)$ as follows:

$$\begin{bmatrix} \underline{F_0} & F_1 & F_2 & F_3 \\ 0 & 0 & 0 & 0 \\ 0 & 0 & 0 & 0 \\ 0 & 0 & 0 & 0 \end{bmatrix} = \begin{bmatrix} \underline{28} & 1-j & -6 & 1+j \\ 0 & 0 & 0 & 0 \\ 0 & 0 & 0 & 0 \\ 0 & 0 & 0 & 0 \end{bmatrix}.$$

We also consider the splitting-signal corresponding to the next generator $(p, s) = (1, 1)$. For this generator, equations $np + ms = t \bmod 4$ result in the following matrix:

$$\|t = (n \cdot 1 + m \cdot 1) \bmod 4\|_{n,m=0:3} = \begin{vmatrix} \underline{0} & 1 & 2 & 3 \\ 1 & 2 & 3 & 0 \\ 2 & 3 & 0 & 1 \\ 3 & 0 & 1 & 2 \end{vmatrix}.$$

Therefore, the first component of this splitting-signal is calculated by

$$f_{1,1,0} = \chi_{1,1,0} \circ f = \begin{bmatrix} \underline{1} & 0 & 0 & 0 \\ 0 & 0 & 0 & 1 \\ 0 & 0 & 1 & 0 \\ 0 & 1 & 0 & 0 \end{bmatrix} \circ \begin{bmatrix} \underline{1} & 2 & 1 & 3 \\ 2 & 0 & 1 & 1 \\ 1 & 3 & 2 & 2 \\ 2 & 4 & 1 & 2 \end{bmatrix} = 1 + 1 + 2 + 4 = 8 \quad (1.47)$$

and similarly we obtain the next three components

$$\begin{aligned} f_{1,1,1} &= \chi_{1,1,1} \circ f = 2 + 2 + 2 + 1 = 7 \\ f_{1,1,2} &= \chi_{1,1,2} \circ f = 1 + 0 + 1 + 2 = 4 \\ f_{1,1,3} &= \chi_{1,1,3} \circ f = 3 + 1 + 3 + 2 = 9. \end{aligned} \quad (1.48)$$

Thus, the splitting-signal $f_{T_{1,1}} = \{8, 7, 4, 9\}$. The four-point DFT of this signal equals $(A_0, A_1, A_2, A_3) = (28, 4 + 2j, -4, 4 - 2j)$, which defines the 2-D DFT of f at the frequency-points $(0,0), (1,1), (2,2)$, and $(3,3)$. At this step, we can write the other three values of the 2-D DFT as follows:

$$\begin{bmatrix} \underline{F_0} & F_1 & F_2 & F_3 \\ 0 & A_1 & 0 & 0 \\ 0 & 0 & A_2 & 0 \\ 0 & 0 & 0 & A_3 \end{bmatrix} = \begin{bmatrix} \underline{28} & 1-j & -6 & 1+j \\ 0 & 4+2j & 0 & 0 \\ 0 & 0 & -4 & 0 \\ 0 & 0 & 0 & 4-2j \end{bmatrix}.$$

The first component, A_0, equals $F_0 = 28$, and could be omitted from the calculations to avoid the redundancy. The redundancy of calculation takes place for other splitting-signals not only at frequency-point $(0,0)$, but at frequency-points with even coordinates, i.e., in the quarter of all frequency-points, and that can be seen in Figure 1.8.

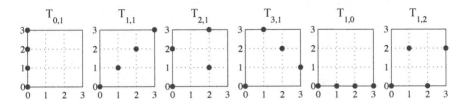

FIGURE 1.8
The disposition of frequency-points of six groups T of the covering σ of $X_{4,4}$.

For instance, the four-point DFT of the splitting-signal $f_{T_{2,1}} = \{4, 8, 7, 9\}$ equals $(B_0, B_1, B_2, B_3) = (28, -3 + j, -6, -3 - j)$. These values define the 2-D DFT at the frequency-points $(0,0), (2,1), (0,2)$, and $(2,3)$. At this step, we can write the other two values of the 2-D DFT as follows:

$$
\begin{bmatrix} F_0 & F_1 & F_2 & F_3 \\ 0 & A_1 & 0 & 0 \\ 0 & B_1 & A_2 & B_3 \\ 0 & 0 & 0 & A_3 \end{bmatrix} = \begin{bmatrix} \underline{28} & 1-j & -6 & 1+j \\ 0 & 4+2j & 0 & 0 \\ 0 & -3+j & -4 & -3-j \\ 0 & 0 & 0 & 4-2j \end{bmatrix}.
$$

It is clear, that the incomplete four-point DFT is required, to avoid calculations for components B_0 and B_2, which have already been calculated.

We continue the calculation of the 2-D DFT. The four-point DFT of the splitting-signal $f_{T_{3,1}} = \{5, 7, 7, 9\}$ equals $(C_0, C_1, C_2, C_3) = (28, -2 + 2j, -4, -2 - 2j)$. It defines the 2-D DFT of f at the frequency-points $(0,0), (3,1), (2,2)$, and $(1,3)$. At this step, we can record two new values of the 2-D DFT as follows:

$$
\begin{bmatrix} F_0 & F_1 & F_2 & F_3 \\ 0 & A_1 & 0 & C_3 \\ 0 & B_1 & A_2 & B_3 \\ 0 & C_1 & 0 & A_3 \end{bmatrix} = \begin{bmatrix} \underline{28} & 1-j & -6 & 1+j \\ 0 & 4+2j & 0 & -2-2j \\ 0 & -3+j & -4 & -3-j \\ 0 & -2+2j & 0 & 4-2j \end{bmatrix}.
$$

For this signal, the incomplete four-point DFT is required to avoid calculations for components C_0 and C_2, which have already been calculated in the second step of our calculations (when (p, s) was $(1, 1)$).

The remaining five values of the 2-D DFT will be calculated by the splitting-signals corresponding to the generators $(1, 0)$ and $(1, 2)$. The splitting-signal $f_{T_{1,0}} = \{7, 4, 8, 8\}$. The four-point DFT of this signal equals $(D_0, D_1, D_2, D_3) = (28, -1 + 5j, 2, -1 - 5j)$ and defines the 2-D DFT of f at the frequency-points $(0,0),(1,0),(2,0)$, and $(3,0)$. At this step, we can record three new values of the 2-D DFT as follows:

$$
\begin{bmatrix} F_0 & F_1 & F_2 & F_3 \\ D_1 & A_1 & 0 & C_3 \\ D_2 & B_1 & A_2 & B_3 \\ D_3 & C_1 & 0 & A_3 \end{bmatrix} = \begin{bmatrix} \underline{28} & 1-j & -6 & 1+j \\ -1+5j & 4+2j & 0 & -2-2j \\ 2 & -3+j & -4 & -3-j \\ -1-5j & -2+2j & 0 & 4-2j \end{bmatrix}.
$$

The redundancy of calculation on this step is only at $(0,0)$. At the last step, the four-point DFT of the splitting-signal $f_{T_{1,2}} = \{7,9,8,4\}$ is calculated. It equals $(E_0, E_1, E_2, E_3) = (28, -1-5j, 2, -1+5j)$ and defines the 2-D DFT of f at the frequency-points $(0,0),(1,2),(2,0)$, and $(3,2)$. We fill the 2-D DFT with the values of E_1 and E_2,

$$
\begin{bmatrix} F_0 & F_1 & F_2 & F_3 \\ D_1 & A_1 & E_1 & C_3 \\ D_2 & B_1 & A_2 & B_3 \\ D_3 & C_1 & E_3 & A_3 \end{bmatrix} = \begin{bmatrix} \underline{28} & 1-j & -6 & 1+j \\ -1+5j & 4+2j & -1-5j & -2-2j \\ 2 & -3+j & -4 & -3-j \\ -1-5j & -2+2j & -1+5j & 4-2j \end{bmatrix} = \begin{bmatrix} F_{0,0} & F_{0,1} & F_{0,2} & F_{0,3} \\ F_{1,0} & F_{1,1} & F_{1,2} & F_{1,3} \\ F_{2,0} & F_{2,1} & F_{2,2} & F_{2,3} \\ F_{3,0} & F_{3,1} & F_{3,2} & F_{3,3} \end{bmatrix}.
$$

For this signal, it is sufficient to perform an incomplete four-point DFT, to avoid the redundancy of calculations in two points. Thus, the 4×4-point 2-D DFT is calculated in the tensor algorithm by six four-point DFTs. Namely, one full four-point DFT and five incomplete four-point DFTs are required in this algorithm.

Example 1.9 (Image 256×256**)** We consider the representation of the image in the group of frequencies, which is generated by $(p_1, p_2) = (16, 1)$. Figure 1.9 shows (a) the clock-and-moon image 256×256 in part a, along with the image-signal $f_{T_{16,1}}$ of length 256 in b, the 1-D DFT of the image-signal (in absolute scale) in c, and 256 samples of this 1-D DFT at frequency-points of the subset $T_{16,1}$ of $X_{256,256}$ at which the 2-D DFT of the image is filled by the 1-D DFT in d. The value of the 1-D DFT at point 0, which is the sum 43885 of the image, has been truncated in parts c and d.

Example 1.10 (Image 256×256**)** Figure 1.10 illustrates the image 256×256 in part a, along with the 1-D DFT over the image-signal $f_{T_{1,3}}$ of length 256 in c, and the magnitude spectrum of the image in d. Three bright parallel lines on the spectrum show the samples at points of the group $T_{1,3}$ at which the 2-DFT of the image is the 1-DFT of the image-signal.

The 2-D DFT at samples at this group has been amplified, in order to see the location of the group and directions of the projection, along which the components of the tensor are calculated as sums of line-integrals. The image after amplifying the 2-DFT at samples of the group $T_{1,3}$ is shown in part b. The effect of amplifying those spectral components illustrates the meaning of Equation (1.21). The projections are calculated at angle $\psi = 18.4349°$ and the 1-D DFT fills the 2-D DFT along three lines at angle $\theta = 90 - \psi = 71.5651°$.

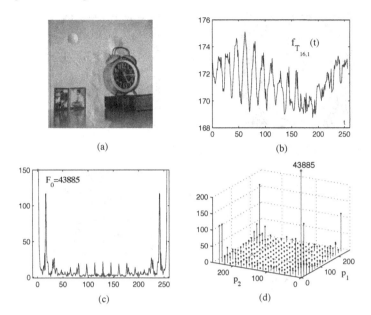

FIGURE 1.9

(a) The image 256×256, (b) image-signal corresponding to the generator $(p_1, p_2) = (16, 1)$, (c) absolute spectrum of the image-signal (with the truncated zero component), and (d) the arrangement of values of the 1-D DFT in the 2-D DFT of the image (in the 3-D view).

Example 1.11 (Image 512×512) Figure 1.11 illustrates the image of size 512×512 in part a, along with the splitting-signal $f_{T_{3,1}}$ of length 512 in b, the 1-D DFT over this splitting-signal in c, and points of the subset $T_{3,1}$ at which the 2-D DFT of the image is filled by the 1-D DFT in d. The image is the max-image of stacked fluorescent in situ hybridization (FISH) images arising from normal glands. In general, the image f is real and components $f_{p,s,t}$ can be considered to be coefficients of decomposition of the 2-D DFT in the group $T_{p,s}$ by the $\cos(\omega x)$- and $\sin(\omega x)$-waves with the frequency $\omega = 2\pi t/N$. In the 3-D representation $f_{n,m} \to f_{p,s,t}$, two dimensions correspond to frequency (p, s) and one dimension refers the time coordinate t of the signal $f_{T_{p,s}}$.

1.4.4 N is a power of an odd prime

We now consider a splitting of the $N \times N$-point DFT, when $N = L^r$, $L > 1$ is an arbitrary odd prime number, and $r > 1$. The irreducible covering $\sigma_J = (T_{p,s})$ of the lattice X_{L^r, L^r} consists of $L^{r-1}(L + 1)$ cyclic groups and can be

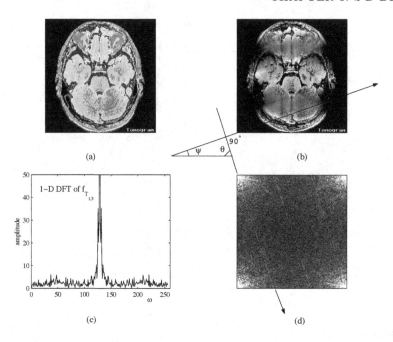

FIGURE 1.10
(a) Image 256×256. (b) Image after amplifying the 2-DFT at samples of the group $T_{1,3}$. (c) Absolute value of the 1-D DFT of the image-signal $f_{T_{1,3}}$ (zero component is shifted to the center and truncated). (d) 2-D DFT of the image with amplified samples at the group $T_{1,3}$. Angles $\psi = 18.4349°$ and $\theta = 71.5651°$.

defined by the following set of generators:

$$J = J_{L^r, L^r} = \bigcup_{p_1=0}^{L^r-1} (p_1, 1) \bigcup \bigcup_{p_2=0}^{L^{r-1}-1} (1, Lp_2). \tag{1.49}$$

Thus, to calculate the $L^r \times L^r$-point DFT, it is sufficient to fulfill $L^{r-1}(L+1)$ L^r-point DFTs of the splitting-signals. The column-row algorithm uses $2(L^r)$ L^r-point DFTs, i.e., $2L/(L+1)$ times more. For instance, the 9×9-point 2-D DFT is calculated in the tensor algorithm by 12 one-dimensional nine-point DFTs. In the column-row algorithm, this 2-D transform requires 18 nine-point DFTs. The tensor algorithm can be improved, when removing the redundancy of calculations in the intersections of the cyclic groups $T_{p,s}$ of the covering. The redundancy of the algorithm is in calculation of the spectral components at all frequency-points (p_1, s_1) with coordinates that are integer multiple to L, i.e., when $(p_1, s_1) = (Lp \mod N, Ls \mod N)$ and $(p, s) \in J$.

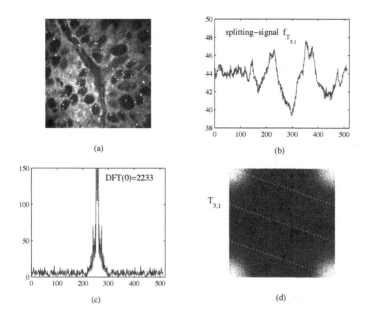

FIGURE 1.11
(a) FISH image, (b) the splitting-signal $f_{T_{3,1}}$, (c) 1-D DFT of the splitting-signal (in absolute scale and shifted), and (d) arrangement of values of the 1-D DFT in the 2-D DFT of the image at frequency-points of the set $T_{3,1}$.

1.4.5 Case $N = L_1 L_2$ $(L_1 \neq L_2 > 1)$

We consider the $N = L_1 L_2$ case, where L_1 and L_2 are arbitrary coprime numbers > 1. The irreducible covering σ_J of the lattice $X_{L_1 L_2, L_1 L_2}$ consists of $(L_1 + 1)(L_2 + 1)$ cyclic groups $T_{p,s}$. Such covering σ_J can be determined by the following set of generators:

$$J = \bigcup_{p_1=0}^{L_1 L_2 - 1} (p_1, 1) \cup \{(1,0)\} \cup \left(\bigcup_{g.c.d.(p_2, L_1 L_2) > 1} (1, p_2) \right) \cup \{(L_1, L_2)\} \cup \{(L_2, L_1)\}. \tag{1.50}$$

To calculate the 2-D $(L_1 L_2) \times (L_1 L_2)$-point DFT, it is sufficient to perform $(L_1 + 1)(L_2 + 1)$ $L_1 L_2$-point 1-D DFTs of splitting-signals $f_{T_{p,s}}$. For instance, the calculation of the 20×20-point 2-D DFT is reduced to calculation of thirty 20-point DFTs, instead of forty 20-point DFTs in the column-row algorithm.

1.4.6 General case

In the general $N \times N$ case, the construction of the irreducible covering

$$\sigma_J = \left(T_{p,s}\right)_{p,s \in J} \tag{1.51}$$

of the square lattice $X_{N,N}$ can be implemented in the following way [48]. We first define the set $B_N = \{n \in X_N; \ g.c.d.(n, N) > 1\}$ and function $\beta(p)$, which equals the number of elements $s \in B_N$ being coprime with p and such that $ps < N$. Then the set of generators can be defined as follows:

$$J = \bigcup_{p=0}^{N-1} (p, 1) \bigcup \left(\bigcup_{s \in B_N} (1, s) \right) \bigcup \left(\bigcup_{p,s \in B_N,\ g.c.d.(p,s)=1,\ p,s \leq N} (p, s) \right). \qquad (1.52)$$

The number of generators in this set, or the number of 1-D DFTs required to calculate the 2-D $N \times N$-point DFT equals

$$card\,\sigma_J = 2N - \phi(N) + \sum_{p \in B_N} \beta(p) \qquad (1.53)$$

where we denote the Euler's function by $\phi(N)$, that is, the number of positive integers that are smaller than N and coprime with N. It is easy to verify, that $\phi(N) \geq \sum\{\beta(p); \ p \in B_N\}$, so that $card\,\sigma \leq 2N$.

1.4.7   Other orders $N_1 \times N_2$

The tensor algorithm, as well as the improved tensor algorithm, can be constructed for other orders $N_1 \times N_2$ of the 2-D DFT, when $N_1 \neq N_2 > 1$. The tensor algorithm is defined by the irreducible covering σ_J of the lattice X_{N_1, N_2} by the cyclic groups $T_{p,s}$. It is not difficult to compose such covering for each case $N_1 \times N_2$ under consideration.

Example 1.12 (The 3×6-point DFT) We consider the 3×6-point discrete Fourier transformation, $\mathcal{F}_{3,6}$, i.e., the case when $N_1 = 3$ and $N_2 = 6$. The irreducible covering σ_J of the lattice $X_{3,6}$ can be defined by the set of four generators; $J = \{(0,1), (1,1), (2,1), (1,3)\}$. All cyclic groups of this covering and their intersections are shown below:

$$T_{0,1} = \{(0,0), (0,1), (0,2), (0,3), (0,4), (0,5)\}$$
$$T_{1,1} = \{(\underline{0,0}), (1,1), (2,2), (\underline{0,3}), (1,4), (2,5)\}$$
$$T_{2,1} = \{(\underline{0,0}), (2,1), (1,2), (\underline{0,3}), (2,4), (1,5)\}$$
$$T_{1,3} = \left\{ \begin{array}{ll} (0,0), & (0,3) \\ (1,0), & (1,3) \\ (2,0), & (2,3) \end{array} \right\}.$$

Figure 1.12 illustrates the locations of all frequency-points of these four groups. The groups are intersected only at two frequency-points $(0,0)$ and $(0,3)$, i.e., when the coordinates of the points are integer multiple to 3. Therefore, the transformation $\mathcal{F}_{3,6}$ can be split by one six-point DFT and three incomplete six-point DFTs. We note for comparison, that in the column-row algorithm, three six-point DFTs with real inputs and six three-point DFTs with complex inputs are used. This algorithm uses therefore more operations of multiplication than the tensor algorithm.

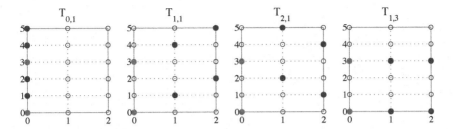

FIGURE 1.12
Arrangement of frequency-points of groups T_{p_1,p_2} covering the lattice $X_{3,6}$.

Example 1.13 (The 6×8-point DFT) We consider the 6×8-point Fourier transformation, $\mathcal{F}_{6,8}$. The irreducible covering σ_J of the lattice $X_{6,8}$ can be defined by the set of generators $J = \{(1,1), (2,1), (1,2), (1,4), (1,0)\}$. Intersections of cyclic groups of this covering are shown below:

$$
\begin{aligned}
T_{1,1} &= \left\{ \begin{array}{l} (0,0), (1,1), (2,2), (3,3), (4,4), (5,5), (0,6), (1,7) \\ (2,0), (3,1), (4,2), (5,3), (0,4), (1,5), (2,6), (3,7) \\ (4,0), (5,1), (0,2), (1,3), (2,4), (3,5), (4,6), (5,7) \end{array} \right\} \\[2mm]
T_{2,1} &= \left\{ \begin{array}{l} (0,0), (2,1), \underline{(4,2)}, (0,3), \underline{(2,4)}, (4,5), \underline{(0,6)}, (2,7) \\ \underline{(4,0)}, (0,1), \underline{(2,2)}, (4,3), \underline{(0,4)}, (2,5), \underline{(4,6)}, (0,7) \\ \underline{(2,0)}, (4,1), \underline{(0,2)}, (2,3), \underline{(4,4)}, (0,5), \underline{(2,6)}, (4,7) \end{array} \right\} \\[2mm]
T_{1,2} &= \left\{ \begin{array}{llll} (0,0), & (1,2), & \underline{(2,4)}, & (3,6) \\ \underline{(4,0)}, & (5,2), & \underline{(0,4)}, & (1,6) \\ \underline{(2,0)}, & (3,2), & \underline{(4,4)}, & (5,6) \end{array} \right\} \\[2mm]
T_{1,4} &= \left\{ \begin{array}{ll} (0,0), & (1,4) \\ \underline{(2,0)}, & (3,4) \\ \underline{(4,0)}, & (5,4) \end{array} \right\} \\[2mm]
T_{1,0} &= \left\{ \begin{array}{l} (0,0) \\ \underline{(1,0)} \\ \underline{(2,0)} \\ \underline{(3,0)} \\ \underline{(4,0)} \\ \underline{(5,0)} \end{array} \right\}.
\end{aligned}
\qquad (1.54)
$$

The locations of frequency-points of the five cyclic groups $T_{p,s}$ of the covering are shown in Figure 1.13.

The groups are intersected only at 12 frequency-points with even coordinates. Therefore, the transformation $\mathcal{F}_{6,8}$ can be split by one 24-point DFT, two incomplete 24- and 12-point DFTs, and two incomplete six-point DFTs. These incomplete DFTs are not calculated for the components with even points, and, therefore, they can be reduced to calculation of the DFTs of twice smaller orders. Thus the 6×8-point DFT can be split by the 24-, 12-,

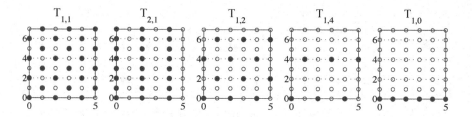

FIGURE 1.13
Arrangement of frequency-points of groups $T_{p,s}$ covering the lattice $X_{6,8}$.

and 6-point DFTs, and two 3-point DFTs, and the redundancy of calculations in the tensor algorithm will be removed.

1.5   Discrete Fourier transform and its geometry

For a better understanding the tensor representation of the 2-D DFT, we first consider the geometry of the 1-D DFT, as a beautiful system rotating the data on the plane around circles. For that, the block structure of the matrix of the N-point discrete Fourier transform (DFT) in the real space R^{2N} is described. All blocks 2×2 of this matrix correspond to the Givens transformations, or elementary rotations describing the multiplications by twiddle coefficients.

The N-point discrete Fourier transform of the signal is defined as a decomposition of the signal by N roots of the unit

$$W^k = W_N^k = e^{-i\frac{2\pi}{N}k} = c_k - is_k = \cos(\frac{2\pi}{N}k) - i\sin(\frac{2\pi}{N}k), \quad k = 0 : (N-1).$$

The roots are located on the unit circle, $(W^k)^N = 1$. The DFT of the vector-signal $\mathbf{f} = (f_0, f_1, f_2, ..., f_{N-1})'$, which is calculated by

$$F_p = R_p + iI_p = \sum_{n=0}^{N-1} W^{np} f_n, \quad p = 0 : (N-1),$$

has the following matrix in the complex space C^N:

$$[\mathcal{F}_N] = \begin{bmatrix} 1 & 1 & 1 & 1 & 1 & 1 \\ 1 & W^1 & W^2 & W^3 & \cdots W^{N-1} \\ 1 & W^2 & W^4 & W^6 & \cdots W^{N-2} \\ 1 & \cdots & \cdots & \cdots & \cdots & \cdots \\ 1 & & & & & \\ 1 & W^{N-1} & W^{N-2} & W^{N-3} & & W^1 \end{bmatrix}.$$

We now describe in matrix form the multiplication of the complex number $\mathbf{x} = x_1 + ix_2$, which is considered as the column-vector $(x_1, x_2)'$ or row-vector (x_1, x_2), by the twiddle coefficients W^k,

$$\mathbf{x} = \begin{pmatrix} x_1 \\ x_2 \end{pmatrix} \to W^k \mathbf{x} = \begin{pmatrix} c_k x_1 + s_k x_2 \\ c_k x_2 - s_k x_1 \end{pmatrix}, \quad k = 0 : (N-1),$$

where $W^k = (c_k, -s_k) = \cos\varphi_k - i\sin\varphi_k$, and the angles $\varphi_k = 2\pi k/N$. In matrix form, this multiplication can be written as

$$T^k \mathbf{x} = \begin{pmatrix} \cos\varphi_k & \sin\varphi_k \\ -\sin\varphi_k & \cos\varphi_k \end{pmatrix} \begin{pmatrix} x_1 \\ x_2 \end{pmatrix}.$$

The matrix of rotation by the angle $\varphi_k = k\varphi_1$ is denoted by T^k. In the $k = 0$ case, $\varphi_0 = 0$, $T = I$, and there is no rotation of the signal-data.

Thus, the complex plane is transferred into the 2-D real space, $C \to R^2$, and each operation of multiplication by the twiddle coefficient is considered as the elementary rotation, or the Givens transformation, $W^k \to T^k$, $k = 0 : (N-1)$. We next consider the inclusion of the complex space C^N into the real space R^{2N} by

$$\mathbf{f} = (f_0, f_1, ..., f_{N-1})' \to \bar{\mathbf{f}} = (r_0, i_0, r_1, i_1, ..., r_{N-1}, i_{N-1})',$$

where we denote $r_k = \text{Re} f_k$ and $i_k = \text{Im} f_k$ for $k = 0 : (N-1)$. The vector $\bar{\mathbf{f}}$ is composed from the original vector, or signal $\mathbf{f}$, and its vector-component is denoted by $\bar{\mathbf{f}}_k = (\bar{f}_{2k}, \bar{f}_{2k+1})' = (r_k, i_k)'$. The N-point DFT of $\mathbf{f}$ is represented in the real space R^{2N} as the $2N$-point transform

$$\bar{\mathbf{F}}_p = \begin{bmatrix} R_p \\ I_p \end{bmatrix} = \sum_{k=0}^{N-1} T^{kp} \begin{bmatrix} r_k \\ i_k \end{bmatrix} = \sum_{k=0}^{N-1} \begin{pmatrix} \cos(k\varphi_p) & \sin(k\varphi_p) \\ -\sin(k\varphi_p) & \cos(k\varphi_p) \end{pmatrix} \begin{bmatrix} r_k \\ i_k \end{bmatrix}, \quad (1.55)$$

where $p = 0 : (N-1)$.

In matrix form, the DFT in the space R^{2N} is described by the following matrix $2N \times 2N$:

$$X = \begin{bmatrix} I & I & I & I & I & I \\ I & T^1 & T^2 & T^3 & \cdots & T^{N-1} \\ I & T^2 & T^4 & T^6 & \cdots & T^{N-2} \\ I & \cdots & & & & \\ I & T^{N-1} & T^{N-2} & T^{N-3} & & T^1 \end{bmatrix}. \quad (1.56)$$

The (n, p)-th blocks 2×2 of this matrix is T^{np}, where $n, p = 0 : (N-1)$. The matrix $I = I_2$ is the identity matrix 2×2. The rotation matrices T^k compose a one-parametric group with period N. In other words

$$T^{k_1 + k_2} = T^{k_1} T^{k_2}, \quad (T^0 = T^N = I),$$

for any $k_1, k_2 = 0 : (N-1)$.

According to (1.55), when calculating the component of the DFT at the frequency-point p, the vector-components, or points (r_k, i_k) are rotated by the corresponding angles $k\varphi_p$. Thus the first point (r_0, i_0) stays on its place, and the second point (r_1, i_1) is rotated around the circle of radius $\sqrt{r_1^2 + i_1^2}$ by angle φ_p. The next point (r_2, i_2) is rotated around the circle of radius $\sqrt{r_2^2 + i_2^2}$ by the twice larger angle $2\varphi_p$, or twice faster, than the point (r_1, i_1), and so on. Then the coordinates of all rotated points are added and one point is defined, which represents the DFT at frequency-point p. For $p = 0$, there is no rotation; the sums of coordinates of the original points define the component F_0.

Example 1.14 Consider the vector-signal of length 10 with the following real and imaginary parts:

$$\{r_k; \ k = 0:9\} = \{ \ 5, \ -3, 2, -5, -1, 3, -4, -2, \ 7, -6\}$$
$$\{i_k; \ k = 0:9\} = \{-2, \ 1, 4, -3, \ 7, 2, -5, \ 5, -1, \ 3\}$$

This signal is plotted in the form of ten points on the plane in Figure 1.14. The points are numbered and ten circles on which they lie are shown. The figure

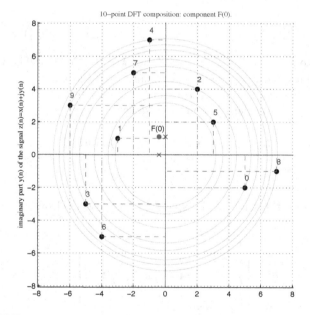

FIGURE 1.14
Calculation of the first component of the DFT, F_0.

also shows projections of all points of the signal on the X-axis and Y-axis (shown by dash lines) and their sums (by x), and the point $F_0 = (-0.4, 1.1)$ (by the bullet with label F_0). The coordinates of F_0 are normalized by the factor of $N = 10$, i.e., $F_0/10$.

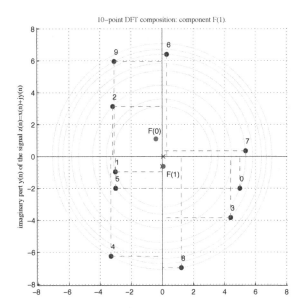

FIGURE 1.15 (See color insert)
Calculation of the second component of the DFT, F_1.

In the next stage of the DFT calculation, points of the signals are rotated as shown in Figure 1.15, and the sums (normalized by 10) of projections of the rotated points on the X-axis and Y-axis define the point $F_1 = (0.0684, -0.6155)$.

Similarly, these ten points are rotated eight more times, and during these rotations the sums of coordinates of ten points overlay a new set of ten points with the following coordinates:

$$\{R_p\} = \{-0.4, \quad 0.0684, \quad 0.7729, -2.0982, \quad 4.0537, \quad 2.2,$$
$$-0.0648, \quad 0.1566, -0.3618, 0.6733\}$$
$$\{I_p\} = \{ \quad 1.1, -0.6155, -0.3367, \quad 0.4982, -0.1407, \quad -0.5,$$
$$-0.5211, -1.5836, -0.1015, 0.2009\}$$

where $p = 0 : 9$. These ten points $F_p = (R_p, I_p)$ are show in Figure 1.16.

1.5.1 Inverse DFT

The inverse discrete Fourier transform (IDFT) is defined as

$$f_n = \frac{1}{N} \sum_{p=0}^{N-1} W^{-np} F_p, \quad n = 0 : (N-1). \tag{1.57}$$

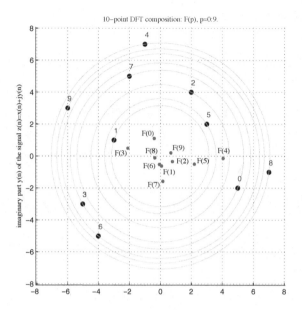

FIGURE 1.16 (See color insert)
Result of ten rotations of the signal-data.

In the real space R^{2N}, the N-point IDFT of $\{F_p; \; p = 0 : (N-1)\}$ is represented as

$$
\bar{\mathbf{f}}_n = \begin{bmatrix} r_n \\ i_n \end{bmatrix} = \frac{1}{N} \sum_{p=0}^{N-1} T^{-np} \begin{bmatrix} R_p \\ I_p \end{bmatrix} = \frac{1}{N} \sum_{p=0}^{N-1} \begin{pmatrix} \cos(n\varphi_p) & -\sin(n\varphi_p) \\ \sin(n\varphi_p) & \cos(n\varphi_p) \end{pmatrix} \begin{bmatrix} R_p \\ I_p \end{bmatrix},
$$

where $n = 0 : (N-1)$.

The illustration of the IDFT is similar to the DFT, but is a clock-wise rotation, i.e., in the direction opposite to the rotation in the DFT. If the projections of all rotated points have been not normalized for the DFT, then after rotating the points, the sums of the coordinates of rotated points are normalized by the factor of N. Thus, if we start with the ten points $\{F_p; \; p = 0 : 9\}$ in the above considered example, as shown in Figure 1.16, the sequential clockwise rotations of these points around the circles, on which they lie, define the original signal, i.e., ten points with coordinates

$$
\{r_k; \; k = 0 : 9\} = \{\; 5, \; -3, 2, -5, -1, 3, -4, -2, \; 7, -6\},
$$
$$
\{i_k; \; k = 0 : 9\} = \{-2, \; 1, 4, -3, \; 7, 2, -5, \; 5, -1, \; 3\}.
$$

Example 1.15 We consider the following symmetric complex signal $z(n)$ of length $N = 201$,

$$
z(n) = x(n) + iy(n) = n + i(n/5), \quad z(-n) = z(n), \quad n = 0 : 100.
$$

Figures 1.17-1.20 illustrate the geometry of a few components of the 201-point DFT. These figures were generated by code `horoscope_1byDFTfiles.m`, which can be found at *http://www.fasttransforms.com* in the folder *Lectures/DSP*. The star in these figures denotes the value of F_0.

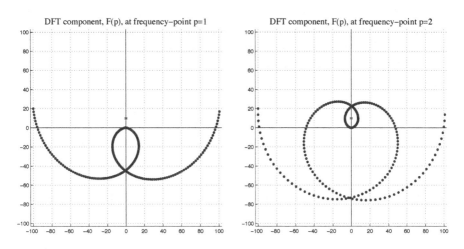

FIGURE 1.17
DFT components F_1 and F_2.

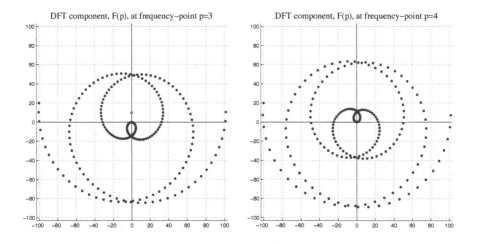

FIGURE 1.18
DFT components F_3 and F_4.

When $f_{n,m}$ is a real image and of size $N \times N$, all its N^2 first points are

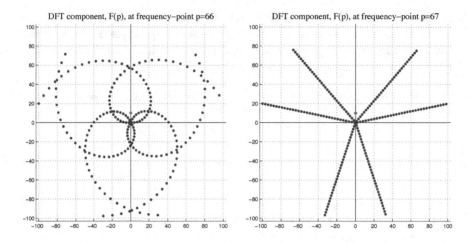

FIGURE 1.19
DFT components F_{66} and F_{67}.

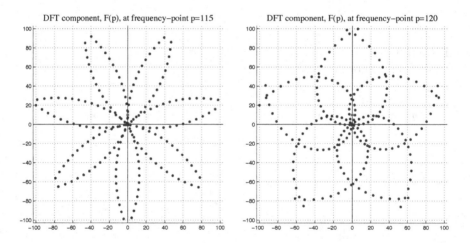

FIGURE 1.20
DFT components F_{115} and F_{120}.

located on the real line and then are rotated, step-by-step, N^2 times in the plane R^2. The process of rotation of all points is more complicated when comparing with the N-point DFT. For instance, the point $f_{1,1}$ is rotated as $W^{p+s}f_{1,1}$, i.e.,

$$f_{1,1} \rightarrow W^1 f_{1,1} \rightarrow W^2 f_{1,1} \rightarrow W^3 f_{1,1} \rightarrow ... \rightarrow W^{N-1} f_{1,1}$$

for components $F_{1,0}, F_{2,0}, F_{3,0}, ..., F_{N-1,0}$, as well as for components $F_{0,1}, F_{0,2}$,

$F_{0,3}, ..., F_{0,N-1}$, and also rotated as $f_{1,1} \to W^2 f_{1,1} \to W^3 f_{1,1} \to W^4 f_{1,1} \to$ ... $\to f_{1,1}$ for components $F_{1,1}$ and $F_{2,1}, F_{3,1}, ..., F_{N-1,1}$, and so on. This process of rotation of point $f_{1,1}$ is fulfilled when combined with other points $f_{n,m}$, that are rotated as $W^{np+ms} f_{n,m}$, where $(p, s) \in X \setminus (1, 1)$.

In tensor representation, one can describe the entire picture of rotation, or the geometry of all components of the 2-D DFT. Indeed, the image $N \times N$ is the set of splitting-signals of length N each, and the process of rotation of the splitting-signal generated by frequency-point (p, s) describes the 2-D DFT at frequency-points of the corresponding group $T_{p,s}$.

Problems

Problem 1.1 Consider the following image: $\{f_{n,m} = (n + m) \bmod 6, n = 0 : 7, m = 0 : 3.\}$ Determine the number of splitting-signals that describe the image in tensor representation. Calculate the tensor transform and plot all splitting-signals of the image.

Problem 1.2 Consider the tree image of size 257×257 in tensor representation. Calculate and plot two splitting-signals generated by frequency-points $(1, 2)$ and $(17, 34)$. Express one signal in terms of another signal.

Problem 1.3 For the tree image of size 256×256, calculate and plot the splitting-signal generated by frequency-point $(1, 7)$. Compute the 1-D DFT of the splitting-signal. In the lattice 256×256, show the frequency-points at which the 2-D DFT of the image is defined by this splitting-signal.

Problem 1.4 Consider the tree image of size 256×256 and its 2-D DFT in tensor representation.

A. Calculate and sketch the splitting-signal of the $(1, 13)$-projection.

B. Assume that the $(1, 12)$-projection is not calculated, or is not available.

(i) How would the tree image be changed if the splitting-signal of the $(1, 12)$-projection was substituted by the splitting-signal of the $(1, 13)$-projection?

(ii) How would the tree image be changed if the splitting-signal of the $(1, 12)$-projection was substituted by the zeros?

Problem 1.5 Consider the tree image of size 256×256 in tensor representation.

A. Remove the horizontal, vertical, and diagonal projections from the set of projections. Calculate and show the image reconstructed by the remaining 255 projections.

B. Calculate and sketch the image reconstructed only by three projections, which are the horizontal, vertical, and diagonal projections.

C. Calculate and sketch the image reconstructed only by two projections, the horizontal and vertical projections.

Problem 1.6 Repeat Problem 1.5 for the tree image extended to the size 257×257.

Problem 1.7 (Circular Convolution) Consider a discrete image $f_{n,m}$ of size $N \times N$, where $N > 1$ is a prime, and the small window 3×3 with the coefficients

$$h_{n,m} = \begin{cases} 1/9, & \text{if } n, m = -1, 0, 1; \\ 0, & \text{otherwise.} \end{cases}$$

A. Determine if the circular 2-D convolution

$$(f \otimes h)_{n,m} = \sum_{n_1} \sum_{m_1} f_{n+n_1 \bmod N, m+m_1 \bmod N} h_{n_1,m_1}$$

can be accomplished by $N + 1$ circular 1-D convolutions of splitting-signals of the image and window.

B. Sketch the results in *A* for the tree image $f_{n,m}$ of size 257×257.

Problem 1.8 (Motion of particles in the ring) Consider the binary image of the following ellipse: $\{f = (x, y)\} = \{(a \sin \phi, b \cos \phi); \ \phi \in [0, 2\pi)\}$, where $a = 2$ and $b = 1$. Compose the complex 1-D signal $f_n = (x_n, y_n) = x_n + jy_n$ of length 64, which represents this ellipse, by selecting 64 values of ϕ uniformly distributed in $[0, 2\pi)$. Use the 1-D DFT-based system to rotate 64 selected points of the ellipse in the ring $b^2 \le x^2 + y^2 \le a^2$, and draw the geometry of such rotation, for rotations number $p = 1, 2, 16, 21, 31, 32, 61, 62$, and 63.

Problem 1.9 (Geometry of the signals) Consider the tree image of size 256×256 and the splitting-signal $f_{T_{2,1}}$, which are shown in Figure 1.21 in parts a and b, respectively.

(a)

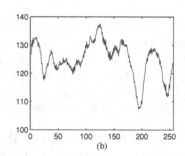
(b)

FIGURE 1.21
(a) Tree image and (b) the splitting-signal generated by $(p, s) = (2, 1)$.

Use the concept of the 1-D DFT in the real space and sketch the geometry of components $F_1, F_2, F_3, F_{83}, F_{110}$, and F_{127} of the 256-point DFT of the splitting-signal. Discuss the reconstruction of the splitting-signal from the geometry of any of these components.

2

Direction Images

This chapter discusses the decomposition of the image by direction images, which is based on the concept of the tensor representation and its advanced form, the paired representation [41]-[48]. The 2-D image is considered to be of the size $N \times N$, where N is prime, a power of two, and a power of odd prime. The tensor and paired representations in the frequency-and-time domain define the image as a set of 1-D splitting-signals. Each of such splitting-signals is calculated as the sum of the image along the parallel lines, and it defines the direction image as a component of the original image. The unique decomposition of the image by direction images is described and formulas for the inverse tensor and paired transforms are given. These formulas can be used for image reconstruction from projections [53]-[56], when splitting-signals or their direction images are calculated directly from the projection data. The number of required projections is uniquely defined by the tensor representation of the image.

2.1 2-D direction images on the lattice

The components $f_{p_1,p_2,t}$, $t = 0 : (N-1)$, of the splitting-signals of the image f_{n_1,n_2} are defined as sums of the image at points lying on the corresponding set $V_{p_1,p_2,t}$ defined in (1.18). To describe these sets, we first consider the case when $N_1 = N_2 = N$. Given a sample $(p_1, p_2) \in X$ and a nonnegative integer $t < N$, the set

$$V_{p_1,p_2,t} = \{(n_1, n_2); \ n_1 p_1 + n_2 p_2 = t \bmod N, \ n_1, n_2 = 0 : (N-1)\},$$

if it is not empty, is the set of points (n_1, n_2) along a family of parallel straight lines at the angle of $\psi = -\arctan(p_2/p_1)$ to the horizontal axis. The equations for these lines are

$$\left. \begin{array}{r} xp_1 + yp_2 = t \\ xp_1 + yp_2 = t + N \\ \cdots \quad \cdots \\ xp_1 + yp_2 = t + kN \end{array} \right\} \tag{2.1}$$

where $k \leq p_1 + p_2$. We denote this family by $\mathcal{L}_{p_1,p_2,t}$. For different values of $t_1 \neq t_2 < N$, the families of lines $\mathcal{L}_{p_1,p_2,t_1}$ and $\mathcal{L}_{p_1,p_2,t_2}$ do not intersect. All together, the sets $V_{p_1,p_2,t}$, $t = 0 : (N - 1)$, compose a partition of the period X. It is interesting to note, that the direction of parallel lines of $\mathcal{L}_{p_1,p_2,t}$ is perpendicular to the direction of frequency-points of the cyclic group T_{p_1,p_2}.

Example 2.1 (Lattice 8×8) On the lattice $X_{8,8}$, we consider two sets of parallel lines $\mathcal{L}_{2,1,1}$ and $\mathcal{L}_{2,1,2}$. Each family contains three parallel lines. For the family $\mathcal{L}_{2,1,1}$, the parallel lines are

$$y_1 : 2x + y = 1, \quad y_9 : 2x + y = 9, \quad y_{17} : 2x + y = 17.$$

One point $(0,1)$ of the set $V_{2,1,1}$ lies on the first line of $\mathcal{L}_{2,1,1}$, four points $(1,7),(2,5),(3,3),(4,1)$ on the second line, and three points $(5,7),(6,5),(7,3)$ on the third one. Therefore, $f_{2,1,1} = (x_{0,1}) + (x_{1,7} + x_{2,5} + x_{3,3} + x_{4,1}) + (x_{5,7} + x_{6,5} + x_{7,3})$. The parallel lines of the family $\mathcal{L}_{2,1,2}$ are defined by

$$y_2 : 2x + y = 2, \quad y_{10} : 2x + y = 10, \quad y_{18} : 2x + y = 18,$$

and the component $f_{2,1,2} = (x_{0,2} + x_{1,0}) + (x_{2,6} + x_{3,4} + x_{4,2} + x_{5,0}) + (x_{6,6} + x_{7,4})$. The disposition of the points lying on the parallel lines of these sets is given

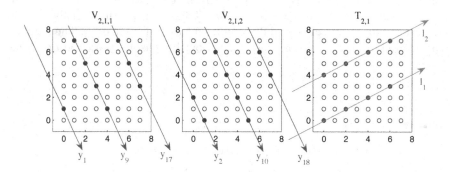

FIGURE 2.1
The locations of points of sets $V_{2,1,1}$ and $V_{2,1,2}$ and frequency-points of the group $T_{2,1}$.

in Figure 2.1. The location of the frequency-points of the group $T_{2,1}$ is also shown in this figure. Two parallel lines pass through these frequency-points, which are defined in the frequency plane (w_1, w_2) as $l_1 : 2w_2 - w_1 = 0$ and $l_2 : 2w_2 - w_1 = 8$. The parallel lines l_1 and l_2 are perpendicular to the parallel lines y_n of $\mathcal{L}_{2,1,1}$ and $\mathcal{L}_{2,1,2}$, as well as all other families $\mathcal{L}_{2,1,t}$, $t = 0, 3 : 7$.

The disposition of the points of all disjoint eight sets $V_{2,1,t}$, when $t = 0 : 7$, is given in Figure 2.2.

We can identify the opposite sides of boundaries of the square $Y = [0, N] \times [0, N]$ and consider Y as a torus and the 2-D lattice X as a net traced on the

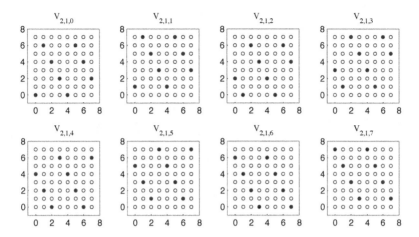

FIGURE 2.2
The disposition of eight sets of points, $V_{2,1,t}$, $t = 0 : 7$ (shown by filled circles).

torus in the 3-D space. Then, the straight lines of $\mathcal{L}_{p_1,p_2,t}$ will compose a closed spiral $S_t = S_{p,s,t}$ on the torus. As an example for $N = 32$, Figure 2.3 shows the location of points of sets $V_{1,1,3}$ and $V_{1,1,7}$ on the square grid 32×32 in parts (a) and (b), respectively. Each set is located on two parallel lines.

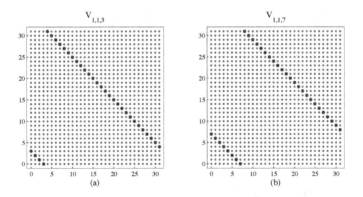

FIGURE 2.3
The sets (a) $V_{1,1,3}$ and (b) $V_{1,1,7}$ on the lattice 32×32.

We now consider the same sets on the discrete torus. Figure 2.4 shows the locus of two spirals S_3 and S_7 on the net, for $(p, s) = (1, 1)$. They correspond respectively to the parallel straight lines of families $\mathcal{L}_{1,1,3}$ and $\mathcal{L}_{1,1,7}$ described by (2.1). The sums $\{f_{p_1,p_2,t}; t = 0 : (N - 1)\}$, calculated on N parallel spirals S_t, $t = 0 : (N-1)$, represent the image f_{n_1,n_2} in the group T_{p_1,p_2} of frequency-

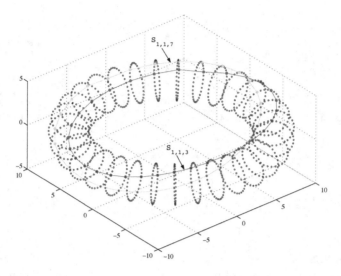

FIGURE 2.4 (See color insert)
The net with knots of the grid 32×32 in the 3-D space with locus of two
spirals $S_{1,1,3}$ and $S_{1,1,7}$.

points. These points lie on the spiral that passes through the initial point $(0,0)$
of the net and make an angle $\pi/2$ with the spirals S_t.

Each image-signal is the set of discrete integrals along the family $\mathcal{L}_{p_1,p_2,t}$
of parallel lines. Therefore, the processing of the image-signal f_T yields the
change in the Fourier transform at frequency-points of the corresponding cyclic
group T_{p_1,p_2}. After performing the inverse 2-D discrete Fourier transform, the
corresponding change might be observed in the image along the parallel lines
of sets $V_{p_1,p_2,t}$, $t = 0 : (N-1)$. As an example, Figure 2.5 shows the tree
image of size 256×256 in part a, along with the results of amplifying only
one image-signal $f_{T_{2,1}}$ by the factor of 4 in b, and signal $f_{T_{5,1}}$ by the factor
of 6 in c. The directions of parallel lines of the corresponding families $\mathcal{L}_{2,1,t}$,
and $\mathcal{L}_{5,1,t}$, $t = 0 : 255$, on the image can easily be observed.

2.1.1   Superposition of directions

The images of Figure 2.5 illustrate the fact that an image $f = f_{n_1,n_2}$ can
be composed by specific collection of direction images. The representation of
the image as the set of splitting-signals leads to the image decomposition by
direction images. To show that, we consider the tensor representation of the
image

$$\{f_{n_1,n_2}\} \to \{f_T; \, T = T_{p_1,p_2} \in \sigma_J\},$$

where $\sigma = \sigma_J$ is the irreducible covering of the period X. We assume that the
size of X is $N \times N$, and J is the set of generators.

(a)   (b)   (c)

FIGURE 2.5
(a) Tree image before and after processing by the image-signals (b) $f_{T_{2,1}}$ and (c) $f_{T_{5,1}}$.

The tensor transform can be used for calculating and processing the 2-D discrete Fourier transform (DFT) of the image, since each of splitting-signal $f_{T_{p,s}}$ defines the 2-D DFT of the image at frequency-points of the cyclic group $T_{p,s}$. The image can thus be calculated from the tensor transform, by using the inverse 2-D DFT. The image can also be reconstructed directly from its tensor transform, without calculating the inverse 2-D DFT. The splitting-signals define unique direction components of the image.

As an example, Figure 2.6 shows the bridge image 256×256 extended to the size 257×257 in part a, along with the image of all 258 splitting-signals in b, the set of 1-D DFTs of these splitting-signals in c, which represent the 2-D DFT of the image shown in d. The splitting-signals and their DFTs are

(a) image   (b) 2–D TT   (c) 1–D DFTs   (d) 2–D DFT

FIGURE 2.6
(a) Image 257×257, (b) image of 258 splitting-signals, (c) image of 258 257-point DFTs of the splitting-signals, and (d) 2-D DFT of the image. (All DFTs are shown in absolute mode.)

written row-wise in images each of size 258×257 in (b) and (c), respectively. These images can be reduced to the original size 257×257. Indeed, the sum of all components of the splitting-signals is equal to the sum of the image, i.e.,

$$\sum_{t=0}^{N-1} f_{p,s,t} = f_{0,0,0} = \sum_{n=0}^{N-1}\sum_{m=0}^{N-1} f_{n,m}.$$

Therefore, we can save one complete splitting-signal, let say, $\{f_{0,1,t}; t = 0 : (N-1)\}$, and only $(N-1)$ first components in the remaining N splitting-signals, $\{f_{1,p,t}; t = 0 : (N-2)\}$, where $p = 0 : (N-1)$. The total number of the required components of all signals equals $N + N(N-1) = N^2$. The DFTs of splitting-signals carry the same information only at the frequency-point $(0,0)$, which equals $f_{0,0,0}$.

For a given generator $(p, s) \neq (0, 0)$, we consider the following incomplete 2-D DFT with values at frequency-points of the group $T_{p,s}$:

$$D_{p_1,p_2} = D_{p_1,p_2}^{p,s} = \begin{cases} F_{\overline{kp},\overline{ks}}, & \text{if } (p_1,p_2) = (\overline{kp},\overline{ks}), \ k = 0 : (N-1), \\ 0, & \text{otherwise,} \end{cases} \qquad (2.2)$$

where $p_1, p_2 = 0 : (N-1)$. Examples of such incomplete 2-D DFTs were given in Example 1.6 for the $N = 5$ case. We define the *direction image* d_{n_1,n_2} as the inverse 2-D DFT of the data D,

$$d_{n_1,n_2} = d_{n_1,n_2}^{(p,s)} = (\mathcal{F}_{N,N}^{-1} \circ D^{p,s})_{n_1,n_2} = \frac{1}{N^2} \sum_{p_1=0}^{N-1}\sum_{p_2=0}^{N-1} D_{p_1,p_2} W^{n_1 p_1 + n_2 p_2}$$

$$n_1, n_2 = 0 : (N-1).$$

$$(2.3)$$

Since the splitting-signal $f_{T_{p,s}}$ determines the 2-D DFT of the image at frequency-points of the group $T = T_{p,s}$,

$$(\mathcal{F}_N \circ f_{T_{p,s}})_k = F_{\overline{kp},\overline{ks}}, \qquad k = 0 : (N-1),$$

the following calculations hold:

$$d_{n_1,n_2} = \frac{1}{N^2} \sum_{p_1=0}^{N-1}\sum_{p_2=0}^{N-1} D_{p_1,p_2} W^{n_1 p_1 + n_2 p_2} = \frac{1}{N^2} \sum_{k=0}^{N-1} F_{\overline{kp},\overline{ks}} W^{n_1(kp)+n_2(ks)}$$

$$= \frac{1}{N}\frac{1}{N} \sum_{k=0}^{N-1} F_{\overline{kp},\overline{ks}} W^{k(n_1 p + n_2 s)} = \frac{1}{N} f_{p,s,(n_1 p + n_2 s) \bmod N} \cdot$$

$$(2.4)$$

Thus, N values of the splitting-signal are placed on all N^2 points of the square lattice $X_{N,N}$. Namely, each value $f_{p,s,t}$ is placed at all points that are located on the parallel lines of the corresponding family $\mathcal{L}_{p,s,t}$.

The interesting property of the tensor transform is derived. The direction image is composed of N values of the splitting-signal, that are placed at all points of the image $N \times N$ along the parallel lines.

As an example, Figure 2.7 shows the tree image of size 257×257 in part

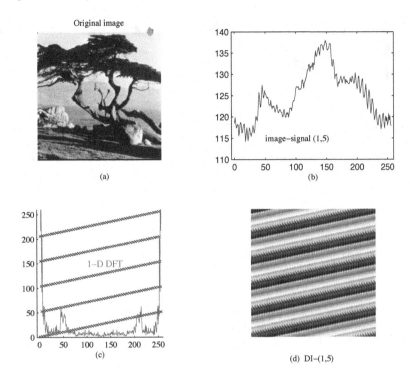

FIGURE 2.7
(a) The image 257×257, (b) the splitting-signal $\{f_{1,5,t};\ t = 0 : 256\}$, and (c) the 1-D DFT of the splitting-signal and frequency-points of $T_{1,5}$, and (d) the corresponding direction image d_{n_1,n_2} (the image has been scaled).

(a), along with the splitting-signal generated by the frequency $(p, s) = (1, 5)$ in b. The original size 256×256 of the tree image was extended to the size 257×257, by adding a zero row and column. The 1-D DFT of the splitting-signal in absolute scale and the location of all frequency-points of the cyclic group $T_{1,5}$, at which the 2-D DFT of the image is defined by this splitting-signal, are shown in c. The direction image $d_{n_1,n_2}^{(1,5)}$ is illustrated in d.

The direction images $d_{n_1,n_2}^{(p,s)}$ of the tree image, which are generated by the frequencies $(p, s) = (0, 1), (1, 1), (2, 1), ...,$ and $(9, 1)$ are shown in Figure 2.8 in parts a-j, respectively.

When N is a prime, $(N+1)$ direction signal-images are required to compose the image $f_{n,m}$. Indeed, the covering $\sigma_J = (T_{p,s};\ (p, s) \in J)$ consists of $(N+1)$ cyclic groups $T_{p,s}$. The groups T of σ_J intersect only at the point $(0, 0)$. Therefore, the sum of all $(N + 1)$ incomplete 2-D DFTs $D_{p_1,p_2}^{(p,s)}$, $(p, s) \in J_{N,N}$, equals the 2-D DFT of the image plus N times single values of the $F_{0,0}$ at

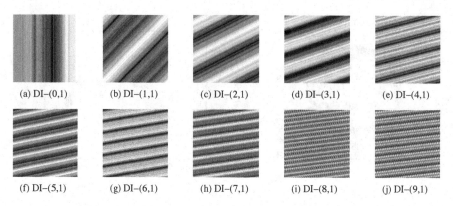

(a) DI–(0,1)   (b) DI–(1,1)   (c) DI–(2,1)   (d) DI–(3,1)   (e) DI–(4,1)

(f) DI–(5,1)   (g) DI–(6,1)   (h) DI–(7,1)   (i) DI–(8,1)   (j) DI–(9,1)

FIGURE 2.8
(a)-(j) The first ten direction images of the tree image 257×257. (All images were scaled.)

point $(0,0)$. We denote by $O_{p_1 p_2}$ the complex matrix with all zero coefficients, except the first coefficient that equals $O_{0,0} = F_{0,0}$.

The inverse 2-D DFT of the sum of all $(N+1)$ incomplete 2-D DFTs can be calculated as follows:

$$
\sum_{(p,s)\in J_{N,N}} d^{(p,s)}_{n_1,n_2} = \sum_{(p,s)\in J_{N,N}} \left[\mathcal{F}^{-1}_{N,N} \circ D^{p,s}\right]_{n_1,n_2} = \left[\mathcal{F}^{-1}_{N,N} \circ \sum_{(p,s)\in J_{N,N}} D^{p,s}\right]_{n_1,n_2}
$$

$$
= \left[\mathcal{F}^{-1}_{N,N} \circ \left(F_{p_1,p_2} + N O_{p_1 p_2}\right)\right]_{n_1,n_2}
$$

$$
= \left(\mathcal{F}^{-1}_{N,N} \circ F_{p_1,p_2}\right)_{n_1,n_2} + N \left(\mathcal{F}^{-1}_{N,N} \circ O_{p_1 p_2}\right)_{n_1,n_2}
$$

$$
= f_{n_1,n_2} + \frac{1}{N} F_{0,0} = f_{n_1,n_2} + N E[f].
$$

$$(2.5)$$

Here $E[f]$ denotes the mean of the image,

$$
E[f] = \frac{1}{N^2} \sum_{n=0}^{N-1} \sum_{m=0}^{N-1} f_{n,m} = \frac{1}{N^2} F_{0,0} = \frac{1}{N^2} f_{0,0,0}.
$$

We obtain the following formula of the inverse tensor transformation:

$$
f_{n_1,n_2} = \sum_{(p,s)\in J_{N,N}} d^{(p,s)}_{n_1,n_2} - N E[f] = \frac{1}{N} \sum_{(p,s)\in J_{N,N}} f_{p,s,(n_1 p + n_2 s)\bmod N} - N E[f].
$$

$$(2.6)$$

Considering the set $J_{N,N}$ as

$$
J_{N,N} = \{(1,0),(1,1),(1,2),(1,3),...,(1,N-1)\} \cup \{(0,1)\},
$$

we obtain the following statement.

Statement 1: (Superposition by direction images) The image f_{n_1,n_2} of size $N \times N$, where $N > 2$ is a prime, can be composed of $(N+1)$ direction images as follows:

$$
\begin{aligned}
f_{n_1,n_2} &= \left[\sum_{s=0}^{N-1} d_{n_1,n_2}^{(1,s)} + d_{n_1,n_2}^{(0,1)} \right] - NE[f] \\
&= \frac{1}{N} \left[\sum_{s=0}^{N-1} f_{1,s,(n_1+sn_2) \bmod N} + f_{0,1,n_2} \right] - NE[f], \\
&\quad n_1, n_2 = 0 : (N-1).
\end{aligned} \tag{2.7}
$$

To simplify the above equation, we could assume that the image is centered $f_{n_1,n_2} \to f_{n_1,n_2} - E[f]$. Then, the sum of direction images equals the image,

$$
f_{n_1,n_2} = \sum_{(p,s) \in J} d_{n_1,n_2}^{(p,s)} = \frac{1}{N} \sum_{(p,s) \in J} f_{p,s,(n_1p+n_2s) \bmod N}. \tag{2.8}
$$

If we define the set of generators to be

$$
J_{N,N} = \{(0,1),(1,1),(2,1),(3,1),...,(N-1,1)\} \cup \{(1,0)\},
$$

Statement 1 would be written as follows:

$$
\begin{aligned}
f_{n_1,n_2} &= \left[\sum_{p=0}^{N-1} d_{n_1,n_2}^{(p,1)} + d_{n_1,n_2}^{(1,0)} \right] - NE[f] \\
&= \frac{1}{N} \left[\sum_{p=0}^{N-1} f_{p,1,(n_1p+n_2) \bmod N} + f_{1,0,n_1} \right] - NE[f].
\end{aligned} \tag{2.9}
$$

The image is the sum of components of $(N+1)$ splitting-signals, or the sum of $(N+1)$ direction images of the tensor representation of the image. The above inverse formulas for the discrete image reconstruction from its tensor transform requires only two operations of multiplication and division by N. The division by N can be considered in the definition of the tensor transform components. Then, the reconstruction requires only N^2 such operations.

Each directional $d^{(p,s)}$ completes the original image with details, or parallel straight lines of different brightness in gray scale. The direction of these lines is defined by the generator (p, s). Equation (2.7) describes the principle of superposition of the direction components in image formation. Each direction image is determined by the corresponding splitting-signal, which is calculated by the discrete linear integrals (sums) of the image f_{n_1,n_2} along the parallel lines of $\mathcal{L}_{p,s,t}$, $t = 0 : (N-1)$. These integrals can be calculated from the projection data along the angle defined by the generator (p, s). Thus we obtain the simple formula of reconstruction of the discrete image by its projection data on the Cartesian lattice, by using the splitting-signals of the tensor representation of the image. The number of projections equals the number of generators, i.e., $(N + 1)$, when N is prime. This is the required number of projections for the exact reconstruction of the image $N \times N$.

As an example, Figure 2.9 shows the tree-image after removing the projection data $f_{p,s,(n_1 p + n_2 s) \bmod N}$ that correspond to the direction defined by the generator $(p, s) = (1, 0)$ in part a, $(p, s) = (1, 1)$ in b, and $(p, s) = (0, 1)$ in c.

(a)　　　　　　　　　　　(b)　　　　　　　　　　　(c)

FIGURE 2.9
Images 257×257 reconstructed by 257 projections by splitting-signals, when the projection data of one generator (p, s) have been removed, for (p, s) equals (a) $(1, 0)$, (b) $(1, 1)$, and (c) $(0, 1)$.

The angles of the required projections for reconstructing the image of size 257×257 by the splitting-signals (or direction images) compose the following set

$$\Phi_{257,257} = \{\arctan(p);\ p = 0 : 257\} \cup \{\pi/2\}.$$

Figure 2.10 illustrates all central angles of this set on the unit circle. One

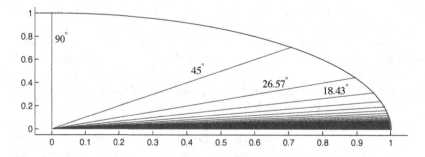

FIGURE 2.10
Central angles of 258 projections for reconstructing the image 257×257.

can see that the points on the circle are not uniformly distributed, and the increment of the angles is not a constant. The main part of the projections are taken along the small angles. For instance, there are no projections for angles in the intervals $(45°, 90°)$ and $(26.57°, 45°)$, as well as $(18.43°, 26.57°)$.

2.2 The inverse tensor transform: Case N is prime

In many practical applications when an image is processed through the frequency domain, the size of the image is considered to be a power of two, such as 256×256, in the case $N \times N$, because of the fast algorithm for the 1-D discrete Fourier transform. Many transformations possess complete systems of basis functions, which are orthogonal, and the 2-D paired transformation is one such transformations. The tensor transform is not orthogonal, but invertible. Therefore, the availability of effective algorithms of the inverse tensor transform would be very useful for many applications, such as image reconstruction from projections, image filtration, and enhancement. The basic functions of the tensor transform define the line-integrals of the discrete image and add them in a specific way, to compose the transform. These basic functions are binary, with values of 0 and 1 only, and therefore not orthogonal. However, such system is complete in the sense that the system defines the inverse tensor transform, as well. We here describe the system of basic functions of the inverse tensor transform, and then show how to calculate these functions by the basic functions of the tensor transform.

Consider the tensor representation of the image $f = \{f_{n,m}\}$ of size $N \times N$,

$$\chi_{N,N} : \{f_{n,m}\} \to \{\{f_{p,s,t}; t = 0 : (N - 1)\}; (p, s) \in J_{N,N}\}, \qquad (2.10)$$

in the case when N is prime. The set of generators is

$$J_{N,N} = \{(1, 0), (1, 1), (1, 2), ..., (1, N - 1)\} \cup \{(0, 1)\}.$$

Given $(p, s) \in J_{N,N}$, the components of the splitting-signal $\{f_{p,s,0}, f_{p,s,1}, ..., f_{p,s,N-1}\}$ are defined as the sums of the image along the parallel lines,

$$f_{p,s,t} = \sum_{(n,m) \in V_{p,s,t}} f_{n,m}, \quad t = 0 : (N - 1), \qquad (2.11)$$

where the sets $V_{p,s,t} = \{(n, m); np + ms = t \bmod N\}$.

2.2.1 Inverse tensor transform

The formula of image reconstruction from its tensor transform, which is given in the principle of superposition, can be described by using the complete system of basic functions of the tensor transform. These functions

$$\chi_{p,s,t}(n, m) = \begin{cases} 1; & \text{if } np + ms = t \bmod N, \\ 0; & \text{otherwise,} \end{cases} \quad t = 0 : (N - 1),$$

define the components $f_{p,s,t} = \chi_{p,s,t} \circ f$.

The number of basic functions, or components in the tensor transform equals $N(N + 1) = N^2 + N$, which exceeds the size of the image. The redundancy can be removed, if we consider that the sum of all components of each splitting-signal $\{f_{p,s,0}, f_{p,s,1}, ..., f_{p,s,N-1}\}$ equals $N^2 E(f)$, and the last component of the signal can be calculated by

$$f_{p,s,N-1} = N^2 E(f) - (f_{p,s,0} + f_{p,s,1} + \cdots + f_{p,s,N-2}).$$

Therefore, we define the tensor transform as the following set of N^2 components:

$$\chi[f] = \begin{cases} \{f_{1,s,0}, f_{1,s,1}, ..., f_{p,s,N-2}\}, & s = 0 : (N-1), \\ \{f_{0,1,0}, f_{0,1,1}, ..., f_{p,s,N-2}, f_{0,1,N-1}\}. \end{cases} \tag{2.12}$$

The system of basic functions of the tensor transform is thus considered to be

$$\chi = \begin{cases} \{\chi_{1,s,0}, \chi_{1,s,1}, ..., \chi_{p,s,N-2}\}, & s = 0 : (N-1), \\ \{\chi_{0,1,0}, \chi_{0,1,1}, ..., \chi_{p,s,N-2}, \chi_{0,1,N-1}\}. \end{cases} \tag{2.13}$$

All basic functions are numbered in the order shown in the above equation, i.e., the first function is $\chi_{1,0,0}$, the second is $\chi_{1,0,1}$, ..., the Nth is $\chi_{1,1,0}$, ..., the $N(N - 1)$th is $\chi_{1,N-1,N-2}$, the $N(N - 1) + 1$ is $\chi_{0,1,0}$, ..., and the last function is $\chi_{0,1,N-1}$. Thus, the triplet-number (p, s, t) corresponds to the following number $k = k(p, s, t)$ from the interval $[0, N^2 - 1]$:

$$k(p, s, t) = \begin{cases} (N-1)s + t, & \text{if } p = 1, \text{ and } s = 0 : (N-1), t = 0 : (N-2), \\ N(N-1) + t, & \text{if } p = 0, \text{ and } s = 1, t = 0 : (N-1). \end{cases}$$

The set of ordered triplet-numbers (p, s, t) is denoted by $U = U_{N,N}$.

The inverse tensor transform is described by N^2 functions that are similar to the functions $\chi_{p,s,t}$. The basic functions in (2.13) can be written row-wise in the matrix $N \times N$ and its inverse matrix can be calculated. Columns of the inverse matrix correspond to the apures of the basis functions of the inverse tensor transform. We denote these two matrices by $[\chi_{N,N}]$ and $[\chi_{N,N}^{-1}]$, respectively.

As an example, consider the $N = 5$ case. The matrix 25×25, composed of the apures of 25 basic functions of the tensor transform, and the transpose inverse matrix equal, respectively,

$$[\chi_{5,5}] =$$

$$
\begin{bmatrix}
1111100000000000000000000\\
0000011111000000000000000\\
0000000000111110000000000\\
0000000000000001111100000\\
1000000001000100010001000\\
0100010000000010001000100\\
0010001000100000000100010\\
0001000100010001000000001\\
1000000100000010100000010\\
0001010000001000000101000\\
0100000010100000010000001\\
0000101000000101000000100\\
1000000010010000000100100\\
0010010000000100100000001\\
0000100100100000001001000\\
0100000001001001000000010\\
1000001000001000001000001\\
0000110000010000010000010\\
0001000001100000100000100\\
0010000010000011000001000\\
1000010000100001000010000\\
0100001000010000100001000\\
0010000100001000010000100\\
0001000010000100001000010\\
0000100001000010000100001
\end{bmatrix}
$$

$$[\chi_{5,5}^{-1}]^{T} =$$

$$
, \frac{1}{5}
\begin{bmatrix}
11111000000000000000 - 1 - 1 - 1 - 1 - 1\\
00000111110000000000 - 1 - 1 - 1 - 1 - 1\\
00000000001111100000 - 1 - 1 - 1 - 1 - 1\\
00000000000000011111 - 1 - 1 - 1 - 1 - 1\\
1000 - 1000 - 1100 - 1100 - 1100 - 11000\\
0100 - 1100 - 1000 - 1010 - 1010 - 10100\\
0010 - 1010 - 1010 - 1000 - 1001 - 10010\\
0001 - 1001 - 1001 - 1001 - 1000 - 10001\\
10 - 1000010 - 10 - 1001010 - 10 - 10010\\
00 - 1101000 - 10 - 1100000 - 11 - 11000\\
01 - 1000001 - 11 - 1000001 - 10 - 10001\\
00 - 1010100 - 10 - 1010100 - 10 - 10100\\
100 - 100 - 10100100 - 100 - 101 - 10100\\
001 - 101 - 10000001 - 101 - 100 - 10001\\
000 - 110 - 11001000 - 100 - 110 - 11000\\
010 - 100 - 10010010 - 110 - 100 - 10010\\
1 - 100001 - 100001 - 100001 - 1 - 10001\\
0 - 100110 - 100010 - 100010 - 1 - 10010\\
0 - 101000 - 101100 - 100100 - 1 - 10100\\
0 - 110000 - 110000 - 111000 - 1 - 11000\\
00000 \quad 00000 \quad 00000 \quad 00000 \quad 50000\\
-11000 - 11000 - 11000 - 11000 \quad 41000\\
-10100 - 10100 - 10100 - 10100 \quad 40100\\
-10010 - 10010 - 10010 - 10010 \quad 40010\\
-10001 - 10001 - 10001 - 10001 \quad 40001
\end{bmatrix}.
$$

The tensor transform $\mathbf{Z} = (Z_0, Z_1, Z_2, ..., Z_{N^2-1})^T$ of the image $\{f_{n,m}\}$ can be considered as the product of the matrix $[\chi_{N,N}]$ on the image-vector $\mathbf{f} = (f_{n+mN} = f_{n,m}; \; n, m = 0 : N - 1)^T,$

$$\mathbf{Z} = [\chi_{N,N}]\mathbf{f},$$

and $f_{p,s,t} = Z_{k(p,s,t)}, \; (p, s, t) \in U.$ If we pack the tensor transform into the matrix $(N \times N)$ by $Z_k \to Z_{k_1,k_2}$, where $k_1 + k_2 N = k, \; k = 0 : (N^2 - 1)$, then the following formula can be considered:

$$
f_{p,s,t} =
\begin{cases}
Z_{(s(N-1)+t) \bmod N, \lfloor s + \frac{t-s}{N} \rfloor}, & \text{if } p = 1, \; s = 0 : (N - 1), \; t = 0 : (N - 2),\\
Z_{N-1,t}, & \text{if } p = 0, \; s = 1, \; t = 0 : (N - 1),
\end{cases}
$$

where $\lfloor . \rfloor$ denotes the floor function.

One can notice that all basic functions of the inverse tensor transform have values of ± 1 and 0, except the last five functions. Each of these functions considered as the 2-D matrix has one coefficient that equals 5 or 4 on the last five rows (in column number 21). Such peculiarity holds for matrices $N \times N$ for other cases of prime $N > 2$; these matrices have coefficients N and $N - 1$ in the last N rows.

For example, the transpose inverse matrix $[\chi_{7,7}^{-1}]^T$ of the 7×7-point tensor transform presents the following table (up to the factor of $1/7$):

```
1 1 1 1 1 1 1 1-0-0 0 0 0 0 0 0 0-0 0-0-0 0-0-0-0-0 0-0-0-0-0 0 0-0-0-0-0 0 0 0-0 0-0-1-1-1-1-1-1-1
0 0 0 0 0 0 0 1 1 1 1 1 1 1 0 0 0 0 0 0 0 0 0 0 0 0 0 0 0 0 0 0 0 0 0 0 0 0 0 0-1-1-1-1-1-1-1
0 0 0 0 0 0 0 0 0 0 0 0 0 1 1 1 1 1 1 1 0 0 0 0 0 0 0 0 0 0 0 0 0 0 0 0 0 0 0 0-1-1-1-1-1-1-1
0 0 0 0 0 0 0 0 0 0 0 0 0 0 0 0 0 0 0 0 1 1 1 1 1 1 1 0 0 0 0 0 0 0 0 0 0 0 0 0-1-1-1-1-1-1-1
0 0 0 0 0 0 0 0 0 0 0 0 0 0 0 0 0 0 0 0 0 0 0 0 0 0 0 1 1 1 1 1 1 1 0 0 0 0 0 0-1-1-1-1-1-1-1
0 0 0 0 0 0 0 0 0 0 0 0 0 0 0 0 0 0 0 0 0 0 0 0 0 0 0 0 0 0 0 0 0 0 1 1 1 1 1 1 1-1-1-1-1-1-1-1
1 0 0 0 0 0-1 0 0 0 0 0-1 1 0 0 0 0-1 1 0 0 0 0-1 1 0 0 0 0-1 1 0 0 0 0-1 1 0 0 0 0-1 1 0 0 0 0 0
0 1 0 0 0 0-1 1 0 0 0 0-1 0 0 0 0 0-1 0 1 0 0 0-1 0 1 0 0 0-1 0 1 0 0 0-1 0 1 0 0 0-1 0 1 0 0 0 0
0 0 1 0 0 0-1 0 1 0 0 0-1 0 1 0 0 0-1 0 1 0 0 0-1 0 0 1 0 0-1 0 0 1 0 0-1 0 0 1 0 0-1 0 0 1 0 0 0
0 0 0 1 0 0-1 0 0 1 0 0-1 0 0 1 0 0-1 0 0 0 0 0-1 0 0 0 1 0-1 0 0 0 1 0-1 0 0 0 1 0-1 0 0 0 1 0 0
0 0 0 0 1 0-1 0 0 0 1 0-1 0 0 0 1 0-1 0 0 0 1 0-1 0 0 0 1 0-1 0 0 0 0-1 0 0 0 0-1 1 0 0 0 1 0
0 0 0 0 0 1-1 0 0 0 0 1-1 0 0 0 0 1-1 0 0 0 0 1-1 0 0 0 0 1-1 0 0 0 0 1-1 0 0 0 0-1 0 0 0 0 0 1
1 0 0-1 0 0 0 0 1 0 0-1 0 0 0 1 0 0-1 0 0 1 0 0-1 0 0 1 0 0-1 0 0 1 0 0 0-1 0 0 0 0 0 1
0 0 0-1 1 0 0 1 0 0-1 0 0 1 1 0 0 0 0-1 1 0-1 1 0 0 0 0 0-1 1 0-1 1 0 0 0 0
0 1 0-1 0 0 0 0 0 1 0-1 1 0-1 0 0 0 0 0 1 0-1 0 0-1 0 1 0 0 0 1 0-1 0 0-1 0 0 1 0
0 0 0-1 0 1 0 0 1 0 0 0-1 0 0-1 0 1 0 0 1 0 0 0-1 0 0-1 0 1 0 0 0 0 0-1 0 1-1 0 1 0 0 0 0
0 0 1-1 0 0 0 0 0 0 1-1 0 1-1 0 0 0 0 1-1 0 0-1 0 1 0 0 0 0 0 1-1 0 0-1 0 1 0 0 0 0 0 1
0 0 0-1 0 0 1 0 0 1 0 0-1 0 0 1 0 0-1 0 0 1 0 0 1 0 0-1 0 0 1 0 0-1 0 0 1 0 0-1 0 0 1 0 0 0
1 0-1 0 0 0 0 0 1 0-1 0 0 0 0 0 1 0-1 0 0 0 1 0-1 0 0 0 0 1 0-1 0 0 0 0 1 0-1 0 0 0 0 1 0
0 0-1 0 0 1 0 1 0 0-1 0 0 0 0 0-1 0 0 1 0 0 0 0-1 0 0 1 0 1 0 0-1 0 0 0 0 0-1 1-1 1 0 0 0 0
0 0-1 1 0 0 0 0 0-1 1 0 1 0 0 0-1 0 1 0 0 0 0 0-1 1-1 1 0 0 0 0 0 0-1 1-1 1 1 0 0 0 0
0 1-1 0 0 0 0 1-1 0 0 0 0 0 0 1-1 0 0 0 0 1-1 0 0 0 0-1 0 0 0 0 0 1-1 0 0 0-1 0 1 0 0 0 0 0
0 0-1 0 0 1 0 1 0-1 0 0 0 0 1 0-1 0 1 0 0 0 0-1 0 1 0 1 0 0-1 0 0 0 0 1-1 0 0 0 0
1 0 0 0-1 0 0 0 0-1 0 1 0 0-1 0 1 0 0 0 0 0-1 0 0 0 0-1 1 0 0 0 0-1 0 1 0 0
0 0 1 0 0-1 0 1 0 0 0-1 0 0 1 0 0 0-1 0 1 0-1 0 0 0-1 0 1 0 0 1-1 0 0 0 1 0
0 0 0 1-1 0 0 0 1-1 0 0 0-1 1 0 0 0 0 1-1 0 1-1 0 0 0 0 0 1-1 0 1-1 0 0 0 0
1 0 0 0-1 0 0 0 0-1 0 1 0 0 0-1 0 1 0 0 0 0 0-1 0 1 0 0-1 0 0 0 0-1 1 0 0 0 0
0 0 1 0 0-1 0 1 0 0 0-1 0 0 1 0 0 0-1 0 1 0-1 0 0 0-1 0 1 0 0 1-1 0 0 0 1 0
0 0 0 1-1 0 0 0 1-1 0 0 0-1 1 0 0 0 0 1-1 0 1-1 0 0 0 0 0 1-1 0 1-1 0 0 0 0 0 0 0 1
0 0 0 0-1 1 0 0-1 1 0 0 0 0 0-1 1 0-1 1 0 0 0 0 0-1 1 0-1 1 0 0-1 1 0 0 0 0
0 1 0 0-1 0 0 1 0 0 0-1 0 0 0 0-1 0 0 0 0 1 0 0 0-1 1 0 0 0 0 0-1 1 0 0 0 0-1 0 0 1 0 0
1-1 0 0 0 0 0 1-1 0 0 0 0 1-1 0 0 0 0 1-1 0 0 0 0 1-1 0 0 0 0 1-1-1 0 0 0 0 1
0-1 0 0 0 1 1 0-1 0 0 0 1 0-1 0 0 0 1 0-1 0 0 0 1 0-1 0 0 0 1 0-1 0 1 0 0 1 0
0-1 0 0 0 1 0 0-1 0 0 1 1 0 0-1 0 0 0 1 0 0-1 0 0 1 0 0 0-1 1 0 0 0 0
0-1 0 0 1 0 0 0-1 0 1 0 0 0-1 0 1 0 0 0-1 0 1 0 0 0-1 0 1 0 0 0-1 0 1 0 0
0-1 0 1 0 0 0 0-1 0 1 0 0 0 0-1 0 1 0 0 0 0-1 0 1 1 0 0 0 0-1 0 1 0 0
0-1 1 0 0 0 0 0-1 1 0 0 0 0 0-1 1 0 0 0 0 0-1 1 0 0 0 0 0-1 1 1 0 0 0 0 0-1 1 0 0 0 0
0 0 0 0 0 0 0 0 0 0 0 0 0 0 0 0 0 0 0 0 0 0 0 0 0 0 0 0 0 0 0 0 0 0 0 0 0 7 0 0 0 0 0 0
-1 1 0 0 0 0 0-1 1 0 0 0 0 0-1 1 0 0 0 0 0-1 1 0 0 0 0 0-1 1 0 0 0 0 0 6 1 0 0 0 0 0
-1 0 1 0 0 0 0-1 0 1 0 0 0 0-1 0 1 0 0 0 0-1 0 1 0 0 0 0-1 0 1 0 0 0 0 6 0 1 0 0 0 0
-1 0 0 1 0 0 0-1 0 0 1 0 0 0-1 0 0 1 0 0 0-1 0 0 1 0 0 0-1 0 0 1 0 0 0 6 0 0 1 0 0 0
-1 0 0 0 1 0 0-1 0 0 0 1 0 0-1 0 0 0 1 0 0-1 0 0 0 1 0 0-1 0 0 0 1 0 0 6 0 0 0 1 0 0
-1 0 0 0 0 1 0-1 0 0 0 0 1 0-1 0 0 0 0 1 0-1 0 0 0 0 1 0-1 0 0 0 0 1 0 6 0 0 0 0 1 0
-1 0 0 0 0 0 1-1 0 0 0 0 0 1-1 0 0 0 0 0 1-1 0 0 0 0 0 1-1 0 0 0 0 0 1 6 0 0 0 0 0 1
```

A simple analytical formula can be derived for basic functions of the inverse tensor transform (ITT). To show that, we number the basic functions of the ITT by the same triplets $(p, s, t) \in U$ and preserve their orders as in the forward TT. Then the following holds:

$$\chi_{p,s,t}^{-1}(n, m) = \frac{1}{N} \begin{cases} \chi_{p,s,t}(n, m) - \chi_{p,s,N-1}(n, m) \\ \quad \text{if } p = 1, \ s = 0 : (N - 1), \ t = 0 : (N - 2), \\ \chi_{p,s,t}(n, m) - \chi_{p,s,0}(n, m) + N\delta_{N-1,0}(n, m) \\ \quad \text{if } p = 0, \ s = 1, \ t = 0 : (N - 1), \end{cases} \quad (2.14)$$

where $n, m = 0 : (N - 1)$. As an example, we first consider the $N = 7$ case and the triplet-number $(p, s, t) = (1, 5, 3)$. This number corresponds to the row with number $k = k(1, 5, 3) = 34$ in the matrix $[\chi_{7,7}]$.

Therefore we consider the 34-th row in the matrix $[\chi_{7,7}^{-1}]^T$. We obtain the following basic functions:

$$[\chi_{1,5,3}] = \begin{bmatrix} 0 & 0 & 0 & 1 & 0 & 0 & 0 \\ 0 & 0 & 0 & 0 & 0 & 1 & 0 \\ 1 & 0 & 0 & 0 & 0 & 0 & 0 \\ 0 & 0 & 1 & 0 & 0 & 0 & 0 \\ 0 & 0 & 0 & 0 & 1 & 0 & 0 \\ 0 & 0 & 0 & 0 & 0 & 0 & 1 \\ 0 & 1 & 0 & 0 & 0 & 0 & 0 \end{bmatrix}, \quad [\chi_{1,5,3}^{-1}] = \frac{1}{7} \begin{bmatrix} 0 & 0 & 0 & 1 & 0 & 0 & -1 \\ 0 & -1 & 0 & 0 & 0 & 1 & 0 \\ 1 & 0 & 0 & -1 & 0 & 0 & 0 \\ 0 & 0 & 1 & 0 & 0 & -1 & 0 \\ -1 & 0 & 0 & 0 & 1 & 0 & 0 \\ 0 & 0 & -1 & 0 & 0 & 0 & 1 \\ 0 & 1 & 0 & -1 & 0 & 0 & 0 \end{bmatrix},$$

and

$$[\chi_{1,5,3}] - 7[\chi_{1,5,3}^{-1}] = [\chi_{1,5,6}] = \begin{bmatrix} 0 & 0 & 0 & 0 & 0 & 0 & 1 \\ 0 & 1 & 0 & 0 & 0 & 0 & 0 \\ 0 & 0 & 0 & 1 & 0 & 0 & 0 \\ 0 & 0 & 0 & 0 & 0 & 1 & 0 \\ 1 & 0 & 0 & 0 & 0 & 0 & 0 \\ 0 & 0 & 1 & 0 & 0 & 0 & 0 \\ 0 & 0 & 0 & 0 & 1 & 0 & 0 \end{bmatrix},$$

or $[\chi_{1,5,3}^{-1}] = \frac{1}{7}([\chi_{1,5,3}] - [\chi_{1,5,6}])$.

One also can see that for the 35-th basic function, we have the following:

$$[\chi_{1,5,4}] - 7[\chi_{1,5,4}^{-1}] = \begin{bmatrix} 0 & 0 & 0 & 0 & 1 & 0 & 0 \\ 0 & 0 & 0 & 0 & 0 & 0 & 1 \\ 0 & 1 & 0 & 0 & 0 & 0 & 0 \\ 0 & 0 & 0 & 1 & 0 & 0 & 0 \\ 0 & 0 & 0 & 0 & 0 & 1 & 0 \\ 1 & 0 & 0 & 0 & 0 & 0 & 0 \\ 0 & 0 & 1 & 0 & 0 & 0 & 0 \end{bmatrix} - \begin{bmatrix} 0 & 0 & 0 & 0 & 1 & 0 & -1 \\ 0 & -1 & 0 & 0 & 0 & 0 & 1 \\ 0 & 1 & 0 & -1 & 0 & 0 & 0 \\ 0 & 0 & 0 & 1 & 0 & -1 & 0 \\ -1 & 0 & 0 & 0 & 0 & 1 & 0 \\ 1 & 0 & -1 & 0 & 0 & 0 & 0 \\ 0 & 0 & 1 & 0 & -1 & 0 & 0 \end{bmatrix} = [\chi_{1,5,6}],$$

or $[\chi_{1,5,4}^{-1}] = \frac{1}{7}([\chi_{1,5,4}] - [\chi_{1,5,6}])$.

Now we consider the triplet-number $(p, s, t) = (0, 1, 2)$. This number corresponds to the row number $k = k(0, 1, 2) = 45$. The basic functions of the direct and inverse tensor transform with number $(0, 1, 2)$ equal, respectively,

$$[\chi_{0,1,2}] = \begin{bmatrix} 0 & 0 & 0 & 0 & 0 & 0 & 0 \\ 0 & 0 & 0 & 0 & 0 & 0 & 0 \\ 1 & 1 & 1 & 1 & 1 & 1 & 1 \\ 0 & 0 & 0 & 0 & 0 & 0 & 0 \\ 0 & 0 & 0 & 0 & 0 & 0 & 0 \\ 0 & 0 & 0 & 0 & 0 & 0 & 0 \\ 0 & 0 & 0 & 0 & 0 & 0 & 0 \end{bmatrix}, \quad [\chi_{0,1,2}^{-1}] = \frac{1}{7} \begin{bmatrix} -1 & -1 & -1 & -1 & -1 & -1 & 6 \\ 0 & 0 & 0 & 0 & 0 & 0 & 0 \\ 1 & 1 & 1 & 1 & 1 & 1 & 1 \\ 0 & 0 & 0 & 0 & 0 & 0 & 0 \\ 0 & 0 & 0 & 0 & 0 & 0 & 0 \\ 0 & 0 & 0 & 0 & 0 & 0 & 0 \\ 0 & 0 & 0 & 0 & 0 & 0 & 0 \end{bmatrix},$$

and $[\chi^{-1}_{0,1,2}] = \frac{1}{7}\big([\chi_{0,1,2}] - [\chi_{0,1,0}]\big) + [\delta_{6,0}]$. Indeed

$$[\chi_{0,1,2}] - 7[\chi^{-1}_{0,1,2}] = [\chi_{0,1,0}] - 7[\delta_{6,0}] = \begin{bmatrix} 1 & 1 & 1 & 1 & 1 & 1 & 1-7 \\ 0 & 0 & 0 & 0 & 0 & 0 & 0 \\ 0 & 0 & 0 & 0 & 0 & 0 & 0 \\ 0 & 0 & 0 & 0 & 0 & 0 & 0 \\ 0 & 0 & 0 & 0 & 0 & 0 & 0 \\ 0 & 0 & 0 & 0 & 0 & 0 & 0 \\ 0 & 0 & 0 & 0 & 0 & 0 & 0 \end{bmatrix}.$$

For the 46-th function, we have $[\chi^{-1}_{0,1,3}] = \frac{1}{7}\big([\chi_{0,1,3}] - [\chi_{0,1,0}]\big) + [\delta_{6,0}]$. Indeed

$$[\chi_{0,1,3}] - 7[\chi^{-1}_{0,1,3}] = \begin{bmatrix} 0 & 0 & 0 & 0 & 0 & 0 & 0 \\ 0 & 0 & 0 & 0 & 0 & 0 & 0 \\ 0 & 0 & 0 & 0 & 0 & 0 & 0 \\ 1 & 1 & 1 & 1 & 1 & 1 & 1 \\ 0 & 0 & 0 & 0 & 0 & 0 & 0 \\ 0 & 0 & 0 & 0 & 0 & 0 & 0 \\ 0 & 0 & 0 & 0 & 0 & 0 & 0 \end{bmatrix} - \begin{bmatrix} -1 & -1 & -1 & -1 & -1 & -1 & 6 \\ 0 & 0 & 0 & 0 & 0 & 0 & 0 \\ 0 & 0 & 0 & 0 & 0 & 0 & 0 \\ 1 & 1 & 1 & 1 & 1 & 1 & 1 \\ 0 & 0 & 0 & 0 & 0 & 0 & 0 \\ 0 & 0 & 0 & 0 & 0 & 0 & 0 \\ 0 & 0 & 0 & 0 & 0 & 0 & 0 \end{bmatrix}$$

$$= [\chi_{0,1,0}] - 7[\delta_{6,0}].$$

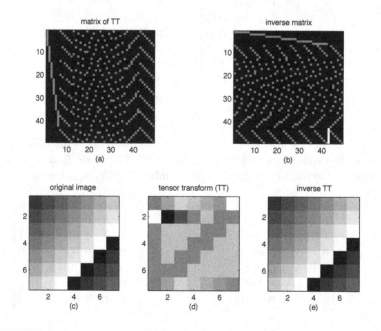

FIGURE 2.11
(a) The matrix 49×49 of the DTT and (b) transpose inverse matrix, (c) the image of size 7×7, (d) DTT of the image, and (e) inverse DTT.

Figure 2.11 shows the binary image of the matrix $[\chi_{7,7}]$ of the tensor transformation in part a, along with the gray-scale image of the transpose inverse matrix in b. The image $f_{n,m} = (n+m-1) \bmod 10$, $n, m = 0:6$, is shown in c, the discrete tensor transform (DTT) of the image in d, and the inverse tensor transform in e. In the image of the inverse matrix in b, the values $-1, 0, 1$, and $6, 7$ are shown in grays from black to white.

2.2.2   Formula of the inverse tensor transform

As follows directly from Equation (2.14), the following calculations hold for the inverse discrete tensor transform:

$$Nf_{n,m} = \sum_{(p,s,t)\in U} f_{p,s,t}\chi_{p,s,t}(n,m) - \sum_{s=0}^{N-1} \left(\sum_{t=0}^{N-2} f_{1,s,t} \right) \chi_{1,s,N-1}(n,m)$$

$$- \left(\sum_{t=0}^{N-1} f_{0,1,t} \right) [\chi_{0,1,0}(n,m) - N\delta_{N-1,0}(n,m)]$$

$$= \sum_{(p,s,t)\in U} f_{p,s,t}\chi_{p,s,t}(n,m) - \sum_{s=0}^{N-1} \left(N^2 E(f) - f_{1,s,N-1} \right) \chi_{1,s,N-1}(n,m)$$

$$- N^2 E(f) [\chi_{0,1,0}(n,m) - N\delta_{N-1,0}(n,m)]$$

$$= \sum_{(p,s,t)\in U} f_{p,s,t}\chi_{p,s,t}(n,m) + \sum_{s=0}^{N-1} f_{1,s,N-1}\chi_{1,s,N-1}(n,m)$$

$$- N^2 E(f) \left(\sum_{s=0}^{N-1} \chi_{1,s,N-1}(n,m) + \chi_{0,1,0}(n,m) - N\delta_{N-1,0}(n,m) \right).$$

It is not difficult to see that the image in the round braces is constant,

$$c(n,m) = \sum_{s=0}^{N-1} \chi_{1,s,N-1}(n,m) + \chi_{0,1,0}(n,m) - N\delta_{N-1,0}(n,m) \equiv 1.$$

Thus, the inversion formula states that the image is composed as the difference

$$f_{n,m} = b_U(n,m) - e(f), \quad n, m = 0:(N-1), \tag{2.15}$$

of the backprojection

$$b_U(n,m) = \frac{1}{N} \left(\sum_{(p,s,t)\in U} f_{p,s,t}\chi_{p,s,t}(n,m) + \sum_{s=0}^{N-1} f_{1,s,N-1}\chi_{1,s,N-1}(n,m) \right) \tag{2.16}$$

and the constant-image, which is the mean of the splitting-signal

$$e(f) = NE(f) = \frac{1}{N} \sum_{n=0}^{N-1}\sum_{m=0}^{N-1} f_{n,m} = \frac{1}{N} \sum_{t=0}^{N-1} f_{1,0,t}. \tag{2.17}$$

Thus, although we have considered the tensor transform in the form of (2.12) with N^2 components, the entire tensor $(N+1) \times N$ is actually used in the image reconstruction, when *backprojecting by spirals* the transform into the points (n, m). Therefore, we denote the set of all $(N+1)N$ triplets in the tensor transform by

$$U' = U'_{N,N} = U_{N,N} \cup \{(1, s, N-1); \ s = 0 : (N-1)\}.$$

The 2nd Inverse formula for DTT (Spiral Backprojection): The image of size $N \times N$, when N is prime, can be calculated from the tensor transform by

$$f_{n,m} = \frac{1}{N} \sum_{(p,s,t) \in U'} f_{p,s,t} \chi_{p,s,t}(n, m) - e(f), \quad n, m = 0 : (N-1). \quad (2.18)$$

We just mentioned that this formula describes the image reconstruction by projecting back the transform data, $f_{p,s,t}$, in each point (n, m) along the spirals, not the line as in the traditional method of backprojection which was originated by considering the polar system of coordinates instead of the Cartesian. Indeed, as shown above, one can identify the opposite sides of boundaries of the square $[0, N] \times [0, N]$ and consider it in the 3-D space as a torus and the 2-D Cartesian lattice $X_{N,N} = \{(n, m); n, m = 0 : (N - 1)\}$ as a net traced on the torus. Then, for a given triplet-number (p, s, t), the straight lines in the set $V_{p,s,t}$ compose one discrete spiral $S_{p,s,t}$ on the torus, because of the equality $(xp + ys) \bmod N = (t + kN) \bmod N = t$ for all integers k. Each component $f_{p,s,t}$ of the tensor transform is the sum of line-integrals, the integral along one spiral on the torus. Equation (2.18) describes the backprojection by spirals, or *spiral backprojection*.

Thus, in the case when N is prime, the image reconstruction from its tensor transform, or the inverse tensor transform, is determined by the same system of basic functions $\{\chi_{p,s,t}\}$. The tensor transformation is a good example of a non-separable and non-orthogonal 2-D transformation with the simple system of non-negative basic functions.

2.2.2.1   Code for inverse tensor transform

Below is the simple code with script demo_7x7inverseTT.m, which is written to calculate the tensor transform of the image, all basic functions of the transform, and the inverse tensor transform in the 7×7 case.

```
%%%%%%%%%%%%%%%%%%%%%%%%% MAIN CODE %%%%%%%%%%%%%%%%%%%%%%%%%%%%%%%%%%%%%
% call: demo_7x7inverseTT.m
% Demo: The inverse formula for the tensor transform 7x7
% Art-Meruzhan Grigoryan, November 12, 2010
   N=7; N49=N*N;
   fprintf('    Tensor transform (%g,%g) \n',N,N);
   TT_matrix=[];   k=1; p=1;
   for s=0:N-1
```

```
        for t=0:N-2
            A=matrix_pst(N,p,s,t); TT_matrix(k,:)=reshape(A,1,N49);
            fprintf(' %g  tensor matrix for (%g,%g,%g) \n',k,p,s,t);
            k=k+1;
        end
    end
    s=1; p=0;
    for t=0:N-1
        A=matrix_pst(N,p,s,t); TT_matrix(k,:)=reshape(A,1,N49);
        fprintf(' %g  tensor matrix for (%g,%g,%g) \n',k,p,s,t);
        k=k+1;
    end
    TT_imatrix=inv(TT_matrix)*N;        % (*N) can be removed  *
    % Calculate TT:
    % Example with the following original image:
    %  1      2     3     4     5     6     7
    %  2      3     4     5     6     7     8
    %  3      4     5     6     7     8     9
    %  4      5     6     7     8     9     0
    %  5      6     7     8     9     0     1
    %  6      7     8     9     0     1     2
    %  7      8     9     0     1     2     3
    for n=1:N
        for m=1:N
            image_1(n,m)=mod(n+m-1,10);
        end
    end
    signal_1=reshape(image_1,N49,1);
    signal_xx=TT_matrix*signal_1;
    XX=reshape(signal_xx,N,N);
    % XX is the TT of the image_1
    %  28     49    29    29    29    29    28
    %  35      9    39    39    39    29    35
    %  42     19    39    39    29    29    42
    %  39     29    39    29    29    39    39
    %  36     39    29    29    39    39    36
    %  33     29    29    39    39    39    33
    %  49     29    39    39    39    39    30 * the last s-signal
    % sum(sum(XX(1:7,7)))=243=sum(sum(image_1))
    % -----------------------------------------------------------------
    % Check if the inverse TT is correct:
    signal_2=round(TT_imatrix*signal_xx/N);  % (/N) can be removed *
    image_2=reshape(signal_2,N,N);           % it equals image_1
    % -----------------------------------------------------------------
    % The image reconstruction by the derived formulas (2.15)-(2.18)
    fprintf(' Inverse Tensor transform (%g,%g) \n',N,N);
    % TT is the vector-signal 'signal_xx' of length 49.
    B=zeros(N);
    p=1; k=1;
```

```
for s=0:N-1
    for t=0:N-2
        A=matrix_pst(N,p,s,t); B=B+signal_xx(k)*A; k=k+1;
    end
end
s=1; p=0;
for t=0:N-1
    A=matrix_pst(N,p,s,t); B=B+signal_xx(k)*A; k=k+1;
end
% Consider additional N matrices (and coefficients)
for s=0:N-1
    A=matrix_pst(N,1,s,N-1);
    TT=sum(sum(A.*image_1));   % i.e. f_{1,s,N-1}
    B=B+TT*A;
end
B=B/N;
C=B-sum(sum(image_1))/N;       % image reconstruction
%%%%%%%%%%%%%%%%%%%%%%%%%%% End of the code %%%%%%%%%%%%%%%%%%%%%%%%%%%%%%%%%
% Call: matrix_pst.m        /    Artyom Grigoryan, 1997.
function A=matrix_pst(N,p,s,t)
  if t>N t=mod(t,N); end
  A=zeros(N,N);
  ms=0;
  for m=1:N
      np=0;
      for n=1:N
        t1=np+ms; t1=mod(t1,N); if t1==t A(m,n)=1; end
        np=np+p;
      end
      ms=ms+s;
  end
```

2.3   3-D paired representation

We introduce a more advanced form of representation of images, than the tensor representation. This form associates with special partitions of the lattice in the frequency domain, which reveal the two-dimensional discrete Fourier transform. The tensor representation of the image is associated with an irreducible covering σ not being partitions of the lattice. This is why there is a redundancy in calculations; the same spectral information is contained in different groups of frequency-points and the splitting-signals together contains more points than the size of the represented image. As an illustration, Figure 2.12 shows the image 256×256 in part a, along with the image of 384 1-D splitting-signals in b, which lie on rows of this image. These 1-D signals describe uniquely the original image and, at the same time, they split the mathematical structure of the 2-D discrete Fourier transform (shown in c) into a set of separate 1-D transforms (shown all together as an image in d). The

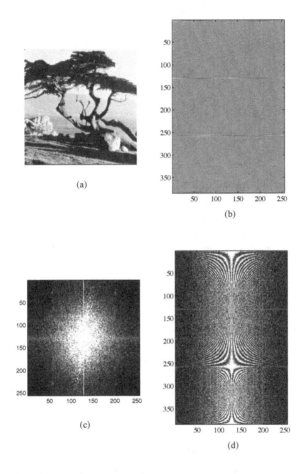

FIGURE 2.12
Conventional and tensor representation of the image and its spectrum. (a) Image 256 × 256, (b) the image of all 384 splitting-signals, (c) the 256 × 256-point DFT of the image, and (d) the image of the 256-point DFTs of splitting-signals. (DFTs are in absolute scale and shifted to the centers.)

2-D DFT is thus considered as a unique set of separate 1-D DFTs. The image in b is the tensor representation of the image, or the tensor transform of the image,

$$\chi = \chi_\sigma : f \to \{f_{T_{p,s}}; T_{p,s} \in \sigma_J\}. \tag{2.19}$$

One can note that a few splitting-signals are well expressed, which can be seen also from the energy plot of all splitting-signals, which is given in Figure 2.13. For this image, the high energy is concentrated in the splitting-

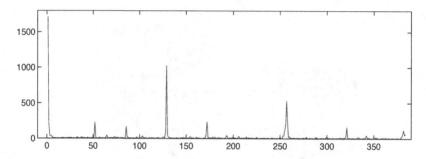

FIGURE 2.13
Energy of all 384 splitting-signals of the tree image.

signals of numbers $1, 129, 257, 172, 52, 258,$ and $2,$ which correspond to the generators $(p, s) = (0, 1), (128, 1), (1, 0), (171, 1), (51, 1), (1, 2),$ and $(1, 1),$ respectively. The processing of only these signals may lead to good results, for instance in image enhancement [49, 50, 52]. One such example is shown in Figure 1.4, when the image is enhanced by one splitting-signal of the generator $(p, s) = (7, 1).$ However, all together the splitting-signals contain 384×256 points, which is much greater than the original size $256 \times 256.$

In order to remove such redundancy, we consider the concept of the paired representation of images. Unlike the tensor representation, the paired representation allows for distributing the spectral information of the image by disjoint sets of frequency-points. This property makes the paired representation the most attractive in image processing, as well as in developing effective algorithms for calculating the 2-D DFT transform.

2.3.1   2D-to-3D paired transform

A complete system of functions can be derived from the two-dimensional discrete Fourier transform as a system that splits the transform [40]-[42],[46, 55]. In other words, there exists a system of functions that reveals completely the mathematical structure of the 2-D DFT and represents it as a minimal composition of short 1-D DFTs. Such a system is three-dimensional, i.e., the system of functions are numbered by three parameters, namely, two parameters for the spatial frequencies and one parameter for the time. The change in time determines a series of functions, and the total number of triples numbering the system of functions, since the system is complete, equals the size of the 2-D DFT, let us say $N \times N.$ The complete systems of such functions which are called *the paired functions* exist also in the one- and multi-dimensional cases.

Let N be a power of two, $N = 2^r, r > 1.$ We consider an irreducible covering σ_J of the lattice $X_{N,N}$, and for each generator $(p, s) \in J$, we determine the characteristic functions of the sets $V_{p,s,t} = \{(n_1, n_2); n_1 p + n_2 s = t \bmod N\}$,

$$\chi_{p,s,t}(n_1, n_2) = \begin{cases} 1; & (n_1, n_2) \in V_{p,s,t}, \\ 0; & \text{otherwise}, \end{cases} \quad t = 0 : (N-1). \tag{2.20}$$

These functions describe the tensor transformation of the 2-D image f,

$$f_{p,s,t} = \chi_{p,s,t} \circ f = \sum_{n_1=0}^{N-1} \sum_{n_2=0}^{N-1} \chi_{p,s,t}(n_1, n_2) f_{n_1,n_2}, \quad t = 0 : (N-1). \tag{2.21}$$

Therefore, the tensor transformation χ_σ is described by the following system of binary functions:

$$\chi_\sigma = \{\chi_{p,s,t}; \, T_{p,s} \in \sigma_J, \, t = 0 : (N-1)\}. \tag{2.22}$$

The tensor transformation is not orthogonal, but, by means of this transformation, one can synthesize unitary transformations.

Definition 2.1 For a given frequency-point $(p, s) \in X_{N,N}$ and integer $t \in \{0, 1, 2, ..., N/2 - 1\}$, the function

$$\chi'_{p,s,t}(n_1, n_2) = \chi_{p,s,t}(n_1, n_2) - \chi_{p,s,t+N/2}(n_1, n_2)$$

$$= \begin{cases} 1, \text{if } (n_1, n_2) \in V_{p,s,t} \\ -1, \text{if } (n_1, n_2) \in V_{p,s,t+N/2} \\ 0, \text{otherwise} \end{cases} \tag{2.23}$$

$$(n_1, n_2) \in X_{N,N},$$

is called *the paired function*.

Example 2.2 We consider the $N = 8$ case. The paired functions are calculated by

$$\chi'_{p,s,t}(n_1, n_2) = \begin{cases} 1; & \text{if } n_1 p + n_2 s = t \bmod 8 \\ -1; & \text{if } n_1 p + n_2 s = (t+4) \bmod 8 \\ 0; & \text{otherwise} \end{cases} \tag{2.24}$$

where $t = 0 : 3$. For instance, the values of the function $\chi'_{3,1,2}(n_1, n_2)$ can be written in the form of the following mask:

$$[\chi'_{3,1,2}] = \begin{bmatrix} 0 & 0 & 1 & 0 & 0 & 0 & -1 & 0 \\ 0 & 0 & 0 & -1 & 0 & 0 & 0 & 1 \\ -1 & 0 & 0 & 0 & 1 & 0 & 0 & 0 \\ 0 & 1 & 0 & 0 & 0 & -1 & 0 & 0 \\ 0 & 0 & -1 & 0 & 0 & 0 & 1 & 0 \\ 0 & 0 & 0 & 1 & 0 & 0 & 0 & -1 \\ 1 & 0 & 0 & 0 & -1 & 0 & 0 & 0 \\ 0 & -1 & 0 & 0 & 0 & 1 & 0 & 0 \end{bmatrix}.$$

The elements $f'_{p,s,t}$ of the paired representation of the image $f_{n,m}$ are defined as

$$f'_{p,s,t} = f_{p,s,t} - f_{p,s,t+N/2} = \chi'_{p,s,t} \circ f_{n,m} = \sum_{n=0}^{N-1}\sum_{m=0}^{N-1} \chi'_{p,s,t}(n,m)f_{n,m} \quad (2.25)$$

where the time parameter t runs integers from the interval $[0, N/2-1]$. Thus, the pair of elements of the tensor representation, $f_{p,s,t}$ and $f_{p,s,t+N/2}$, defines $f'_{p,s,t}$.

As an example, Figure 2.14 shows the image in part a, along with the tensor splitting-signal $\{f_{1,4,0}, f_{1,4,1}, f_{1,4,2}, f_{1,4,3}, \dots f_{1,4,255}\}$ of length 256 in b, and paired splitting-signal $\{f'_{1,4,0}, f'_{1,4,1}, f'_{1,4,2}, \dots f'_{1,4,127}\}$ of length 128 in c. The signals are generated by the frequency-point $(p, s) = (1, 4)$.

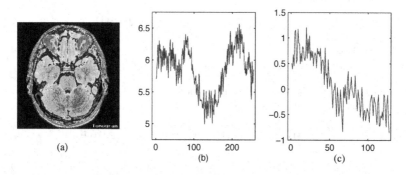

(a)   (b)   (c)

FIGURE 2.14

(a) Image 256×256, (b) splitting-signal $f_{T_{1,4}}$, and (c) splitting-signal $f_{T'_{1,4}}$.

The 2-D paired transform is defined as a set of N^2 such components

$$\{f'_{p,s,t}; \; (p,s,t) \in U\}, \quad\quad (2.26)$$

where the set U of triplet-numbers (p, s, t) can be taken as follows:

$$U = \bigcup_{n=0}^{r-1} \left\{ (p, s, 2^n t); \; (p, s) \in 2^n J_{N/2^n, N/2^n}, t = 0 : (N/2^{n+1} - 1) \right\} \cup \{(0,0,0)\}.$$

$$(2.27)$$

The selection of this set U allows us to consider the 2-D paired transform as the representation of the 2-D image as the set of complete splitting-signals. It means the paired transform is written as

$$\{f'_{p,s,t}; \; (p,s,t) \in U\} = \bigcup_{n=0}^{r-1} \bigcup_{(p,s)\in 2^n J_{N/2^n, N/2^n}} \left\{ f'_{p,s,2^n t}; \; t = 0 : (N/2^{n+1} - 1) \right\}$$

$$\cup \{f_{0,0,0}\},$$

$$(2.28)$$

and *the paired splitting-signals* generated by frequencies (p, s) are

$$f_{T'_{p,s}} = \{f'_{p,s,0}, f'_{p,s,2^n}, f'_{p,s,2\cdot 2^n}, \ldots, f'_{p,s,N/2-2^n}\}. \qquad (2.29)$$

The higher the frequency (p, s), i.e., $2^n = \text{g.c.d.}(p, s)$, the shorter the splitting-signal, and the time $2^n t$ runs the interval $[0, N/2 - 1]$ faster; with the step 2^n. In the collection of the splitting-signals, there are $3N/2$ signals of length $N/2$ each, $3N/4$ signals of length $N/4$ each, and so on. The total number of splitting-signals, or generators (p, s) equals $3N - 2$.

It follows directly from the definition of the set U, that the set of frequency-points (p, s) of splitting-signals of the paired transform is defined as

$$J'_{N,N} = \left\{ \bigcup_{n=0}^{r-1} 2^n J_{N/2^n, N/2^n} \right\} \cup \{(0,0)\},$$

and its subsets are defined as follows:

$$J_{N/2^n, N/2^n} = \{2^n(p, 1); p = 0 : (N/2^n - 1)\} \cup \{2^n(1, 2s); s = 0 : (N/2^{n+1} - 1)\}.$$

The set $J'_{N,N}$ is the projection of the set of triplets U on the lattice $X_{N,N}$. For instance, when $N = 8$, the set $J'_{8,8}$ equals

$$\begin{cases} (0,1), (1,1), (2,1), (3,1), (4,1), (5,1), (6,1), (7,1), (1,0), (1,2), (1,4), (1,6) \\ (0,2), (2,2), (4,2), (6,2), (2,0), (2,4) \\ (0,4), (4,4), (4,0) \\ (0,0) \end{cases}$$

Figure 2.15 shows the location (marked by filled circles) of frequency-points of this set, as well as the set $J'_{16,16}$ on the grid $X_{16,16}$.

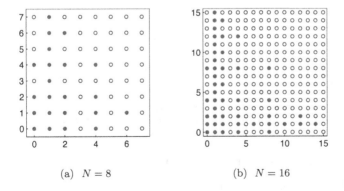

(a) $N = 8$ (b) $N = 16$

FIGURE 2.15
Sets of $(3N - 2)$ generators (in red) of paired splitting-signals.

Figure 2.16 shows the tomo image 256×256 in part a, along with the

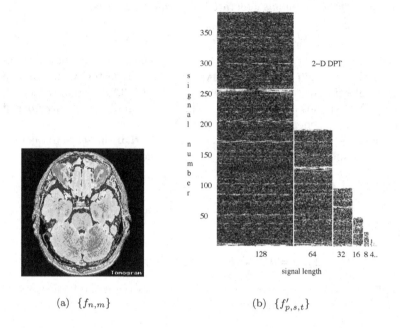

(a) $\{f_{n,m}\}$ (b) $\{f'_{p,s,t}\}$

FIGURE 2.16
(a) Image and (b) splitting-signals of lengths $128, 64, 32, 16, 8, 4, 2, 1, 1$.

totality of 766 paired splitting-signals of the image in b. The first 384 splitting-signals $\{f_{T'_{p,s}}; (p,s) \in J_{256,256}\}$ of length 128 each are shown in the form of the image 384×128. The next 192 splitting-signals $\{f_{T'_{p,s}}; (p,s) \in 2J_{128,128}\}$ of length 64 each are shown in the form of the image 192×64, and so on. The splitting-signals of the same length are united and separated from others according to the order of the paired transform components, which is given in the definition of the set U. The whole picture represents the paired transform of the tomo image. The set of all splitting-signals composing the paired transform of the image can be packed into the original form, the square matrix 256×256, as shown in Figure 2.17. All splitting-signals of length 128 are written in three squares 128×128 of the matrix. All splitting-signals of length 64 are written in three squares 64×64 of the remaining square 128×128, and so on.

2.3.2   The splitting of the 2-D DFT

The paired splitting-signal defines the 2-D DFT of the image at specific frequency-points. To show that, we first consider the case when $g.c.d.(p,s) = 1$, for instance $(1,s)$, or $(p,1)$. The splitting-signal is of length $N/2$,

$$f_{T'_{p,s}} = \{f'_{p,s,0}, f'_{p,s,1}, f'_{p,s,2}, \ldots, f'_{p,s,N/2-1}\}.$$

FIGURE 2.17
766 splitting-signals of lengths $128, 64, 32, 16, 8, 4, 2, 1, 1$.

If we apply the shifted $N/2$ -point DFT over this signal,

$$F_m^s = \left(\mathcal{F}_{N/2}^s \circ f_{T'_{p,s}} \right)_m = \sum_{t=0}^{N/2-1} f'_{p,s,t} W_{N/2}^{(m+\frac{1}{2})t}, \quad m = 0 : N/2 - 1,$$

we obtain the following [40, 41, 46]:

$$F_m^s = \sum_{t=0}^{N/2-1} \left(f'_{p,s,t} W^t \right) W_{N/2}^{mt} = F_{\overline{(2m+1)p}, \overline{(2m+1)s}} \tag{2.30}$$

for $m = 0 : (N/2 - 1)$. The sum in this equation represents the $N/2$-point DFT of the modified splitting-signal

$$g_{T'_{p,s}} = \{ f'_{p,s,0}, f'_{p,s,1} W^1, f'_{p,s,2} W^2, \ldots, f'_{p,s,N/2-1} W^{N/2-1} \}.$$

Thus the 2-D DFT of the image $f_{n,m}$ at frequency-points of the subset

$$T'_{p,s} = \left\{ (p,s), (\overline{3p}, \overline{3s}), , (\overline{5p}, \overline{5s}) \ldots, (\overline{(N-1)p}, \overline{(N-1)s}) \right\} \tag{2.31}$$

of the cyclic group $T_{p,s}$ is defined by the splitting-signal of length $N/2$.

As an example, Figure 2.18 shows the tree image of size 256×256 in part a, along with the splitting-signal $f_{T'_{3,1}}$ of length 128 together with the real part of the modified signal $f_{T'_{3,1}}$ in b, the 128-point DFT of the modified signal in c, and the samples of the 2-D DFT of the image at frequency-points of the subset $T'_{3,1}$ on the lattice $X_{256,256}$, at which the 2-D DFT can be filled by the 1-D DFT of the modified splitting-signal.

If g.c.d.$(p,s) = 2^n$, where $n \geq 0$, then sets $V_{p,s,t} = \emptyset$ and components $f'_{p,s,t} = 0$ if t is not divisible by 2^n. Therefore, the splitting-signal $f_{T'_{p,s}}$ can be considered as the following signal of length $N/2^{n+1}$:

$$f_{T'_{p,s}} = \{ f'_{p,s,0}, f'_{p,s,2^n}, f'_{p,s,2 \cdot 2^n}, \ldots, f'_{p,s,N/2-2^n} \}.$$

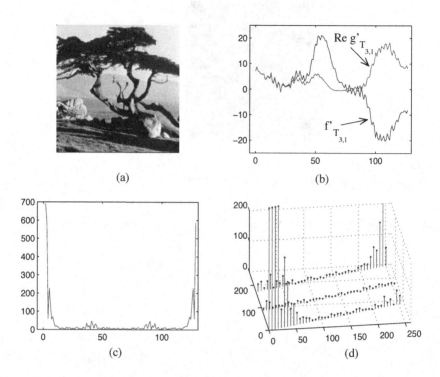

(a)   (b)

(c)   (d)

FIGURE 2.18 (See color insert)
(a) Tree image 256×256, (b) the paired splitting-signal $f_{T'_{3,1}}$ and the real part of the modified signal, (c) the 1-D DFT of the modified signal, and (d) the arrangement of values of the 1-D DFT at frequency-points of the subset $T'_{3,1}$. (The 1-D DFT is shown in the absolute scale.)

The $N/2$-point DFT in the right side of (2.30) represents the $N/2^{n+1}$-point DFT of the splitting-signal $f_{T'_{p,s}}$ modified by the vector of rotated coefficients $\{W^t_{N/2^n}; t = 0 : (N/2^{n+1} - 1)\}$ as follows:

$$F_{\overline{(2m+1)p}, \overline{(2m+1)s}} = \sum_{t=0}^{N/2^{n+1}-1} \left(f'_{p,s,2^n t} W^t_{N/2^n} \right) W^{mt}_{N/2^{n+1}} \qquad (2.32)$$

when $m = 0 : (N/2^{n+1} - 1)$.

As an example, Figure 2.19 shows the 256×256-point DFT of the tomo image in the form of the 1-D DFTs of the modified paired splitting-signals in part a. This totality of 1-D DFTs is shown in a form that is similar to the figure of splitting-signals in Figure 2.16(b). These 1-D DFTs represent the splitting of the 2-D DFT. The set of 1-D DFTs in the form of the square matrix 256×256 is shown in part b.

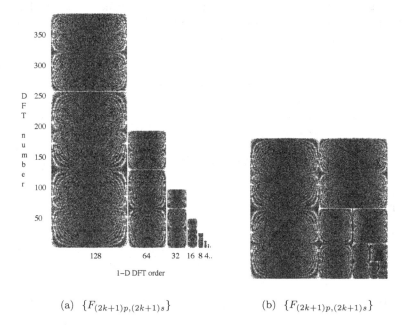

(a) $\{F_{(2k+1)p,(2k+1)s}\}$ (b) $\{F_{(2k+1)p,(2k+1)s}\}$

FIGURE 2.19
(a), (b) 2-D DFT of the image composed of the 1-D DFT of modified splitting-signals of lengths $128, 64, 32, 16, 8, 4, 2, 1, 1$.

As was mentioned above, the projection of frequency-time points of the 3-D set U on the 2-D frequency domain $X_{N,N}$ can be taken as the required set of generators (p, s) of subsets $T'_{p,s}$ covering $X_{N,N}$. Indeed, the family of $2^r 3 - 2$ subsets

$$\sigma' = \left(\left(\left(T'_{2^n p, 2^n s} \right)_{(p,s) \in J_{2^{r-n}, 2^{r-n}}} \right)_{n=0:(r-1)}, \{(0,0)\} \right) \qquad (2.33)$$

is the partition of X, which reveals the transformation $\mathcal{F}_{2^r, 2^r}$. The $2^r \times 2^r$-point DFT is thus split by this partition σ' into $2^r 3 - 2$ short DFTs, namely, $3 \cdot 2^{r-1}$ 2^{r-1}-point DFTs, $3 \cdot 2^{r-2}$ 2^{r-2}-point DFTs, ..., and six 2-point DFTs,

$$N \times N\text{-point 2-D DFT} \rightarrow \begin{cases} 3N/2 & N/2\text{-point DFTs} \\ 3N/4 & N/4\text{-point DFTs} \\ 3N/8 & N/8\text{-point DFTs} \\ \dots\dots & \dots\dots \\ 6 & 2\text{-point DFTs} \\ 3 & 1\text{-point DFTs} \\ 1 & 1\text{-point DFTs} \end{cases}$$

The totality of $(3N - 2)$ splitting-signals $\{f_{T'}, \ T' \in \sigma'\}$ consists of $2^{r-1}3$

signals of length 2^{r-1} each, $2^{r-2}3$ signals of length 2^{r-2} each, and so on. The summary length of all splitting-signals equals N^2. This set of splitting-signals is referred to as the σ'-representation (or the *paired* representation) of the image $f_{n,m}$.

Example 2.3 (Case 8×8) To compose a partition of the lattice $X_{8,8}$, we first note that each subset $T'_{p,s}$ is the orbit of the point (p, s) with respect to the movement group $G = \{1, 3, 5, 7\}$. For instance, the subset $T'_{1,1} = \{(1,1), (3,3), (5,5), (7,7)\}$ is the orbit of the point $(1,1)$. The point $(1,1)$ "is moving" starting (at time $t = 0$) from itself and "returning" back in four units of time (at time $t = 4$) as $(1,1) \rightarrow (3,3) \rightarrow (5,5) \rightarrow (7,7) \rightarrow (1,1)$. In the orbit $T'_{3,1}$, the point "is moving" as follows: $(3,1) \rightarrow (1,3) \rightarrow (7,5) \rightarrow (5,7) \rightarrow (3,1)$. The set of all orbits dividing the lattice $X_{8,8}$ is unique and given in Table 2.1.

TABLE 2.1
Set of orbits in the grid 8×8

$T'_{0,1}$	$T'_{1,1}$	$T'_{2,1}$	$T'_{3,1}$	$T'_{4,1}$	$T'_{5,1}$	$T'_{6,1}$	$T'_{7,1}$	$T'_{1,0}$	$T'_{1,2}$	$T'_{1,4}$	$T'_{1,6}$	4
$T'_{0,2}$	$T'_{2,2}$	$T'_{4,2}$	$T'_{6,2}$	$T'_{2,0}$	$T'_{2,4}$							2
$T'_{0,4}$	$T'_{4,4}$	$T'_{4,0}$										1
$T'_{0,0}$												1

The lattice is divided by 22 orbits and the orbits that are shown on the same row have equal cardinalities, which are shown in the right column of this table. For instance, the point $(0,1)$ runs around its orbit for time $t = 4$, and the orbit of the point $(0,2)$ is twice shorter, $(0,2) \rightarrow (0,6) \rightarrow (0,2)$. The set of generators of all orbits is shown in Figure 2.20 in part a, by the filled circles. The locations of frequency-points of orbits $T'_{p,s}$ for generators $(p, s) = (1,1), (3,1)$, and $(6,1)$ are shown in parts b, c, and d, respectively.

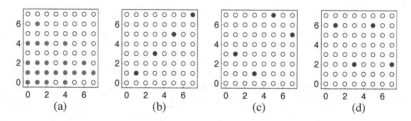

FIGURE 2.20
(a) 22 generators on the lattice 8×8, and the orbits of points (b) $(1,1)$, (c) $(3,1)$, and (d) $(6,1)$. (The generators and elements of the orbits are shown by the filled circles.)

The 22 splitting-signals that are defined for the generators of subsets, or orbits T' of the partition σ' determine values of the 2-D DFT at frequency-points of these subsets. For instance, the signal $f_{T'_{1,1}}$ carries the spectral information of the image at frequency-points $(1,1), (3,3), (5,5)$, and $(7,7)$. According to Table 2.1, the 8×8-point DFT is split by twelve 4-point DFTs, six 2-point DFTs, and four 1-point DFTs (which are the identity transformations).

This table also shows how to compose the complete set of basis paired functions. The 64 triplets (p, s, t), or numbers of the complete system of paired functions $\chi'_{p,s,t}$ are composed from the generators (p, s) and time variable t, which runs numbers $0, 1, 2, 3$, for the orbits in the first row in the table. For short orbits in the second row, t runs only values of 0 and 2. Indeed, both coordinates of points on these orbits are even; therefore conditions $(n_1 p + n_2 s = t \bmod 8)$ and $(n_1 p + n_2 s = t + 4 \bmod 8)$ do not hold for odd $t = 1$ and 3. In other words, for these triples (p, s, t), we have $\chi'_{p,s,t} \equiv 0$, and we do not consider such functions. There is no movement (or $t = 0$) in single-point orbits for $(p, s) = (0, 4), (4, 4), (4, 0)$, and $(0, 0)$. The set of 64 triplet-numbers for the complete system of paired functions is given in Table 2.2.

TABLE 2.2
Set of triplets (p, s, t)

(p, s)	t
$(0,1), (1,1), (2,1), (3,1), (4,1), (5,1), (6,1), (7,1), (1,0), (1,2), (1,4), (1,6)$	$0, 1, 2, 3$
$(0,2), (2,2), (4,2), (6,2), (2,0), (2,4)$	$0, 2$
$(0,4), (4,4), (4,0)$	0
$(0,0)$	0

Figure 2.21 shows the first 36 basis paired functions $\chi'_{p,s,t}$, which are placed in the order given in the above table, starting from the top on the left to the bottom on the right. For instance, the first four functions have numbers $(0, 1, t)$, where $t = 0 : 3$. The filled circle is used for value 1 and the open circle for -1. It is not difficult to see that all these functions are orthogonal. The complete set of paired functions defines *the paired representation* of the image,

$$f \to \{f_{T'_{p,s}}; \; T'_{p,s} \in \sigma'\}. \tag{2.34}$$

The components of the paired splitting-signals $f_{T'_{p,s}}$ are calculated from the components of the splitting-signals $f_{T_{p,s}}$ of the tensor representation of f by $f'_{p,s,t} = f_{p,s,t} - f_{p,s,t+4}$, $t \in \{0, 1, 2, 3\}$.

The transformation

$$\chi' : f \to \{f'_{p,s,2^n t}; \; T'_{p,s} \in \sigma', \; 2^n = \text{g.c.d.}(p, s), \; t = 0 : (\text{card}(T'_{p,s}) - 1)\} \tag{2.35}$$

is called *the paired transformation*. Thus the paired transform is the representation of the image in the form of 1-D splitting-signals; components of all splitting-signals together compose the 2-D discrete paired transform.

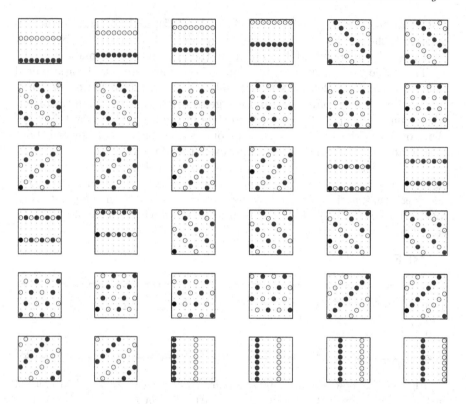

FIGURE 2.21

The first 36 basic functions of the complete system of the 8 × 8-paired transformation. (Values of 1 are shown by the filled circles, and −1 by open circles.)

Let $(p, s) = (3, 1)$ and let f be the following image of size 8 × 8 :

$$f = \begin{bmatrix} \underline{1} & 2 & 1 & 3 & 1 & 2 & 1 & 3 \\ 2 & 0 & 1 & 2 & 2 & 4 & 2 & 1 \\ 1 & 3 & 2 & 2 & 1 & 1 & 1 & 2 \\ 4 & 1 & 0 & 1 & 3 & 1 & 3 & 1 \\ 2 & 4 & 1 & 2 & 1 & 3 & 2 & 2 \\ 2 & 4 & 1 & 2 & 1 & 2 & 2 & 1 \\ 2 & 4 & 1 & 2 & 1 & 5 & 2 & 1 \\ 2 & 4 & 1 & 2 & 1 & 1 & 3 & 2 \end{bmatrix}.$$

We will calculate the 2-D DFT of this image at frequency-points of the orbit of the point $(3, 1)$. For that, we first write all values of t in the equations

$3n_1 + n_2 = t \mod 8$ in the form of the table

$$||t = (3n_1 + n_2) \mod 8||_{n_2,n_1=0:7} = \begin{bmatrix} 0 & 1 & 2 & 3 & 4 & 5 & 6 & 7 \\ 3 & 4 & 5 & 6 & 7 & 0 & 1 & 2 \\ 6 & 7 & 0 & 1 & 2 & 3 & 4 & 5 \\ 1 & 2 & 3 & 4 & 5 & 6 & 7 & 0 \\ 4 & 5 & 6 & 7 & 0 & 1 & 2 & 3 \\ 7 & 0 & 1 & 2 & 3 & 4 & 5 & 6 \\ 2 & 3 & 4 & 5 & 6 & 7 & 0 & 1 \\ 5 & 6 & 7 & 0 & 1 & 2 & 3 & 4 \end{bmatrix}.$$

Then, the components of the splitting-signal $f_{T'_{3,1}}$ are defined as follows:

$$f_{T'_{3,1}} = \begin{cases} f'_{3,1,0} = (f_{0,0} + f_{1,5} + f_{2,2} + f_{3,7} + f_{4,4} + f_{5,1} + f_{6,6} + f_{7,3}) - \\ \qquad (f_{0,4} + f_{1,1} + f_{2,6} + f_{3,3} + f_{4,0} + f_{5,5} + f_{6,2} + f_{7,7}) \\ \quad = (1 + 4 + 2 + 1 + 1 + 4 + 2 + 2) - \\ \qquad -(1 + 0 + 1 + 1 + 2 + 2 + 1 + 2) = 7, \\ f'_{3,1,1} = (f_{0,1} + f_{1,6} + f_{2,3} + f_{3,0} + f_{4,5} + f_{5,2} + f_{6,7} + f_{7,4}) - \\ \qquad (f_{0,5} + f_{1,2} + f_{2,7} + f_{3,4} + f_{4,1} + f_{5,6} + f_{6,3} + f_{7,0}) \\ \quad = (2 + 2 + 2 + 4 + 3 + 1 + 1 + 1) - \\ \qquad -(2 + 1 + 2 + 3 + 4 + 2 + 2 + 2) = -2, \\ f'_{3,1,2} = (f_{0,2} + f_{1,7} + f_{2,4} + f_{3,1} + f_{4,6} + f_{5,3} + f_{6,0} + f_{7,5}) - \\ \qquad (f_{0,6} + f_{1,3} + f_{2,0} + f_{3,5} + f_{4,2} + f_{5,7} + f_{6,4} + f_{7,1}) \\ \quad = (1 + 1 + 1 + 1 + 2 + 2 + 2 + 1) - \\ \qquad -(1 + 2 + 1 + 1 + 1 + 1 + 1 + 4) = -1, \\ f'_{3,1,3} = (f_{0,3} + f_{1,0} + f_{2,5} + f_{3,2} + f_{4,7} + f_{5,4} + f_{6,1} + f_{7,6}) - \\ \qquad (f_{0,7} + f_{1,4} + f_{2,1} + f_{3,6} + f_{4,3} + f_{5,0} + f_{6,5} + f_{7,2}) \\ \quad = (3 + 2 + 1 + 0 + 2 + 1 + 4 + 3) - \\ \qquad -(3 + 2 + 3 + 3 + 2 + 2 + 5 + 1) = -5. \end{cases}$$

This splitting-signal is modified as $f_{T'_{3,1}} \mathbf{W}$, where

$$\mathbf{W} = \text{diag}\{1, e^{-j\pi/4}, e^{-j2\pi/4}, e^{-j3\pi/4}\} = \text{diag}\{1, a(1-j), -j, -a(1+j)\}$$

and $a = \sqrt{2}/2 = 0.7071$. The four-point DFT of the modified splitting-signal

$$g_{T'_{3,1}} = f_{T'_{3,1}} \mathbf{W} = \{7, -1.4142 + j1.4142, j, 3.5355 + j3.5355\}$$

is calculated as follows:

$$\begin{bmatrix} 1 & 1 & 1 & 1 \\ 1 & -j & -1 & j \\ 1 & -1 & 1 & -1 \\ 1 & j & -1 & -j \end{bmatrix} \begin{bmatrix} 1 & 0 & 0 & 0 \\ 0 & a(1-j) & 0 & 0 \\ 0 & 0 & -j & 0 \\ 0 & 0 & 0 & -a(1+j) \end{bmatrix} \begin{bmatrix} 7 \\ -2 \\ -1 \\ -5 \end{bmatrix} = \begin{bmatrix} 9.1213 + j5.9497 \\ 4.8787 + j3.9497 \\ 4.8787 - j3.9497 \\ 9.1213 - j5.9497 \end{bmatrix}.$$

The obtained four-point DFT coincides with the 8×8-point DFT of the image

f at four frequency-points of the orbit $T'_{3,1}$, i.e.

$$\begin{bmatrix} F_{3,1} \\ F_{1,3} \\ F_{7,5} \\ F_{5,7} \end{bmatrix} = \begin{bmatrix} 9.1213 + j5.9497 \\ 4.8787 + j3.9497 \\ 4.8787 - j3.9497 \\ 9.1213 - j5.9497 \end{bmatrix}.$$

One can notice from this example, that the pairs of complex conjugate components $F_{3,1}$ and $F_{5,7}$, $F_{1,3}$ and $F_{7,5}$, are obtained from the same splitting-signal. In general, if (p_1, p_2) is on the orbit $T'_{p,s}$, then the frequency-point $(N - p_1, N - p_2)$ is also on this orbit.

Twelve subsets T' of the partition σ' of the lattice $X_{8,8}$ consist of four points each, six subsets of two points each, and the remaining four subsets are one-point sets. The splitting of the 8×8-point DFT thus equals

$$\mathcal{R}(\mathcal{F}_{8,8}; \sigma') = \Big\{ \underbrace{\mathcal{F}_4, \ldots, \mathcal{F}_4}_{12 \text{ times}}, \underbrace{\mathcal{F}_2, \ldots, \mathcal{F}_2}_{6 \text{ times}}, \underbrace{\mathcal{F}_1, \ldots, \mathcal{F}_1}_{4 \text{ times}} \Big\}. \qquad (2.36)$$

The first twelve four-point DFTs are calculated over the modified splitting-signals

$$\begin{aligned} g_{T'_{p,s}} &= \{ f'_{p,s,0}, f'_{p,s,1} W, f'_{p,s,2} W^2, f'_{p,s,3} W^3 \} \\ &= \{ f'_{p,s,0}, a(1-j) f'_{p,s,1}, -j f'_{p,s,2}, -a(1+j) f'_{p,s,3} \} \end{aligned} \qquad (2.37)$$

where $a = \sqrt{2}/2 = 0.7071$ and $W = \exp(-j2\pi/8) = a(1-j)$. The generators (p, s) of these signals lie on the first row in Table 2.2. The calculation in (2.37) uses two operations of multiplication by the factor of a, for each splitting-signal. The next six two-point DFTs are calculated over the short modified splitting-signals

$$g_{T'_{p,s}} = \{ f'_{p,s,0}, f'_{p,s,2} W^2 \} = \{ f'_{p,s,0}, -j f'_{p,s,2} \}.$$

The generators (p, s) of these signals are given in the second row in Table 2.2. Thus, the total number of operations of multiplication for calculating the 8×8-point DFT through the splitting (2.36) equals $m'_{8,8} = 12 \cdot 2 = 24$.

Example 2.4 (Case 16×16) The splitting of the 16×16-point DFT by the partition $\sigma' = (T')$ equals

$$\mathcal{R}(\mathcal{F}_{16,16}; \sigma') = \Big\{ \underbrace{\mathcal{F}_8, \ldots, \mathcal{F}_8}_{24 \text{ times}}, \underbrace{\mathcal{F}_4, \ldots, \mathcal{F}_4}_{12 \text{ times}}, \underbrace{\mathcal{F}_2, \ldots, \mathcal{F}_2}_{6 \text{ times}}, \underbrace{\mathcal{F}_1, \ldots, \mathcal{F}_1}_{4 \text{ times}} \Big\}.$$

The cardinalities of the largest subsets $T'_{p,s}$ of the partition σ' of the lattice $X_{16,16}$ equal 8. The generators for these subsets are taken from the set $J_{16,16}$, and the corresponding modified splitting-signals have the following form:

$$g_{T'_{p,s}} = f_{T'_{p,s}} \mathbf{W} = \{ f'_{p,s,0}, f'_{p,s,1} W, f'_{p,s,2} W^2, f'_{p,s,3} W^3, \ldots, f'_{p,s,7} W^7 \} \qquad (2.38)$$

where $W = \exp(-j2\pi/16) = 0.9239 - j0.3827$. This product uses six non-trivial operations of multiplication by the twiddle factors W^t, $t = 1, 2, 3, 5, 6, 7$. The eight-point DFT requires two operations of multiplication. Therefore the calculation of 24 eight-point DFTs in the splitting of the 16×16-point DFT requires $24 \cdot 8$ operations of multiplication. The calculation of the four-point DFTs over the modified splitting-signals of length four was described in Example 2.3. Each such transform uses two operations of multiplication. The total number of operations of multiplication required to calculate the 16×16-point DFT by the paired transform is calculated by

$$m'_{16,16} = 24(m_8 + 8 - 2) + 12(m_4 + 4 - 2) = 24(2 + 6) + 12(2) = 192 + 24 = 216.$$

We note for comparison, that the tensor algorithm of the 16×16-point DFT uses $24m_{16} = 24 \cdot 10 = 240$ operations of multiplication.

In the general $N = 2^r$ case, the number of operations of multiplication required for calculating the $2^r \times 2^r$-point DFT by the paired transforms can be estimated as

$$m_{2^r,2^r} = \sum_{n=1}^{r-2} 2^{r-n} 3\big(m_{2^{r-n}} + 2^{r-n} - 2\big) \le 2 \cdot 4^{r-1}(r - 7/3) + 8/3, \quad (r > 2).$$

When estimating this number, we use the fact that the 2^n-point fast paired DFT requires $m_{2^n} = 2^{n-1}(n - 3) + 2$ multiplications, when $n \ge 3$ [41, 45].

2.4 Complete system of 2-D paired functions

The paired functions compose a basis in the linear space of discrete images of size $N \times N$, and the system of functions

$$\chi' = \{\chi'_{p,s,t}; \ (p, s, t) \in U\}$$

is complete [42]. We can construct other such sets of triples, U. For instance, we can substitute all or a part of triplets (p, s, t) by (s, p, t).

The 2-D discrete paired transform does not require multiplications. The complete system of paired functions is defined as

$$\chi'_{p,s,2^k t}(n, m) = \begin{cases} 1, \text{if } np + ms = 2^k t \bmod N, \\ -1, \text{if } np + ms = (2^k t + N/2) \bmod N, \\ 0, \text{otherwise}, \end{cases} \quad (2.39)$$

which is the difference of two characteristic functions $\chi_{p,s,2^k t}$ and $\chi_{p,s,2^k t + N/2}$. When $(p, s) = (0, 0)$, the paired function is defined as $\chi'_{0,0,0}(n, m) \equiv 1$. The paired function $\chi'_{p,s,2^k t}$ itself represents a 2-D plane wave, and the decomposition of the image by the paired functions is the decomposition of the image by plane waves. All basic paired functions are orthogonal. In the masks of paired functions, all "1"s and all "−1"s lie on the different but parallel lines, the directions of which are defined by (p, s). Inside each series of functions generated by frequency (p, s),

$$\{\chi'_{p,s,2^k t};\ t = 0, 1, 2, ..., N/2^{k+1} - 1,\ (2^k = \text{g.c.d.}(p, s))\},$$

the paired functions have the same frequency, direction, and they are parallel shifted functions with respect to each other. As an example, Figure 2.22 shows, for the $N = 32$ case, the gray-scale images of basic paired functions $\chi'_{1,4,1}$ and $\chi'_{1,4,6}$ in parts a and b, and functions $\chi'_{2,2,6}$ and $\chi'_{2,2,12}$ in c and d, respectively.

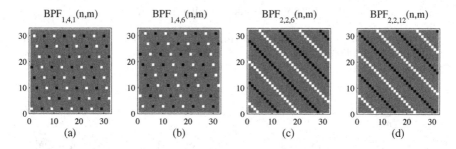

FIGURE 2.22

Images 32×32 of the paired functions (a) $\chi'_{1,4,1}$, (b) $\chi'_{1,4,6}$, (c) $\chi'_{2,2,6}$, and (d) $\chi'_{2,2,12}$. (Values $1, -1$, and 0 correspond to the white, black, and gray levels, respectively.)

The paired function can be considered as the quantized 2-D cosine wave

$$\chi'_{p,s,2^k t}(n, m) = M\left[c_{p,s,2^k t}(n, m)\right] = M\left[\cos\left(\frac{2\pi(np + ms - 2^k t)}{2^r}\right)\right],$$
(2.40)

where the quantization operation, M, is defined as $M[1] = 1$, $M[-1] = -1$, and $M[x] = 0$, if $x \neq \pm 1$.

For the $N = 256$ case, Figure 2.23 shows the gray-scale images of the 2-D cosine wave with number $(4, 8, 8)$ and its quantized function (or the paired function) in parts a and b, respectively.

Similar to paired functions, inside each series of functions generated by frequency (p, s) the cosine waves are shifted in parallel and defined by the same direction. Figure 2.24 shows the images of the 2-D cosine waves with numbers $(1, 4, 1)$, $(1, 8, 1)$, and $(12, 4, 4)$ in parts a, b, and c, respectively. Four periods of the wave $c_{1,4,1}$ cover the discrete lattice $X_{256,256}$; the wave $c_{1,8,1}$ covers it with eight periods, and the wave $c_{12,4,4}$ with twelve periods. Thus, the length of each wave depends on the frequency (p, s) that generates this

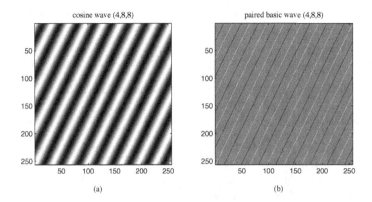

FIGURE 2.23
The images of (a) the 2-D cosine wave $c_{4,8,8}(n,m)$ and (b) the paired function $\chi'_{4,8,8}(n,m)$.

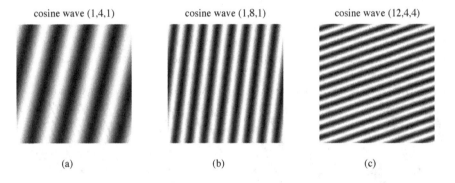

FIGURE 2.24
The images 256×256 of the 2-D cosine waves (a) $c_{1,4,1}$, (b) $c_{1,8,1}$, and (c) $c_{12,4,4}$.

wave, and the wave is oriented in the 3-D space. It covers the period 256×256 in the direction that is also defined by the frequency.

In addition we observe the following property of the multiresolution of the image, which is defined along directions. We call this property *the directed multiresolution* of the image by the paired transform. The paired functions generated by the frequencies (p,s) and $(2p,2s)$, as well as $(4p,4s),(8p,8s),\ldots$, are defined by the same direction. Indeed, the angles along which the coefficients "1" and "−1" are located in the masks equal to the angle $\vartheta = \text{arctg}(p/s) = \text{arctg}(2p/2s) = \text{arctg}(4p/4s)$. The paired function $\chi'_{2p,2s,2t}$ can be calculated by

$$\chi'_{2p,2s,2t} = \chi_{2p,2s,2t} - \chi_{2p,2s,2t+N/2} = \chi_{p,s,t} - \chi_{p,s,t+N/4} + \chi_{p,s,t+N/2} - \chi_{p,s,t+3N/4},$$

and the components of the paired transform can be calculated through the
tensor transform as follows:

$$f'_{2p,2s,2t} = f'_{\overline{2p},\overline{2s},2t} = f_{p,s,t} - f_{p,s,t+N/4} + f_{p,s,t+N/2} - f_{p,s,t+3N/4}.$$

Indeed, the following relation holds for tensor characteristic functions:

$$\chi_{2p,2s,2t} = \chi_{\overline{2p},\overline{2s},2t} = \chi_{p,s,t} + \chi_{p,s,t+N/2}, \quad t = 0 : (N/2 - 1),$$

where subscripts are considered modulo N. Similarly, the components of the
paired splitting-signals generated by frequencies $(\overline{4p}, \overline{4s})$, $(\overline{8p}, \overline{8s})$, ... can be
calculated from the same splitting-signal $\{f_{p,s,t}; t = 0 : (N - 1)\}$. Although
the complete family of paired functions are generated by $(3N - 2)$ frequencies
(p, s), all functions can be calculated from $3N/2$ characteristic functions of the
tensor transformation. In other words, all N^2 paired functions are determined
by directions at $3N/2$ angles of the set

$$\Psi_N = \{\arctan(p); p = 0 : (N - 1)\} \cup \{\arctan(1/(2s)); s = 0 : (N/2 - 1)\}.$$

As an example, Figure 2.25 shows the images of the 2-D cosine waves with
numbers $(1, 2, 1)$, $(2, 4, 2)$, $(4, 8, 4)$, and $(8, 16, 8)$ from $J'_{256,256}$ in parts a-d,
respectively. These waves have different frequencies, but they are oriented at

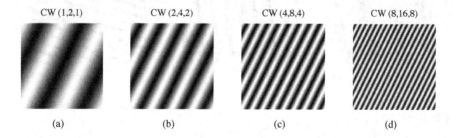

CW (1,2,1)   CW (2,4,2)   CW (4,8,4)   CW (8,16,8)

(a)   (b)   (c)   (d)

FIGURE 2.25
The images 256×256 of the 2-D cosine waves (a) $c_{1,2,1}$, (b) $c_{2,4,2}$, (c) $c_{4,8,4}$,
and (d) $c_{8,16,8}$.

the same angle. There are three other series of cosine waves with frequency-
numbers $(16, 32)$, $(32, 64)$, and $(64, 128)$, which are defined by the direction at
this angle. Thus, seven series of the basic paired functions have the same orien-
tation, but define different resolutions. The triplet-numbers of these functions
and the number of parallel shifts of functions inside each series are given in
Table 2.3. The total number of these paired functions equals 254. In all triplet-
numbers of the above functions, the first two components are integer multiple
to $(p_0, s_0) = (1, 2)$ in modulo N. Therefore, we call this set of seven series of
cosine functions (and paired functions) *the $(1, 2)$-generated set of functions*.
We now consider the $(1, 5)$-generated set of functions. Figure 2.26 shows the

TABLE 2.3
Numbers of basic paired functions defined along the direction at angle
$\text{arctg}(1/2)$

(p,s,t)	$(1,2,t)$	$(2,4,2t)$	$(4,8,4t)$	$(8,16,8t)$	$(16,32,16t)$	$(32,64,32t)$	$(64,128,64t)$
t	$0:127$	$0:63$	$0:31$	$0:15$	$0:7$	$0:3$	$0,1$
#	128	64	32	16	8	4	2
shift by	1	2	4	8	16	32	64

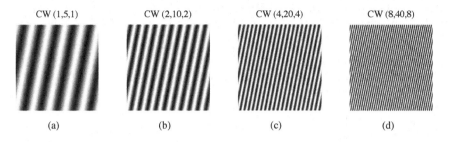

CW (1,5,1) CW (2,10,2) CW (4,20,4) CW (8,40,8)

(a)　　　　(b)　　　　(c)　　　　(d)

FIGURE 2.26
The images 256×256 of the 2-D cosine waves (a) $c_{1,5,1}$, (b) $c_{2,10,2}$, (c) $c_{4,20,4}$,
and (d) $c_{8,40,8}$.

images for the 2-D cosine waves with numbers $(1,5,1)$, $(2,10,2)$, $(4,20,4)$,
and $(8,40,8)$ in parts a-d, respectively. The other two series of cosine waves
with frequency-numbers $(16,80)$ and $(32,160)$ are also defined by the same
direction. The triplet-numbers of these functions and the number of parallel
shifts of functions inside each series of the $(1,5)$-generated set are given in
Table 2.4. In the general case of $(p_0,s_0) \in J_{N,N}$, the (p_0,s_0)-generated set of

TABLE 2.4
Numbers of basic paired functions defined along the direction at angle
$\text{arctg}(1/5)$

(p,s,t)	$(1,5,t)$	$(2,10,2t)$	$(4,20,4t)$	$(8,40,8t)$	$(16,80,16t)$	$(32,160,32t)$	$(64,64,64t)$
t	$0:127$	$0:63$	$0:31$	$0:15$	$0:7$	$0:3$	$0,1$
#	128	64	32	16	8	4	2
shift by	1	2	4	8	16	32	64

functions is defined similarly.

It should be noted that the series of functions with high frequencies $2^k(p,s)$
have very dense structures with small periods $N/2^k \times N/2^k$. Such functions describe high-resolution components of the image and they can be determined by
different directions. As an example, we consider the cosine wave with triplet-
number $(8,40,8) = 8 \cdot (1,5,1)$ from Table 2.4. This wave has the period
32×32 and is directionally defined at angle $\arctan(1/5)$. The same triplet
can be defined as $(8,40,8) = 8 \cdot (1,5+32n,1)$ in modulo 256, for any integer n. We consider, for instance, the cases when $n = 3,4$, and 5, i.e., the
triplets $(1,101,1)$, $(1,133,1)$, and $(1,165,1)$. The corresponding frequency-
points $(1,101)$, $(1,133)$, and $(1,165)$ are from the set of generators $J_{256,256}$.

The cosine wave with number $(8, 40, 8)$ can thus be referred to the $(1, 101)$-oriented set of functions, as well as to the $(1, 133)$- and $(1, 165)$-oriented sets. Directions of these three sets are determined by the angles $\arctan(1/101)$, $\arctan(1/133)$, and $\arctan(1/165)$, respectively.

2.4.0.1 Code: System of basic paired functions

The complete set of basic paired functions can be calculated by the program with script matrices_2Dpaired.m. It prints all N^2 matrices of paired functions.

```
% -----------------------------------------------
% call: matrices_2Dpaired.m
N=input('   Set the value of N (power of 2) = ');
r=log2(N); L2=N/2; p=1;
for nn=1:r
    for s=0:p:N-1
        for t=0:p:L2-1
            fprintf('   #(%g,%g,%g) ',p,s,t);
            A=pmatrix_pst(N,p,s,t);
        end
    end
    s=p; step=2*s;
    for p=0:step:N-1
        for t=0:s:L2-1
            fprintf('   #(%g,%g,%g) ',p,s,t);
            A=pmatrix_pst(N,p,s,t);
        end
    end
    p=2*s;
end
Alast=pmatrix_pst(N,0,0,0)
% ----------------------------------------------
% Call: pmatrix_pst.m
function A=pmatrix_pst(N,p,s,t)
    if (t>N) | (t<0) t=mod(t,N); end
    A=zeros(N,N);
    ms=0; t2=t+N/2;
    for m=1:N
        np=0;
        for n=1:N
            t1=mod(np+ms,N);
            if (t1==t) A(m,n)=1;
            elseif (t1==t2) A(m,n)=-1;
            end
            np=np+p;
        end
        ms=ms+s;
    end
```

The function pmatrix_pst.m calculates one matrix with triplet (p, s, t).

2.4.1   1-D DFT and paired transform

The 2-D discrete paired transformation is not separable, but its basic functions can be calculated by means of the convolution from the basic functions of the 1-D paired transform [46, 48]. The 1-D paired transform reveals the structure of the 1-D DFT in a way similar to the 2-D case, and this transform can also be defined from the cosine transform.

Given $N = 2^r$, $r > 1$, the 1-D N-point discrete paired transform χ'_N is determined by the following complete system of the paired functions [41, 46]:

$$\chi'_{2^k,2^k t}(n) = M\left(\cos\left(\frac{2\pi(n - t)}{2^{r-k}}\right)\right), \quad t = 0 : (2^{r-k-1} - 1),\ k = 0 : (r - 1),$$

$$\chi'_{0,0}(n) \equiv 1, \quad n = 0 : (N - 1),$$

$$(2.41)$$

where M is the real function which differs from zero only on the bounds of the interval $[-1, 1]$ and takes values $M(-1) = -1$, $M(1) = 1$. The double numbering of the paired functions refers to the frequency $(p = 2^k)$ and time (t). The paired transform is a transform of the discrete-time signal f_n to the set of frequency-time signals,

$$f_n \to \left\{ \{f'_{2^k,0}, f'_{2^k,2^k}, f'_{2^k,2^k 2}, ..., f'_{2^k,N/2^{k+1}-1}\},\ k = 0 : (r - 1) \right\},$$

that splits the 2^r-point DFT by the 2^{r-k-1}-point DFTs, $k = 0 : (r - 1)$. The components of these splitting-signals are calculated by

$$f'_{2^k,2^k t} = \chi'_{2^k,2^k t} \circ f_n = \sum_{n=0}^{2^r-1} \chi'_{2^k,2^k t}(n) f_n, \quad t = 0 : (2^{r-k-1} - 1).$$

Example 2.5 (Case $N = 8$) The matrix of the eight-point discrete paired transform (DPT) is defined as

$$[\chi'_8] = \begin{bmatrix} [\chi'_{1,0}] \\ [\chi'_{1,1}] \\ [\chi'_{1,2}] \\ [\chi'_{1,3}] \\ [\chi'_{2,0}] \\ [\chi'_{2,2}] \\ [\chi'_{4,0}] \\ [\chi'_{0,0}] \end{bmatrix} = \begin{bmatrix} 1 & 0 & 0 & 0 & -1 & 0 & 0 & 0 \\ 0 & 1 & 0 & 0 & 0 & -1 & 0 & 0 \\ 0 & 0 & 1 & 0 & 0 & 0 & -1 & 0 \\ 0 & 0 & 0 & 1 & 0 & 0 & 0 & -1 \\ 1 & 0 & -1 & 0 & 1 & 0 & -1 & 0 \\ 0 & 1 & 0 & -1 & 0 & 1 & 0 & -1 \\ 1 & -1 & 1 & -1 & 1 & -1 & 1 & -1 \\ 1 & 1 & 1 & 1 & 1 & 1 & 1 & 1 \end{bmatrix}.$$

The first four basis paired functions correspond to the frequency $p = 1$, the next two functions correspond to frequency $p = 2$, and the last two functions correspond to the frequencies $p = 4$ and 0, respectively. The process of composition of these functions from the corresponding cosine waves defined in the interval $[0, 7]$ is illustrated in Figure 2.27.

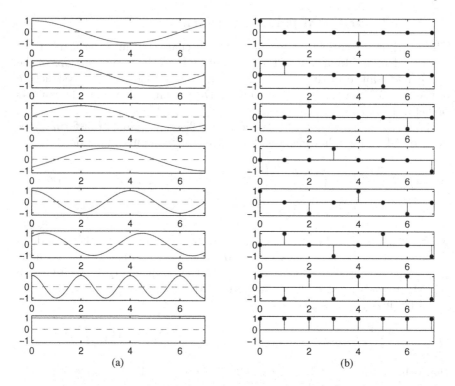

FIGURE 2.27

(a) Cosine waves and (b) discrete paired functions of the eight-point DPT.

Let f_n be the signal $\{1, 2, 2, 4, 5, 3, 1, 3\}$. The paired transform of this signals results in four splitting-signals as follows:

$$\{f_n\} \rightarrow \begin{cases} \{f'_{1,0}, f'_{1,1}, f'_{1,2}, f'_{1,3}\} = \{-4, -1, 1, 1\}, \\ \{f'_{2,0}, f'_{2,2}\} = \{3, -2\}, \\ \{f'_{4,0}\} = \{-3\}, \\ \{f'_{0,0}\} = \{21\}. \end{cases}$$

The splitting of the 2^r-point DFT into $(r+1)$ short DFTs is described by

$$F_{\overline{(2m+1)2^k}} = \sum_{t=0}^{2^{r-k-1}-1} \left(f'_{p,t} W^t_{2^{r-k}} \right) W^{mt}_{2^{r-k-1}}, \quad m = 0 : (2^{r-k-1} - 1). \quad (2.42)$$

The set of 2^r frequency-points $X_{2^r} = \{0, 1, 2, ..., 2^r - 1\}$ is divided by $(r + 1)$ subsets, or orbits

$$T'_p = \{(2m + 1)p \bmod 2^r; m = 0 : (2^{r-1}/p - 1)\},$$

where $p = 2^k$, $k = 0 : (r - 1)$, and $T'_0 = \{0\}$. These subsets compose a partition

σ' of X_{2^r}, and the 2^r-point DFT is split as follows:

$$\mathcal{R}(\mathcal{F}_{2^r};\sigma') = \{\mathcal{F}_{2^{r-1}}, \mathcal{F}_{2^{r-2}}, \mathcal{F}_{2^{r-3}}, ..., \mathcal{F}_2, 1, 1\}. \qquad (2.43)$$

Using the similar splitting for each short transform $\mathcal{F}_{2^{r-k-1}}$ of this splitting, we obtain the full decomposition of the 2^r-point DFT, by the paired transforms.

Example 2.6 (8-point DFT) Let $N = 8$ and let $\{f_n\}$ be the signal of Example 2.5. According to (2.42), the calculation of the eight-point DFT of f_n can be written in matrix form as

$$
\begin{bmatrix} F_7 \\ F_3 \\ F_5 \\ F_1 \\ F_6 \\ F_2 \\ F_4 \\ F_0 \end{bmatrix} =
\begin{bmatrix} \begin{bmatrix}[\mathcal{F}_2]\end{bmatrix} & & \\ & 1 & \\ & & 1 \end{bmatrix}
\operatorname{diag}\left\{\begin{matrix}1\\-j\\1\\1\end{matrix}\right\} [\chi_4']
\begin{bmatrix} & & \\ & [\mathcal{F}_2] & \\ & & 1 \\ & & & 1 \end{bmatrix}
\operatorname{diag}\left\{\begin{matrix}1\\W\\-j\\W^3\\1\\-j\\1\\1\end{matrix}\right\} [\chi_8']
\begin{bmatrix}1\\2\\2\\4\\5\\3\\1\\3\end{bmatrix} =
\begin{bmatrix} -5.4142+j \\ -2.5858+j \\ -2.5858-j \\ -5.4142-j \\ 3-2j \\ 3+2j \\ -3 \\ 21 \end{bmatrix}
$$

where the twiddle factors $W = \exp(-2\pi j/8) = 0.7071(1-j)$ and $W^3 = -0.7071(1+j)$.

In the general $N = 2^r$ case, where $r \geq 2$, the following matrix decomposition holds for the 1-D DFT [45, 48, 51]:

$$[\mathcal{F}_{2^r}] = \left[\left(\bigoplus_{k=0}^{r-1}[\mathcal{F}_{2^{r-k-1}}]\right) \oplus 1\right] D_{2^r}[\chi_{2^r}], \qquad (2.44)$$

where $\oplus$ denotes the operation of the Kronecker sum of matrices and the diagonal matrix

$$D_{2^r} = \operatorname{diag}\{1, W, W^2, W^3, ..., W^{2^r-1}, 1, W^2, W^4, ..., W^{2^r-2},$$
$$1, W^4, W^8, ..., W^{2^r-4}, 1, ..., 1, 1\}.$$

2.5 Paired transform direction images

In this section, we describe the principle of superposition, which states that the image $\{f_{n,m}\}$ can be composed with the direction images, which are determined by splitting-signals in paired representation. Let (p, s) be the generator from the set $J'_{N,N}$ and let $2^k = \text{g.c.d.}(p, s)$, where $k \geq 0$. We denote by D_{p_1,s_1} the incomplete 2-D DFT of the image, which is defined only at frequency-points of the subset $T'_{p,s}$, i.e.,

$$D_{p_1,s_1} = D^{(p,s)}_{p_1,s_1} = \begin{cases} F_{p_1,s_1}; \text{if } (p_1, s_1) \in T'_{p,s}, \\ 0; \qquad \text{otherwise.} \end{cases} \qquad (2.45)$$

We start with the first series of generators, when g.c.d.$(p,s) = 1$. There are $3N/2$ such generators that compose the subset $J_{N,N}$. The inverse transform of the above incomplete 2-D DFT can be calculated as follows:

$$d_{n,m} = \frac{1}{N^2} \sum_{p_1=0}^{N-1} \sum_{s_1=0}^{N-1} D_{p_1,s_1} W^{-(np_1+ms_1)}$$

$$= \frac{1}{N^2} \sum_{(p_1,s_1)\in T'_{p,s}} F_{p_1,s_1} W^{-(np_1+ms_1)}$$

$$= \frac{1}{N^2} \sum_{k=0}^{N/2-1} F_{\overline{(2k+1)p},\overline{(2k+1)s}} W^{-(n\overline{(2k+1)p}+m\overline{(2k+1)s})}$$

$$= \frac{1}{N^2} \sum_{k=0}^{N/2-1} F_{\overline{(2k+1)p},\overline{(2k+1)s}} W^{-(2k+1)\overline{(np+ms)}} \qquad (2.46)$$

$$= \frac{1}{2N} \left(\frac{2}{N} \sum_{k=0}^{N/2-1} F_{\overline{(2k+1)p},\overline{(2k+1)s}} W_{N/2}^{-kt} \right) W^{-t}$$

$$= \frac{1}{2N} \left(f'_{p,s,t} W^t \right) W^{-t} = \frac{1}{2N} f'_{p,s,(np+ms) \bmod N} ,$$

where we denote $t = (np + ms) \bmod N$.

Thus we obtain the following direction image of the 1st series:

$$d_{n,m} = d_{n,m}^{(p,s)} = \frac{1}{2N} f'_{p,s,(np+ms) \bmod N}, \qquad n,m = 0 : (N-1).$$

The direction image $N \times N$ is composed of $N/2$ values of the splitting-signal $f_{T'_{p,s}}$, which are placed on the image along the set of parallel lines $np + ms = t \bmod N$, $t = 0 : (N-1)$. As an example, Figure 2.28 shows for the bridge image 256×256 the first nine direction images defined by frequencies $(p,s) = (0,1), (1,1), \ldots, (8,1)$ in parts a-i, respectively.

According to the definition of the paired representation, the following property holds for its components: $f'_{p,s,t+N/2} = -f'_{p,s,t}$ when $t = 0 : (N/2-1)$. When $(np + ms) \bmod N \geq N/2$, the values of $f'_{p,s,t}$ are placed on the image with the minus sign. The composition of the direction image from the splitting-signal is simple. All values of this image can be calculated, for instance, by the following method of permutation of values of the splitting-signal. Let us assume that a value of the direction image at point (n,m) has been calculated. Then the next three neighbor values of the image can be calculated as follows:

$$
d_{n,m} = f'_{p,s,t} \rightarrow
\begin{aligned}
d_{n+1,m} &= \begin{cases} f'_{p,s,t+p}; & \text{if } \overline{t+p} < N/2, \\ -f'_{p,s,t+p}; & \text{otherwise,} \end{cases} \\
d_{n,m+1} &= \begin{cases} f'_{p,s,t+s}; & \text{if } \overline{t+s} < N/2, \\ -f'_{p,s,t+s}; & \text{otherwise,} \end{cases} \\
d_{n+1,m+1} &= \begin{cases} f'_{p,s,t+p+s}; & \text{if } \overline{t+p+s} < N/2, \\ -f'_{p,s,t+p+s}; & \text{otherwise,} \end{cases}
\end{aligned}
\qquad (2.47)
$$

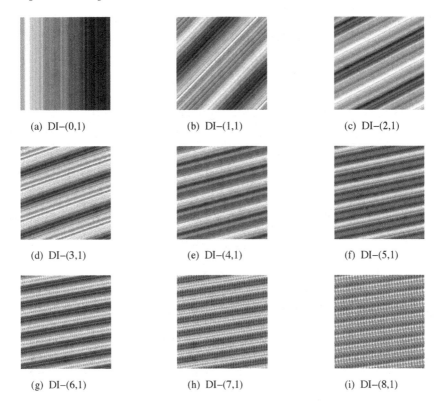

(a) DI–(0,1)   (b) DI–(1,1)   (c) DI–(2,1)

(d) DI–(3,1)   (e) DI–(4,1)   (f) DI–(5,1)

(g) DI–(6,1)   (h) DI–(7,1)   (i) DI–(8,1)

FIGURE 2.28
Nine first direction image components of the bridge image. (All images have been scaled.)

where we omitted the factor $1/(2N)$. Thus, the direction image can be constructed right after the corresponding projection data at the angle $\text{arctg}(p/s)$ have been obtained.

We now consider the $(k + 1)$th series of generators, i.e., such that g.c.d.$(p, s) = 2^k$, when $k \in \{1, 2, ..., r - 1\}$. The number of such generators in the set $J'_{N,N}$ equals $3N/2^{k+1}$. The calculation of the inverse transform of the incomplete 2-D DFT, D_{p_1,s_1}, results in the following direction image:

$$
\begin{aligned}
d_{n,m} = d_{n,m}^{(p,s)} &= \frac{1}{N^2} \sum_{p_1=0}^{N-1} \sum_{s_1=0}^{N-1} D_{p_1,s_1} W^{-(np_1+ms_1)} \\
&= \frac{1}{N^2} \sum_{l=0}^{N/2^{k+1}-1} F_{\overline{(2l+1)p},\overline{(2l+1)s}} W^{-(2l+1)\overline{(np+ms)}} \\
&= \frac{1}{2^{k+1}N} \left(\frac{2^{k+1}}{N} \sum_{l=0}^{N/2^{k+1}-1} F_{\overline{(2l+1)p},\overline{(2l+1)s}} W_{N/2^{k+1}}^{-lt} \right) W_{N/2^k}^{-t} \\
&= \frac{1}{2^{k+1}N} \left(f'_{p,s,2^kt} W_{N/2^k}^t \right) W_{N/2^k}^{-t} \\
&= \frac{1}{2^{k+1}N} f'_{p,s,(np+ms) \bmod N},
\end{aligned}
\tag{2.48}
$$

where we denoted $2^k t = (np + ms) \bmod N$.

For the bridge image, Figure 2.29 shows the first nine direction images of the 2nd series, for the generators $(p,s) = (0,2),(2,2),\ldots,(16,2)$ in parts a-i, respectively.

The last series of generators contains only the frequency $(0,0)$ and the set $T'_{0,0} = \{(0,0)\}$. The transform D_{p_1,s_1} is zero at all frequency-points except the point $(0,0)$, where the transform equals $F_{0,0}$. In this case, we have the following constant-image:

$$
d_{n,m} = d_{n,m}^{(0,0)} = \frac{1}{N^2} F_{0,0} = \frac{1}{N^2} f'_{0,0,0}.
$$

All $(3N - 2)$ subsets $T'_{p,s}$, with generators $(p,s) \in J'_{N,N}$, compose a partition of the lattice $N \times N$. It means the sum of corresponding $(3N - 2)$ incomplete 2-D DFTs equals the 2-D DFT of the image. It also means the sum of $(3N - 2)$ direction images $d_{n,m}^{(p,s)}$ equals the image $f_{n,m}$.

Statement 2: (Superposition by paired direction images) The discrete image of size $N \times N$, where $N = 2^r$, $(r > 1)$, can be composed from $(3N - 2)$ direction images by

$$
\begin{aligned}
f_{n,m} &= \sum_{(p,s) \in J'_{N,N}} d_{n,m}^{(p,s)} \\
&= \frac{1}{2N} \sum_{k=0}^{r-1} \frac{1}{2^k} \sum_{(p,s) \in J_{2^{r-k},2^{r-k}}} f'_{p,s,(np+ms) \bmod N} + \frac{1}{N^2} f'_{0,0,0},
\end{aligned}
\tag{2.49}
$$

$n, m = 0 : (N - 1)$.

This is the formula of composition of the image by direction images and the formula of reconstruction of the image from its 2-D discrete paired transform.

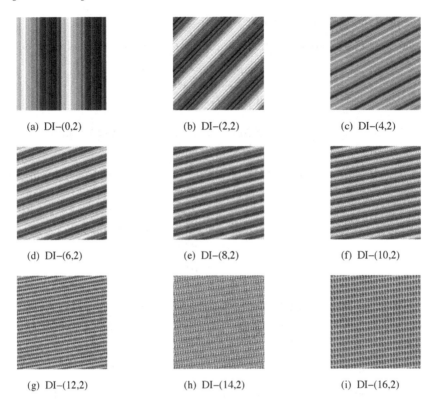

(a) DI–(0,2) (b) DI–(2,2) (c) DI–(4,2)

(d) DI–(6,2) (e) DI–(8,2) (f) DI–(10,2)

(g) DI–(12,2) (h) DI–(14,2) (i) DI–(16,2)

FIGURE 2.29
Nine direction image components of the bridge image. (All images have been scaled.)

2.6 *L*-paired representation of the image

Consider the case when $N = L^r$ and $L > 1$ is odd prime. The paired functions are defined similarly to the $L = 2$ case. The components of the tensor transform are united with the twiddle factors, which divide the unit circle into L parts. Below is the definition of the paired transform in the general case of $N \times N$.

Definition 2.2 Let L be a non-trivial divisor of the number N and let $W_L = \exp(-2\pi i/L)$. Given frequency-point $(p, s) \in X_{N,N}$ and integer $t \in \{0, 1, ..., N-1\}$, the function

$$\chi'_{p,s,t}(n, m) = \chi'_{p,s,t;L}(n, m) = \sum_{k=0}^{L-1} W_L^k \, \chi_{p,s,t+kN/L}(n, m) \qquad (2.50)$$

is called *the L-paired function.*

The L-paired function, or simply the paired-function is defined thus as

$$\chi'_{p,s,t}(n,m) = \begin{cases} 1, & \text{if } np+ms = t \bmod N \\ W_L, & \text{if } np+ms = (t+N/L) \bmod N \\ W_L^2, & \text{if } np+ms = (t+2N/L) \bmod N \\ \quad\dots, \quad\dots \\ W_L^{L-1}, & \text{if } np+ms = (t+(L-1)N/L) \bmod N \\ 0, & \text{otherwise.} \end{cases} \tag{2.51}$$

Given triplet (p,s,t), we denote by $f'_{p,s,t}$ the coefficient of the representation of the image $f = \{f_{n,m}\}$ by the L-paired function $\chi'_{p,s,t}$

$$f'_{p,s,t} = \chi'_{p,s,t} \circ f = \sum_{n=0}^{N-1}\sum_{m=0}^{N-1} \chi'_{p,s,t}(n,m)f_{n,m}$$

$$= \sum_{k=0}^{L-1} W_L^k f_{p,s,t+kN/L}. \tag{2.52}$$

We now consider the $N = L^r$ case, when $L > 2$ is prime and $r > 1$. As in the $N = 2^r$ case, we construct a partition $\sigma' = (T')$ of the lattice $X_{N,N}$, which is composed of the following subsets:

$$T'_{p,s;L} = \left\{ \left(\overline{(mL+1)p}, \overline{(mL+1)s}\right); \; m = 0:(N/L-1) \right\}. \tag{2.53}$$

To construct such a partition σ', we first consider the subset of generators (p,s),

$$J = J_{L^r,L^r} = \bigcup_{s=0}^{L^r-1}(1,s) \bigcup_{p=0}^{L^{r-1}-1}(Lp,1), \tag{2.54}$$

which is used for the tensor representation of the image.

Theorem 2.1 For an arbitrary prime number $L > 1$ and $r > 1$, the following is valid:

1) The totality of $(L+1)(L^r-1)+1$ subsets

$$\sigma' = \left(\left(\left(\left(T'_{jL^np,jL^ns;L} \right)_{(p,s)\in J_{L^{r-n},L^{r-n}}} \right)_{j=1:(L-1)} \right)_{n=0:(r-1)}, \{(0,0)\} \right)$$

is the partition of the lattice X_{L^r,L^r}.

2) The system of the L-paired functions

$$\chi' = \{\chi'_{p,s,t}; \; (p,s,t) \in U\}$$

is the complete set of functions, where the set U is defined as

$$U = \bigcup_{n=0}^{r-1}\bigcup_{j=1}^{L-1} \left\{ (L^n(jp,js,t); \; (p,s) \in J_{L^{r-n},L^{r-n}} , t = 0:(L^{r-n-1}-1) \right\}$$

$$\bigcup \{(0,0,0)\} \qquad \square$$

This theorem can be proved from the following statement [43, 46]:

Theorem 2.2 Let $L > 1$ be a prime number and let $r > 1$. Then, the following is valid:

1) For each sample (p, s) of X and $m = 0 : (L^{r-n-1} - 1)$,

$$F_{\overline{(mL+1)p},\overline{(mL+1)s}} = \sum_{t=0}^{L^{r-n-1}-1} \left(f'_{p,s,L^n t} W^t_{L^{r-n}} \right) W^{mt}_{L^{r-n-1}} . \tag{2.55}$$

The 2-D DFT at frequency-points of the subset $T'_{p,s}$ is defined by the N/L^{n+1}-point DFT of the splitting-signal modified by twiddle coefficients. Here L^n is defined as g.c.d.$(p, s) = jL^n$, where $j \in \{1, ..., L - 1\}$.

2) The $L^r \times L^r$-point DFT is split into $(L^2 - 1)L^{r-1}$ L^{r-1}-point DFTs, $(L^2 - 1)L^{r-2}$ L^{r-2}-point DFTs, ..., and $(L^2 - 1)L$ L-point DFTs.

It should be noted that the sum on the right side of Equation (2.55) describes the shifted 1-D DFT of the splitting-signal. For instance, when $n = 0$, we have

$$F_{m+\frac{1}{L}} = \sum_{t=0}^{L^{r-1}-1} f'_{p,s,t} W^{(m+\frac{1}{L})t}_{L^{r-1}} = \sum_{t=0}^{L^{r-1}-1} \left(f'_{p,s,t} W^t_{L^r} \right) W^{mt}_{L^{r-1}}$$

$$m = 0 : (L^{r-1} - 1),$$

which is the shifted L^{r-1}-point DFT in the frequency domain, $(m \to m+1/L)$.

The 2-D paired transform, or the L-paired transform is the representation of the 2-D image in the 2-D frequency and 1-D time domain in the form of a family of paired splitting-signals,

$$\{f_{n,m}\} \to \bigcup_{n=0}^{r-1} \bigcup_{j=1}^{L-1} \bigcup_{(p,s) \in jL^n J_{N/L^n, N/L^n}} \left\{ f'_{p,s,L^n t}; \ t = 0 : (N/L^{n+1} - 1) \right\}$$

$$\bigcup \{f_{0,0,0}\},$$

and *the L-paired splitting-signals* generated by the frequencies (p, s) are

$$f_{T'_{p,s}} = \{f'_{p,s,0}, f'_{p,s,L^n}, f'_{p,s,2 \cdot L^n}, \dots, f'_{p,s,N/L-L^n}\}. \tag{2.56}$$

Figure 2.30 shows the bridge image of size 243×243 in part a, and the 3-paired transform of the image in b. The first two sets of $2(324) = 648$ splitting-signals $\{f_{T'_{p,s}}; (p, s) \in j J_{243,243}\}$, $j = 1, 2$, of length 81 each are shown in the form of two images of size 324×81 each in part a. The next two sets of $2(108) = 216$ splitting-signals $\{f_{T'_{p,s}}; (p, s) \in j3 J_{81,81}\}$, $j = 1, 2$, of length 27 each are shown in the form of two images of size 108×27 each, and so on. The whole picture of these 968 splitting-signals represents the paired transform of the tree image.

The 2-D DFT of the tree image as the set of 968 1-D DFTs of all 3-paired splitting-signals is shown in Figure 2.31 in a similar way.

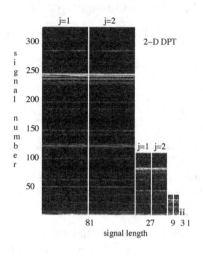

(a) $\{f_{n,m}\}$ (b) $\{f'_{p,s,t}\}$

FIGURE 2.30
(a) Bridge image 243×243 and (b) the splitting-signals of lengths $81, 27, 9, 3, 1, 1$.

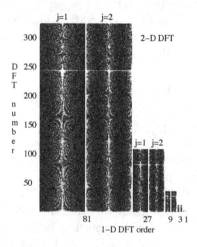

FIGURE 2.31
1-D DFTs of the modified splitting-signals. (All transforms are in absolute scale and shifted to the center).

2.6.1   Principle of superposition: General case

In L-paired representation, the image $f_{n,m}$ of size $N \times N$, when $N = L^r$, $r > 1$, is the set of $(L+1)(L^r - 1) + 1$ splitting-signals,

$$f_{T'_{p,s}} = \{f'_{p,s,0}, f'_{p,s,L^n}, f'_{p,s,2 \cdot L^n}, \ldots, f'_{p,s,N/L - L^n}\}, \tag{2.57}$$

which define the 2-D DFT at the corresponding subsets $T'_{p,s} = T'_{p,s;L}$. The set of all generators (p,s) of the paired transform components is defined as

$$J'_{N,N} = \bigcup_{k=0}^{r-1} \bigcup_{j=1}^{L-1} \{jL^k J_{L^{r-k},L^{r-k}}\} \cup \{(0,0)\},$$

and the subsets J_{L^k,L^k}, for $k = 1 : r$, are

$$J_{L^k,L^k} = \bigcup_{s=0}^{L^k-1} (1,s) \bigcup_{p=0}^{L^{k-1}-1} (Lp,1). \tag{2.58}$$

According to (2.55), the 1-D DFT of the splitting-signals weighted by the twiddle factors carry the spectral information of the image at frequency-points of the subsets $T'_{p,s}$,

$$F_{\overline{(mL+1)p},\overline{(mL+1)s}} = G_m = \sum_{t=0}^{L^{r-n-1}-1} g'_t W^{mt}_{L^{r-n-1}}, \tag{2.59}$$

where we define

$$g'_t = g'_{p,s,t} = f'_{p,s,L^n t} W^t_{L^{r-n}}, \quad t = 0 : L^{r-n-1} - 1.$$

Note that these components are periodic,

$$g'_t = g'_{t+L^{r-n-1}} = g'_{t \bmod L^{r-n-1}}. \tag{2.60}$$

For instance, when g.c.d.$(p,s) = j$, or $n = 0$, the following calculations are valid:

$$g'_{t+N/L} = f'_{p,s,t+N/L} W^{t+N/L} = (f'_{p,s,t+N/L} W_L) W^t$$

$$= \left(\sum_{k=0}^{L-1} f_{p,s,(t+N/L)+kN/L} W^k_L W_L \right) W^t$$

$$= \left(\sum_{k=0}^{L-1} f_{p,s,t+(k+1)N/L} W^{k+1}_L \right) W^t$$

$$= \left(\sum_{k=1}^{L-1} f_{p,s,t+kN/L} W^k_L + f_{p,s,t+N} W^L_L \right) W^t$$

$$= \left(\sum_{k=0}^{L-1} f_{p,s,t+kN/L} W^k_L \right) W^t = f'_{p,s,t} W^t = g'_t,$$

since the tensor transform is periodic, i.e., $f_{p,s,t+N} = f_{p,s,t}$.

The direction images are constructed similar to the $N = 2^r$ case. The incomplete 2-D DFT is composed only from the components of the 2-D DFT with frequency-points of the subset $T'_{p,s}$,

$$D_{p_1,s_1} = \begin{cases} F_{p_1,s_1}; \text{if } (p_1, s_1) \in T'_{p,s;L} \\ 0; \quad \text{otherwise.} \end{cases} \tag{2.61}$$

The image of this incomplete 2-D transform is composed only of values of the splitting-signal generated by (p, s).

We first consider this image for a generator (p, s) from the first series of generators in $J'_{N,N}$, i.e., when g.c.d.$(p, s) \in \{1, 2, ..., L - 1\}$. The following calculations hold for the inverse 2-D DFT:

$$d_{n,m}^{(p,s)} = \frac{1}{N^2} \sum_{p_1=0}^{N-1} \sum_{s_1=0}^{N-1} D_{p_1,s_1} W^{-(np_1+ms_1)}$$

$$= \frac{1}{N^2} \sum_{(p_1,s_1)\in T'_{p,s}} F_{p_1,s_1} W^{-(np_1+ms_1)}$$

$$= \frac{1}{N^2} \sum_{k=0}^{N/L-1} F_{\overline{(kL+1)p},\overline{(kL+1)s}} W^{-(n(kL+1)p+m(kL+1)s)}$$

$$= \frac{1}{N^2} \sum_{k=0}^{N/L-1} F_{\overline{(Lk+1)p},\overline{(Lk+1)s}} W^{-(Lk+1)(np+ms)}$$

$$= \frac{1}{LN} \left(\frac{L}{N} \sum_{k=0}^{N/L-1} F_{\overline{(Lk+1)p},\overline{(kL+1)s}} W_{N/L}^{-kt} \right) W^{-t}$$

where we denote $t = (np + ms) \bmod N$. According to (2.59), the sum in the brackets is the N/L-point inverse DFT, which represents the data of g'. Therefore,

$$d_{n,m}^{(p,s)} = \frac{1}{NL} g'_{p,s,t \bmod (N/L)} W^{-t}$$

$$= \frac{1}{NL} f'_{p,s,t \bmod (N/L)} W^{t \bmod (N/L)} W^{-t} \qquad .$$

$$= \frac{1}{NL} f'_{p,s,t \bmod (N/L)} W^{-t+t \bmod (N/L)}. \tag{2.62}$$

Since the image is real, we can write this direction image as

$$d_{n,m}^{(p,s)} = \frac{1}{NL} \left(\sum_{n=0}^{L-1} f_{p,s,\tilde{t}+nN/L} W_L^n \right) W^{-(t-\tilde{t})}$$

$$= \frac{1}{NL} \sum_{n=0}^{L-1} f_{p,s,\tilde{t}+nN/L} W_L^{n-\frac{t-\tilde{t}}{N/L}} \tag{2.63}$$

$$= \frac{1}{NL} \sum_{n=0}^{L-1} \cos\left(\frac{2\pi}{L} \left[n - \frac{1}{L^{r-1}}(t - \tilde{t}) \right] \right) f_{p,s,\tilde{t}+nN/L}$$

where $\tilde{t} = t \bmod (N/L)$.

For (p, s) from the second series of generators in $J'_{N,N}$, i.e., when g.c.d.$(p, s) = jL$ and $j \in \{1, 2, ..., L-1\}$, the following calculations are valid:

$$d^{(p,s)}_{n,m} = \frac{1}{N^2} \sum_{k=0}^{N/L^2-1} F_{\overline{(kL+1)p},\overline{(kL+1)s}} W^{-(kL+1)(np/L+ms/L)}_{N/L}$$

$$= \frac{1}{NL^2} \left(\frac{L^2}{N} \sum_{k=0}^{N/L^2-1} F_{\overline{(kL+1)p},\overline{(kL+1)s}} W^{-k(np/L+ms/L)}_{N/L^2} \right) W^{-(np/L+ms/L)}_{N/L}$$

(we denote $t = np/L + ms/L \bmod N/L$)

$$= \frac{1}{NL^2} g'_{p,s,t \bmod N/L^2} \cdot W^{-t}_{N/L}$$

$$= \frac{1}{NL^2} f'_{p,s,L(t \bmod N/L^2)} W^{t \bmod N/L^2}_{N/L} \cdot W^{-t}_{N/L}$$

$$= \frac{1}{NL^2} f'_{p,s,L(t \bmod N/L^2)} W^{-t+t \bmod N/L^2}_{N/L}. \tag{2.64}$$

Similarly, for (p, s) from the remaining series of generators in $J'_{N,N}$, that is, when g.c.d.$(p, s) = jL^k$, where $j \in \{1, 2, ..., L-1\}$ and $k = 2 : (r-1)$, the direction images are calculated by:

$$d^{(p,s)}_{n,m} = \frac{1}{NL^{k+1}} f'_{p,s,L^k(t \bmod N/L^{k+1})} W^{-t+t \bmod N/L^{k+1}}_{N/L^k} \tag{2.65}$$

where $t = (np/L^k + ms/L^k) \bmod N/L^k$.

We can also write this image as

$$d^{(p,s)}_{n,m} = \frac{1}{NL^{k+1}} f'_{p,s,L^k\tilde{t}} W^{-t+\tilde{t}}_{N/L^k}$$

$$= \frac{1}{NL^{k+1}} \sum_{n=0}^{L-1} f_{p,s,L^k\tilde{t}+nN/L} W^n_L W^{-t+\tilde{t}}_{N/L^k}$$

$$= \frac{1}{NL^{k+1}} \sum_{n=0}^{L-1} f_{p,s,L^k\tilde{t}+nN/L} W^{n-\frac{t-\tilde{t}}{L^{r-k-1}}}_L \tag{2.66}$$

$$= \frac{1}{NL^{k+1}} \sum_{n=0}^{L-1} \cos\left(\frac{2\pi}{L} \left[n - \frac{1}{L^{r-k-1}}(t - \tilde{t}) \right] \right) f_{p,s,L^k\tilde{t}+nN/L}$$

where $\tilde{t} = t \bmod (N/L^{k+1})$.

Statement 3: Paired Transform Slice Theorem II: The discrete image $\{f_{n,m}\}$ of size $N \times N$, where $N = L^r$, L is prime, and $r > 1$, can be decomposed by $(L+1)(N-1)+1$ direction images as

$$
\begin{aligned}
f_{n,m} &= \sum_{(p,s)\in J'_{N,N}} d_{n,m}^{(p,s)} \\
&= \sum_{k=0}^{r-1}\sum_{j=1}^{L-1}\sum_{(p,s)\in jL^k J_{N/L^k,N/L^k}} d_{n,m}^{(p,s)} + d_{n,m}^{(0,0)}.
\end{aligned}
\tag{2.67}
$$

Each component of the paired transform can be calculated from the projection data. To reconstruct the discrete image on the Cartesian lattice, only $(L+1)N/L$ projections are needed at the angles of the set

$$
\begin{aligned}
\Phi &= \{\varphi(p,s) = \arctan(p/s);\ (p,s)\in J_{N,N}\} \\
&= \{\arctan(p);\ p = 0 : (N-1)\}\cup\{\arctan(1/(Ls));\ s = 0 : (N/L-1)\}.
\end{aligned}
$$

Codes for discrete image decomposition by direction images, or image reconstruction from its projections by the tensor and paired transforms can be found in Image Reconstruction at http://www.fasttransforms.com.

Problems

Problem 2.1 On the lattice 16×16, sketch the parallel lines, or rays that pass through the points of the sets $\{V_{1,4,t};\ t = 0 : 3\}$. Determine and sketch parallel lines that pass through the points of the group $T_{1,4}$.

Problem 2.2 Consider the lattice 7×7 and the frequency-point $(1,3)$. Show that the set of parallel rays that pass all points of the sets $V_{1,3,t}$, $t = 0 : 6$, is not unique. Determine at least two different sets of such parallel rays and for each of these sets, determine the total number of rays.

Problem 2.3 Calculate all basic functions of the 2-D tensor transformation for the $N = 11$ case and pack them into the matrix 121×121. Calculate also the matrix of the inverse tensor transformation. Use MATLAB® function `imagesc` to sketch these matrices as images.

Problem 2.4 For the tree image 256×256, calculate the direction images $d_{n,m}^{(1,3)}$ and $d_{n,m}^{(1,16)}$ from the corresponding splitting-signals generated by frequencies $(p, s) = (1, 3)$ and $(1, 16)$, respectively.

Problem 2.5 Consider the following discrete image of size 7×7 :

$$[f_{n,m};\ n, m = 0 : 6] = \begin{array}{|c|c|c|c|c|c|c|} \hline 1 & 1 & 0 & 1 & 0 & 1 & 1 \\ \hline 0 & 1 & 1 & 0 & 1 & 1 & 0 \\ \hline 0 & 1 & 2 & 1 & 2 & 1 & 0 \\ \hline 1 & 0 & 1 & 6 & 1 & 0 & 1 \\ \hline 0 & 1 & 2 & 1 & 2 & 1 & 0 \\ \hline 0 & 1 & 1 & 0 & 1 & 1 & 0 \\ \hline 1 & 1 & 0 & 1 & 0 & 1 & 1 \\ \hline \end{array} .$$

Calculate and sketch eight direction components of this image in tensor representation. Calculate the original image from the sum of these eight direction images.

Problem 2.6 Accomplish the subsampling of the tree image down to the size 32×32 and consider this image in paired representation.

A. Splitting-signals of the image have different energies. Determine and sketch the five paired splitting-signals of the subsampled image that have the highest energies.

B. Sketch the direction images defined by the splitting-signals in *A*. The values of the direction image across all parallel rays are periodically repeated. Determine these periods for each of the above five direction images.

C. The number of splitting-signals for the subsampled image 32×32 is 94, and they gave different lengths. It is possible to pack all splitting-signals into a matrix 32×32. Develop an algorithm to pack all splitting-signals in such matrix.

Problem 2.7 The 2-D paired transformation $\chi'_{N,N}$ is not separable. Prove that the 2-D basic paired functions can be obtained from the basic functions of the 1-D paired transformation.

Problem 2.8 Suppose that you are given an image of size 256×256. Develop an algorithm for calculating the 2-D separable paired transform of this image. Compare your result with the 2-D non-separable paired transform of the image.

Problem 2.9 Assume that the image you are given has the size 243×243. The number 243 is the power of three, $243 = 3^5$. Develop a fast algorithm of the 3-paired transform of this image. You may calculate first the tensor transform of this image, and then use it to calculate all 3-paired splitting-signals. Sketch the image and the magnitude of the 3-paired transform.

Problem 2.10 The given image 512×512 can be extended to the size 513×513. Discuss the effect of using the image 513×513 in the frequency domain, instead of the original image.

3

Image Sampling Along Directions

The concept of the CT or computed X-ray tomography is illustrated in Figure 3.1, which shows the process of projection data collection. On both sides of the analyzed object or image, which is assumed to be immovable, the X-ray set and detectors are disposed and revolved around the object by a small angle in certain time intervals, and the measurement data of radiation and detection of X-rays are collected along other directions. This set of measurements, or

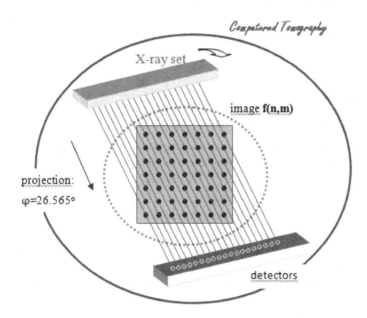

FIGURE 3.1
Method of computed X-ray tomography.

projections, taken at different positions of X-ray sources and detectors, is used by specific mathematical methods to reconstruct the two-dimensional (slice), or in general, the three-dimensional image of the observed object or tissue. The methods of image reconstruction must be fast and accurate because of the desired high quality of images for diagnostic purposes. And more important, we need methods that reconstruct the image directly on the discrete Cartesian lattice and with a minimal number of projections, in order to not over-radiate the body in CT.

In this chapter, we analyze solutions of the problem of reconstruction of the discrete image on the Cartesian lattice from projections of the image, which are based on the concepts of the tensor and paired transformations. In the framework of the constructed model, we show a way of using the ray-sums, or line-integrals of the image, or real projection data, for exact reconstructing the discrete image. Our goal is to show the way to transfer the geometry from the image plane to the Cartesian lattice. The sampling theorem of Kotelnikov is not used [35] to obtain the discrete image, but rather the sampling along the projections. We need to reconstruct the image from a finite number of its projections. The approach we propose for image reconstruction differs from the well-known methods of backprojection, iterative reconstruction, Fourier filtering, Radon filtering, and convolution filtering [25]-[32]. Each projection is processed by a system of linear equations, or linear convolutions, to calculate the corresponding part of the 2-D paired transform of the image, and then the inverse paired transform is calculated to obtain the discrete image. The model described for reconstruction is simple and the reconstruction is exact.

3.1   Image reconstruction: Model I

The image $f(x, y)$ is considered in the form of a digital image $f_d(x, y)$ which is represented by a finite number of elements, or image elements (IE) with constant values and of size $(\Delta x) \times (\Delta y)$ each,

$$f_d(x, y) = \frac{1}{(\Delta x)^2} \int_{\text{IE}} f(x, y) dx dy, \quad \text{if } (x, y) \in \text{IE}. \qquad (3.1)$$

We assume that the original image $f(x, y)$ occupies the square region $[0, 1] \times [0, 1]$, which is divided into N^2 image elements by the Cartesian lattice $N \times N$, where $N > 1$; therefore $\Delta x = \Delta y = 1/N$. All image elements are numbered by (n, m), where $n, m = 0 : (N - 1)$, and the image is considered in matrix form $f_{n,m}$ with values

$$f_{n,m} = \int_{(n,m)\text{-th IE}} f(x, y) dx dy = (\Delta x)^2 f_d(x_0, y_0), \quad \forall (x_0, y_0) \in (\text{n,m})\text{-th IE},$$
$$(3.2)$$

which are placed in the center of the corresponding (n, m)-th IE (see Figure 3.2). Thus, the point (x_0, y_0) is considered to be the center of the (n, m)-th IE. Thus the discrete image is considered on the square grid, or the Cartesian lattice $X_{N,N} = \{(n, m); n, m = 0 : (N - 1)\}$ inside the region $[0, 1] \times [0, 1]$. In such a model, the ray-sum, or line-integral along the ray l in the (n, m)-th IE

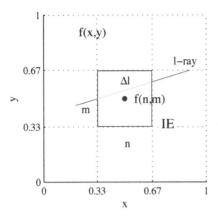

FIGURE 3.2
Model of the discrete image.

is defined as follows:

$$w_l^{(n,m)} = \int_{(n,m)\text{-th IE}} f(x,y)dl = (\Delta l)f_d(x_0,y_0) = (\Delta l)\frac{1}{(\Delta x)^2}f_{n,m} = (\Delta l)N^2 f_{n,m}$$

where $\Delta l = \Delta l_{n,m}$ is the length of the ray l in the (n,m)-th IE. Therefore, if $\Delta l \neq 0$, i.e., the ray passes through the IE, then

$$f_{n,m} = \frac{1}{N^2 \Delta l_{n,m}} w_l^{(n,m)}, \quad n,m = 0 : (N-1). \tag{3.3}$$

The line-integral of the image $f(x,y)$ along the ray l is denoted by w_l,

$$w_l = \int_l f(x,y)dl = \sum_{m=0}^{N-1}\sum_{n=0}^{N-1} w_l^{(n,m)} = N^2 \sum_{m=0}^{N-1}\sum_{n=0}^{N-1} \Delta l_{n,m} f_{n,m}. \tag{3.4}$$

We also can write that

$$w_l = \sum_{(n,m)\text{-th IE}\,\cap\, l} w_l^{(n,m)}$$

where the summation is performed by those IE which lie in the direction of the ray l. Here, the idealized rays are considered, i.e., rays of width 0 each.

The sum of the discrete image along the ray l is

$$v_l = \sum_{(n,m)\in l} f_{n,m} = \frac{1}{N^2}\sum_{(n,m)\in l} \frac{1}{\Delta l_{n,m}} w_l^{(n,m)}. \tag{3.5}$$

We will call v_l the line-sum.

If the number of projections is L, and $\{l\}$ is the set of rays passing through the (n, m)-th IE, then we can write the following:

$$f_{n,m} = \frac{1}{N^2} \frac{1}{L} \left[\sum_{l \in \{l\}} \frac{1}{\Delta l_{n,m}} w_l^{(n,m)} \right], \quad n, m = 0 : (N-1). \quad (3.6)$$

Here, we can probably approximate each normalized integral $w_l^{(n,m)} / \Delta l_{n,m}$ by the full integral of the image along the ray that passes through the point (n, m), after normalizing this integral by the length of the ray. As a result, we obtain an approximation

$$f_{n,m} \approx \frac{1}{N^2} \frac{1}{L} \sum_{l \in \{l\}} \frac{w_l}{\Delta l}, \quad n, m = 0 : (N-1), \quad (3.7)$$

where Δl are the lengths of the rays inside the square region $[0, 1] \times [0, 1]$. This approximation, which can be referred to as backprojection [29, 30], will be considered in detail in Chapter 7. Now we discuss a method of exact reconstruction of the image.

3.1.1   Coordinate systems and rays

We consider two co-ordinate systems on the plane. The first system X-Y of coordinates (x, y) is for the image $f(x, y)$ on the square $[0, 1] \times [0, 1]$. The second coordinate system (n, m), where n and m are integers, is for the lattice $X_{N,N}$ placed in the center of the square. This system is used for the discrete image $f_{n,m}$. Parameters x and n run from left to right, and parameters y and m run from the top down, as shown in Figure 3.3, for the $N = 4$ case. The first point of the image, $f_{0,0}$, is in the point with coordinates $(x, y) = (1/8, 1/8)$.

Given the frequency-point $(p, s) \in X_{N,N}$, such that $g.c.d.(p, s) = 1$, we consider the lines

$$l(t) = l_{p,s}(t) = \{(n, m); \, pn + sm = t\}, \quad t = 0 : (p+s)(N-1),$$

on the square grid $X_{N,N}$. These lines are referred to as *the arithmetical rays*. The equation of these lines on the square $[0, 1] \times [0, 1]$ is

$$l(t) = l_{p,s}(t) = \left\{ (x, y); \, px + sy = \frac{t}{N} + \frac{p+s}{2N} \right\}, \quad t = 0 : (p+s)(N-1).$$

These lines are referred to as *the geometrical rays*, to distinguish the discrete and continuous cases. These two types of rays are denoted by $l(t)$, and the same set of t is considered for the rays, $t = 0 : (p+s)(N-1)$. This set of t for the geometrical rays may be changed for effective calculation of sums v_l for different (p, s), as will be shown later.

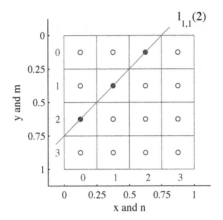

FIGURE 3.3
Two coordinate systems for images $f(x, y)$ and $f(n, m)$.

The generator (p, s) defines the slope, $-\tan^{-1}(p/s)$, of these rays. The sums of the image along the arithmetical rays are denoted by

$$v_{p,s}(t) = v_{l(t)} = \sum_{l_{p,s,t}} f_{n,m}.$$

Line-integrals w_l are considered to be the physical measurements of the projection, or measurements of geometrical rays, and the sums v_l are measurements of arithmetical rays. The main problem that we are going to solve is formulated simply: Determine the sums v_l from the line-integrals w_l. Since the discrete tensor and paired transforms of the image are completely defined by the direction binary 2-D functions, these transforms can be used to solve this problem and reconstruct the image from its projections. We first describe in detail the case in which the discrete image is considered of size $N \times N$ and N is a power of two. In this case, the tensor representation $\{f_{p,s,t}\}$ is modified into the unitary paired transform $\{f'_{p,s,t}\}$, which does not require operations of multiplication. The case where N is prime is considered in Chapter 5.

3.2 Inverse paired transform

Our task is to obtain the discrete image $f_{n,m}$ of size $N \times N$, by using the 2-D discrete paired transform (DPT), which can be calculated from the integrals of the image $f(x, y)$ along the rays, which correspond to the projections to be defined. If we find a way of calculating the 2-D DPT from the physical measurements, the problem of image reconstruction in discrete form will then be solved by using the inverse 2-D DPT. We here recall briefly the main results of the paired representation of the image, which defines a unique image superposition by direction images.

The discrete image $\{f_{n,m}\}$ of size $N \times N$, where $N = 2^r$, $r > 1$, can be composed of $(3N - 2)$ splitting-signals as

$$f_{n,m} = \frac{1}{2N} \sum_{k=0}^{r-1} \frac{1}{2^k} \sum_{(p,s)\in 2^k J_{2^{r-k},2^{r-k}}} f'_{p,s,(np+ms)\bmod N} + \frac{1}{N^2} f'_{0,0,0}. \qquad (3.8)$$

The components of the paired transform

$$f'_{p,s,2^k t} = f_{p,s,2^k t} - f_{p,s,2^k t+N/2}, \quad t = 0 : (N/2^{k+1} - 1),$$

where $2^k = \text{g.c.d.}(p, s)$, are calculated by the system of orthogonal 2-D paired functions $\chi'_{p,s,2^k t}(n, m)$ defined in (2.39). The set of N^2 triplet-numbers of paired functions is defined as

$$U_{N,N} = \bigcup_{k=0}^{r-1} \left\{ (p, s, 2^k t); (p, s) \in 2^k J_{N/2^k, N/2^k}, t = 0 : (N/2^{k+1}-1) \right\} \cup \{(0,0,0)\},$$

$$(3.9)$$

where the subsets of generators (p, s) are calculated by

$$J_{N/2^k,N/2^k} = \left\{ (1, s); s = 0 : (N/2^k - 1) \right\} \cup \left\{ (2p, 1); p = 0 : (N/2^{k+1} - 1) \right\}.$$

If we normalize the components of the 2-D DPT as

$$f'_{p,s,t} \rightarrow \frac{1}{2^{k+1}} f'_{p,s,t} = \frac{1}{2 \cdot \text{g.c.d}(p, s)} f'_{p,s,t}, \quad (f'_{0,0,0} \rightarrow f'_{0,0,0}/N),$$

then the decomposition of the image in (3.8) can be written as

$$f_{n,m} = \frac{1}{N} \left[\sum_{k=0}^{r-1} \sum_{(p,s)\in 2^k J_{2^{r-k},2^{r-k}}} f'_{p,s,(np+ms)\bmod N} + f'_{0,0,0} \right]. \qquad (3.10)$$

Due to this formula, the value of the discrete image at each point (n, m) is defined by $3N - 2$ values of the 2-D DPT, which are calculated along arithmetical rays of $3N/2$ projections that pass through points (n, m) on the Cartesian grid on the image. The number of operations of addition which are required to reconstruct the image by this equation equals

$$a(N) = N^2[(3N - 2) - 1] = 3N^2(N - 1).$$

Additional operations are required for calculating the splitting-signals from the physical measurements, or the line-integrals w_l.

3.3 Example: Image 4×4

We consider the masks of all basic paired functions $\chi'_{p,s,t}$ in detail, for the $N = 4$ case. Sixteen triplets (p, s, t) of the 4×4-point paired transform are defined by the set

$$U_{4,4} = \begin{cases} (1,0,0), (1,0,1), (1,1,0), (1,1,1), (1,2,0), (1,2,1), (1,3,0), (1,3,1), \\ (0,1,0), (0,1,1), (2,1,0), (2,1,1), \\ (2,0,0), (2,2,0), (0,2,0), \\ (0,0,0). \end{cases}$$

3.3.1 Horizontal and vertical projections

First, we separate seven paired functions which are generated by $(p, s) = (1,0), (0,1), (0,2), (2,0)$, and $(0,0)$ and defined by the horizontal and vertical projections. For the generator $(p, s) = (1,0)$, the masks of two paired basic functions are

$$[\chi'_{1,0,0}] = \begin{bmatrix} 1 & 0 & -1 & 0 \\ 1 & 0 & -1 & 0 \\ 1 & 0 & -1 & 0 \\ 1 & 0 & -1 & 0 \end{bmatrix}, \qquad [\chi'_{1,0,1}] = \begin{bmatrix} 0 & 1 & 0 & -1 \\ 0 & 1 & 0 & -1 \\ 0 & 1 & 0 & -1 \\ 0 & 1 & 0 & -1 \end{bmatrix}. \tag{3.11}$$

The summation of the image is performed along the geometrical rays

$$l(t) = l_{1,0}(t) = \{(x, y); 1x + 0y = t/4 + 1/8\}, \quad t = 0, 1, 2, 3,$$

as shown in Figure 3.4. Note that along the horizontal axis, the integer value

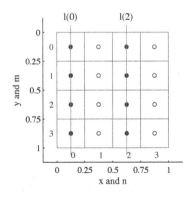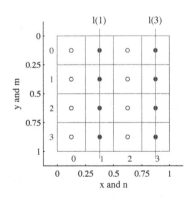

FIGURE 3.4

Schemes of calculating four line-integrals for $(p, s) = (1, 0)$.

n runs from 0 to 3, and x runs the interval $[0, 1]$. Along the vertical axis, the integer value m runs from 0 to 3, and y runs the interval $[0, 1]$. The same system of coordinates is used for all masks of basic functions of the paired transform.

We denote the sums of the discrete image along these lines by $v_{1,0}(t)$,

$$v_{1,0}(t) = v_{l(t)} = \sum_{l_{1,0}(t)} f_{n,m} = \sum_{m=0}^{3} f_{t,m}, \quad t = 0, 1, 2, 3.$$

They are four measurements of the projection by the angle $\varphi = \varphi_{1,0} = 0$ to the horizontal axis. This horizontal projection defines the first two components of the 2-D paired transform of the image,

$$f'_{1,0,0} = v_{1,0}(0) - v_{1,0}(2), \quad f'_{1,0,1} = v_{1,0}(1) - v_{1,0}(3).$$

All sums $v_{1,0}(t)$ can be calculated from the horizontal projection of the image $f(x, y)$,

$$w_{1,0}(t) = w_{l(t)} = \int_{l(t)} f(x, y)dl, \quad t = 0, 1, 2, 3.$$

The length of intersection of the geometrical ray $l(t)$ with the (t, m)-th IE equals $\Delta l = \Delta l_{t,m} = 1/N$, and

$$v_{1,0}(t) = \frac{1}{N^2 \Delta l} \sum_{m=0}^{N-1} w_l^{(t,m)} = \frac{1}{N^2 \frac{1}{N}} w_{l(t)} = \frac{1}{N} w_{l(t)} = \frac{1}{N} w_{1,0}(t). \quad (3.12)$$

Therefore

$$f'_{1,0,t} = \frac{1}{N}[w_{1,0}(t) - w_{1,0}(t+2)], \quad t = 0, 1.$$

The next two paired functions generated by $(p, s) = (0, 1)$ have the following masks:

$$[\chi'_{0,1,0}] = \begin{bmatrix} 1 & 1 & 1 & 1 \\ 0 & 0 & 0 & 0 \\ -1 & -1 & -1 & -1 \\ 0 & 0 & 0 & 0 \end{bmatrix}, \quad [\chi'_{0,1,1}] = \begin{bmatrix} 0 & 0 & 0 & 0 \\ 1 & 1 & 1 & 1 \\ 0 & 0 & 0 & 0 \\ -1 & -1 & -1 & -1 \end{bmatrix}. \quad (3.13)$$

The summation of the image is performed along the geometrical rays

$$l(t) = l_{0,1}(t) = \{(x, y); 0x + 1y = t/4 + 1/8\}, \quad t = 0, 1, 2, 3,$$

as shown in Figure 3.5.

Here, four measurements of the projection by the angle $\varphi = \varphi_{0,1} = \pi/2$ to the horizontal axis are used. We denote the sums of the discrete image along these lines by $v_{0,1}(t)$,

$$v_{0,1}(t) = \sum_{l_{0,1}(t)} f_{n,m} = \sum_{n=0}^{3} f_{n,t}, \quad t = 0 : 3.$$

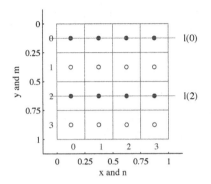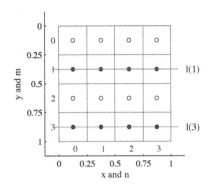

FIGURE 3.5
Schemes of calculating four line-integrals for $(p, s) = (0, 1)$.

This vertical projection defines two components of the 2-D paired transform,

$$f'_{0,1,0} = v_{0,1}(0) - v_{0,1}(2), \quad f'_{0,1,1} = v_{0,1}(1) - v_{0,1}(3).$$

All sums $v_{1,0}(t)$ can be calculated from the vertical projection of the image $f(x, y)$,

$$w_{0,1}(t) = w_{l(t)} = \int_{l(t)} f(x, y) dl, \quad t = 01, 2, 3,$$

in a way similar to the horizontal projection,

$$v_{0,1}(t) = \frac{1}{N^2 \Delta l} \sum_{n=0}^{N-1} w_l^{n,t} = \frac{1}{N^2 \frac{1}{N}} w_{l(t)} = \frac{1}{N} w_{l(t)} = \frac{1}{N} w_{0,1}(t) \qquad (3.14)$$

and, therefore,

$$f'_{1,0,t} = \frac{1}{N} [w_{1,0}(t) - w_{1,0}(t + 2)], \quad t = 0, 1.$$

The next three basic functions have the following masks:

$$[\chi'_{0,2,0}] = \begin{bmatrix} 1 & 1 & 1 & 1 \\ -1 & -1 & -1 & -1 \\ 1 & 1 & 1 & 1 \\ -1 & -1 & -1 & -1 \end{bmatrix}, \quad [\chi'_{2,0,0}] = \begin{bmatrix} 1 & -1 & 1 & -1 \\ 1 & -1 & 1 & -1 \\ 1 & -1 & 1 & -1 \\ 1 & -1 & 1 & -1 \end{bmatrix}, \quad [\chi'_{0,0,0}] = \begin{bmatrix} 1 & 1 & 1 & 1 \\ 1 & 1 & 1 & 1 \\ 1 & 1 & 1 & 1 \\ 1 & 1 & 1 & 1 \end{bmatrix},$$

and they are defined by the vertical and horizontal projections. The corresponding components of the 2-D paired transform can be calculated by

$$f'_{0,2,0} = v_{0,1}(0) - v_{0,1}(1) + v_{0,1}(2) - v_{0,1}(3)$$
$$= \tfrac{1}{N}\left[w_{0,1}(0) - w_{0,1}(1) + w_{0,1}(2) - w_{0,1}(3)\right],$$
$$f'_{2,0,0} = v_{1,0}(0) - v_{1,0}(1) + v_{1,0}(2) - v_{1,0}(3)$$
$$= \tfrac{1}{N}\left[w_{1,0}(0) - w_{1,0}(1) + w_{1,0}(2) - w_{1,0}(3)\right],$$
$$f'_{0,0,0} = v_{0,1}(0) + v_{0,1}(1) + v_{0,1}(2) + v_{0,1}(3)$$
$$= \tfrac{1}{N}\left[w_{0,1}(0) + w_{0,1}(1) + w_{0,1}(2) + w_{0,1}(3)\right],$$
$$= v_{1,0}(0) + v_{1,0}(1) + v_{1,0}(2) + v_{1,0}(3)$$
$$= \tfrac{1}{N}\left[w_{1,0}(0) + w_{1,0}(1) + w_{1,0}(2) + w_{1,0}(3)\right].$$

To calculate the component $f'_{0,0,0}$, both horizontal and vertical projections can be used.

Summarizing the above calculations for the vertical and horizontal projections, we can write the above equations in matrix form

$$\begin{bmatrix} f'_{0,1,0} \\ f'_{0,1,1} \\ f'_{0,2,0} \\ f'_{0,0,0} \end{bmatrix} = \frac{1}{4} \begin{bmatrix} 1 & 0 & -1 & 0 \\ 0 & 1 & 0 & -1 \\ 1 & -1 & 1 & -1 \\ 1 & 1 & 1 & 1 \end{bmatrix} \begin{bmatrix} w_{0,1}(0) \\ w_{0,1}(1) \\ w_{0,1}(2) \\ w_{0,1}(3) \end{bmatrix},$$

and

$$\begin{bmatrix} f'_{1,0,0} \\ f'_{1,0,1} \\ f'_{2,0,0} \\ f'_{0,0,0} \end{bmatrix} = \frac{1}{4} \begin{bmatrix} 1 & 0 & -1 & 0 \\ 0 & 1 & 0 & -1 \\ 1 & -1 & 1 & -1 \\ 1 & 1 & 1 & 1 \end{bmatrix} \begin{bmatrix} w_{1,0}(0) \\ w_{1,0}(1) \\ w_{1,0}(2) \\ w_{1,0}(3) \end{bmatrix},$$

where

$$[\chi'_{4,4}] = \begin{bmatrix} 1 & 0 & -1 & 0 \\ 0 & 1 & 0 & -1 \\ 1 & -1 & 1 & -1 \\ 1 & 1 & 1 & 1 \end{bmatrix}$$

is the matrix of the 4-point 1-D paired transformation, χ'_4. The calculation of the component $f'_{0,0,0}$ is repeated and, therefore, can be omitted from one of these equations, or considered as the mean of calculations by both projections,

$$f'_{0,0,0} = \frac{1}{2}\left[\frac{1}{4}[w_{0,1}(0) + w_{0,1}(1) + w_{0,1}(2) + w_{0,1}(3)] + \right.$$
$$\left. + \frac{1}{4}[w_{1,0}(0) + w_{1,0}(1) + w_{1,0}(2) + w_{1,0}(3)]\right].$$

3.3.2 Diagonal projections

Now we consider the second set of five paired functions, which are generated by $(p, s) = (1, 1), (1, 3)$, and $(2, 2)$. The masks of the first two functions are

$$[\chi'_{1,1,0}] = \begin{bmatrix} 1 & 0 & -1 & 0 \\ 0 & -1 & 0 & 1 \\ -1 & 0 & 1 & 0 \\ 0 & 1 & 0 & -1 \end{bmatrix}, \quad [\chi'_{1,1,1}] = \begin{bmatrix} 0 & 1 & 0 & -1 \\ 1 & 0 & -1 & 0 \\ 0 & -1 & 0 & 1 \\ -1 & 0 & 1 & 0 \end{bmatrix}.$$

Figure 3.6 shows two masks of the paired functions for the frequency-point $(p, s) = (1, 1)$ and seven parallel rays along which the coefficients ± 1 are located. These rays define the diagonal projection of the image when the

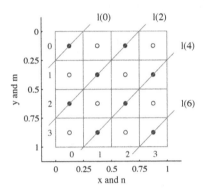 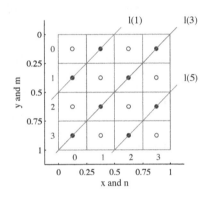

FIGURE 3.6
Schemes of calculating seven line-integrals for $(p, s) = (1, 1)$.

angle of projection equals $\pi/4$. The sums of the image along the arithmetical rays

$$l(t) = l_{1,1,t} = \{(x, y); \, 1x + 1y = t/4 + 1/4\}, \quad t = 0, 1, ..., 6,$$

define the following components of the 2-D paired transform:

$$f'_{1,1,0} = v_{1,1}(0) - v_{1,1}(2) + v_{1,1}(4) - v_{1,1}(6),$$
$$f'_{1,1,1} = v_{1,1}(1) - v_{1,1}(3) + v_{1,1}(5).$$

In this case, $\Delta l = \Delta l_{1,1} = \sqrt{2}\Delta x$ and the sums $v_{1,1}(t)$ are defined by the line-integrals $w_{1,1}(t)$ as

$$w_{1,1}(t) = \Delta l \frac{v_{1,1}(t)}{(\Delta x)^2} = \sqrt{2}\frac{v_{1,1}(t)}{\Delta x} = \sqrt{2}Nv_{1,1}(t), \quad (3.15)$$

or $v_{1,1}(t) = w_{1,1}(t)/(\sqrt{2}N)$, $t = 0, 1, 2, 3$. Therefore these two components of the 2-D paired transform can be calculated by

$$f'_{1,1,0} = \frac{1}{4\sqrt{2}}[w_{1,1}(0) - w_{1,1}(2) + w_{1,1}(4) - w_{1,1}(6)],$$

$$f'_{1,1,1} = \frac{1}{4\sqrt{2}}[w_{1,1}(1) - w_{1,1}(3) + w_{1,1}(5)].$$

The mask of the paired function $\chi'_{2,2,0}$ contains only coefficients ± 1,

$$[\chi'_{2,2,0}] = \begin{bmatrix} 1 & -1 & 1 & -1 \\ -1 & 1 & -1 & 1 \\ 1 & -1 & 1 & -1 \\ -1 & 1 & -1 & 1 \end{bmatrix},$$

which can be seen on the seven parallel lines $1x + 1y = t/4 + 1/4$, $t = 0 : 6$,

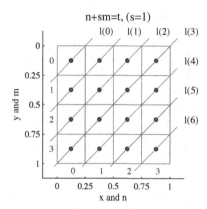 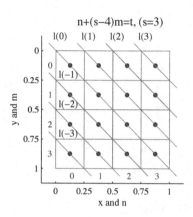

FIGURE 3.7
Two schemes of calculating seven line-integrals for $(p, s) = (2, 2)$.

which are shown in Figure 3.7 (on the left). Therefore the component $f'_{2,2,0}$ of the paired transform can be calculated by

$$f'_{2,2,0} = \sum_{t=0}^{6}(-1)^n v_{1,1}(t) = \frac{1}{4\sqrt{2}}\sum_{t=0}^{6}(-1)^n w_{1,1}(t).$$

The above calculations can be written in the following matrix form:

$$\begin{bmatrix} f'_{1,1,0} \\ f'_{1,1,1} \\ f'_{2,2,0} \\ f'_{0,0,0}/\sqrt{2} \end{bmatrix} = \frac{1}{4\sqrt{2}}\begin{bmatrix} 1 & 0 & -1 & 0 & 1 & 0 & -1 \\ 0 & 1 & 0 & -1 & 0 & 1 & 0 \\ 1 & -1 & 1 & -1 & 1 & -1 & 1 \\ 1 & 1 & 1 & 1 & 1 & 1 & 1 \end{bmatrix}\begin{bmatrix} w_{1,1}(0) \\ w_{1,1}(1) \\ w_{1,1}(2) \\ w_{1,1}(3) \\ w_{1,1}(4) \\ w_{1,1}(5) \\ w_{1,1}(6) \end{bmatrix},$$

or

$$
\begin{bmatrix} f'_{1,1,0} \\ f'_{1,1,1} \\ f'_{2,2,0} \\ f'_{0,0,0}/\sqrt{2} \end{bmatrix} = \frac{1}{4\sqrt{2}} \begin{bmatrix} 1 & 0 & -1 & 0 \\ 0 & 1 & 0 & -1 \\ 1 & -1 & 1 & -1 \\ 1 & 1 & 1 & 1 \end{bmatrix} \begin{bmatrix} w_{1,1}(0) + w_{1,1}(4) \\ w_{1,1}(1) + w_{1,1}(5) \\ w_{1,1}(2) + w_{1,1}(6) \\ w_{1,1}(3) \end{bmatrix}.
$$

The matrix 4×4 in this equation is the matrix of the four-point DPT.

3.3.3 Other projections

The remaining three projections that are generated by the frequency-points $(p, s) = (1, 3), (1, 2)$, and $(2, 1)$ are described similarly to the projections generated by $(p, s) = (1, 0), (0, 1)$, and $(1, 1)$. We will start with the $(1, 3)$ case, which corresponds to the diagonal projection, and then consider two symmetric cases $(1, 2)$ and $(2, 1)$.

3.3.3.1 Generator $(1, 3)$

Consider two paired functions that are generated by the frequency-point $(1, 3)$ and have the following masks:

$$
[\chi'_{1,3,0}] = \begin{bmatrix} 1 & 0 & -1 & 0 \\ 0 & 1 & 0 & -1 \\ -1 & 0 & 1 & 0 \\ 0 & -1 & 0 & 1 \end{bmatrix}, \quad [\chi'_{1,3,1}] = \begin{bmatrix} 0 & 1 & 0 & -1 \\ -1 & 0 & 1 & 0 \\ 0 & -1 & 0 & 1 \\ 1 & 0 & -1 & 0 \end{bmatrix}.
$$

Coefficients ± 1 on these masks are situated on the parallel rays

$$
l(t) = l_{1,3}(t) = \{(x, y); 1 \cdot x + 3 \cdot y = t/4 + 1/2\},
$$

where t runs the integer numbers from 0 to 12. The rays compose the angle $\varphi_{1,3} = \pi - \tan^{-1}(1/3)$ to the horizontal axis, as shown in the following matrix:

$$
[1 \cdot n + 3 \cdot m = t] = \begin{bmatrix} 0 & 1 & 2 & 3 \\ 3 & 4 & 5 & 6 \\ 6 & 7 & 8 & 9 \\ 9 & 10 & 11 & 12 \end{bmatrix}. \tag{3.16}
$$

The line connecting two 6's shows the ray $l(6)$ and the direction of the parallel rays in the projection. This large number of rays can be replaced with another set of parallel rays with number less than 13. To show that, we first consider the location of four coefficients -1 in the mask of the function $\chi'_{1,3,0}(n, m)$. The locations of these units are determined by the above matrix, where coefficients $t = 2, 6$, and 10 equal 2 mod 4. These four -1 are located on three parallel rays by the angle $\varphi_{1,3}$ to the horizontal. One ray passes through two points with number 6, and one ray passes through the points with numbers 2 and 10 each. Similarly, one can notice that four coefficients 1 in the mask $[\chi'_{1,3,0}]$, which are shown by coefficients $t = 0, 4, 8$, and 12 which equal 0 mod 4, are on the different four parallel rays by the angle $\varphi_{1,3}$ to the horizontal. Thus, all units of this mask are situated on seven rays.

Consider the similar matrix $[t]$ which is determined by the point $(1, 3-4) = (1, -1)$,

$$[1 \cdot n - 1 \cdot m = t] = \begin{bmatrix} 0 & 1 & \underline{2} & 3 \\ -1 & 0 & 1 & \underline{2} \\ -\underline{2} & -1 & 0 & 1 \\ -3 & -\underline{2} & -1 & 0 \end{bmatrix}.$$

One can see that coefficients ± 2 are located on two parallel rays, and all coefficients 0 are on one diagonal ray. These rays compose the angle $\varphi_{1,-1} = \tan^{-1}(1) = \pi/4$ with the horizontal axis,

$$l(t) = l_{1,-1}(t) = \{(x, y); \ 1 \cdot x - 1 \cdot y = t/4\},$$

and the total number of them equals $(1 + |-1|)3 + 1 = 7$, as shown in Figure 3.7 (on the right).

The sums of the image along these arithmetical rays $l(t)$ define the following components of the 2-D paired transform:

$$f'_{1,3,0} = -v_{1,-1}(-2) + v_{1,-1}(0) - v_{1,-1}(2),$$
$$f'_{1,3,1} = v_{1,-1}(-3) - v_{1,-1}(-1) + v_{1,-1}(1) - v_{1,-1}(3),$$

as shown in Figure 3.8.

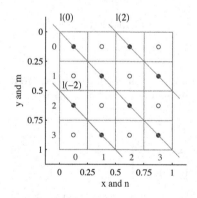 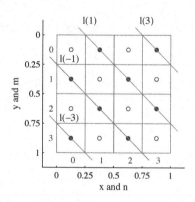

FIGURE 3.8
Schemes of calculating seven line-integrals for $(p, s) = (1, -1)$.

Similar to the $(p, s) = (1, 1)$ case, the sums $v_{1,-1}(t)$ of the image are defined by the line-integrals $w_{1,-1}(t)$ as follows:

$$w_{1,-1}(t) = \Delta l \frac{v_{1,-1}(t)}{(\Delta x)^2} = \sqrt{2} \frac{v_{1,-1}(t)}{\Delta x} = N\sqrt{2} v_{1,-1}(t), \qquad (3.17)$$

or $v_{1,-1}(t) = w_{1,-1}(t)/(\sqrt{2}N)$, $t = -3 : 3$. Therefore we obtain

$$f'_{1,3,0} = \frac{1}{4\sqrt{2}}[-w_{1,-1}(-2) + w_{1,-1}(0) - w_{1,-1}(2)],$$

$$f'_{1,3,1} = \frac{1}{4\sqrt{2}}[w_{1,-1}(-3) - w_{1,-1}(-1) + w_{1,-1}(1) - w_{1,-1}(3)].$$

As follows from Figure 3.7 (on the right), one can also calculate the component $f'_{2,2,0}$, by using these line-integrals,

$$f'_{2,2,0} = \sum_{t=-3}^{3} (-1)^n v_{1,-1}(t) = \frac{1}{4\sqrt{2}} \sum_{t=-3}^{3} (-1)^n w_{1,-1}(t).$$

The calculations of these three components of the paired transform together with $f'_{0,0,0}$ can be written in matrix form

$$
\begin{bmatrix} f'_{1,3,0} \\ f'_{1,3,1} \\ f'_{2,2,0} \\ f'_{0,0,0}/\sqrt{2} \end{bmatrix} = \frac{1}{4\sqrt{2}}
\begin{bmatrix} 0 & -1 & 0 & 1 & 0 & -1 & 0 \\ 1 & 0 & -1 & 0 & 1 & 0 & -1 \\ -1 & 1 & -1 & 1 & -1 & 1 & -1 \\ 1 & 1 & 1 & 1 & 1 & 1 & 1 \end{bmatrix}
\begin{bmatrix} w_{1,-1}(-3) \\ w_{1,-1}(-2) \\ w_{1,-1}(-1) \\ w_{1,-1}(0) \\ w_{1,-1}(1) \\ w_{1,-1}(2) \\ w_{1,-1}(3) \end{bmatrix},
$$

which can also be written as

$$
\begin{bmatrix} f'_{1,3,0} \\ f'_{1,3,1} \\ f'_{2,2,0} \\ f'_{0,0,0}/\sqrt{2} \end{bmatrix} = \frac{1}{4\sqrt{2}}
\begin{bmatrix} 1 & 0 & -1 & 0 \\ 0 & 1 & 0 & -1 \\ 1 & -1 & 1 & -1 \\ 1 & 1 & 1 & 1 \end{bmatrix}
\begin{bmatrix} w_{1,-1}(0) \\ w_{1,-1}(1) + w_{1,-1}(-3) \\ w_{1,-1}(2) + w_{1,-1}(-2) \\ w_{1,-1}(3) + w_{1,-1}(-1) \end{bmatrix}.
$$

3.3.3.2 Generator $(1,2)$

We now consider the masks of the paired functions for $(p, s) = (1, 2)$,

$$
[\chi'_{1,2,0}] = \begin{bmatrix} 1 & 0 & -1 & 0 \\ -1 & 0 & 1 & 0 \\ 1 & 0 & -1 & 0 \\ -1 & 0 & 1 & 0 \end{bmatrix}, \quad
[\chi'_{1,2,1}] = \begin{bmatrix} 0 & 1 & 0 & -1 \\ 0 & -1 & 0 & 1 \\ 0 & 1 & 0 & -1 \\ 0 & -1 & 0 & 1 \end{bmatrix}.
$$

The coefficients ± 1 on these masks are situated on the parallel rays at the angle $\varphi_{1,2} = \pi - \tan^{-1}(1/2)$ with the horizontal axis (see Figure 3.9),

$$l(t) = l_{1,2}(t) = \{(x, y); \ 1 \cdot x + 2 \cdot y = t/4 + 3/8\},$$

where t runs the integer numbers from 0 to 9. The sums of the image along

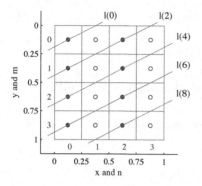 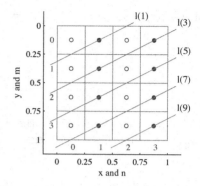

FIGURE 3.9

Schemes of calculating seven line-integrals for $(p, s) = (1, 2)$.

the corresponding arithmetical rays define the following components of the paired transform:

$$f'_{1,2,0} = v_{1,2}(0) - v_{1,2}(2) + v_{1,2}(4) - v_{1,2}(6) + v_{1,2}(8),$$
$$f'_{1,2,1} = v_{1,2}(1) - v_{1,2}(3) + v_{1,2}(5) - v_{1,2}(7) + v_{1,2}(9).$$

One can see from Figure 3.9 and Figure 3.10 (on the left), that the geometrical rays $l(t)$ pass through the additional squares, or image elements on which the arithmetical rays are not defined. For example, the arithmetical ray $l(6)$ passes through only two image elements with numbers $(0, 3)$ and $(2, 2)$, and the geometrical ray $l(6)$ passes through six image elements with numbers $(0, 3), (1, 3), (1, 2), (2, 2), (3, 2)$, and $(3, 1)$. Each sum $v_{1,2}(t)$ cannot be calculated from only the integral $w_{1,2}(t)$, which is defined along the geometrical rays $l(t)$, as was the case for the $(p, s) = (1, 1)$ and $(1, 3)$ cases. To derive a simple expression between the measurements along the arithmetical and geometrical rays for this projection, we first consider the arithmetical rays $l(6)$ and $l(7)$ and the geometrical ray $l = \{(x, y); x + 2y = 2\}$ between them, which can be considered as the ray $\tilde{l}(7) = l(7 - 0.5)$ shifted in the vertical direction. This geometrical ray intersects four image elements with intersection of length $\sqrt{5}/2\Delta x = \sqrt{5}/8$ in each. Therefore, the line-integral along this ray can be expressed by two sums of the image along the neighbor rays, but arithmetical, as follows:

$$w = w_{1,2}(7) = w(\tilde{l}(7)) = \frac{\sqrt{5}}{8} \frac{1}{(\Delta x)^2} [f_{0,3} + f_{1,3} + f_{2,2} + f_{3,2}]$$
$$= \frac{\sqrt{5}}{8} \frac{1}{(\Delta x)^2} [(f_{0,3} + f_{2,2}) + (f_{1,3} + f_{3,2})]$$
$$= \frac{\sqrt{5}}{8} \left[\frac{v_{1,2}(6) + v_{1,2}(7)}{(\Delta x)^2} \right] = 2\sqrt{5} [v_{1,2}(6) + v_{1,2}(7)].$$

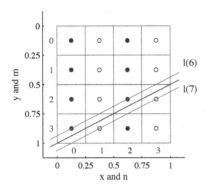
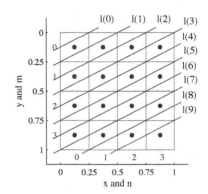

FIGURE 3.10
Schemes of calculating seven line-integrals for $(p, s) = (1, 2)$.

Case 1: Consider the set of 10 shifted geometrical rays

$$\tilde{l}(t) = l_{1,2}(t - 0.5) = \{(x, y);\ 1 \cdot x + 2 \cdot y = t/4 + 1/8\},$$

which are shown in Figure 3.10 (on the right) and numbered from left to right and top to bottom (and without the fraction $1/2$ in the numbers). In the figure, rays $\tilde{l}(t)$ are denoted by $l(t)$. For all line-integrals of the image along these rays, we can write the following:

$$w_{1,2}(t) = 2\sqrt{5}\,[v_{1,2}(t - 1) + v_{1,2}(t)], \quad t = 0:9, \tag{3.18}$$

where it is assumed that $v_{1,2}(-1) = 0$. These equations can be written in matrix form

$$\mathbf{w} = \begin{bmatrix} w_{1,2}(0) \\ w_{1,2}(1) \\ w_{1,2}(2) \\ w_{1,2}(3) \\ w_{1,2}(4) \\ w_{1,2}(5) \\ w_{1,2}(6) \\ w_{1,2}(7) \\ w_{1,2}(8) \\ w_{1,2}(9) \end{bmatrix} = 2\sqrt{5} \begin{bmatrix} 1000000000 \\ 1100000000 \\ 0110000000 \\ 0011000000 \\ 0001100000 \\ 0000110000 \\ 0000011000 \\ 0000001100 \\ 0000000110 \\ 0000000011 \end{bmatrix} \begin{bmatrix} v_{1,2}(0) \\ v_{1,2}(1) \\ v_{1,2}(2) \\ v_{1,2}(3) \\ v_{1,2}(4) \\ v_{1,2}(5) \\ v_{1,2}(6) \\ v_{1,2}(7) \\ v_{1,2}(8) \\ v_{1,2}(9) \end{bmatrix}. \tag{3.19}$$

The above Toeplitz matrix 10×10 with a left bandwidth of 1 has the triangle inverse matrix, and all sums along the arithmetical rays are calculated by

$$
\begin{bmatrix} v_{1,2}(0) \\ v_{1,2}(1) \\ v_{1,2}(2) \\ v_{1,2}(3) \\ v_{1,2}(4) \\ v_{1,2}(5) \\ v_{1,2}(6) \\ v_{1,2}(7) \\ v_{1,2}(8) \\ v_{1,2}(9) \end{bmatrix}
= \frac{1}{2\sqrt{5}}
\begin{bmatrix}
1 & 0 & 0 & 0 & 0 & 0 & 0 & 0 & 0 & 0 \\
-1 & 1 & 0 & 0 & 0 & 0 & 0 & 0 & 0 & 0 \\
1 & -1 & 1 & 0 & 0 & 0 & 0 & 0 & 0 & 0 \\
-1 & 1 & -1 & 1 & 0 & 0 & 0 & 0 & 0 & 0 \\
1 & -1 & 1 & -1 & 1 & 0 & 0 & 0 & 0 & 0 \\
-1 & 1 & -1 & 1 & -1 & 1 & 0 & 0 & 0 & 0 \\
1 & -1 & 1 & 0 & 1 & -1 & 1 & 0 & 0 & 0 \\
-1 & 1 & -1 & 1 & -1 & 1 & -1 & 1 & 0 & 0 \\
1 & -1 & 1 & -1 & 1 & -1 & 1 & -1 & 1 & 0 \\
-1 & 1 & -1 & 1 & -1 & 1 & -1 & 1 & -1 & 1
\end{bmatrix}
\begin{bmatrix} w_{1,2}(0) \\ w_{1,2}(1) \\ w_{1,2}(2) \\ w_{1,2}(3) \\ w_{1,2}(4) \\ w_{1,2}(5) \\ w_{1,2}(6) \\ w_{1,2}(7) \\ w_{1,2}(8) \\ w_{1,2}(9) \end{bmatrix} .
\tag{3.20}
$$

This equation can also be written in the following simple recurrent form:

$$
\begin{aligned}
v_{1,2}(t) &= (-1)^t [w_{1,2}(0) - w_{1,2}(1) + w_{1,2}(2) - \dots + (-1)^t w_{1,2}(t)] \\
&= w_{1,2}(t) - v_{1,2}(t-1), \quad (v_{1,2}(-1) = 0) \\
& t = 0, 1, 2, \dots, 9,
\end{aligned}
\tag{3.21}
$$

where the factor of $1/2\sqrt{5}$ is omitted for simplicity.

Case 2: We now consider another set of geometrical rays,

$$
\tilde{l}(t) = l_{1,2}(t + 0.5) = \{(x, y); 1 \cdot x + 2 \cdot y = t/4 + 1/2\},
$$

which are defined by shifting the arithmetical rays to the left, as shown in Figure 3.11. The geometrical ray $l = \{(x, y); x + 2y = 2\}$ between the arith-

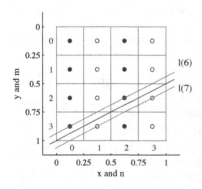
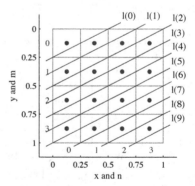

FIGURE 3.11

The 2nd set of rays for calculating seven line-integrals for $(p, s) = (1, 2)$.

metical rays $l(6)$ and $l(7)$ is considered as the shifted ray $\tilde{l}(6) = l(6 + 0.5)$. The line-integral along this ray can be expressed by two sums of the image along these arithmetical rays, as follows:

$$
w = w_{1,2}(6) = w(\tilde{l}(6)) = 2\sqrt{5}\,[v_{1,2}(6) + v_{1,2}(7)].
$$

For all line-integrals of the image along the rays $\tilde{l}(t)$, we can write the following:

$$w_{1,2}(t) = 2\sqrt{5}\,[v_{1,2}(t) + v_{1,2}(t+1)], \quad t = 0:9, \tag{3.22}$$

where it is assumed that $v_{1,2}(10) = 0$. These equations can be written in matrix form

$$\mathbf{w} = \begin{bmatrix} w_{1,2}(0) \\ w_{1,2}(1) \\ w_{1,2}(2) \\ w_{1,2}(3) \\ w_{1,2}(4) \\ w_{1,2}(5) \\ w_{1,2}(6) \\ w_{1,2}(7) \\ w_{1,2}(8) \\ w_{1,2}(9) \end{bmatrix} = 2\sqrt{5} \begin{bmatrix} 1100000000 \\ 0110000000 \\ 0011000000 \\ 0001100000 \\ 0000110000 \\ 0000011000 \\ 0000001100 \\ 0000000110 \\ 0000000011 \\ 0000000001 \end{bmatrix} \begin{bmatrix} v_{1,2}(0) \\ v_{1,2}(1) \\ v_{1,2}(2) \\ v_{1,2}(3) \\ v_{1,2}(4) \\ v_{1,2}(5) \\ v_{1,2}(6) \\ v_{1,2}(7) \\ v_{1,2}(8) \\ v_{1,2}(9) \end{bmatrix}. \tag{3.23}$$

The Toeplitz matrix 10×10 with a right bandwidth of 1 has the upper triangle inverse matrix, and all sums along the arithmetical rays are calculated by

$$\begin{bmatrix} v_{1,2}(0) \\ v_{1,2}(1) \\ v_{1,2}(2) \\ v_{1,2}(3) \\ v_{1,2}(4) \\ v_{1,2}(5) \\ v_{1,2}(6) \\ v_{1,2}(7) \\ v_{1,2}(8) \\ v_{1,2}(9) \end{bmatrix} = \frac{1}{2\sqrt{5}} \begin{bmatrix} 1 & -1 & 1 & -1 & 1 & -1 & 1 & -1 & 1 & -1 \\ 0 & 1 & -1 & 1 & -1 & 1 & -1 & 1 & -1 & 1 \\ 0 & 0 & 1 & -1 & 1 & -1 & 1 & -1 & 1 & -1 \\ 0 & 0 & 0 & 1 & -1 & 1 & -1 & 1 & -1 & 1 \\ 0 & 0 & 0 & 0 & 1 & -1 & 1 & -1 & 1 & -1 \\ 0 & 0 & 0 & 0 & 0 & 1 & -1 & 1 & -1 & 1 \\ 0 & 0 & 0 & 0 & 0 & 0 & 1 & -1 & 1 & -1 \\ 0 & 0 & 0 & 0 & 0 & 0 & 0 & 1 & -1 & 1 \\ 0 & 0 & 0 & 0 & 0 & 0 & 0 & 0 & 1 & -1 \\ 0 & 0 & 0 & 0 & 0 & 0 & 0 & 0 & 0 & 1 \end{bmatrix} \begin{bmatrix} w_{1,2}(0) \\ w_{1,2}(1) \\ w_{1,2}(2) \\ w_{1,2}(3) \\ w_{1,2}(4) \\ w_{1,2}(5) \\ w_{1,2}(6) \\ w_{1,2}(7) \\ w_{1,2}(8) \\ w_{1,2}(9) \end{bmatrix}. \tag{3.24}$$

This equation can also be written in the following recurrent form:

$$\begin{aligned} v_{1,2}(9) &= w_{1,2}(9) \\ v_{1,2}(t) &= w_{1,2}(t) - v_{1,2}(t+1), \quad t = 8,7,6,\ldots,1,0, \end{aligned} \tag{3.25}$$

where the factor of $1/(2\sqrt{5})$ is omitted for simplicity.

3.3.3.3 Generator $(2,1)$

The last two paired functions are defined by the generator $(p,s) = (2,1)$, and their masks are

$$[\chi'_{2,1,0}] = \begin{bmatrix} 1 & -1 & 1 & -1 \\ 0 & 0 & 0 & 0 \\ -1 & 1 & -1 & 1 \\ 0 & 0 & 0 & 0 \end{bmatrix}, \quad [\chi'_{2,1,1}] = \begin{bmatrix} 0 & 0 & 0 & 0 \\ 1 & -1 & 1 & -1 \\ 0 & 0 & 0 & 0 \\ -1 & 1 & -1 & 1 \end{bmatrix}.$$

These masks are the transposition of the corresponding matrices considered above for the generator $(1, 2)$. Ten parallel rays

$$l(t) = l_{2,1,t} = \{(x, y); \ 2 \cdot x + 1 \cdot y = t/4 + 3/8\}, \quad t = 0 : 9,$$

are shown separately in Figure 3.12. The sums of the image along such arith-

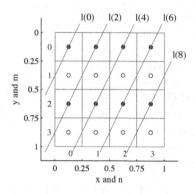 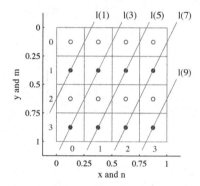

FIGURE 3.12
Schemes of calculating 11 line-integrals for $(p, s) = (2, 1)$.

metical rays define the following components of the paired transform:

$$f'_{2,1,0} = v_{2,1}(0) - v_{2,1}(2) + v_{2,1}(4) - v_{2,1}(6) + v_{2,1}(8),$$
$$f'_{2,1,1} = v_{2,1}(1) - v_{2,1}(3) + v_{2,1}(5) - v_{2,1}(7) + v_{2,1}(9).$$

Case 1: As in the $(p, s) = (1, 2)$ case, the geometrical rays $l(t)$ pass through some image elements which are not intersected by arithmetical rays of this projection. Therefore we consider a new set of shifted geometrical rays

$$\tilde{l}(t) = l_{2,1}(t - 0.5) = \{(x, y); \ 2 \cdot x + 1 \cdot y = t/4 + 1/4\}.$$

The example with arithmetical rays $l(4)$ and $l(5)$ given in Figure 3.13 (on the left) shows that the geometrical ray between them intersects four image elements with intersection of length $\sqrt{5}/8$ in each. Therefore, the line-integral along this ray can be expressed by two neighbor rays, but arithmetical, as follows:

$$w = w(5) = \frac{\sqrt{5}}{8} \frac{1}{(\Delta x)^2} \left[(f_{1,3} + f_{2,1}) + (f_{1,2} + f_{2,0}) \right]$$
$$= \frac{\sqrt{5}}{8} \left[\frac{v_{2,1}(5) + v_{1,2}(4)}{(\Delta x)^2} \right] = 2\sqrt{5} \left[v_{2,1}(5) + v_{2,1}(4) \right].$$

Ten such shifted geometrical rays are shown in Figure 3.13 (on the right). For all line-integrals of the image along these rays, we can write the following:

$$w_{2,1}(t) = 2\sqrt{5} \left[v_{2,1}(t) + v_{2,1}(t - 1) \right], \quad t = 0 : 9, \qquad (3.26)$$

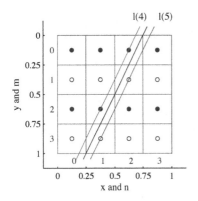 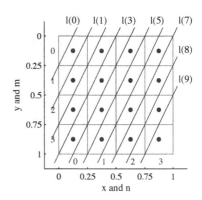

FIGURE 3.13
Schemes of calculating 11 line-integrals for $(p, s) = (2, 1)$.

where $v_{2,1}(-1) = 0$. The sums of the image along the arithmetical rays can be calculated from ten line-integrals along the geometrical rays, by using the following equation:

$$
\begin{aligned}
v_{2,1}(t) &= (-1)^t [w_{2,1}(0) - w_{2,1}(1) + w_{2,1}(2) - \ldots + (-1)^t w_{2,1}(t)] \\
&= w_{2,1}(t) - v_{2,1}(t-1), \quad (v_{2,1}(-1) = 0) \\
t &= 0, 1, 2, \ldots, 9,
\end{aligned} \tag{3.27}
$$

where the factor of $1/(2\sqrt{5})$ is omitted.

We also can define two new binary orthogonal matrices

$$
[\chi'_{2,1,0+1}] = \begin{bmatrix} 1 & -1 & 1 & -1 \\ 1 & -1 & 1 & -1 \\ -1 & 1 & -1 & 1 \\ -1 & 1 & -1 & 1 \end{bmatrix}, \quad [\chi'_{2,1,0-1}] = \begin{bmatrix} 1 & -1 & 1 & -1 \\ -1 & 1 & -1 & 1 \\ -1 & 1 & -1 & 1 \\ 1 & -1 & 1 & -1 \end{bmatrix}
$$

and consider new components of superposition of these functions with the image,

$$
f'_{2,1,0+1} = \sum_{m=0}^{3} \sum_{n=0}^{3} \chi'_{2,1,0+1}(n, m) f_{n,m}
$$

and

$$
f'_{2,1,0-1} = \sum_{m=0}^{3} \sum_{n=0}^{3} \chi'_{2,1,0-1}(n, m) f_{n,m}.
$$

These components can be calculated from the line-integrals along the geometrical rays as

$$f'_{2,1,0+1} = (v_{2,1,0} - v_{2,1,2} + v_{2,1,4} - v_{2,1,6} + v_{2,1,8})$$
$$+ (v_{2,1,1} - v_{2,1,3} + v_{2,1,5} - v_{2,1,7} + v_{2,1,9})$$
$$= (v_{2,1,0} + v_{2,1,1}) - (v_{2,1,2} + v_{2,1,3}) + (v_{2,1,4} + v_{2,1,5})$$
$$- (v_{2,1,6} + v_{2,1,7}) + (v_{2,1,8} + v_{2,1,9})$$
$$= \frac{1}{2\sqrt{5}}(w_{2,1,1} - w_{2,1,3} + w_{2,1,5} - w_{2,1,7} + w_{2,1,9}),$$

where $v_{2,1,t} = v_{2,1}(t)$, $w_{2,1,t} = w_{2,1}(t)$, and

$$f'_{2,1,0-1} = (v_{2,1,0} - v_{2,1,2} + v_{2,1,4} - v_{2,1,6} + v_{2,1,8})$$
$$- (v_{2,1,1} - v_{2,1,3} + v_{2,1,5} - v_{2,1,7} + v_{2,1,9})$$
$$= v_{2,1,0} - (v_{2,1,2} + v_{2,1,1}) + (v_{2,1,4} + v_{2,1,3})$$
$$- (v_{2,1,6} + v_{2,1,5}) + (v_{2,1,8} + v_{2,1,7}) - v_{2,1,9}$$
$$= \frac{1}{2\sqrt{5}}(w_{2,1,0} - w_{2,1,2} + w_{2,1,4} - w_{2,1,6} + w_{2,1,8} - w_{2,1,10}),$$

when considering $v_{2,1,-1} = v_{2,1,10} = 0$.

Case 2: Let us consider another set of geometrical rays, by shifting the arithmetical rays to the left,

$$\tilde{l}(t) = l_{2,1}(t + 0.5) = \{(x, y); 2 \cdot x + 1 \cdot y = t/4 + 1/2\},$$

as shown in Figure 3.14. The geometrical ray $l = \{(x, y); x + 2y = 2\}$ between

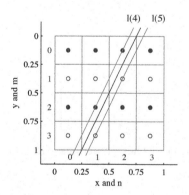 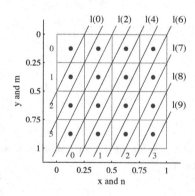

FIGURE 3.14
The 2nd set of rays for calculating seven line-integrals for $(p, s) = (2, 1)$.

the arithmetical rays $l(4)$ and $l(5)$ is considered as the shifted ray $\tilde{l}(4) = l(4 + 0.5)$. The line-integral along this ray can be expressed by two sums of the image along these arithmetical rays, as follows:

$$w = w_{2,1}(4) = w(\tilde{l}(4)) = 2\sqrt{5}\left[v_{1,2}(4) + v_{1,2}(5)\right].$$

Ten geometrical rays are defined as $\tilde{l}(t) = l(t+0.5)$, $t = 0:9$. All line-integrals of the image along these rays can be written as

$$w_{2,1}(t) = 2\sqrt{5}\left[v_{2,1}(t) + v_{2,1}(t+1)\right], \quad t = 0:9, \tag{3.28}$$

where it is assumed that $v_{2,1}(10) = 0$. The solution of this system of equations can be calculated by using the following recurrent procedure:

$$\begin{aligned} b(9) &= w_{2,1}(9), \\ b(t) &= w_{2,1}(t) - b(t+1), \quad t = 8,7,6,\ldots,1,0, \end{aligned} \tag{3.29}$$

and $v_{2,1}(t) = b(t)/(2\sqrt{5})$, for $t = 0:8$.

Thus for reconstruction of the image 4×4, six projections are used with total $4 + 4 + 7 + 7 + 10 + 10 = 42$ measurements along the geometrical rays, as shown in Table 3.1.

TABLE 3.1
Data of the 4×4 discrete image reconstruction

Projection	Number (p, s, t) of basic paired functions	# measurements
$0°$	$(1,0,0), (1,0,1), (2,0,0), (0,0,0)$	4
$90°$	$(0,1,0), (0,1,1), (0,2,0)$	4
$45°$	$(1,1,0), (1,1,1), (2,2,0)$	7
$-45°$ $(135°)$	$(1,3,0), (1,3,1)$	7
$63.43°$	$(1,2,0), (1,2,1)$	10
$26.57°$	$(2,1,0), (2,1,1)$	10

All equations of the A-rays $l(t)$, G-rays $\tilde{l}(t) = l(t - t_0)$ with shifts t_0, and the sums $v_{p,s}(t)$ considered above for the example $N = 4$, are given in Table 3.2. In the last column, the normalized factors are given for the sums $v_{p,s}(t)$, so that $v_{p,s}(t) = v_{p,s}(t)/K$. For generators $(p, s) = (1, 2)$ and $(2, 1)$, both cases of G-rays $\tilde{l}(t)$ are given, when the corresponding set of A-rays are shifted to the left and right.

TABLE 3.2
Equations of G-rays for the image 4×4

(p, s)	A-rays	t_0	G-rays $l(t - t_0)$	t	$v_{p,s} = A^{-1}w_{p,s}$	K
$(1,0)$	$l_{1,0}(t)$	0	$x = \frac{t}{4} + \frac{1}{8}$	$0:3$	$w_{1,0}(t)$	4
$(1,1)$	$l_{1,1}(t)$	0	$x + y = \frac{t}{4} + \frac{1}{4}$	$0:6$	$w_{1,1}(t)$	$4\sqrt{2}$
$(1,2)$	$l_{1,2}(t)$	$1/2$	$x + 2y = \frac{t}{4} + \frac{1}{8}$	$0:9$	$w_{1,2}(t) - v_{1,2}(t-1)$	$2\sqrt{5}$
	$l_{1,2}(t)$	$-1/2$	$x + 2y = \frac{t}{4} + \frac{1}{2}$	$9:0$	$w_{1,2}(t) - v_{1,2}(t+1)$	$2\sqrt{5}$
$(1,3)$	$l_{1,-1}(t)$	0	$x - y = \frac{t}{4}$	$3:-3$	$w_{1,-1}(t)$	$4\sqrt{2}$
$(0,1)$	$l_{0,1}(t)$	0	$y = \frac{t}{4} + \frac{1}{8}$	$0:7$	$w_{0,1}(t)$	4
$(2,1)$	$l_{2,1}(t)$	$1/2$	$2x + y = \frac{t}{4} + \frac{1}{4}$	$0:9$	$w_{2,1}(t) - v_{2,1}(t-1)$	$2\sqrt{5}$
	$l_{2,1}(t)$	$-1/2$	$2x + y = \frac{t}{4} + \frac{t}{2}$	$9:0$	$w_{2,1}(t) - v_{2,1}(t+1)$	$2\sqrt{5}$

Model is original: equations are $np + ms = t$ and $xp + ys = t_1$.

3.4   Property of the directed multiresolution

The paired functions generated by frequency-points (p, s) and $(2p, 2s)$, as well as $(4p, 4s), (8p, 8s), \ldots$, are defined by the same direction. We denote these set of points by $U(p, s)$. This property holds because of the relation

$$\chi_{2p,2s,2t} = \chi_{\overline{2p},\overline{2s},2t} = \chi_{p,s,t} + \chi_{p,s,t+N/2}, \quad t = 0 : (N/2 - 1),$$

where subscripts are considered modulo N. The paired function $\chi'_{2p,2s,2t}$ can thus be calculated by

$$\chi'_{2p,2s,2t} = \chi_{2p,2s,2t} - \chi_{2p,2s,2t+N/2} = \chi_{p,s,t} - \chi_{p,s,t+N/4} + \chi_{p,s,t+N/2} - \chi_{p,s,t+3N/4}.$$

Components of the 2-D paired transform are defined as $f'_{p,s,t} = f_{p,s,t} - f_{p,s,t+N/2}$, and the functions $f'_{2p,2s,2t}$ can be calculated by

$$f'_{2p,2s,2t} = f'_{\overline{2p},\overline{2s},2t} = f_{p,s,t} - f_{p,s,t+N/4} + f_{p,s,t+N/2} - f_{p,s,t+3N/4}.$$

Similarly, for other triplet-numbers of the set $U_{p,s}$, the components of the paired splitting-signals generated by frequencies $(\overline{4p}, \overline{4s})$, $(\overline{8p}, \overline{8s})$, ... can be calculated from the same splitting-signal $\{f_{p,s,t}; t = 0 : (N-1)\}$. This means the following: For each generator $(p, s) \in J_{N,N}$, the set $U(p, s)$ consists of N triplet-numbers, and the 2-D paired transform of the image on this set is defined by the 1-D paired transform of the tensor splitting-signal $f_{T_{p,s}} = \{f_{p,s,t}; t = 0 : (N-1)\}$. Thus

$$\chi'_N[f_{T_{p,s}}] = \begin{cases} \{f'_{p,s,t}; t = 0 : (N/2 - 1)\}, \\ \{f'_{2p,2s,2t}; t = 0 : (N/4 - 1)\}, \\ \{f'_{4p,4s,4t}; t = 0 : (N/8 - 1)\}, \\ \quad \cdots \quad \cdots \quad \cdots, \\ \{f'_{N/2,N/2,0}\}, \\ \{f'_{0,0,0}\}. \end{cases}$$

This property will be used in the next $N = 8$ example, as well as in the general case when $N = 2^r$, $r > 1$.

3.5 Example: Image 8×8

The 64 triplets (p, s, t) of the 8×8-point paired transform are defined by the set

$$
U_{8,8} = \begin{cases}
(1,0,0), (1,0,1), (1,0,2), (1,0,3), (1,1,0), (1,1,1), (1,1,2), (1,1,3), \\
(1,2,0), (1,2,1), (1,2,2), (1,2,3), (1,3,0), (1,3,1), (1,3,2), (1,3,3), \\
(1,4,0), (1,4,1), (1,4,2), (1,4,3), (1,5,0), (1,5,1), (1,5,2), (1,5,3), \\
(1,6,0), (1,6,1), (1,6,2), (1,6,3), (1,7,0), (1,7,1), (1,7,2), (1,7,3), \\
(0,1,0), (0,1,1), (0,1,2), (0,1,3), (2,1,0), (2,1,1), (2,1,2), (2,1,3), \\
(4,1,0), (4,1,1), (4,1,2), (4,1,3), (6,1,0), (6,1,1), (6,1,2), (6,1,3), \\
(2,0,0), (2,0,2), (2,2,0), (2,2,2), (2,4,0), (2,4,2), (2,6,0), (2,6,2), \\
(0,2,0), (0,2,2), (4,2,0), (4,2,2), \\
(4,0,0), (4,4,0), (0,4,0), \\
(0,0,0).
\end{cases}
$$

We separate these 64 triplets by subsets wherein the triplets correspond to the components of the 2-D paired transform, which are defined by the same directions, or projections. The number of these directions is 12 and they are calculated by $\vartheta = \operatorname{arctg}(p/s)$, where $(p, s) \in J_{8,8}$, and

$$
J_{8,8} = \{(1, s); \ s = 0 : 7\} \cup \{(2p, 1); \ p = 0 : 3\}.
$$

Given generator $(p, s) \in J_{N,N}$, the corresponding subset of triplets $2^k(p, s, t)$, where $t = 0 : N/2^{k+1} - 1$, when $k = 0 : (r - 1)$, is called $U(p, s)$. The triplet $(0, 0, 0)$ is also an element of this set.

3.5.1 Horizontal projection

Consider the first subset of triplets (p, s, t),

$$
U(1, 0) = \begin{cases}
(1,0,0), (1,0,1), (1,0,2), (1,0,3), \\
(2,0,0), (2,0,2), \\
(4,0,0), \\
(0,0,0).
\end{cases}
$$

Masks of the corresponding eight 2-D paired functions equal

$$
[\chi'_{1,0,0}] = \begin{bmatrix}
1 & 0 & 0 & 0 & -1 & 0 & 0 & 0 \\
1 & 0 & 0 & 0 & -1 & 0 & 0 & 0 \\
1 & 0 & 0 & 0 & -1 & 0 & 0 & 0 \\
1 & 0 & 0 & 0 & -1 & 0 & 0 & 0 \\
1 & 0 & 0 & 0 & -1 & 0 & 0 & 0 \\
1 & 0 & 0 & 0 & -1 & 0 & 0 & 0 \\
1 & 0 & 0 & 0 & -1 & 0 & 0 & 0 \\
1 & 0 & 0 & 0 & -1 & 0 & 0 & 0
\end{bmatrix}, \quad
[\chi'_{1,0,1}] = \begin{bmatrix}
0 & 1 & 0 & 0 & 0 & -1 & 0 & 0 \\
0 & 1 & 0 & 0 & 0 & -1 & 0 & 0 \\
0 & 1 & 0 & 0 & 0 & -1 & 0 & 0 \\
0 & 1 & 0 & 0 & 0 & -1 & 0 & 0 \\
0 & 1 & 0 & 0 & 0 & -1 & 0 & 0 \\
0 & 1 & 0 & 0 & 0 & -1 & 0 & 0 \\
0 & 1 & 0 & 0 & 0 & -1 & 0 & 0 \\
0 & 1 & 0 & 0 & 0 & -1 & 0 & 0
\end{bmatrix},
$$

$$[\chi'_{1,0,2}] = \begin{bmatrix} 0 & 0 & 1 & 0 & 0 & 0 & -1 & 0 \\ 0 & 0 & 1 & 0 & 0 & 0 & -1 & 0 \\ 0 & 0 & 1 & 0 & 0 & 0 & -1 & 0 \\ 0 & 0 & 1 & 0 & 0 & 0 & -1 & 0 \\ 0 & 0 & 1 & 0 & 0 & 0 & -1 & 0 \\ 0 & 0 & 1 & 0 & 0 & 0 & -1 & 0 \\ 0 & 0 & 1 & 0 & 0 & 0 & -1 & 0 \\ 0 & 0 & 1 & 0 & 0 & 0 & -1 & 0 \end{bmatrix}, \quad [\chi'_{1,0,3}] = \begin{bmatrix} 0 & 0 & 0 & 1 & 0 & 0 & 0 & -1 \\ 0 & 0 & 0 & 1 & 0 & 0 & 0 & -1 \\ 0 & 0 & 0 & 1 & 0 & 0 & 0 & -1 \\ 0 & 0 & 0 & 1 & 0 & 0 & 0 & -1 \\ 0 & 0 & 0 & 1 & 0 & 0 & 0 & -1 \\ 0 & 0 & 0 & 1 & 0 & 0 & 0 & -1 \\ 0 & 0 & 0 & 1 & 0 & 0 & 0 & -1 \\ 0 & 0 & 0 & 1 & 0 & 0 & 0 & -1 \end{bmatrix},$$

$$[\chi'_{2,0,0}] = \begin{bmatrix} 1 & 0 & -1 & 0 & 1 & 0 & -1 & 0 \\ 1 & 0 & -1 & 0 & 1 & 0 & -1 & 0 \\ 1 & 0 & -1 & 0 & 1 & 0 & -1 & 0 \\ 1 & 0 & -1 & 0 & 1 & 0 & -1 & 0 \\ 1 & 0 & -1 & 0 & 1 & 0 & -1 & 0 \\ 1 & 0 & -1 & 0 & 1 & 0 & -1 & 0 \\ 1 & 0 & -1 & 0 & 1 & 0 & -1 & 0 \\ 1 & 0 & -1 & 0 & 1 & 0 & -1 & 0 \end{bmatrix}, \quad [\chi'_{2,0,2}] = \begin{bmatrix} 0 & 1 & 0 & -1 & 0 & 1 & 0 & -1 \\ 0 & 1 & 0 & -1 & 0 & 1 & 0 & -1 \\ 0 & 1 & 0 & -1 & 0 & 1 & 0 & -1 \\ 0 & 1 & 0 & -1 & 0 & 1 & 0 & -1 \\ 0 & 1 & 0 & -1 & 0 & 1 & 0 & -1 \\ 0 & 1 & 0 & -1 & 0 & 1 & 0 & -1 \\ 0 & 1 & 0 & -1 & 0 & 1 & 0 & -1 \\ 0 & 1 & 0 & -1 & 0 & 1 & 0 & -1 \end{bmatrix},$$

$$[\chi'_{4,0,0}] = \begin{bmatrix} 1 & -1 & 1 & -1 & 1 & -1 & 1 & -1 \\ 1 & -1 & 1 & -1 & 1 & -1 & 1 & -1 \\ 1 & -1 & 1 & -1 & 1 & -1 & 1 & -1 \\ 1 & -1 & 1 & -1 & 1 & -1 & 1 & -1 \\ 1 & -1 & 1 & -1 & 1 & -1 & 1 & -1 \\ 1 & -1 & 1 & -1 & 1 & -1 & 1 & -1 \\ 1 & -1 & 1 & -1 & 1 & -1 & 1 & -1 \\ 1 & -1 & 1 & -1 & 1 & -1 & 1 & -1 \end{bmatrix}, \quad [\chi'_{0,0,0}] = \begin{bmatrix} 1 & 1 & 1 & 1 & 1 & 1 & 1 & 1 \\ 1 & 1 & 1 & 1 & 1 & 1 & 1 & 1 \\ 1 & 1 & 1 & 1 & 1 & 1 & 1 & 1 \\ 1 & 1 & 1 & 1 & 1 & 1 & 1 & 1 \\ 1 & 1 & 1 & 1 & 1 & 1 & 1 & 1 \\ 1 & 1 & 1 & 1 & 1 & 1 & 1 & 1 \\ 1 & 1 & 1 & 1 & 1 & 1 & 1 & 1 \\ 1 & 1 & 1 & 1 & 1 & 1 & 1 & 1 \end{bmatrix}.$$

Eight rays of the horizontal projection are used to define all components of the 2-D paired transform with triplet-numbers of $U(1,0)$. We define the control points for these rays by the bullets, from which the rays pass. These control points are numbered as shown below:

$$\begin{bmatrix} \cdot & \cdot & \cdot & \cdot & \cdot & \cdot & \cdot & \cdot \\ \vdots & \vdots & \vdots & \vdots & \vdots & \vdots & \vdots & \vdots \\ 0_{\bullet} & 1_{\bullet} & 2_{\bullet} & 3_{\bullet} & 4_{\bullet} & 5_{\bullet} & 6_{\bullet} & 7_{\bullet} \end{bmatrix} \qquad \text{(angle of rays is } 90°\text{)}.$$

We denote the sums of the image along these vertical rays by $v_t = v_{1,0}(t)$, where $t = 0 : 7$. It is not difficult to notice that the components of the 2-D paired transform with these triplet-numbers can be calculated as follows:

$$\begin{aligned} f'_{1,0,0} &= v_0 - v_4 \\ f'_{1,0,1} &= v_1 - v_5 \\ f'_{1,0,2} &= v_2 - v_6 \\ f'_{1,0,3} &= v_3 - v_7 \\ f'_{2,0,0} &= (v_0 + v_4) - (v_2 + v_6) \\ f'_{2,0,2} &= (v_1 + v_5) - (v_3 + v_7) \\ f'_{4,0,0} &= (v_0 + v_4) - (v_1 + v_5) + (v_2 + v_6) - (v_3 + v_7) \\ f'_{0,0,0} &= (v_0 + v_4) + (v_1 + v_5) + (v_2 + v_6) + (v_3 + v_7). \end{aligned} \qquad (3.30)$$

These calculations correspond to the 8-point discrete paired transform (DPT) of the vector-measurement $\mathbf{v} = (v_0, v_1, v_2, ..., v_7)'$, which requires only $2N - 2 = 14$ operations of addition and subtraction [48]. Thus, we obtain $\mathbf{P}(1,0) = [\chi_8']\mathbf{v}$. Here $\mathbf{P}(1,0)$ denotes the 2-D discrete paired transform data on the left side of the linear system of equations in (3.30), i.e., the vector $(f_{1,0,0}', f_{1,0,1}', \cdots, f_{2,0,2}', f_{4,0,0}', f_{0,0,0}')'$. The length of intersection of the geometrical ray $l(t)$ with the image element is $\Delta x = 1/8$ (see Figure 3.15). Therefore, the line-integrals

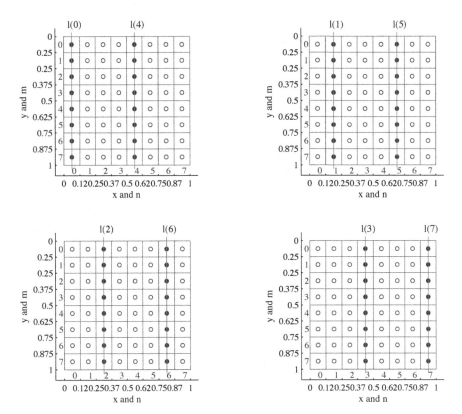

FIGURE 3.15
Schemes of calculating the eight line-integrals of the $(1,0)$-projection.

equal

$$w(t) = w_{1,0}(t) = \frac{1}{\Delta x}v(t) = 8v(t), \quad t = 0:7,$$

and the 2-D paired transform data in the subset $U(1,0)$ is calculated by

$$\mathbf{P}(1,0) = \frac{1}{8}[\chi_8']\mathbf{w},$$

where $\mathbf{w}$ denotes the vector-projection $(w(0), w(1), w(2), ..., w(7))'$.

3.5.2   Vertical projection

The vertical projection is considered similarly. The subset of triplet-numbers (p, s, t) for this projection is

$$U(0,1) = \begin{cases} (0,1,0), (0,1,1), (0,1,2), (0,1,3), \\ (0,2,0), (0,2,2), \\ (0,4,0), \\ (0,0,0). \end{cases}$$

We first consider seven masks of the paired functions with triplet-numbers of $U(0,1)$,

$$[\chi'_{0,1,0}] = \begin{bmatrix} 1 & 1 & 1 & 1 & 1 & 1 & 1 & 1 \\ 0 & 0 & 0 & 0 & 0 & 0 & 0 & 0 \\ 0 & 0 & 0 & 0 & 0 & 0 & 0 & 0 \\ 0 & 0 & 0 & 0 & 0 & 0 & 0 & 0 \\ -1 & -1 & -1 & -1 & -1 & -1 & -1 & -1 \\ 0 & 0 & 0 & 0 & 0 & 0 & 0 & 0 \\ 0 & 0 & 0 & 0 & 0 & 0 & 0 & 0 \\ 0 & 0 & 0 & 0 & 0 & 0 & 0 & 0 \end{bmatrix}, \; [\chi'_{0,1,1}] = \begin{bmatrix} 0 & 0 & 0 & 0 & 0 & 0 & 0 & 0 \\ 1 & 1 & 1 & 1 & 1 & 1 & 1 & 1 \\ 0 & 0 & 0 & 0 & 0 & 0 & 0 & 0 \\ 0 & 0 & 0 & 0 & 0 & 0 & 0 & 0 \\ 0 & 0 & 0 & 0 & 0 & 0 & 0 & 0 \\ -1 & -1 & -1 & -1 & -1 & -1 & -1 & -1 \\ 0 & 0 & 0 & 0 & 0 & 0 & 0 & 0 \\ 0 & 0 & 0 & 0 & 0 & 0 & 0 & 0 \end{bmatrix}$$

$$[\chi'_{0,1,2}] = \begin{bmatrix} 0 & 0 & 0 & 0 & 0 & 0 & 0 & 0 \\ 0 & 0 & 0 & 0 & 0 & 0 & 0 & 0 \\ 1 & 1 & 1 & 1 & 1 & 1 & 1 & 1 \\ 0 & 0 & 0 & 0 & 0 & 0 & 0 & 0 \\ 0 & 0 & 0 & 0 & 0 & 0 & 0 & 0 \\ 0 & 0 & 0 & 0 & 0 & 0 & 0 & 0 \\ -1 & -1 & -1 & -1 & -1 & -1 & -1 & -1 \\ 0 & 0 & 0 & 0 & 0 & 0 & 0 & 0 \end{bmatrix}, \; [\chi'_{0,1,3}] = \begin{bmatrix} 0 & 0 & 0 & 0 & 0 & 0 & 0 & 0 \\ 0 & 0 & 0 & 0 & 0 & 0 & 0 & 0 \\ 0 & 0 & 0 & 0 & 0 & 0 & 0 & 0 \\ 1 & 1 & 1 & 1 & 1 & 1 & 1 & 1 \\ 0 & 0 & 0 & 0 & 0 & 0 & 0 & 0 \\ 0 & 0 & 0 & 0 & 0 & 0 & 0 & 0 \\ 0 & 0 & 0 & 0 & 0 & 0 & 0 & 0 \\ -1 & -1 & -1 & -1 & -1 & -1 & -1 & -1 \end{bmatrix}$$

$$[\chi'_{0,2,0}] = \begin{bmatrix} 1 & 1 & 1 & 1 & 1 & 1 & 1 & 1 \\ 0 & 0 & 0 & 0 & 0 & 0 & 0 & 0 \\ -1 & -1 & -1 & -1 & -1 & -1 & -1 & -1 \\ 0 & 0 & 0 & 0 & 0 & 0 & 0 & 0 \\ 1 & 1 & 1 & 1 & 1 & 1 & 1 & 1 \\ 0 & 0 & 0 & 0 & 0 & 0 & 0 & 0 \\ -1 & -1 & -1 & -1 & -1 & -1 & -1 & -1 \\ 0 & 0 & 0 & 0 & 0 & 0 & 0 & 0 \end{bmatrix}, \; [\chi'_{0,2,2}] = \begin{bmatrix} 0 & 0 & 0 & 0 & 0 & 0 & 0 & 0 \\ 1 & 1 & 1 & 1 & 1 & 1 & 1 & 1 \\ 0 & 0 & 0 & 0 & 0 & 0 & 0 & 0 \\ -1 & -1 & -1 & -1 & -1 & -1 & -1 & -1 \\ 0 & 0 & 0 & 0 & 0 & 0 & 0 & 0 \\ 1 & 1 & 1 & 1 & 1 & 1 & 1 & 1 \\ 0 & 0 & 0 & 0 & 0 & 0 & 0 & 0 \\ -1 & -1 & -1 & -1 & -1 & -1 & -1 & -1 \end{bmatrix}$$

$$[\chi'_{0,4,0}] = \begin{bmatrix} 1 & 1 & 1 & 1 & 1 & 1 & 1 & 1 \\ -1 & -1 & -1 & -1 & -1 & -1 & -1 & -1 \\ 1 & 1 & 1 & 1 & 1 & 1 & 1 & 1 \\ -1 & -1 & -1 & -1 & -1 & -1 & -1 & -1 \\ 1 & 1 & 1 & 1 & 1 & 1 & 1 & 1 \\ -1 & -1 & -1 & -1 & -1 & -1 & -1 & -1 \\ 1 & 1 & 1 & 1 & 1 & 1 & 1 & 1 \\ -1 & -1 & -1 & -1 & -1 & -1 & -1 & -1 \end{bmatrix}.$$

The numbered control points of eight rays for this projection are denoted by the bullets as shown below:

$$
\begin{bmatrix}
0_\bullet & \cdot & \cdot & \cdot & \cdot & \cdot & \cdot & \cdot \\
1_\bullet & \cdot & \cdot & \cdot & \cdot & \cdot & \cdot & \cdot \\
2_\bullet & \cdot & \cdot & \cdot & \cdot & \cdot & \cdot & \cdot \\
\vdots & \vdots & \vdots & \vdots & \vdots & \vdots & \vdots & \vdots \\
6_\bullet & \cdot & \cdot & \cdot & \cdot & \cdot & \cdot & \cdot \\
7_\bullet & \cdot & \cdot & \cdot & \cdot & \cdot & \cdot & \cdot
\end{bmatrix}
\qquad \text{(angle of rays is } 0°\text{)}.
$$

The 2-D paired transform in the subset of triplets $(p, s, t) \in U(0, 1)$ can be calculated by the 8-point DPT of the vector $\mathbf{w}$, which is composed from the projection data $w(t) = w_{1,0}(t)$, $t = 0 : 7$,

$$
\mathbf{P}(0, 1) = \frac{1}{8}[\chi_8']\mathbf{w}.
$$

3.5.3 Diagonal projection

Now we consider the third subset of triplet-numbers

$$
U(1, 1) = \begin{cases}
(1, 1, 0), (1, 1, 1), (1, 1, 2), (1, 1, 3), \\
(2, 2, 0), (2, 2, 2), \\
(4, 4, 0), \\
(0, 0, 0),
\end{cases}
$$

and the masks of the corresponding first seven 2-D paired functions

$$
[\chi_{1,1,0}'] = \begin{bmatrix}
1 & 0 & 0 & 0 & -1 & 0 & 0 & 0 \\
0 & 0 & 0 & -1 & 0 & 0 & 0 & 1 \\
0 & 0 & -1 & 0 & 0 & 0 & 1 & 0 \\
0 & -1 & 0 & 0 & 0 & 1 & 0 & 0 \\
-1 & 0 & 0 & 0 & 1 & 0 & 0 & 0 \\
0 & 0 & 0 & 1 & 0 & 0 & 0 & -1 \\
0 & 0 & 1 & 0 & 0 & -1 & 0 \\
0 & 1 & 0 & 0 & -1 & 0 & 0 & 0
\end{bmatrix}, \quad
[\chi_{1,1,1}'] = \begin{bmatrix}
0 & 1 & 0 & 0 & 0 & -1 & 0 & 0 \\
1 & 0 & 0 & 0 & -1 & 0 & 0 & 0 \\
0 & 0 & 0 & -1 & 0 & 0 & 0 & 1 \\
0 & 0 & -1 & 0 & 0 & 0 & 1 & 0 \\
0 & -1 & 0 & 0 & 0 & 1 & 0 & 0 \\
-1 & 0 & 0 & 0 & 1 & 0 & 0 & 0 \\
0 & 0 & 0 & 1 & 0 & 0 & 0 & -1 \\
0 & 0 & 1 & 0 & 0 & 0 & -1 & 0
\end{bmatrix}
$$

$$
[\chi_{1,1,2}'] = \begin{bmatrix}
0 & 0 & 1 & 0 & 0 & 0 & -1 & 0 \\
0 & 1 & 0 & 0 & 0 & -1 & 0 & 0 \\
1 & 0 & 0 & 0 & -1 & 0 & 0 & 0 \\
0 & 0 & 0 & -1 & 0 & 0 & 0 & 1 \\
0 & 0 & -1 & 0 & 0 & 0 & 1 & 0 \\
0 & -1 & 0 & 0 & 0 & 1 & 0 & 0 \\
-1 & 0 & 0 & 0 & 1 & 0 & 0 & 0 \\
0 & 0 & 0 & 1 & 0 & 0 & 0 & -1
\end{bmatrix}, \quad
[\chi_{1,1,3}'] = \begin{bmatrix}
0 & 0 & 0 & 1 & 0 & 0 & 0 & -1 \\
0 & 0 & 1 & 0 & 0 & 0 & -1 & 0 \\
0 & 1 & 0 & 0 & 0 & -1 & 0 & 0 \\
1 & 0 & 0 & 0 & -1 & 0 & 0 & 0 \\
0 & 0 & 0 & -1 & 0 & 0 & 0 & 1 \\
0 & 0 & -1 & 0 & 0 & 0 & 1 & 0 \\
0 & -1 & 0 & 0 & 0 & 1 & 0 & 0 \\
-1 & 0 & 0 & 0 & 1 & 0 & 0 & 0
\end{bmatrix}
$$

$$[\chi'_{2,2,0}] = \begin{bmatrix} 1 & 0 & -1 & 0 & 1 & 0 & -1 & 0 \\ 0 & -1 & 0 & 1 & 0 & -1 & 0 & 1 \\ -1 & 0 & 1 & 0 & -1 & 0 & 1 & 0 \\ 0 & 1 & 0 & -1 & 0 & 1 & 0 & -1 \\ 1 & 0 & -1 & 0 & 1 & 0 & -1 & 0 \\ 0 & -1 & 0 & 1 & 0 & -1 & 0 & 1 \\ -1 & 0 & 1 & 0 & -1 & 0 & 1 & 0 \\ 0 & 1 & 0 & -1 & 0 & 1 & 0 & -1 \end{bmatrix}, \quad [\chi'_{2,2,2}] = \begin{bmatrix} 0 & 1 & 0 & -1 & 0 & 1 & 0 & -1 \\ 1 & 0 & -1 & 0 & 1 & 0 & -1 & 0 \\ 0 & -1 & 0 & 1 & 0 & -1 & 0 & 1 \\ -1 & 0 & 1 & 0 & -1 & 0 & 1 & 0 \\ 0 & 1 & 0 & -1 & 0 & 1 & 0 & -1 \\ 1 & 0 & -1 & 0 & 1 & 0 & -1 & 0 \\ 0 & -1 & 0 & 1 & 0 & -1 & 0 & 1 \\ -1 & 0 & 1 & 0 & -1 & 0 & 1 & 0 \end{bmatrix}$$

$$[\chi'_{4,4,0}] = \begin{bmatrix} 1 & -1 & 1 & -1 & 1 & -1 & 1 & -1 \\ -1 & 1 & -1 & 1 & -1 & 1 & -1 & 1 \\ 1 & -1 & 1 & -1 & 1 & -1 & 1 & -1 \\ -1 & 1 & -1 & 1 & -1 & 1 & -1 & 1 \\ 1 & -1 & 1 & -1 & 1 & -1 & 1 & -1 \\ -1 & 1 & -1 & 1 & -1 & 1 & -1 & 1 \\ 1 & -1 & 1 & -1 & 1 & -1 & 1 & -1 \\ -1 & 1 & -1 & 1 & -1 & 1 & -1 & 1 \end{bmatrix}.$$

Fifteen parallel rays of the diagonal projection are used to define components of the 2-D paired transforms with triplet-numbers of $U(1,1)$. The number of rays is calculated by $(p+s)(N-1)+1 = 15$, where $(p,s) = (1,1)$. We define the control points of these rays by the bullets which are numbered as

$$\begin{bmatrix} 0_{\bullet} & \cdot & \cdot & \cdot & \cdot & \cdot & \cdot & \cdot \\ 1_{\bullet} & \cdot & \cdot & \cdot & \cdot & \cdot & \cdot & \cdot \\ 2_{\bullet} & \cdot & \cdot & \cdot & \cdot & \cdot & \cdot & \cdot \\ \vdots & \vdots & \vdots & \vdots & \vdots & \vdots & \vdots & \vdots \\ 6_{\bullet} & \cdot & \cdot & \cdot & \cdot & \cdot & \cdot & \cdot \\ 7_{\bullet} & 8_{\bullet} & 9_{\bullet} & 10_{\bullet} & 11_{\bullet} & 12_{\bullet} & 13_{\bullet} & 14_{\bullet} \end{bmatrix} \qquad \text{(angle of rays is } -45°\text{)}.$$

These masks with the rays, on which the coefficients 1 and -1 are situated, are shown in Figures 3.16-3.18.

The sums of the discrete image along these rays are denoted by $v_t = v_{1,1}(t)$, where $t = 0 : 14$. It is not difficult to notice that the components of the 8×8-point 2-D paired transform with the triplet-numbers $(p,s,t) \in U(1,1)$ can be calculated by

$$\begin{aligned}
f'_{1,1,0} &= v_0 - v_4 + v_8 - v_{12} \\
f'_{1,1,1} &= v_1 - v_5 + v_9 - v_{13} \\
f'_{1,1,2} &= v_2 - v_6 + v_{10} - v_{14} \\
f'_{1,1,3} &= v_3 - v_7 + v_{11} - 0 \\
f'_{2,2,0} &= (v_0 + v_4 + v_8 + v_{12}) - (v_2 + v_6 + v_{10} + v_{14}) \\
f'_{2,2,0} &= (v_1 + v_5 + v_9 + v_{13}) - (v_3 + v_7 + v_{11} + 0) \\
f'_{4,4,0} &= v_0 - v_1 + v_2 - v_3 + \cdots - v_{11} + v_{12} - v_{13} + v_{14} \\
f'_{0,0,0} &= v_0 + v_1 + v_2 + v_3 + \cdots + v_{11} + v_{12} + v_{13} + v_{14}.
\end{aligned} \tag{3.31}$$

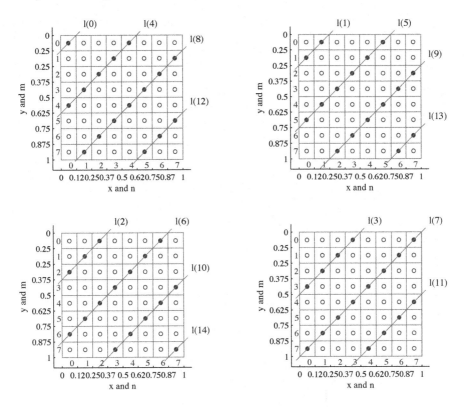

FIGURE 3.16

Rays along which the line-integrals are calculated for $(p, s) = (1, 1)$.

Introducing the new variables

$$a_n = \sum_m v_{n+8m} = \sum_m v_{1,1,n+8m}, \qquad n = 0 : 7,$$

the above system of linear equations can be written as follows:

$$
\begin{aligned}
f'_{1,1,0} &= a_0 - a_4 \\
f'_{1,1,1} &= a_1 - a_5 \\
f'_{1,1,2} &= a_2 - a_6 \\
f'_{1,1,3} &= a_3 - a_7 \\
f'_{2,2,0} &= (a_0 + a_4) - (a_2 + a_6) \\
f'_{2,2,2} &= (a_1 + a_5) - (a_3 + a_7) \\
f'_{4,4,0} &= (a_0 + a_4) - (a_1 + a_5) + (a_2 + a_6) - (a_3 + a_7) \\
f'_{0,0,0} &= (a_0 + a_4) + (a_1 + a_5) + (a_2 + a_6) + (a_3 + a_7).
\end{aligned}
\tag{3.32}
$$

These calculations correspond to the 8-point discrete paired transform of the

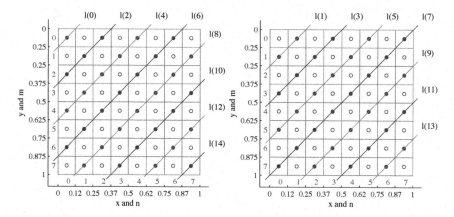

FIGURE 3.17
Rays along which the line-integrals are calculated for $(p, s) = (2, 2)$.

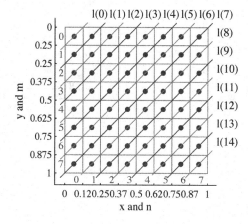

FIGURE 3.18
Rays along which the line-integrals are calculated for $(p, s) = (4, 4)$.

vector $\mathbf{a} = (a_0, a_1, a_2, ..., a_7)'$,

$$\mathbf{P}(1, 1) = [\chi'_8]\mathbf{a},$$

where $\mathbf{P}(1, 1)$ denotes the 2-D paired transform data on the left side of system of equations in (3.32). The calculation of the component $f'_{0,0,0}$ is repeated, since it is obtained from the subsets $U(0, 1)$ and $U(1, 0)$ as well. Thus, the procedure of calculating the above components of the 2-D paired transform of the discrete image $f_{n,m}$ from 15 arithmetical rays is accomplished by the 8-point paired transform. The sums of the image $v(t) = v_{1,1}(t)$ can be calculated from the line-integrals of the image $f(x, y)$ along the corresponding geometrical rays

$$l(t) = l_{1,1}(t) = \left\{ (x,y); \ x+y = \frac{t}{8} + \frac{1}{8} \right\}, \quad t = 0 : 14,$$

where $(x,y) \in [0,1] \times [0,1]$. Indeed, as shown in Figure 3.16, the length of intersection of the geometrical ray with the image element equals $\Delta l = \sqrt{2}\Delta x = \sqrt{2}/N$. Each integral along the geometrical ray is proportional to the sum of the image along the corresponding arithmetical ray,

$$w(t) = w_{1,1}(t) = \Delta l \frac{v(t)}{(\Delta x)^2} = \sqrt{2} \cdot 8v(t), \quad t = 0 : 14.$$

Therefore, $v(t) = w(t)/(8\sqrt{2})$, $t = 0 : 14$.

3.5.4 $(2,1)$- and $(1,2)$-projections

The next subset of triplet-numbers

$$U(2,1) = \begin{cases} (2,1,0), (2,1,1), (2,1,2), (2,1,3), \\ (4,2,0), (4,2,2), \\ (0,4,0), \\ (0,0,0) \end{cases}$$

can be considered together with the subset

$$U(1,2) = \begin{cases} (1,2,0), (1,2,1), (1,2,2), (1,2,3), \\ (2,4,0), (2,4,2), \\ (4,0,0), \\ (0,0,0). \end{cases}$$

These two symmetric cases with generators $(2,1)$ and $(1,2)$ correspond to the projections by angles $\pi - \tan^{-1}(1/2)$ and $\pi - \tan^{-1}(2/1)$, respectively, and are described by similar equations. We consider in detail the case $(p,s) = (2,1)$.

3.5.4.1 $(2,1)$-projection

Twenty-two rays of the projection are used to define all components of the 2-D paired transforms with triplet-numbers of $U(2,1)$. The number of rays is calculated as $(p+s)(N-1)+1 = 3 \cdot 7 + 1 = 22$. We define the control points of these rays by the bullets and we number them as shown below:

$$\begin{bmatrix} 0_\bullet & \cdot & \cdot & \cdot & \cdot & \cdot & \cdot & \cdot \\ 1_\bullet & \cdot & \cdot & \cdot & \cdot & \cdot & \cdot & \cdot \\ 2_\bullet & \cdot & \cdot & \cdot & \cdot & \cdot & \cdot & \cdot \\ \vdots & \vdots & \vdots & \vdots & \vdots & \vdots & \vdots & \vdots \\ 5_\bullet & \cdot & \cdot & \cdot & \cdot & \cdot & \cdot & \cdot \\ 6_\bullet {}_8_\bullet {}_{10}_\bullet {}_{12}_\bullet {}_{14}_\bullet {}_{16}_\bullet {}_{18}_\bullet {}_{20}_\bullet \\ 7_\bullet {}_9_\bullet {}_{11}_\bullet {}_{13}_\bullet {}_{15}_\bullet {}_{17}_\bullet {}_{19}_\bullet {}_{21}_\bullet \end{bmatrix} \qquad \text{(angle of rays is } -\tan^{-1} 2 = -63.44°\text{).}$$

The equations of the rays $l_{2,1}(t)$ on the square $[0,1] \times [0,1]$ are

$$l(t) = l_{2,1}(t) = \left\{ (x,y); \; 2x + y = \frac{t}{8} + \frac{3}{16} \right\}, \quad t = 0:21.$$

These rays are shown in Figure 3.19.

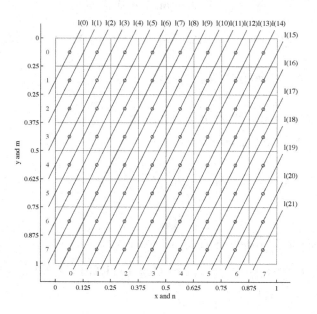

FIGURE 3.19
Rays along which the sums of the image are calculated for $(p,s) = (2,1)$.

The paired functions with numbers $(0,4,0)$ and $(0,0,0)$ were considered for the horizontal and vertical projections. The masks of the paired functions with the first six triplet-numbers of $U(2,1)$ are

$$[\chi'_{2,1,0}] = \begin{bmatrix} 1 & 0 & -1 & 0 & 1 & 0 & -1 & 0 \\ 0 & 0 & 0 & 0 & 0 & 0 & 0 & 0 \\ 0 & -1 & 0 & 1 & 0 & -1 & 0 & 1 \\ 0 & 0 & 0 & 0 & 0 & 0 & 0 & 0 \\ -1 & 0 & 1 & 0 & -1 & 0 & 1 & 0 \\ 0 & 0 & 0 & 0 & 0 & 0 & 0 & 0 \\ 0 & 1 & 0 & -1 & 0 & 1 & 0 & -1 \\ 0 & 0 & 0 & 0 & 0 & 0 & 0 & 0 \end{bmatrix}, \quad [\chi'_{2,1,1}] = \begin{bmatrix} 0 & 0 & 0 & 0 & 0 & 0 & 0 & 0 \\ 1 & 0 & -1 & 0 & 1 & 0 & -1 & 0 \\ 0 & 0 & 0 & 0 & 0 & 0 & 0 & 0 \\ 0 & -1 & 0 & 1 & 0 & -1 & 0 & 1 \\ 0 & 0 & 0 & 0 & 0 & 0 & 0 & 0 \\ -1 & 0 & 1 & 0 & -1 & 0 & 1 & 0 \\ 0 & 0 & 0 & 0 & 0 & 0 & 0 & 0 \\ 0 & 1 & 0 & -1 & 0 & 1 & 0 & -1 \end{bmatrix}$$

$$[\chi'_{2,1,2}] = \begin{bmatrix} 0 & 1 & 0 & -1 & 0 & 1 & 0 & -1 \\ 0 & 0 & 0 & 0 & 0 & 0 & 0 & 0 \\ 1 & 0 & -1 & 0 & 1 & 0 & -1 & 0 \\ 0 & 0 & 0 & 0 & 0 & 0 & 0 & 0 \\ 0 & -1 & 0 & 1 & 0 & -1 & 0 & 1 \\ 0 & 0 & 0 & 0 & 0 & 0 & 0 & 0 \\ -1 & 0 & 1 & 0 & -1 & 0 & 1 & 0 \\ 0 & 0 & 0 & 0 & 0 & 0 & 0 & 0 \end{bmatrix}, \quad [\chi'_{2,1,3}] = \begin{bmatrix} 0 & 0 & 0 & 0 & 0 & 0 & 0 & 0 \\ 0 & 1 & 0 & -1 & 0 & 1 & 0 & -1 \\ 0 & 0 & 0 & 0 & 0 & 0 & 0 & 0 \\ 1 & 0 & -1 & 0 & 1 & 0 & -1 & 0 \\ 0 & 0 & 0 & 0 & 0 & 0 & 0 & 0 \\ 0 & -1 & 0 & 1 & 0 & -1 & 0 & 1 \\ 0 & 0 & 0 & 0 & 0 & 0 & 0 & 0 \\ -1 & 0 & 1 & 0 & -1 & 0 & 1 & 0 \end{bmatrix}$$

$$[\chi'_{4,2,0}] = \begin{bmatrix} 1 & -1 & 1 & -1 & 1 & -1 & 1 & -1 \\ 0 & 0 & 0 & 0 & 0 & 0 & 0 & 0 \\ -1 & 1 & -1 & 1 & -1 & 1 & -1 & 1 \\ 0 & 0 & 0 & 0 & 0 & 0 & 0 & 0 \\ 1 & -1 & 1 & -1 & 1 & -1 & 1 & -1 \\ 0 & 0 & 0 & 0 & 0 & 0 & 0 & 0 \\ -1 & 1 & -1 & 1 & -1 & 1 & -1 & 1 \\ 0 & 0 & 0 & 0 & 0 & 0 & 0 & 0 \end{bmatrix}, \quad [\chi'_{4,2,2}] = \begin{bmatrix} 0 & 0 & 0 & 0 & 0 & 0 & 0 & 0 \\ 1 & -1 & 1 & -1 & 1 & -1 & 1 & -1 \\ 0 & 0 & 0 & 0 & 0 & 0 & 0 & 0 \\ -1 & 1 & -1 & 1 & -1 & 1 & -1 & 1 \\ 0 & 0 & 0 & 0 & 0 & 0 & 0 & 0 \\ 1 & -1 & 1 & -1 & 1 & -1 & 1 & -1 \\ 0 & 0 & 0 & 0 & 0 & 0 & 0 & 0 \\ -1 & 1 & -1 & 1 & -1 & 1 & -1 & 1 \end{bmatrix}.$$

These masks with the rays, on which the coefficients 1 and -1 are situated, are shown in Figures 3.20 and 3.21.

The sums of the image along these rays are denoted by $v_t = v_{2,1}(t)$, where $t = 0 : 21$. It is not difficult to notice that the components of the 2-D paired transform of the image with the triplets of $U(2,1)$ can be calculated by

$$
\begin{aligned}
f'_{2,1,0} &= v_0 - v_4 + v_8 - v_{12} + v_{16} - v_{20} \\
f'_{2,1,1} &= v_1 - v_5 + v_9 - v_{13} + v_{17} - v_{21} \\
f'_{2,1,2} &= v_3 - v_7 + v_{11} - v_{15} + v_{19} - 0 \\
f'_{2,1,3} &= v_2 - v_6 + v_{10} - v_{14} + v_{18} - 0 \\
f'_{4,2,0} &= (v_0 + v_4 + v_8 + v_{12} + v_{16} + v_{20}) - (v_2 + v_6 + v_{10} + v_{14} + v_{18}) \\
f'_{4,2,2} &= (v_1 + v_5 + v_9 + v_{13} + v_{17} + v_{21}) - (v_3 + v_7 + v_{11} + v_{15} + v_{19}) \\
f'_{0,4,0} &= v_0 - v_1 + v_2 - v_3 + \cdots + v_{18} - v_{19} + v_{20} - v_{21} \\
f'_{0,0,0} &= v_0 + v_1 + v_2 + v_3 + \cdots + v_{18} + v_{19} + v_{20} + v_{21}.
\end{aligned} \tag{3.33}
$$

Introducing the new variables

$$a_n = \sum_m v_{n+8m} = \sum_m v_{2,1,n+8m}, \qquad n = 0 : 7,$$

the above system of equations can be written as follows:

$$
\begin{aligned}
f'_{2,1,0} &= a_0 - a_4 \\
f'_{2,1,1} &= a_1 - a_5 \\
f'_{2,1,2} &= a_2 - a_6 \\
f'_{2,1,3} &= a_3 - a_7 \\
f'_{4,2,0} &= (a_0 + a_4) - (a_2 + a_6) \\
f'_{4,2,2} &= (a_1 + a_5) - (a_3 + a_7) \\
f'_{0,4,0} &= (a_0 + a_4) - (a_1 + a_5) + (a_2 + a_6) - (a_3 + a_7) \\
f'_{0,0,0} &= (a_0 + a_4) + (a_1 + a_5) + (a_2 + a_6) + (a_3 + a_7).
\end{aligned} \tag{3.34}
$$

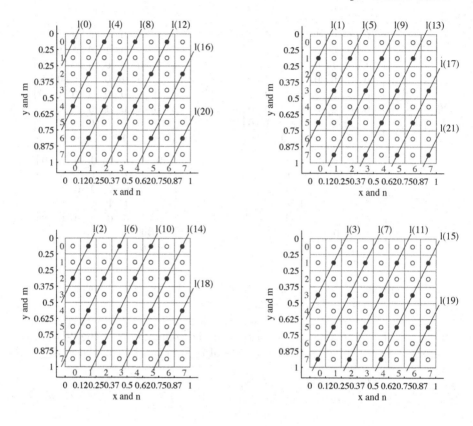

FIGURE 3.20

Rays along which the sums of the image are calculated for $(p, s) = (2, 1)$.

Again we can see that the calculation of the 2-D paired transform data $\mathbf{P}(2, 1)$ corresponds to the 8-point discrete paired transform of the vector $\mathbf{a} = (a_0, a_1, a_2, ..., a_7)'$, i.e., $\mathbf{P}(2, 1) = [\chi'_8]\mathbf{a}$.

Now we express the sums $v(t)$ of the discrete image $f_{n,m}$ through the line-integrals $w(t)$, assuming first that these integrals are calculated along the geometrical rays that coincide with the arithmetical rays $l(t)$, i.e., $w(t) = w_{2,1}(t)$, $t = 0 : 21$. As an example, we consider the ray number seven, $l(7)$. As can be seen from Figure 3.22, this ray crosses the IE of number $(0, 7)$ and then two IE of numbers $(0, 6)$ and $(1, 6)$, and such intersections are repeated on its way to the IE of number $(4, 0)$. The length of the intersection $\Delta l_{0,7}$ of the ray with the IE of number $(0, 7)$ is twice the length of intersection with each IE of number $(0, 6)$ and $(1, 6)$.

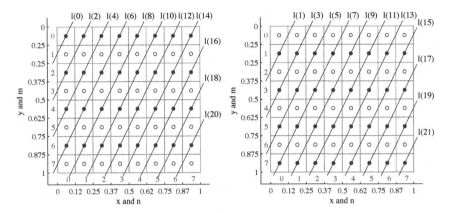

FIGURE 3.21
Rays along which the sums of the image are calculated for $(p, s) = (4, 2)$.

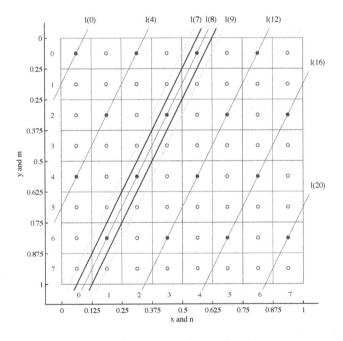

FIGURE 3.22 (See color insert)
Mask of the paired function $\chi'_{2,1,0}(n, m)$ and rays.

Since $\Delta l_{0,7} = \sqrt{5}/2\Delta x$, we can write the following:

$$w(7) = w_{2,1}(7) = \frac{\Delta l_{0,7}}{(\Delta x)^2}\left[v(7) + \frac{1}{2}v(6) + \frac{1}{2}v(8)\right] = \frac{\sqrt{5}}{2}8\left[\frac{1}{2}v(6) + v(7) + \frac{1}{2}v(8)\right].$$

For the line-integral along the ray $l(8)$, we also can write

$$w(8) = w_{2,1}(8) = 4\sqrt{5}\left[\frac{1}{2}v(7) + \frac{1}{2}v(9) + v(8)\right] = 4\sqrt{5}\left[\frac{1}{2}v(7) + v(8) + \frac{1}{2}v(9)\right].$$

Similar equations can be used for the integrals along all these geometrical rays,

$$w(t) = w_{2,1}(t) = 4\sqrt{5}\left[\frac{1}{2}v(t-1) + v(t) + \frac{1}{2}v(t+1)\right], \quad t = 0:21.$$

We consider two rays which are between rays $l(7), l(8)$ and $l(8), l(9)$ and call them $\tilde{l}(8)$ and $\tilde{l}(9)$, respectively. One can notice that new ray number 8 intersects equally the IE of number $(0,7)$, the IE of number $(1,6)$ and other the IE, i.e., the length of intersection of this ray with all these IEs is the same and equals $\Delta l_{0,7} = \sqrt{5}/2\Delta x$. New ray number 9 also equally intersects image elements. The line-integrals along these two rays can be defined as

$$w(8) = w(\tilde{l}(8)) = 4\sqrt{5}\left[v(7) + v(8)\right], \quad w(9) = w(\tilde{l}(9)) = 4\sqrt{5}\left[v(8) + v(9)\right].$$

Thus we can simplify the calculations, by considering the shifting, $t \to t - 1/2$, and the following new set of geometrical rays:

$$\tilde{l}(t) = l_{2,1}(t - 1/2) = \left\{(x,y); 2x + y = \frac{t}{8} + \frac{1}{8}\right\}, \quad t = 0:21,$$

which are shown in Figure 3.23. For this set of geometrical rays, we can define the line-integrals along the rays as follows:

$$w(t) = w(\tilde{l}(t)) = 4\sqrt{5}\left[v(t-1) + v(t)\right], \quad t = 0:21,$$

where $v(-1) = 0$. This system of equations can be written in matrix form

$$\mathbf{w} = 4\sqrt{5}\mathbf{A}\mathbf{v} = 4\sqrt{5}\begin{bmatrix} 1 & 0 & 0 & 0 & \dots & 0 & 0 & 0 \\ 1 & 1 & 0 & 0 & \dots & 0 & 0 & 0 \\ 0 & 1 & 1 & 0 & \dots & 0 & 0 & 0 \\ 0 & 0 & 1 & 1 & \dots & 0 & 0 & 0 \\ \dots & \dots & \dots & \dots & \dots & \dots & \dots & \dots \\ 0 & 0 & 0 & 0 & \dots & 1 & 1 & 0 \\ 0 & 0 & 0 & 0 & \dots & 0 & 1 & 1 \end{bmatrix}\mathbf{v}, \tag{3.35}$$

where $\mathbf{w} = (w(0), w(1), ..., w(21))'$ and $\mathbf{v} = (v_0, v_1, ..., v_{21})'$. The above Toeplitz matrix 22×22 has the triangle inverse matrix, which is similar to the matrix described in (3.19) for the $N = 4$ example. The inverse matrix is triangular and consists of unit diagonals whose signs alternate,

$$\mathbf{A}^{-1} = \begin{bmatrix} 1 & 0 & 0 & 0 & \dots & 0 & 0 & 0 \\ -1 & 1 & 0 & 0 & \dots & 0 & 0 & 0 \\ 1 & -1 & 1 & 0 & \dots & 0 & 0 & 0 \\ -1 & 1 & -1 & 1 & \dots & 0 & 0 & 0 \\ \dots & \dots & \dots & \dots & \dots & \dots & \dots & \dots \\ 1 & -1 & 1 & -1 & \dots & -1 & 1 & 0 \\ -1 & 1 & -1 & 1 & \dots & 1 & -1 & 1 \end{bmatrix}. \tag{3.36}$$

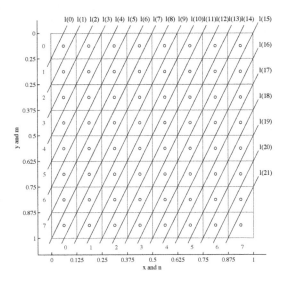

FIGURE 3.23
New set of geometrical rays for calculating line-integrals for $(p,s) = (2,1)$.

The equation $\mathbf{b} = \mathbf{A}^{-1}\mathbf{w}$ can thus be written in the following recurrent form:

$$\begin{aligned} b_0 &= w(0), \\ b_t &= w(t) - b_{t-1}, \quad t = 1, 2, \ldots, 21, \end{aligned} \qquad (3.37)$$

and, therefore, $\mathbf{v} = \mathbf{b}/(4\sqrt{5})$, $t = 0 : 21$. In other words, the solution $\mathbf{v}$ is calculated by the method of forward substitution.

Remark 3.1 We also can consider another set of geometrical rays, by using the shifting, $t \to t + 1/2$, of the original arithmetical rays,

$$\tilde{l}(t) = l_{2,1}(t + 1/2) = \left\{ (x, y); \ 2x + y = \frac{t}{8} + \frac{1}{4} \right\}, \quad t = 0 : 21,$$

which is shown in Figure 3.24. One can notice that geometrical ray number 7 intersects equally another pair of IEs of number $(0, 7)$ and $(1, 6)$. Therefore, we can define the integral along this ray as $w(7) = w(\tilde{l}(7)) = 4\sqrt{5}\,[v(7) + v(8)]$. In general, the integrals along the geometrical rays are defined as follows:

$$w(t) = w(\tilde{l}(t)) = 4\sqrt{5}\,[v(t) + v(t + 1)], \quad t = 0 : 21,$$

considering $v(22) = 0$. With the exception of the IE of number $(0, 0)$, each other IE is intersected by two parallel geometrical rays. This system of linear

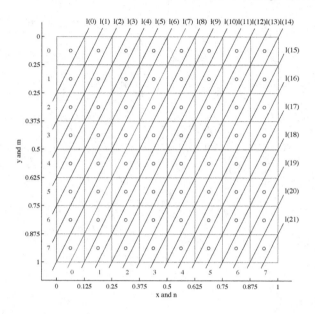

FIGURE 3.24
New set of geometrical rays for calculating line-integrals for $(p, s) = (2, 1)$.

equations can be written in matrix form

$$\mathbf{w} = 4\sqrt{5}\mathbf{A}\mathbf{v} = 4\sqrt{5} \begin{bmatrix} 1 & 1 & 0 & 0 & \dots & 0 & 0 & 0 \\ 0 & 1 & 1 & 0 & \dots & 0 & 0 & 0 \\ 0 & 0 & 1 & 1 & \dots & 0 & 0 & 0 \\ 0 & 0 & 0 & 1 & \dots & 0 & 0 & 0 \\ \multicolumn{8}{c}{\dotfill} \\ 0 & 0 & 0 & 0 & \dots & 0 & 1 & 1 \\ 0 & 0 & 0 & 0 & \dots & 0 & 0 & 1 \end{bmatrix} \mathbf{v}. \tag{3.38}$$

The above Toeplitz matrix 22×22 has an upper triangle inverse matrix that is the transposition of the inverse matrix (3.36), which was used for the first set of geometrical rays. The matrix is triangular and consists of unit diagonals whose signs alternate,

$$\mathbf{A}^{-1} = \begin{bmatrix} 1 & -1 & 1 & -1 & \dots & -1 & \dots & 1 & -1 \\ 0 & 1 & -1 & 1 & -1 & \dots & \dots & -1 & 1 \\ 0 & 0 & 1 & -1 & 1 & -1 & \dots & 1 & -1 \\ 0 & 0 & 0 & 1 & -1 & 1 & \dots & -1 & 1 \\ 0 & 0 & 0 & 0 & 1 & -1 & \dots & \dots & -1 \\ \dots & \dots & \dots & \dots & \dots & \dots & \dots & \dots & \dots \\ 0 & 0 & 0 & 0 & 0 & 0 & \dots & 1 & -1 \\ 0 & 0 & 0 & 0 & 0 & 0 & \dots & 0 & 1 \end{bmatrix}. \tag{3.39}$$

The coefficients of this matrix are

$$A_{m,n} = \sum_{k=0}^{n} (-1)^{m-n} \delta_{m-n,t} \, .$$

The equation $\mathbf{b} = \mathbf{A}^{-1}\mathbf{w}$ can be written in the following recurrent form:

$$\begin{aligned} b_{21} &= w(21), \\ b_t &= w(t) - b_{t+1}, \quad t = 20, 19, \ldots, 0, \end{aligned} \tag{3.40}$$

and, therefore, $\mathbf{v} = \mathbf{b}/(4\sqrt{5})$, $t = 0 : 21$. Thus, the solution $\mathbf{v}$ is calculated by the method of back substitution.

3.5.4.2 $(1, 2)$-projection

We now consider the rays of the projection that corresponds to the generator $(p, s) = (1, 2)$. Twenty-two arithmetical rays of this projection which are used to define all components of the 2-D paired transforms with triplet-numbers of $U(1, 2)$ are shown in Figure 3.25. The set of 22 geometrical rays that coincide

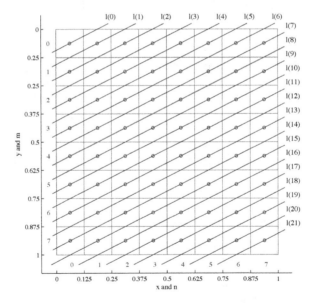

FIGURE 3.25
Set of arithmetical rays for calculating line-sums for $(p, s) = (1, 2)$.

with these arithmetical rays are defined as

$$l(t) = l_{1,2}(t) = \left\{ (x, y); \ x + 2y = \frac{t}{8} + \frac{3}{16} \right\}, \quad t = 0 : 21,$$

where $(x, y) \in [0, 1] \times [0, 1]$.

The masks of the paired functions with the first six triplet-numbers of $U(1, 2)$ are

$$[\chi'_{1,2,0}] = \begin{bmatrix} 1\,0 & 0\,0 & -1\,0 & 0\,0 \\ 0\,0 & -1\,0 & 0\,0 & 1\,0 \\ -1\,0 & 0\,0 & 1\,0 & 0\,0 \\ 0\,0 & 1\,0 & 0\,0 & -1\,0 \\ 1\,0 & 0\,0 & -1\,0 & 0\,0 \\ 0\,0 & -1\,0 & 0\,0 & 1\,0 \\ -1\,0 & 0\,0 & 1\,0 & 0\,0 \\ 0\,0 & 1\,0 & 0\,0 & -1\,0 \end{bmatrix}, \quad [\chi'_{1,2,1}] = \begin{bmatrix} 0 & 1\,0 & 0\,0 & -1\,0 & 0 \\ 0 & 0\,0 & -1\,0 & 0\,0 & 1 \\ 0 & -1\,0 & 0\,0 & 1\,0 & 0 \\ 0 & 0\,0 & 1\,0 & 0\,0 & -1 \\ 0 & 1\,0 & 0\,0 & -1\,0 & 0 \\ 0 & 0\,0 & -1\,0 & 0\,0 & 1 \\ 0 & -1\,0 & 0\,0 & 1\,0 & 0 \\ 0 & 0\,0 & 1\,0 & 0\,0 & -1 \end{bmatrix},$$

$$[\chi'_{1,2,2}] = \begin{bmatrix} 0\,0 & 1\,0 & 0\,0 & -1\,0 \\ 1\,0 & 0\,0 & -1\,0 & 0\,0 \\ 0\,0 & -1\,0 & 0\,0 & 1\,0 \\ -1\,0 & 0\,0 & 1\,0 & 0\,0 \\ 0\,0 & 1\,0 & 0\,0 & -1\,0 \\ 1\,0 & 0\,0 & -1\,0 & 0\,0 \\ 0\,0 & -1\,0 & 0\,0 & 1\,0 \\ -1\,0 & 0\,0 & 1\,0 & 0\,0 \end{bmatrix}, \quad [\chi'_{1,2,3}] = \begin{bmatrix} 0 & 0\,0 & 1\,0 & 0\,0 & -1 \\ 0 & 1\,0 & 0\,0 & -1\,0 & 0 \\ 0 & 0\,0 & -1\,0 & 0\,0 & 1 \\ 0 & -1\,0 & 0\,0 & 1\,0 & 0 \\ 0 & 0\,0 & 1\,0 & 0\,0 & -1 \\ 0 & 1\,0 & 0\,0 & -1\,0 & 0 \\ 0 & 0\,0 & -1\,0 & 0\,0 & 1 \\ 0 & -1\,0 & 0\,0 & 1\,0 & 0 \end{bmatrix},$$

$$[\chi'_{2,4,0}] = \begin{bmatrix} 1\,0 & -1\,0 & 1\,0 & -1\,0 \\ -1\,0 & 1\,0 & -1\,0 & 1\,0 \\ 1\,0 & -1\,0 & 1\,0 & -1\,0 \\ -1\,0 & 1\,0 & -1\,0 & 1\,0 \\ 1\,0 & -1\,0 & 1\,0 & -1\,0 \\ -1\,0 & 1\,0 & -1\,0 & 1\,0 \\ 1\,0 & -1\,0 & 1\,0 & -1\,0 \\ -1\,0 & 1\,0 & -1\,0 & 1\,0 \end{bmatrix}, \quad [\chi'_{2,4,2}] = \begin{bmatrix} 0 & 1\,0 & -1\,0 & 1\,0 & -1 \\ 0 & -1\,0 & 1\,0 & -1\,0 & 1 \\ 0 & 1\,0 & -1\,0 & 1\,0 & -1 \\ 0 & -1\,0 & 1\,0 & -1\,0 & 1 \\ 0 & 1\,0 & -1\,0 & 1\,0 & -1 \\ 0 & -1\,0 & 1\,0 & -1\,0 & 1 \\ 0 & 1\,0 & -1\,0 & 1\,0 & -1 \\ 0 & -1\,0 & 1\,0 & -1\,0 & 1 \end{bmatrix}.$$

These masks with the rays on which the coefficients 1 and -1 are situated, are shown in Figures 3.26 and 3.27. We define and number the control points for these rays by the bullets as shown below:

$$\begin{bmatrix} 0\bullet^1\bullet^2\bullet^3\bullet^4\bullet^5\bullet & 6\bullet & 7\bullet \\ \cdot\;\;\cdot\;\;\cdot\;\;\cdot\;\;\cdot\;\;\cdot\;\; & 8\bullet & 9\bullet \\ \cdot\;\;\cdot\;\;\cdot\;\;\cdot\;\;\cdot\; & 10\bullet & 11\bullet \\ \cdot\;\;\cdot\;\;\cdot\;\;\cdot\;\;\cdot\; & 12\bullet & 13\bullet \\ \cdot\;\;\cdot\;\;\cdot\;\;\cdot\;\;\cdot\; & 14\bullet & 15\bullet \\ \cdot\;\;\cdot\;\;\cdot\;\;\cdot\;\;\cdot\; & 16\bullet & 17\bullet \\ \cdot\;\;\cdot\;\;\cdot\;\;\cdot\;\;\cdot\; & 18\bullet & 19\bullet \\ \cdot\;\;\cdot\;\;\cdot\;\;\cdot\;\;\cdot\; & 20\bullet & 21\bullet \end{bmatrix} \quad \text{(angle of rays is } -\tan^{-1} 1/2 = -26.565°\text{).}$$

The components of the paired transform with the triplet-numbers of the

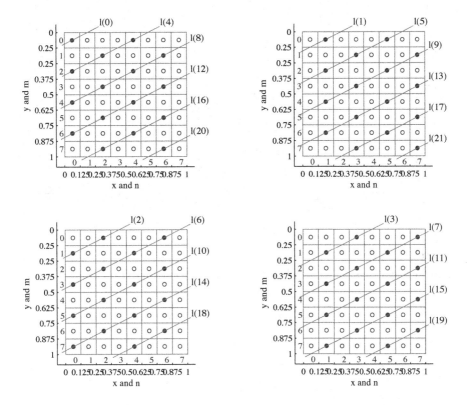

FIGURE 3.26

Arithmetical rays along which the line-sums are calculated for $(p, s) = (1, 2)$.

subset $U(1, 2)$ are calculated by

$$
\begin{aligned}
f'_{1,2,0} &= v_0 - v_4 + v_8 - v_{12} + v_{16} - v_{20} \\
f'_{1,2,1} &= v_1 - v_5 + v_9 - v_{13} + v_{17} - v_{21} \\
f'_{1,2,2} &= v_2 - v_6 + v_{10} - v_{14} + v_{18} \\
f'_{1,2,3} &= v_3 - v_7 + v_{11} - v_{15} + v_{19} \\
f'_{2,4,0} &= (v_0 + v_4 + v_8 + v_{12} + v_{16} + v_{20}) - (v_2 + v_6 + v_{10} + v_{14} + v_{18}) \\
f'_{2,4,2} &= (v_1 + v_5 + v_9 + v_{13} + v_{17} + v_{21}) - (v_3 + v_7 + v_{11} + v_{15} + v_{19}) \\
f'_{4,0,0} &= v_0 - v_1 + v_2 - v_3 + \cdots + v_{18} - v_{19} + v_{20} - v_{21} \\
f'_{0,0,0} &= v_0 + v_1 + v_2 + v_3 + \cdots + v_{18} + v_{19} + v_{20} + v_{21},
\end{aligned}
\tag{3.41}
$$

where variables $v_t = v_{1,2}(t)$ denote the sums of the discrete image along the

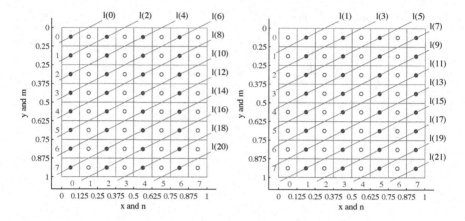

FIGURE 3.27
Arithmetical rays along which the line-sums are calculated for $(p, s) = (2, 4)$.

rays $l(t)$, $t = 0 : 21$. The system of equations can also be written as

$$
\begin{aligned}
f'_{1,2,0} &= a_0 - a_4 \\
f'_{1,2,1} &= a_1 - a_5 \\
f'_{1,2,2} &= a_2 - a_6 \\
f'_{1,2,3} &= a_3 - a_7 \\
f'_{2,4,0} &= (a_0 + a_4) - (a_2 + a_6) \\
f'_{2,4,2} &= (a_1 + a_5) - (a_3 + a_7) \\
f'_{4,0,0} &= (a_0 + a_4) - (a_1 + a_5) + (a_2 + a_6) - (a_3 + a_7) \\
f'_{0,0,0} &= (a_0 + a_4) + (a_1 + a_5) + (a_2 + a_6) + (a_3 + a_7),
\end{aligned}
\tag{3.42}
$$

where a_n are calculated from the sums $v_t = v_{1,2}(t)$ of the discrete image by

$$
a_n = \sum_m v_{n+8m} = \sum_m v_{1,2,n+8m}, \qquad n = 0 : 7.
$$

The 8-point discrete paired transform of the vector $\mathbf{a} = (a_0, a_1, a_2, ..., a_7)'$ results in the 2-D paired transform data at points of $U(1, 2)$,

$$
\mathbf{P}(1, 2) = [\chi'_8]\mathbf{a}.
$$

As in the $(p, s) = (2, 1)$ case, to calculate the vector $\mathbf{a}$ from the projection-data, we will simplify the equation describing the linear relation between the line-integrals $w(t)$ and sums $v(t)$, $t = 0 : 21$, by considering another set of 22 geometric rays. For that, we first consider the mask of the paired function $\chi'_{1,2,0}(n, m)$ with coefficients ± 1 of the rays $l(0), l(4), l(8), l(12), l(16)$, and $l(20)$ which are shown in Figure 3.28. One can see that the geometrical ray $l(8)$ intersects three image elements when x runs two intervals, $2\Delta x$, along

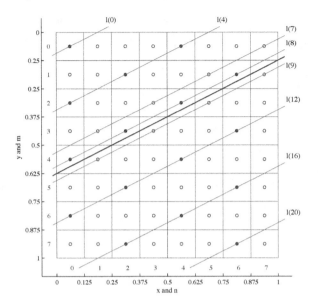

FIGURE 3.28
The mask of number $(1, 2, 0)$ and rays.

the horizontal. Therefore, the integral $w(8) = w_{1,2}(8)$ can be calculated by using three sums, $v(7), v(8),$ and $v(9)$. On other hand, the integral of the image along the geometrical ray between the rays $l(8)$ and $l(9)$ can be calculated from the sums of the image along two arithmetical rays by

$$w = \sqrt{5}/2 \cdot 8 \left[v(8) + v(9) \right].$$

This geometric ray is considered as ray number 8 in the new set of parallel geometrical rays, which are defined as

$$\tilde{l}(t) = l_{1,2}(t + 0.5) = \left\{ (x, y); \; x + 2y = \frac{t}{8} + \frac{1}{4} \right\}, \quad t = 0 : 21,$$

and illustrated in Figure 3.29.

The relation between the measurements along the geometrical rays $\tilde{l}(t)$ and arithmetical rays $l(t)$ is described by the following system of linear equations:

$$w(t) = w(\tilde{l}(t)) = 4\sqrt{5} \left[v(t) + v(t + 1) \right], \quad t = 0 : 21,$$

where $v(22) = 0$. This system of equations was considered in matrix form in (3.38) and (3.39), for the $(p, s) = (2, 1)$ case. The required sums $v(t)$ of the discrete image $f_{n,m}$ are calculated by $v(t) = b_t / 4\sqrt{5}$, where the coefficients b_t can be calculated in the following recurrent form:

$$\begin{aligned} b_{21} &= w(21), \\ b_t &= w(t) - b_{t+1}, \quad t = 20, 19, \ldots, 1, 0. \end{aligned} \tag{3.43}$$

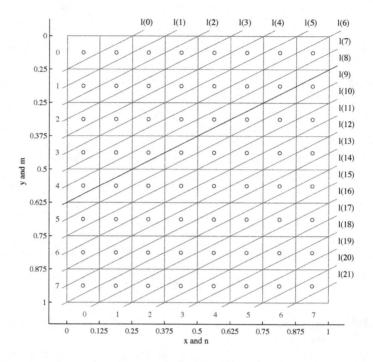

FIGURE 3.29
New set of geometrical rays for calculating line-integrals for $(p, s) = (1, 2)$.

Remark 3.2 Another set of shifted parallel geometrical rays can also be considered. As an example, we consider the rays defined as

$$\tilde{l}(t) = l_{1,2}(t - 0.5) = \left\{ (x, y); \ x + 2y = \frac{t}{8} + \frac{1}{8} \right\}, \quad t = 0 : 21,$$

which are illustrated in Figure 3.30.

The relation between the measurements along the geometrical rays $\tilde{l}(t)$ and arithmetical rays $l(t)$ is described by the following system of linear equations:

$$w(t) = w(\tilde{l}(t)) = \frac{\sqrt{5}}{2} \cdot 8 \left[v(t) + v(t - 1) \right], \quad t = 0 : 21,$$

where $v(-1) = 0$. This system of equations was considered in matrix form in (3.35) and (3.36), for the $(p, s) = (2, 1)$ case. Therefore, the required sums $v(t)$ of the image are calculated by $v(t) = b_t / 4\sqrt{5}$, and the coefficients b_t can be calculated in the following recurrent form:

$$\begin{aligned} b_0 &= w(0), \\ b_t &= w(t) - b_{t-1}, \quad t = 1, 2, \ldots, 20, 21. \end{aligned} \tag{3.44}$$

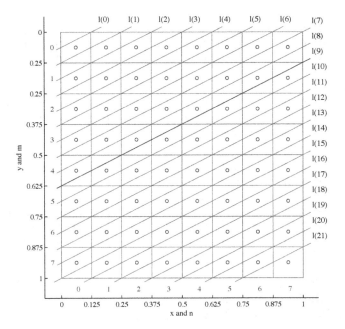

FIGURE 3.30
New set of geometrical rays for calculating line-integrals for $(p, s) = (1, 2)$.

3.5.5 $(1, 3)$-projection

We now consider the subset of triplet-numbers $U(1, 3)$,

$$
U(1, 3) = \begin{cases} (1, 3, 0), (1, 3, 1), (1, 3, 2), (1, 3, 3), \\ (2, 6, 0), (2, 6, 2), \\ (4, 4, 0), \\ (0, 0, 0). \end{cases}
$$

Since the masks of the paired functions $\chi'_{4,4,0}$ and $\chi'_{0,0,0}$ were given above, we consider functions with the first six triplet-numbers of the subset $U(1, 3)$,

$$
[\chi'_{1,3,0}] = \begin{bmatrix} 1 & 0 & 0 & 0 & -1 & 0 & 0 & 0 \\ 0 & -1 & 0 & 0 & 0 & 1 & 0 & 0 \\ 0 & 0 & 1 & 0 & 0 & 0 & -1 & 0 \\ 0 & 0 & 0 & -1 & 0 & 0 & 0 & 1 \\ -1 & 0 & 0 & 0 & 1 & 0 & 0 & 0 \\ 0 & 1 & 0 & 0 & 0 & -1 & 0 & 0 \\ 0 & -1 & 0 & 0 & 0 & 1 & 0 \\ 0 & 0 & 0 & 1 & 0 & 0 & 0 & -1 \end{bmatrix}, \quad
[\chi'_{1,3,1}] = \begin{bmatrix} 0 & 1 & 0 & 0 & 0 & -1 & 0 & 0 \\ 0 & 0 & -1 & 0 & 0 & 0 & 1 & 0 \\ 0 & 0 & 0 & 1 & 0 & 0 & 0 & -1 \\ 1 & 0 & 0 & 0 & -1 & 0 & 0 & 0 \\ 0 & -1 & 0 & 0 & 0 & 1 & 0 & 0 \\ 0 & 0 & 1 & 0 & 0 & 0 & -1 & 0 \\ 0 & 0 & 0 & -1 & 0 & 0 & 0 & 1 \\ -1 & 0 & 0 & 0 & 1 & 0 & 0 & 0 \end{bmatrix}
$$

$$[\chi'_{1,3,2}] = \begin{bmatrix} 0 & 0 & 1 & 0 & 0 & 0 & -1 & 0 \\ 0 & 0 & 0 & -1 & 0 & 0 & 0 & 1 \\ -1 & 0 & 0 & 0 & 1 & 0 & 0 & 0 \\ 0 & 1 & 0 & 0 & 0 & -1 & 0 & 0 \\ 0 & 0 & -1 & 0 & 0 & 0 & 1 & 0 \\ 0 & 0 & 0 & 1 & 0 & 0 & 0 & -1 \\ 1 & 0 & 0 & 0 & -1 & 0 & 0 & 0 \\ 0 & -1 & 0 & 0 & 0 & 1 & 0 & 0 \end{bmatrix}, \quad [\chi'_{1,3,3}] = \begin{bmatrix} 0 & 0 & 0 & 1 & 0 & 0 & 0 & -1 \\ 1 & 0 & 0 & 0 & -1 & 0 & 0 & 0 \\ 0 & -1 & 0 & 0 & 0 & 1 & 0 & 0 \\ 0 & 0 & 1 & 0 & 0 & 0 & -1 & 0 \\ 0 & 0 & 0 & -1 & 0 & 0 & 0 & 1 \\ -1 & 0 & 0 & 0 & 1 & 0 & 0 & 0 \\ 0 & 1 & 0 & 0 & 0 & -1 & 0 & 0 \\ 0 & -1 & 0 & 0 & 0 & 1 & 0 & 0 \end{bmatrix}$$

$$[\chi'_{2,6,0}] = \begin{bmatrix} 1 & 0 & -1 & 0 & 1 & 0 & -1 & 0 \\ 0 & 1 & 0 & -1 & 0 & 1 & 0 & -1 \\ -1 & 0 & 1 & 0 & -1 & 0 & 1 & 0 \\ 0 & -1 & 0 & 1 & 0 & -1 & 0 & 1 \\ 1 & 0 & -1 & 0 & 1 & 0 & -1 & 0 \\ 0 & 1 & 0 & -1 & 0 & 1 & 0 & -1 \\ -1 & 0 & 1 & 0 & -1 & 0 & 1 & 0 \\ 0 & -1 & 0 & 1 & 0 & -1 & 0 & 1 \end{bmatrix}, \quad [\chi'_{2,6,2}] = \begin{bmatrix} 0 & -1 & 0 & 1 & 0 & -1 & 0 & 1 \\ 1 & 0 & -1 & 0 & 1 & 0 & -1 & 0 \\ 0 & 1 & 0 & -1 & 0 & 1 & 0 & -1 \\ -1 & 0 & 1 & 0 & -1 & 0 & 1 & 0 \\ 0 & -1 & 0 & 1 & 0 & -1 & 0 & 1 \\ 1 & 0 & -1 & 0 & 1 & 0 & -1 & 0 \\ 0 & 1 & 0 & -1 & 0 & 1 & 0 & -1 \\ -1 & 0 & 1 & 0 & -1 & 0 & 1 & 0 \end{bmatrix}$$

The masks of these paired functions are shown in Figures 3.31 and 3.32, together with the rays the pass through the coefficients ± 1 of the masks.

The set of 29 geometrical rays of this projection,

$$l(t) = l_{1,3}(t) = \left\{ (x,y); \; x + 3y = \frac{t}{8} + \frac{1}{4} \right\}, \quad t = 0 : 28,$$

where $(x, y) \in [0,1] \times [0,1]$, is shown in Figures 3.33.

We define the control points for these rays as shown below:

$$\begin{bmatrix} 0 & 1 & 2 & 3 & 4 & 5 & 6 & 7 \\ & & & & & 8 & 9 & 10 \\ \cdot & \cdot & \cdot & \cdot & 11 & 12 & 13 \\ \cdot & \cdot & \cdot & \cdot & 14 & 15 & 16 \\ \cdot & \cdot & \cdot & \cdot & 17 & 18 & 19 \\ \cdot & \cdot & \cdot & \cdot & 20 & 21 & 22 \\ \cdot & \cdot & \cdot & \cdot & 23 & 24 & 25 \\ \cdot & \cdot & \cdot & \cdot & 26 & 27 & 28 \end{bmatrix}$$ (angle of rays is $-\tan^{-1}(1/3) = -18.435°$).

The components of the paired transform with the triplet-numbers of the subset $U(1,3)$ are calculated as follows:

$$\begin{aligned} f'_{1,3,0} &= v_0 - v_4 + v_8 - v_{12} + v_{16} - v_{20} + v_{24} - v_{28}, \\ f'_{1,3,1} &= v_1 - v_5 + v_9 - v_{13} + v_{17} - v_{21} + v_{25}, \\ f'_{1,3,2} &= v_2 - v_6 + v_{10} - v_{14} + v_{18} - v_{22} + v_{26}, \\ f'_{1,3,3} &= v_3 - v_7 + v_{11} - v_{15} + v_{19} + v_{23} + v_{27}, \end{aligned}$$

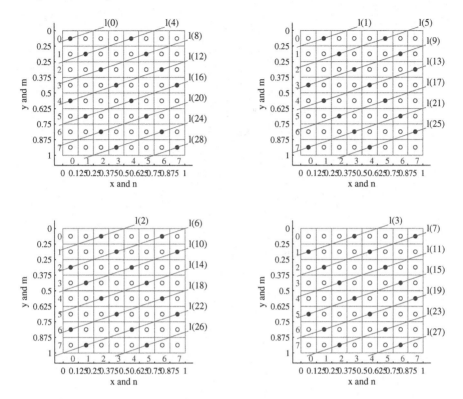

FIGURE 3.31
Masks of the paired functions with triplet-numbers $(1, 3, t)$, $t = 0, 1, 2, 3$.

$$
\begin{aligned}
f'_{2,6,0} &= (v_0 + v_4 + v_8 + v_{12} + v_{16} + v_{20} + v_{24} + v_{28}) \\
&\quad - (v_2 + v_6 + v_{10} + v_{14} + v_{18} + v_{22} + v_{26}), \\
f'_{2,6,2} &= (v_1 + v_5 + v_9 + v_{13} + v_{17} + v_{21} + v_{25}) \\
&\quad - (v_3 + v_7 + v_{11} + v_{15} + v_{19} + v_{23} + v_{27}), \\
f'_{4,4,0} &= v_0 - v_1 + v_2 - v_3 + \cdots + v_{18} - v_{19} + v_{20} - v_{21}, \\
f'_{0,0,0} &= v_0 + v_1 + v_2 + v_3 + \cdots + v_{18} + v_{19} + v_{20} + v_{21},
\end{aligned}
$$

where variables $v_t = v_{1,3}(t)$ denote the sums of the discrete image along the rays $l(t)$, $t = 0 : 28$. The system of equations can also be written as

$$
\begin{aligned}
f'_{1,3,0} &= a_0 - a_4 \\
f'_{1,3,1} &= a_1 - a_5 \\
f'_{1,3,2} &= a_2 - a_6 \\
f'_{1,3,3} &= a_3 - a_7 \\
f'_{2,6,0} &= (a_0 + a_4) - (a_2 + a_6) \\
f'_{2,6,2} &= (a_1 + a_5) - (a_3 + a_7) \\
f'_{4,4,0} &= (a_0 + a_4) - (a_1 + a_5) + (a_2 + a_6) - (a_3 + a_7) \\
f'_{0,0,0} &= (a_0 + a_4) + (a_1 + a_5) + (a_2 + a_6) + (a_3 + a_7)
\end{aligned}
\tag{3.45}
$$

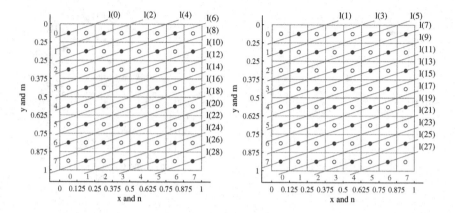

FIGURE 3.32
Masks of the paired functions with triplet-numbers $(2, 6, t)$, $t = 0, 2$.

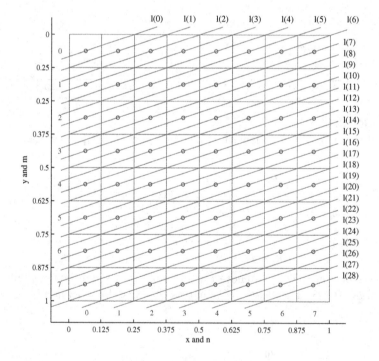

FIGURE 3.33
The set of arithmetical rays for the $(1, 3)$-projection.

where a_n are calculated from the sums $v_t = v_{1,3}(t)$ of the discrete image by

$$a_n = \sum_m v_{n+8m} = \sum_m v_{1,3,n+8m}, \qquad n = 0 : 7.$$

The 8-point discrete paired transform of the vector $\mathbf{a} = (a_0, a_1, a_2, ..., a_7)'$ results in the 2-D paired transform data at points of $U(1,3)$, i.e., $\mathbf{P}(1,3) = [\chi'_8]\mathbf{a}$. To calculate the vector $\mathbf{a}$ from the projection data, we will define a set of geometrical rays that allow calculation of the integrals $w(t)$ by means of the sums $v(t)$, $t = 0 : 28$. For that, we first consider a few arithmetical rays as the geometrical rays in the part of the square $[0,1] \times [0,1]$, which are shown in Figure 3.34. One can notice that each of five rays equally intersects image

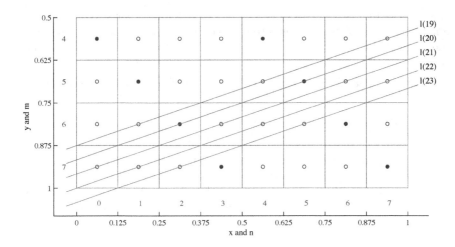

FIGURE 3.34

Five rays of the projection corresponding $(1,3)$.

elements, namely the length of intersection of the ray with each IE equals $\Delta l_{0,7} = \sqrt{10}/3\Delta x$. The line-integrals along the geometrical rays $20, 21$, and 22 can be calculated by the sums of the discrete image as follows:

$$w(20) = w_{1,3}(20) = 8\sqrt{10}/3\,[v(21) + v(19) + v(20)]$$
$$= 8\sqrt{10}/3\,[v(19) + v(20) + v(21)]$$
$$w(21) = w_{1,3}(21) = 8\sqrt{10}/3\,[v(21) + v(22) + v(20)]$$
$$= 8\sqrt{10}/3\,[v(20) + v(21) + v(22)]$$
$$w(22) = w_{1,3}(22) = 8\sqrt{10}/3\,[v(21) + v(22) + v(23)].$$

Similar equations hold for all integrals $w(t)$ of the considered projection

$$w(t) = w_{1,3}(t) = 8\sqrt{10}/3 \left[v(t-1) + v(t) + v(t+1)\right], \quad t = 0 : 28.$$

In matrix form, this system of linear equations is determined by the Toeplitz matrix 29×29,

$$A = \begin{bmatrix} 1 & 1 & 0 & 0 & 0 & \dots & 0 & 0 & 0 \\ 1 & 1 & 1 & 0 & 0 & \dots & 0 & 0 & 0 \\ 0 & 1 & 1 & 1 & 0 & \dots & 0 & 0 & 0 \\ 0 & 0 & 1 & 1 & 1 & \dots & 0 & 0 & 0 \\ \multicolumn{9}{c}{\dots\dots\dots\dots\dots\dots\dots\dots\dots} \\ 0 & 0 & 0 & 0 & 0 & \dots & 1 & 1 & 1 \\ 0 & 0 & 0 & 0 & 0 & \dots & 0 & 1 & 1 \end{bmatrix}, \tag{3.46}$$

which is not triangular and has determinant 0. In other words, the given set of geometrical rays cannot be used to define all required sums $v(t)$, $t = 0 : 21$, for the considered projection. Therefore, as for the projections defined by frequency-points $(p, s) = (2, 1)$ and $(1, 2)$, we consider other sets of rays. The simple way to select such a set is to remove the first ray from the set and add one parallel new ray at the end, or remove the last ray from the set and add one new parallel ray at the beginning. We consider the first case.

Case 1: The set of 29 geometrical rays is defined as (see Figure 3.35)

$$\tilde{l}(t) = l_{1,3}(t+1) = \left\{ (x, y); \; x + 3y = \frac{t}{8} + \frac{3}{8} \right\}, \quad t = 0 : 28,$$

and the integrals along these rays are denoted by $w(t) = w(\tilde{l}(t))$.

One can see, for instance, that $w(\tilde{l}(1)) = w(l(2)) = v(1) + v(2) + v(3)$, up to the normalized coefficient $8\sqrt{10}/3$. The system of equations for all new integrals $w(t)$ of the considered projection can be defined as follows:

$$w(t) = w(\tilde{l}(t)) = 8\sqrt{10}/3 \left[v(t) + v(t+1) + v(t+2)\right], \quad t = 0 : 28,$$

where it is assumed that $v(29) = v(30) = 0$. This system of equations can be written in matrix form as

$$\mathbf{w} = \frac{8\sqrt{10}}{3} \mathbf{A}\mathbf{v} = \frac{8\sqrt{10}}{3} \begin{bmatrix} 1 & 1 & 1 & 0 & \dots & 0 & 0 & 0 \\ 0 & 1 & 1 & 1 & \dots & 0 & 0 & 0 \\ 0 & 0 & 1 & 1 & \dots & 0 & 0 & 0 \\ 0 & 0 & 0 & 1 & \dots & 0 & 0 & 0 \\ \multicolumn{8}{c}{\dots\dots\dots\dots\dots\dots\dots} \\ 0 & 0 & 0 & 0 & \dots & 0 & 1 & 1 \\ 0 & 0 & 0 & 0 & \dots & 0 & 0 & 1 \end{bmatrix} \mathbf{v}. \tag{3.47}$$

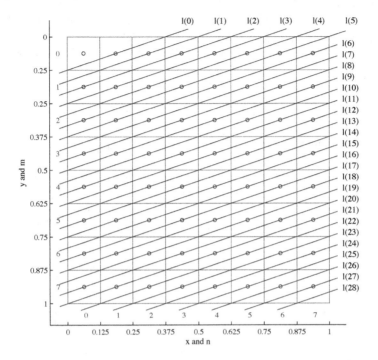

FIGURE 3.35
The set of geometrical rays for the projection defined by $(1,3)$.

The above Toeplitz matrix 29×29 has the upper triangle inverse matrix

$$\mathbf{A}^{-1} = \begin{bmatrix} 1 & -1 & 0 & 1 & -1 & 0 & 1 & -1 & \ldots & 0 & 1 & -1 \\ 0 & 1 & -1 & 0 & 1 & -1 & 0 & 1 & \ldots & -1 & 0 & 1 \\ 0 & 0 & 1 & -1 & 0 & 1 & -1 & 0 & \ldots & 1 & -1 & 0 \\ 0 & 0 & 0 & 1 & -1 & 0 & 1 & -1 & \ldots & 0 & 1 & -1 \\ 0 & 0 & 0 & 0 & 1 & -1 & 0 & 1 & \ldots & -1 & 0 & 1 \\ \cdot & \cdot & \cdot & \cdot & \cdot & \cdot & \cdot & & \ldots & \cdot & \cdot & \cdot \\ 0 & 0 & 0 & 0 & 0 & 0 & 0 & 1 & \ldots & -1 & 0 & 1 \\ 0 & 0 & 0 & 0 & 0 & 0 & 0 & 0 & \ldots & 1 & -1 & 0 \\ 0 & 0 & 0 & 0 & 0 & 0 & 0 & 0 & \ldots & 0 & 1 & -1 \\ 0 & 0 & 0 & 0 & 0 & 0 & 0 & 0 & \ldots & 0 & 0 & 1 \end{bmatrix}. \tag{3.48}$$

The inverse transform $\mathbf{b} = \mathbf{A}^{-1}\mathbf{w}$ can be calculated by the following recurrent form:

$$\begin{aligned} b_{28} &= w(28), \\ b_{27} &= w(27) - b_{28}, \\ b_{26} &= w(26) - (b_{27} + b_{28}), \\ b_{25} &= w(25) - (b_{26} + b_{27}), \\ b_t &= w(t) - (b_{t+1} + b_{t+2}), \quad t = 24, 23, \ldots, 1, 0. \end{aligned} \tag{3.49}$$

The required sums $v(t)$ of the discrete image are calculated by $v(t) = b_t/(8\sqrt{10}/3)$, where $t = 0 : 28$.

   Case 2: The set of 29 geometrical rays can also be defined as (see Figure 3.36)

$$\tilde{l}(t) = l_{1,3}(t-1) = \left\{ (x,y); \; x + 3y = \frac{t}{8} + \frac{1}{8} \right\}, \quad t = 0 : 28.$$

The integrals along these rays are denoted by $w(t) = w(\tilde{l}(t))$.

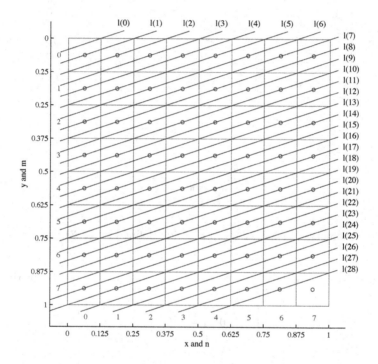

FIGURE 3.36
Second set of geometrical rays for the $(1,3)$-projection.

   In this case, the line-integrals $w(t)$ are defined by the sums $v(t)$ as

$$w(t) = w(\tilde{l}(t)) = 8\sqrt{10}/3 \left[v(t) + v(t-1) + v(t-2) \right], \quad t = 0 : 28. \quad (3.50)$$

In matrix form, this system of linear equations is described by the triangular

Toeplitz matrix 29×29 with left band width 2,

$$
A = \begin{bmatrix}
1\ 0\ 0\ 0\ 0... \ 0\ 0\ 0 \\
1\ 1\ 0\ 0\ 0... \ 0\ 0\ 0 \\
1\ 1\ 1\ 0\ 0... \ 0\ 0\ 0 \\
0\ 1\ 1\ 1\ 0... \ 0\ 0\ 0 \\
\hdotsfor{1} \\
0\ 0\ 0\ 0\ 0... \ 1\ 1\ 0 \\
0\ 0\ 0\ 0\ 0... \ 1\ 1\ 1
\end{bmatrix}, \quad (\det(A) = 1). \tag{3.51}
$$

The inverse matrix equals

$$
\mathbf{A}^{-1} = \begin{bmatrix}
1 & 0 & 0 & 0 & 0 & 0... & 0 & 0 & 0 & 0 & 0 \\
-1 & 1 & 0 & 0 & 0 & 0... & 0 & 0 & 0 & 0 & 0 \\
0 & -1 & 1 & 0 & 0 & 0... & 0 & 0 & 0 & 0 & 0 \\
1 & 0 & -1 & 1 & 0 & 0... & 0 & 0 & 0 & 0 & 0 \\
-1 & 1 & 0 & -1 & 1 & 0... & 0 & 0 & 0 & 0 & 0 \\
0 & -1 & 1 & 0 & -1 & 1... & 0 & 0 & 0 & 0 & 0 \\
 . & . & . & . & . & ... & . & . & . & . & . \\
1 & 0 & -1 & 1 & 0 & -1... & 1 & 0 & -1 & 1 & 0 \\
-1 & 1 & 0 & -1 & 1 & 0... & -1 & 1 & 0 & -1 & 1
\end{bmatrix}. \tag{3.52}
$$

The inverse transform $\mathbf{b} = \mathbf{A}^{-1}\mathbf{w}$ can be calculated in the following recurrent form:

$$
\begin{aligned}
b_0 &= w(0), \\
b_1 &= w(1) - b_0, \\
b_2 &= w(2) - (b_1 + b_0), \\
b_3 &= w(3) - (b_2 + b_1), \\
b_t &= w(t) - (b_{t-1} + b_{t-2}), \quad t = 4, 5, \ldots, 27, 28.
\end{aligned} \tag{3.53}
$$

The required sums $v(t)$ of the discrete image are calculated by $v(t) = 3b_t/(8\sqrt{10})$, where $t = 0 : 28$.

Remark 3.3 Instead of the subset $U(1,3)$, we can consider the subset $U(3,1)$,

$$
U(3,1) = \begin{cases}
(3,1,0), (3,1,1), (3,1,2), (3,1,3), \\
(6,2,0), (6,2,2), \\
(4,4,0), \\
(0,0,0).
\end{cases}
$$

Indeed, it is not difficult to see that the paired functions generated by the frequency $(p,s) = (3,1)$ are $\chi'_{3,1,t} = \pm\chi'_{1,3,t \bmod 4}$, where the sign is determined by $\text{sign}(4 - t \bmod 8)$, for $t = 0, 1, 2, 3$. These two points $(1,3)$ and $(3,1)$ generate the same group $T_{1,3} = T_{3,1}$ in the frequency domain. It also means that all 1's and all -1's are located on the parallel rays at the angles $\pi - \tan^{-1}(1/3)$ and $\pi - \tan^{-1}(3/1)$ to the horizontal. As an example, Figure 3.37 shows the mask of the function $\chi'_{3,1,0} = \chi'_{1,3,0}$ and two sets of parallel rays, along which the coefficients ±1 of the function are located. For the $(3,1)$-projection, the

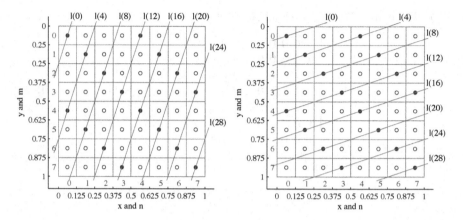

FIGURE 3.37
Two sets of rays on the mask of the paired function with number $(3, 1, 0)$.

control points of the second set of rays are defined as shown below:

$$
\begin{bmatrix}
0 \; \bullet & \cdot & \cdot & \cdot & \cdot & \cdot & \cdot & \cdot \\
1 \; \bullet & \cdot & \cdot & \cdot & \cdot & \cdot & \cdot & \cdot \\
2 \; \bullet & \cdot & \cdot & \cdot & \cdot & \cdot & \cdot & \cdot \\
3 \; \bullet & \cdot & \cdot & \cdot & \cdot & \cdot & \cdot & \cdot \\
4 \; \bullet & \cdot & \cdot & \cdot & \cdot & \cdot & \cdot & \cdot \\
5 \; \bullet & 8 \; \bullet & 11 \; \bullet & 14 \; \bullet & 17 \; \bullet & 20 \; \bullet & 23 \; \bullet & 26 \; \bullet \\
6 \; \bullet & 9 \; \bullet & 12 \; \bullet & 15 \; \bullet & 18 \; \bullet & 21 \; \bullet & 24 \; \bullet & 27 \; \bullet \\
7 \; \bullet & 10 \; \bullet & 13 \; \bullet & 16 \; \bullet & 19 \; \bullet & 22 \; \bullet & 25 \; \bullet & 28 \; \bullet
\end{bmatrix}
$$

(angle of rays is $\pi - \tan^{-1}(3) = 108.435°$).

The set of the 29 geometrical rays of this projection,

$$
l(t) = l_{3,1}(t) = \left\{ (x, y); \; 3x + y = \frac{t}{8} + \frac{1}{4} \right\}, \quad t = 0:28,
$$

where $(x, y) \in [0, 1] \times [0, 1]$, is shown in Figures 3.38.
Below are six masks for the first six triplet-numbers of the subset $U(3, 1)$,

$$
[\chi'_{3,1,0}] =
\begin{bmatrix}
1 & 0 & 0 & 0 & -1 & 0 & 0 & 0 \\
0 & -1 & 0 & 0 & 0 & 1 & 0 & 0 \\
0 & 0 & 1 & 0 & 0 & 0 & -1 & 0 \\
0 & 0 & 0 & -1 & 0 & 0 & 0 & 1 \\
-1 & 0 & 0 & 0 & 1 & 0 & 0 & 0 \\
0 & 1 & 0 & 0 & 0 & -1 & 0 & 0 \\
0 & 0 & -1 & 0 & 0 & 0 & 1 & 0 \\
0 & 0 & 0 & 1 & 0 & 0 & 0 & -1
\end{bmatrix},
\quad
[\chi'_{3,1,1}] =
\begin{bmatrix}
0 & 0 & 0 & 1 & 0 & 0 & 0 & -1 \\
1 & 0 & 0 & 0 & -1 & 0 & 0 & 0 \\
0 & -1 & 0 & 0 & 0 & 1 & 0 & 0 \\
0 & 0 & 1 & 0 & 0 & 0 & -1 & 0 \\
0 & 0 & 0 & -1 & 0 & 0 & 0 & 1 \\
-1 & 0 & 0 & 0 & 1 & 0 & 0 & 0 \\
0 & 1 & 0 & 0 & 0 & -1 & 0 & 0 \\
0 & 0 & -1 & 0 & 0 & 0 & 1 & 0
\end{bmatrix}
$$

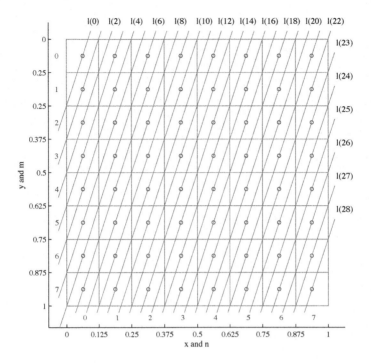

FIGURE 3.38
The set of arithmetical rays for the $(3, 1)$-projection.

$$
[\chi'_{3,1,2}] = \begin{bmatrix}
0 & 0 & -1 & 0 & 0 & 0 & 1 & 0 \\
0 & 0 & 0 & 1 & 0 & 0 & 0 & -1 \\
1 & 0 & 0 & 0 & -1 & 0 & 0 & 0 \\
0 & -1 & 0 & 0 & 0 & 1 & 0 & 0 \\
0 & 0 & 1 & 0 & 0 & 0 & -1 & 0 \\
0 & 0 & 0 & -1 & 0 & 0 & 0 & 1 \\
-1 & 0 & 0 & 0 & 1 & 0 & 0 & 0 \\
0 & 1 & 0 & 0 & 0 & -1 & 0 & 0
\end{bmatrix}, \quad
[\chi'_{3,1,3}] = \begin{bmatrix}
0 & 1 & 0 & 0 & 0 & -1 & 0 & 0 \\
0 & 0 & -1 & 0 & 0 & 0 & 1 & 0 \\
0 & 0 & 0 & 1 & 0 & 0 & 0 & -1 \\
1 & 0 & 0 & 0 & -1 & 0 & 0 & 0 \\
0 & -1 & 0 & 0 & 0 & 1 & 0 & 0 \\
0 & 0 & 1 & 0 & 0 & 0 & -1 & 0 \\
0 & 0 & 0 & -1 & 0 & 0 & 0 & 1 \\
-1 & 0 & 0 & 0 & 1 & 0 & 0 & 0
\end{bmatrix}
$$

$$
[\chi'_{6,2,0}] = \begin{bmatrix}
1 & 0 & -1 & 0 & 1 & 0 & -1 & 0 \\
0 & 1 & 0 & -1 & 0 & 1 & 0 & -1 \\
-1 & 0 & 1 & 0 & -1 & 0 & 1 & 0 \\
0 & -1 & 0 & 1 & 0 & -1 & 0 & 1 \\
1 & 0 & -1 & 0 & 1 & 0 & -1 & 0 \\
0 & 1 & 0 & -1 & 0 & 1 & 0 & -1 \\
-1 & 0 & 1 & 0 & -1 & 0 & 1 & 0 \\
0 & -1 & 0 & 1 & 0 & -1 & 0 & 1
\end{bmatrix}, \quad
[\chi'_{6,2,2}] = \begin{bmatrix}
0 & -1 & 0 & 1 & 0 & -1 & 0 & 1 \\
1 & 0 & -1 & 0 & 1 & 0 & -1 & 0 \\
0 & 1 & 0 & -1 & 0 & 1 & 0 & -1 \\
-1 & 0 & 1 & 0 & -1 & 0 & 1 & 0 \\
0 & -1 & 0 & 1 & 0 & -1 & 0 & 1 \\
1 & 0 & -1 & 0 & 1 & 0 & -1 & 0 \\
0 & 1 & 0 & -1 & 0 & 1 & 0 & -1 \\
-1 & 0 & 1 & 0 & -1 & 0 & 1 & 0
\end{bmatrix}
$$

The masks of these paired functions are shown in Figures 3.39 and 3.40, together with the rays passing through the coefficients ± 1 of the masks.

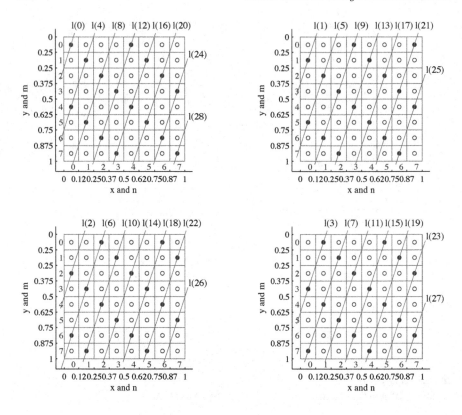

FIGURE 3.39
Masks of the paired functions with triplet-numbers $(3,1,t)$, $t = 0,1,2,3$.

The components of the 2-D paired transform with the triplet-numbers of the subset $U(1,3)$ are calculated by

$$
\begin{aligned}
f'_{3,1,0} &= v_0 - v_4 + v_8 - v_{12} + v_{16} - v_{20} + v_{24} - v_{28}, \\
f'_{3,1,1} &= v_1 - v_5 + v_9 - v_{13} + v_{17} - v_{21} + v_{25}, \\
f'_{3,1,2} &= v_2 - v_6 + v_{10} - v_{14} + v_{18} - v_{22} + v_{26}, \\
f'_{3,1,3} &= v_3 - v_7 + v_{11} - v_{15} + v_{19} + v_{23} + v_{27}, \\
f'_{6,2,0} &= (v_0 + v_4 + v_8 + v_{12} + v_{16} + v_{20} + v_{24} + v_{28}) \\
&\quad - (v_2 + v_6 + v_{10} + v_{14} + v_{18} + v_{22} + v_{26}), \\
f'_{6,2,2} &= (v_1 + v_5 + v_9 + v_{13} + v_{17} + v_{21} + v_{25}) \\
&\quad - (v_3 + v_7 + v_{11} + v_{15} + v_{19} + v_{23} + v_{27}), \\
f'_{4,4,0} &= v_0 - v_1 + v_2 - v_3 + \cdots + v_{18} - v_{19} + v_{20} - v_{21}, \\
f'_{0,0,0} &= v_0 + v_1 + v_2 + v_3 + \cdots + v_{18} + v_{19} + v_{20} + v_{21},
\end{aligned}
\tag{3.54}
$$

where variables $v_t = v_{3,1}(t)$ denote the sums of the discrete image along the

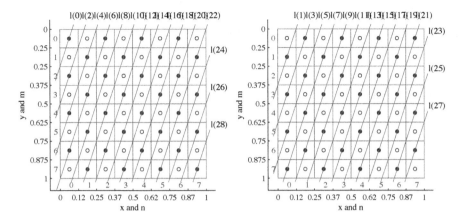

FIGURE 3.40

Masks of the paired functions with triplet-numbers $(6, 2, t)$, $t = 0, 2$.

rays $l(t)$, $t = 0 : 28$. The system of equations can be written as

$$
\begin{aligned}
f'_{3,1,0} &= a_0 - a_4 \\
f'_{3,1,1} &= a_1 - a_5 \\
f'_{3,1,2} &= a_2 - a_6 \\
f'_{3,1,3} &= a_3 - a_7 \\
f'_{6,2,0} &= (a_0 + a_4) - (a_2 + a_6) \\
f'_{6,2,2} &= (a_1 + a_5) - (a_3 + a_7) \\
f'_{4,4,0} &= (a_0 + a_4) - (a_1 + a_5) + (a_2 + a_6) - (a_3 + a_7) \\
f'_{0,0,0} &= (a_0 + a_4) + (a_1 + a_5) + (a_2 + a_6) + (a_3 + a_7)
\end{aligned}
\tag{3.55}
$$

where a_n are calculated from the sums $v_t = v_{3,1}(t)$ of the discrete image by

$$
a_n = \sum_m v_{n+8m} = \sum_m v_{3,1,n+8m}, \qquad n = 0 : 7.
$$

The 8-point discrete paired transform of the vector $\mathbf{a} = (a_0, a_1, a_2, ..., a_7)'$ results in the 2-D paired transform data at points of $U(3, 1)$,

$$
\mathbf{P}(3, 1) = [\chi'_{8,8}]\mathbf{a}.
$$

To calculate the vector $\mathbf{a}$ from the projection data, we will define a set of geometrical rays that allows calculation of the integrals $w(t)$ by means of the sums $v(t)$, $t = 0 : 28$. For that, we first consider a few arithmetical rays as the geometrical rays in the square $[0, 1] \times [0, 1]$, which are shown in Figure 3.41. We define two geometrical rays $\tilde{l}(8) = l(7)$ and $\tilde{l}(9) = l(8)$. One can see that the line-integrals along these rays can be calculated as follows:

$$
\begin{aligned}
w(8) &= w(\tilde{l}(8)) = 8\sqrt{10}/3 \left[v(6) + v(7) + v(8) \right], \\
w(9) &= w(\tilde{l}(9)) = 8\sqrt{10}/3 \left[v(7) + v(8) + v(9) \right].
\end{aligned}
\tag{3.56}
$$

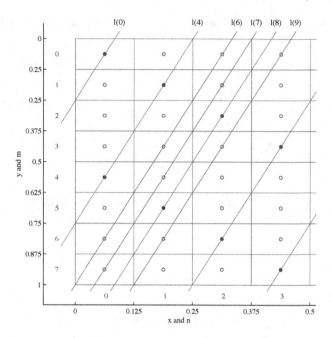

FIGURE 3.41
Eight rays of the $(3, 1)$-projection.

To work with similar equations for all line-integrals $w(t)$ of the considered projection,

$$w(t) = w(\tilde{l}(t)) = 8\sqrt{10}/3\,[v(t) + v(t-1) + v(t-2)], \quad t = 0:28, \quad (3.57)$$

we define the following set of 29 geometrical rays:

$$\tilde{l}(t) = l_{3,1}(t-1) = \left\{ (x, y);\ 3x + y = \frac{t}{8} + \frac{1}{8} \right\}, \quad t = 0:28.$$

A similar system of equations was described in (3.50)-(3.53) for the $(1, 3)$-projection. This system has the following solution:

$$v(t) = 3b_t/(8\sqrt{10}), \qquad t = 0:28,$$

where coefficients $b(t)$ are calculated recursively as

$$
\begin{aligned}
b_0 &= w(0),\\
b_1 &= w(1) - b_0,\\
b_2 &= w(2) - (b_1 + b_0),\\
b_3 &= w(3) - (b_2 + b_1),\\
b_t &= w(t) - (b_{t-1} + b_{t-2}), \quad t = 4, 5, \ldots, 27, 28.
\end{aligned}
\qquad (3.58)
$$

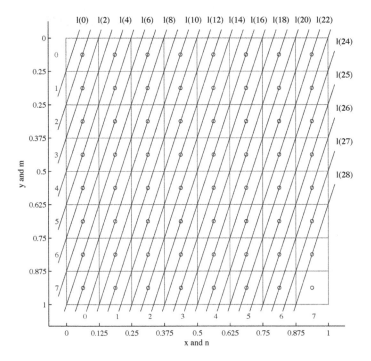

FIGURE 3.42

The set of geometrical rays for the $(3, 1)$-projection.

Figure 3.42 shows the new set of geometrical rays.

The following should be mentioned. If we denote the lines $l(7)$ and $l(8)$ in Figure 3.41 by $\tilde{l}(6)$ and $\tilde{l}(7)$, respectively, the line-integrals along these rays will be defined as

$$w(6) = w(\tilde{l}(6)) = 8\sqrt{10}/3 \left[v(6) + v(7) + v(8) \right],$$
$$w(7) = w(\tilde{l}(7)) = 8\sqrt{10}/3 \left[v(7) + v(8) + v(9) \right].$$

We obtain the system of equations

$$w(t) = w(\tilde{l}(t)) = 8\sqrt{10}/3 \left[v(t) + v(t+1) + v(t+2) \right], \quad t = 0 : 28, \quad (3.59)$$

for the set of 29 geometrical rays which are defined as

$$\tilde{l}(t) = l_{3,1}(t+1) = \left\{ (x, y); \ 3x + y = \frac{t}{8} + \frac{3}{8} \right\}, \quad t = 0 : 28.$$

This set of geometrical rays is shown in Figure 3.43. The system of equations in (3.59) is described by the same Toeplitz matrix as for the subset $U(1, 3)$ in (3.47). Therefore, the sums $v(t)$ of the discrete image can be calculated by $v(t) = b_t/(8\sqrt{10}/3)$, where $t = 0 : 28$, and components b_t are calculated in

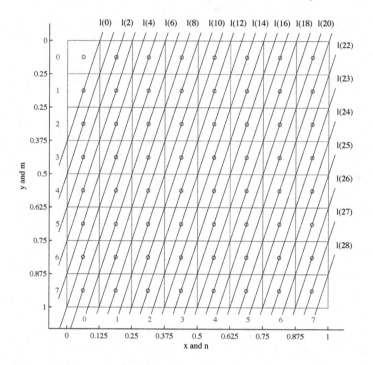

FIGURE 3.43
The second set of geometrical rays for the $(3, 1)$-projection.

recurrent form,

$$
\begin{aligned}
b_{28} &= w(28), \\
b_{27} &= w(27) - b_{28}, \\
b_{26} &= w(26) - (b_{27} + b_{28}), \\
b_t &= w(t) - (b_{t+1} + b_{t+2}), \quad t = 25, 24, \ldots, 1, 0.
\end{aligned}
\tag{3.60}
$$

3.5.6 $(1, 4)$- and $(4, 1)$-projections

Consider two subsets of triplet-numbers

$$
U(1, 4) = \begin{cases}
(1, 4, 0), (1, 4, 1), (1, 4, 2), (1, 4, 3), \\
(2, 0, 0), (2, 0, 2), \\
(4, 0, 0), \\
(0, 0, 0),
\end{cases}
$$

and

$$
U(4, 1) = \begin{cases}
(4, 1, 0), (4, 1, 1), (4, 1, 2), (4, 1, 3), \\
(0, 2, 0), (0, 2, 2), \\
(0, 4, 0), \\
(0, 0, 0).
\end{cases}
$$

The paired functions with numbers $(2,0,0), (0,2,2), (0,2,0), (0,2,2)$, $(4,0,0)$, $(0,4,0)$, and $(0,0,0)$ were described above. We thus consider the paired functions with other triplet-numbers. Four masks of the paired functions with the triplet-numbers $(1,4,t)$, $t = 0:3$, are

$$[\chi'_{1,4,0}] = \begin{bmatrix} 1\ 0\ 0\ 0\ -1\ 0\ 0\ 0 \\ -1\ 0\ 0\ 0\ \ 1\ 0\ 0\ 0 \\ 1\ 0\ 0\ 0\ -1\ 0\ 0\ 0 \\ -1\ 0\ 0\ 0\ \ 1\ 0\ 0\ 0 \\ 1\ 0\ 0\ 0\ -1\ 0\ 0\ 0 \\ -1\ 0\ 0\ 0\ \ 1\ 0\ 0\ 0 \\ 1\ 0\ 0\ 0\ -1\ 0\ 0\ 0 \\ -1\ 0\ 0\ 0\ \ 1\ 0\ 0\ 0 \end{bmatrix}, \quad [\chi'_{1,4,1}] = \begin{bmatrix} 0\ \ 1\ 0\ 0\ 0\ -1\ 0\ 0 \\ 0\ -1\ 0\ 0\ 0\ \ 1\ 0\ 0 \\ 0\ \ 1\ 0\ 0\ 0\ -1\ 0\ 0 \\ 0\ -1\ 0\ 0\ 0\ \ 1\ 0\ 0 \\ 0\ \ 1\ 0\ 0\ 0\ -1\ 0\ 0 \\ 0\ -1\ 0\ 0\ 0\ \ 1\ 0\ 0 \\ 0\ \ 1\ 0\ 0\ 0\ -1\ 0\ 0 \\ 0\ -1\ 0\ 0\ 0\ \ 1\ 0\ 0 \end{bmatrix},$$

$$[\chi'_{1,4,2}] = \begin{bmatrix} 0\ 0\ \ 1\ 0\ 0\ 0\ -1\ 0 \\ 0\ 0\ -1\ 0\ 0\ 0\ \ 1\ 0 \\ 0\ 0\ \ 1\ 0\ 0\ 0\ -1\ 0 \\ 0\ 0\ -1\ 0\ 0\ 0\ \ 1\ 0 \\ 0\ 0\ \ 1\ 0\ 0\ 0\ -1\ 0 \\ 0\ 0\ -1\ 0\ 0\ 0\ \ 1\ 0 \\ 0\ 0\ \ 1\ 0\ 0\ 0\ -1\ 0 \\ 0\ 0\ -1\ 0\ 0\ 0\ \ 1\ 0 \end{bmatrix}, \quad [\chi'_{1,4,3}] = \begin{bmatrix} 0\ 0\ 0\ \ 1\ 0\ 0\ 0\ -1 \\ 0\ 0\ 0\ -1\ 0\ 0\ 0\ \ 1 \\ 0\ 0\ 0\ \ 1\ 0\ 0\ 0\ -1 \\ 0\ 0\ 0\ -1\ 0\ 0\ 0\ \ 1 \\ 0\ 0\ 0\ \ 1\ 0\ 0\ 0\ -1 \\ 0\ 0\ 0\ -1\ 0\ 0\ 0\ \ 1 \\ 0\ 0\ 0\ \ 1\ 0\ 0\ 0\ -1 \\ 0\ 0\ 0\ -1\ 0\ 0\ 0\ \ 1 \end{bmatrix}.$$

$(1+4)7 + 1 = 36$ parallel rays are considered for the $(1,4)$-projection. The control points of the set of rays are defined as shown below:

$$\begin{bmatrix} 0\ \ 1\ \ 2\ \ 3\ \ 4\ \ 5\ \ 6\ \ 7 \\ \ \ \ \ \ \ \ \ \ \ \ \ 8\ \ 9\ 10\ 11 \\ \ \ \ \ \ \ \ \ 12\ 13\ 14\ 15 \\ \ \ \ \ \ \ \ \ 16\ 17\ 18\ 19 \\ \ \ \ \ \ \ \ \ 20\ 21\ 22\ 23 \\ \ \ \ \ \ \ \ \ 24\ 25\ 26\ 27 \\ \ \ \ \ \ \ \ \ 28\ 29\ 30\ 31 \\ \ \ \ \ \ \ \ \ 32\ 33\ 34\ 35 \end{bmatrix} \qquad \text{(angle of rays is } -\tan^{-1}(1/4) = -14.036°\text{)}.$$

The set of the 36 parallel rays of this projection,

$$l(t) = l_{1,4}(t) = \left\{ (x,y); x + 4y = \frac{t}{8} + \frac{5}{16} \right\}, \quad t = 0:35,$$

where $(x,y) \in [0,1] \times [0,1]$, is shown in Figure 3.44. These rays are also shown separately on the masks of the paired functions $\chi'_{1,4,t}$, $t = 0:3$, in Figure 3.45.

We denote by $v_t = v_{1,4}(t)$, where $t = 0:35$, the sums of the discrete image $f_{n,m}$ along these rays. The components of the paired transform with

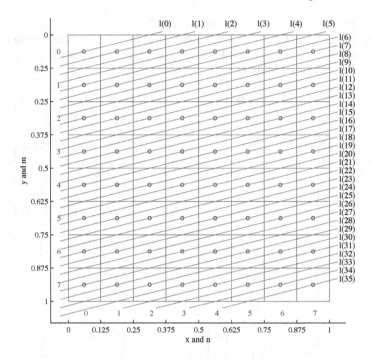

FIGURE 3.44
The sets of rays for the $(1,4)$-projection.

the triplet-numbers $(1,4,t)$ can be calculated as follows:

$$
\begin{aligned}
f'_{1,4,0} &= v_0 - v_4 + v_8 - v_{12} + v_{16} - v_{20} + v_{24} - v_{28} + v_{32}, \\
f'_{1,4,1} &= v_1 - v_5 + v_9 - v_{13} + v_{17} - v_{21} + v_{25} - v_{29} + v_{33}, \\
f'_{1,4,2} &= v_2 - v_6 + v_{10} - v_{14} + v_{18} - v_{22} + v_{26} - v_{30} + v_{34}, \\
f'_{1,4,3} &= v_3 - v_7 + v_{11} - v_{15} + v_{19} - v_{23} + v_{27} - v_{31} + v_{35}.
\end{aligned}
\tag{3.61}
$$

Two masks for the paired functions $\chi'_{2,0,t}$, $t = 0,2$, are shown in Figure 3.46. The corresponding components of the 2-D paired transform are calculated by

$$
\begin{aligned}
f'_{2,0,0} &= (v_0 + v_4 + v_8 + v_{12} + v_{16} + v_{20} + v_{24} + v_{28} + v_{32}) \\
&\quad - (v_2 + v_6 + v_{10} + v_{14} + v_{18} + v_{22} + v_{26} + v_{30} + v_{34}), \\
f'_{2,0,2} &= (v_1 + v_5 + v_9 + v_{13} + v_{17} + v_{21} + v_{25} + v_{29} + v_{33}) \\
&\quad - (v_3 + v_7 + v_{11} + v_{15} + v_{19} + v_{23} + v_{27} + v_{31} + v_{35}).
\end{aligned}
\tag{3.62}
$$

Similarly, for the remaining two components, we have the following: $f'_{4,0,0} = v_0 - v_1 + v_2 - v_3 + \cdots + v_{34} - v_{35}$ and $f'_{0,0,0} = v_0 + v_1 + v_2 + v_3 + \cdots + v_{34} + v_{35}$. Introducing the new variables

$$
a_n = \sum_m v_{n+8m} = \sum_m v_{1,4,n+8m}, \qquad n = 0 : 7,
$$

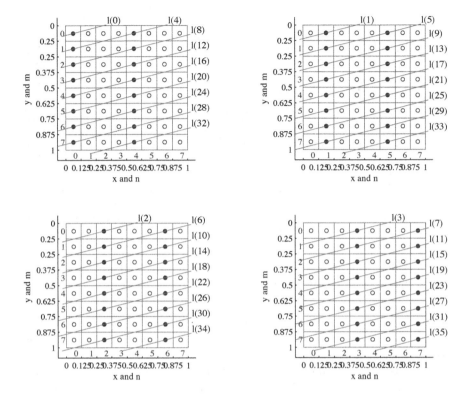

FIGURE 3.45

Four masks with the set of arithmetical rays for the paired functions with triplet-numbers $(1, 4, t)$, $t = 0 : 3$.

the complete system of linear equations for the set $\mathbf{P}(1, 4)$ with eight components $f'_{p_1, s_1, t}$, $(p_1, s_1, t) \in U(1, 4)$, can be written as follows:

$$
\begin{aligned}
f'_{1,4,0} &= a_0 - a_4 \\
f'_{1,4,1} &= a_1 - a_5 \\
f'_{1,4,2} &= a_2 - a_6 \\
f'_{1,4,3} &= a_3 - a_7 \\
f'_{2,0,0} &= (a_0 + a_4) - (a_2 + a_6) \\
f'_{2,0,2} &= (a_1 + a_5) - (a_3 + a_7) \\
f'_{4,0,0} &= (a_0 + a_4) - (a_1 + a_5) + (a_2 + a_6) - (a_3 + a_7) \\
f'_{0,0,0} &= (a_0 + a_4) + (a_1 + a_5) + (a_2 + a_6) + (a_3 + a_7).
\end{aligned}
\tag{3.63}
$$

These calculations correspond to the 8-point discrete paired transform of the vector $\mathbf{a} = (a_0, a_1, a_2, ..., a_7)'$, i.e., $\mathbf{P}(1, 4) = [\chi'_8]\mathbf{a}$.

To express the sums $v(t)$ of the discrete image $f_{n,m}$ through the line-integrals $w(t) = w_{1,4}(t)$, we consider the part of the mask of the paired func-

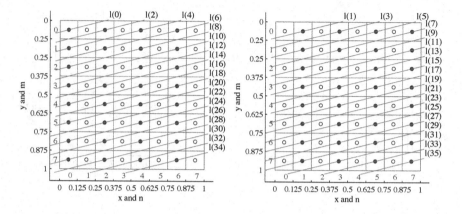

FIGURE 3.46

Two masks with the set of arithmetical rays for the paired functions with triplet-numbers $(2, 0, t)$, $t = 0, 1$.

tion with number $(1, 4, 0)$, which is shown in Figure 3.47. The geometrical ray $l(29)$ between two points of the discrete image, $(0, 7)$ and $(5, 6)$ passes through six image elements of numbers $(0, 7), (1, 7), (2, 7), (3, 7), (3, 6)$, and $(4, 6)$. The length of the intersection $\Delta l_{0,7}$ of the ray with the IE of number $(0, 7)$ is twice the length of intersection with each IE of number $(3, 7)$ and $(3, 6)$, but equal to the length of intersection with the IE of numbers $(1, 7)$ and $(2, 7)$. Since $\Delta l_{0,7} = \sqrt{17}/4\Delta x$, we can write the following:

$$w(29) = w_{1,4}(29) = \frac{\Delta l_{0,7}}{(\Delta x)^2} \left[v(28) + v(29) + v(30) + \frac{1}{2}v(31) + \frac{1}{2}v(27) \right].$$

Similar equations can be used for calculating other integrals $w(t)$, $t = 0 : 35$, and then the sums $v(t)$ can be found.

Now we consider the geometrical rays l_1 and l_2 between rays $28, 29$, and $29, 30$, respectively. The length of the intersection of ray l_1 with the image elements of numbers $(0, 7), (1, 7), (2, 7)$, and $(3, 6)$ is the same, $\sqrt{17}/4\Delta x$. The length of the intersection of ray l_2 with each image element of number $(0, 7), (1, 7), (2, 7)$, and $(3, 7)$ is also $\sqrt{17}/4\Delta x$. Therefore, the integrals w_1 and w_2 of the image along the shifted rays l_1 and l_2, respectively, can be written as follows:

$$w_1 = \frac{\sqrt{17}/4\Delta x}{(\Delta x)^2} \left[v(28) + v(29) + v(30) + v(27) \right]$$

$$= 2\sqrt{17} \left[v(27) + v(28) + v(29) + v(30) \right],$$

$$w_2 = 2\sqrt{17} \left[v(28) + v(29) + v(30) + v(31) \right].$$

These equations can be generalized and we consider two ways to do that.

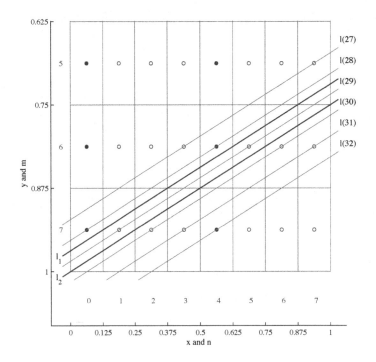

FIGURE 3.47
Set of geometrical rays for calculating line-integrals when $(p, s) = (1, 4)$.

Case 1: The integral w_1 can be denoted by $w(30)$, and w_2 by $w(31)$.
Case 2: The integral w_1 can be denoted by $w(27)$, and w_2 by $w(28)$.

In both cases, a simple solution for sums $v(t)$ can be obtained. Indeed, in the first case, we consider a new set of parallel rays defined by the shifting, $t \to t - 3/2$,

$$\tilde{l}(t) = l_{1,4}(t - 3/2) = \left\{ (x, y); \ x + 4y = \frac{t}{8} + \frac{1}{8} \right\}, \quad t = 0 : 35,$$

which is shown in Figure 3.48. In the second case, we consider the set of parallel rays defined by the shifting, $t \to t + 3/2$,

$$\tilde{l}(t) = l_{1,4}(t + 3/2) = \left\{ (x, y); \ x + 4y = \frac{t}{8} + \frac{1}{2} \right\}, \quad t = 0 : 35,$$

which is shown in Figure 3.49. The line-integrals along the new rays are defined as $w(t) = w(\tilde{l}(t))$, where $t = 0 : 35$.

In the first case, for the integrals along the set of geometrical rays, we can write the following equations:

$$w(t) = 2\sqrt{17} \left[v(t) + v(t - 1) + v(t - 2) + v(t - 3) \right], \quad t = 0 : 35,$$

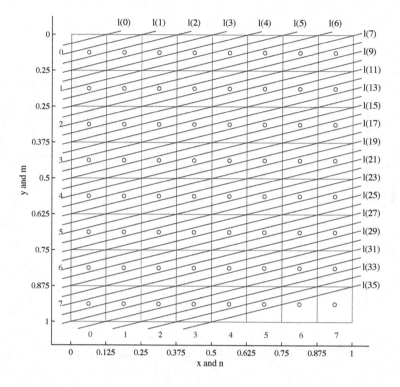

FIGURE 3.48
The first set of geometrical rays for the $(1, 4)$-projection.

where $v(-1) = v(-2) = v(-3) = 0$. This system of linear equations can be written in matrix form as

$$\mathbf{w} = 2\sqrt{17}\mathbf{A}\mathbf{v} = 2\sqrt{17}
\begin{bmatrix}
1\ 0\ 0\ 0\ 0\ 0\ \dots\ 0\ 0\ 0\ 0\ 0 \\
1\ 1\ 0\ 0\ 0\ 0\ \dots\ 0\ 0\ 0\ 0\ 0 \\
1\ 1\ 1\ 0\ 0\ 0\ \dots\ 0\ 0\ 0\ 0\ 0 \\
1\ 1\ 1\ 1\ 0\ 0\ \dots\ 0\ 0\ 0\ 0\ 0 \\
0\ 1\ 1\ 1\ 1\ 0\ \dots\ 0\ 0\ 0\ 0\ 0 \\
0\ 0\ 1\ 1\ 1\ 1\ \dots\ 0\ 0\ 0\ 0\ 0 \\
\cdot\ \cdot\ \cdot\ \cdot\ \cdot\ \cdot\ \dots\ \cdot\ \cdot\ \cdot\ \cdot\ \cdot \\
\cdot\ \cdot\ \cdot\ \cdot\ \cdot\ \cdot\ \dots\ \cdot\ \cdot\ \cdot\ 0\ 0 \\
0\ 0\ 0\ 0\ 0\ 0\ \dots\ 1\ 1\ 1\ 1\ 0 \\
0\ 0\ 0\ 0\ 0\ 0\ \dots\ 0\ 1\ 1\ 1\ 1
\end{bmatrix}
\mathbf{v}, \qquad (3.64)$$

where $\mathbf{w} = (w_0, w_1, ..., w_{35})'$ and $\mathbf{v} = (v_0, v_1, ..., v_{35})'$. The above Toeplitz

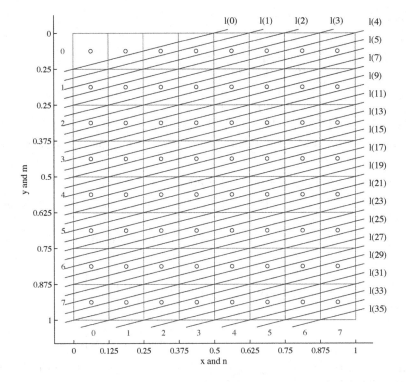

FIGURE 3.49
The second set of geometrical rays for the $(1, 4)$-projection.

matrix 36×36 has the following inverse lower triangle Toeplitz matrix:

$$
\mathbf{A}^{-1} =
\begin{bmatrix}
1 & 0 & 0 & 0 & 0 & 0 & 0... & 0 & 0 & 0 & 0 \\
-1 & 1 & 0 & 0 & 0 & 0 & 0... & 0 & 0 & 0 & 0 \\
0 & -1 & 1 & 0 & 0 & 0 & 0... & 0 & 0 & 0 & 0 \\
0 & 0 & -1 & 1 & 0 & 0 & 0... & 0 & 0 & 0 & 0 \\
1 & 0 & 0 & -1 & 1 & 0 & 0... & 0 & 0 & 0 & 0 \\
-1 & 1 & 0 & 0 & -1 & 1 & 0... & 0 & 0 & 0 & 0 \\
0 & -1 & 1 & 0 & 0 & -1 & 1... & 0 & 0 & 0 & 0 \\
0 & 0 & -1 & 1 & 0 & 0 & -1... & 0 & 0 & 0 & 0 \\
... & ... & ... & ... & ... & ... & ...... & ... & ... & ...... \\
1 & 0 & 0 & -1 & 1 & 0 & 0... & 1 & 0 & 0 & 0 \\
-1 & 1 & 0 & 0 & -1 & 1 & 0... & -1 & 1 & 0 & 0 \\
0 & -1 & 1 & 0 & 0 & -1 & 1... & 0 & -1 & 1 & 0 \\
0 & 0 & -1 & 1 & 0 & 0 & -1... & 0 & 0 & -1 & 1
\end{bmatrix}. \qquad (3.65)
$$

From the solution of the equation $\mathbf{b} = \mathbf{A}^{-1}\mathbf{w}$, we obtain the sums $\mathbf{v} = \mathbf{b}/(2\sqrt{17})$. The vector $\mathbf{b}$ can be calculated by using the following recurrent procedure:

$$
\begin{aligned}
b_0 &= w(0),\\
b_1 &= w(1) - b_0,\\
b_2 &= w(2) - (b_1 + b_0),\\
b_3 &= w(3) - (b_2 + b_1 + b_0),\\
b_4 &= w(4) - (b_3 + b_2 + b_1),\\
b_t &= w(t) - (b_{t-1} + b_{t-2} + b_{t-3}), \quad t = 5, 6, \ldots, 35.
\end{aligned}
\tag{3.66}
$$

In the second case, the integrals along the set of geometrical rays can be defined by equations

$$
w(t) = 2\sqrt{17}\,[v(t) + v(t+1) + v(t+2) + v(t+3)], \quad t = 0 : 35,
$$

where $v(36) = v(37) = v(38) = 0$. This system of linear equations can be written in matrix form as

$$
\mathbf{w} = 2\sqrt{17}\mathbf{A}\mathbf{v} = 2\sqrt{17}
\begin{bmatrix}
1 & 1 & 1 & 1 & 0 & 0 & \ldots & 0 & 0 & 0\\
0 & 1 & 1 & 1 & 1 & 0 & \ldots & 0 & 0 & 0\\
0 & 0 & 1 & 1 & 1 & 0 & \ldots & 0 & 0 & 0\\
0 & 0 & 0 & 1 & 1 & 1 & \ldots & 0 & 0 & 0\\
\vdots & \vdots & \vdots & \vdots & \vdots & \vdots & \ldots & \vdots & \vdots\\
0 & 0 & 0 & 0 & 0 & 0 & \ldots & 0 & 1 & 1\\
0 & 0 & 0 & 0 & 0 & 0 & \ldots & 0 & 0 & 1
\end{bmatrix}
\mathbf{v},
\tag{3.67}
$$

where $\mathbf{w} = (w_0, w_1, \ldots, w_{35})'$ and $\mathbf{v} = (v_0, v_1, \ldots, v_{35})'$. The above Toeplitz matrix 36×36 has the following triangle inverse matrix:

$$
\mathbf{A}^{-1} =
\begin{bmatrix}
1 & -1 & 0 & 0 & 1 & -1 & 0 & 0 & 1 & -1 & 0 & \ldots & 1 & -1 & 0 & 0\\
0 & 1 & -1 & 0 & 0 & 1 & -1 & 0 & 0 & 1 & -1 & \ldots & 0 & 1 & -1 & 0\\
0 & 0 & 1 & -1 & 0 & 0 & 1 & -1 & 0 & 0 & 1 & \ldots & 0 & 0 & 1 & -1\\
0 & 0 & 0 & 1 & -1 & 0 & 0 & 1 & -1 & 0 & 0 & \ldots & -1 & 0 & 0 & 1\\
0 & 0 & 0 & 0 & 1 & -1 & 0 & 0 & 1 & -1 & 0 & \ldots & 1 & -1 & 0 & 0\\
0 & 0 & 0 & 0 & 0 & 1 & -1 & 0 & 0 & 1 & -1 & \ldots & 0 & 1 & -1 & 0\\
. & . & . & . & . & . & . & . & . & . & . & \ldots & . & . & . & .\\
0 & 0 & 0 & 0 & 0 & 0 & 0 & 0 & 0 & 0 & 1 & \ldots & 0 & 0 & 1 & -1\\
0 & 0 & 0 & 0 & 0 & 0 & 0 & 0 & 0 & 0 & 0 & \ldots & -1 & 0 & 0 & 1\\
0 & 0 & 0 & 0 & 0 & 0 & 0 & 0 & 0 & 0 & 0 & \ldots & 1 & -1 & 0 & 0\\
0 & 0 & 0 & 0 & 0 & 0 & 0 & 0 & 0 & 0 & 0 & \ldots & 0 & 1 & -1 & 0\\
0 & 0 & 0 & 0 & 0 & 0 & 0 & 0 & 0 & 0 & 0 & \ldots & 0 & 0 & 1 & -1\\
0 & 0 & 0 & 0 & 0 & 0 & 0 & 0 & 0 & 0 & 0 & \ldots & 0 & 0 & 0 & 1
\end{bmatrix}.
\tag{3.68}
$$

The equation $\mathbf{b} = \mathbf{A}^{-1}\mathbf{w}$ can be written in the following recurrent form:

$$
\begin{aligned}
b_{35} &= w(35),\\
b_{34} &= w(34) - b_{35},\\
b_{33} &= w(33) - (b_{34} + b_{35}),\\
b_{32} &= w(32) - (b_{33} + b_{34} + b_{35}),\\
b_{31} &= w(31) - (b_{32} + b_{33} + b_{34}),\\
b_t &= w(t) - (b_{t+1} + b_{t+2} + b_{t+3}), \quad t = 30, 29, \ldots, 1, 0.
\end{aligned}
\tag{3.69}
$$

Therefore, the sums are calculated by $\mathbf{v} = \mathbf{b}/(2\sqrt{17})$.

Now we consider the first four matrices of the subset $U(4,1)$,

$$[\chi'_{4,1,0}] = \begin{bmatrix} 1 & -1 & 1 & -1 & 1 & -1 & 1 & -1 \\ 0 & 0 & 0 & 0 & 0 & 0 & 0 & 0 \\ 0 & 0 & 0 & 0 & 0 & 0 & 0 & 0 \\ 0 & 0 & 0 & 0 & 0 & 0 & 0 & 0 \\ -1 & 1 & -1 & 1 & -1 & 1 & -1 & 1 \\ 0 & 0 & 0 & 0 & 0 & 0 & 0 & 0 \\ 0 & 0 & 0 & 0 & 0 & 0 & 0 & 0 \\ 0 & 0 & 0 & 0 & 0 & 0 & 0 & 0 \end{bmatrix}, \quad [\chi'_{4,1,1}] = \begin{bmatrix} 0 & 0 & 0 & 0 & 0 & 0 & 0 & 0 \\ 1 & -1 & 1 & -1 & 1 & -1 & 1 & -1 \\ 0 & 0 & 0 & 0 & 0 & 0 & 0 & 0 \\ 0 & 0 & 0 & 0 & 0 & 0 & 0 & 0 \\ 0 & 0 & 0 & 0 & 0 & 0 & 0 & 0 \\ -1 & 1 & -1 & 1 & -1 & 1 & -1 & 1 \\ 0 & 0 & 0 & 0 & 0 & 0 & 0 & 0 \\ 0 & 0 & 0 & 0 & 0 & 0 & 0 & 0 \end{bmatrix}$$

$$[\chi'_{4,1,2}] = \begin{bmatrix} 0 & 0 & 0 & 0 & 0 & 0 & 0 & 0 \\ 0 & 0 & 0 & 0 & 0 & 0 & 0 & 0 \\ 1 & -1 & 1 & -1 & 1 & -1 & 1 & -1 \\ 0 & 0 & 0 & 0 & 0 & 0 & 0 & 0 \\ 0 & 0 & 0 & 0 & 0 & 0 & 0 & 0 \\ 0 & 0 & 0 & 0 & 0 & 0 & 0 & 0 \\ -1 & 1 & -1 & 1 & -1 & 1 & -1 & 1 \\ 0 & 0 & 0 & 0 & 0 & 0 & 0 & 0 \end{bmatrix}, \quad [\chi'_{4,1,3}] = \begin{bmatrix} 0 & 0 & 0 & 0 & 0 & 0 & 0 & 0 \\ 0 & 0 & 0 & 0 & 0 & 0 & 0 & 0 \\ 0 & 0 & 0 & 0 & 0 & 0 & 0 & 0 \\ 1 & -1 & 1 & -1 & 1 & -1 & 1 & -1 \\ 0 & 0 & 0 & 0 & 0 & 0 & 0 & 0 \\ 0 & 0 & 0 & 0 & 0 & 0 & 0 & 0 \\ 0 & 0 & 0 & 0 & 0 & 0 & 0 & 0 \\ -1 & 1 & -1 & 1 & -1 & 1 & -1 & 1 \end{bmatrix}$$

Thirty-six rays of this projection are used to define all components of the 2-D paired transforms with triplet-numbers of $U(4,1)$. We define the control points of the set of rays as shown below:

$$\begin{bmatrix} 0 & \cdot & \cdot & \cdot & \cdot & \cdot & \cdot & \cdot \\ 1 & \cdot & \cdot & \cdot & \cdot & \cdot & \cdot & \cdot \\ 2 & \cdot & \cdot & \cdot & \cdot & \cdot & \cdot & \cdot \\ 3 & \cdot & \cdot & \cdot & \cdot & \cdot & \cdot & \cdot \\ 5 & 8 & 12 & 16 & 20 & 24 & 28 & 32 \\ 5 & 9 & 13 & 17 & 21 & 25 & 29 & 33 \\ 6 & 10 & 14 & 18 & 22 & 26 & 30 & 34 \\ 7 & 11 & 15 & 19 & 23 & 27 & 31 & 35 \end{bmatrix}$$
(angle of rays is $-\tan^{-1}(4) = -75.964°$).

The set of the 36 rays of this projection,

$$l(t) = l_{4,1}(t) = \left\{ (x,y); \ 4x + y = \frac{t}{8} + \frac{5}{16} \right\}, \quad t = 0:35,$$

where $(x,y) \in [0,1] \times [0,1]$, is shown in Figure 3.50 on the masks of the paired functions $\chi'_{4,1,t}$, $t = 0:3$.

The components of the 2-D paired transform with the triplet-numbers $(4,1,t)$ are calculated by

$$\begin{aligned}
f'_{4,1,0} &= v_0 - v_4 + v_8 - v_{12} + v_{16} - v_{20} + v_{24} - v_{28} + v_{32} \\
f'_{4,1,1} &= v_1 - v_5 + v_9 - v_{13} + v_{17} - v_{21} + v_{25} - v_{29} + v_{33} \\
f'_{4,1,2} &= v_2 - v_6 + v_{10} - v_{14} + v_{18} - v_{22} + v_{26} - v_{30} + v_{34} \\
f'_{4,1,3} &= v_3 - v_7 + v_{11} - v_{15} + v_{19} - v_{23} + v_{27} - v_{31} + v_{35}
\end{aligned} \tag{3.70}$$

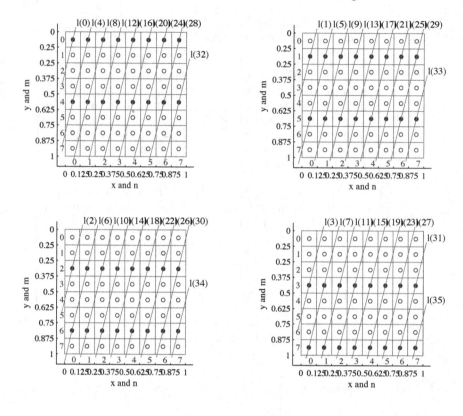

FIGURE 3.50
Masks and rays for calculating line-integrals for the $(4, 1)$-projection.

where the sums of the discrete image along the rays are denoted by $v_t = v_{4,1}(t)$, $t = 0 : 35$. With the new variables

$$a_n = \sum_m v_{n+8m} = \sum_m v_{4,1,n+8m}, \qquad n = 0 : 7,$$

the complete system of linear equations, which defines the subset $\mathbf{P}(4,1)$ of components of the 2-D paired transform with the triplet-numbers of the subset

$U(4, 1)$, can be written as follows:

$$\begin{aligned}
f'_{4,1,0} &= a_0 - a_4 \\
f'_{4,1,1} &= a_1 - a_5 \\
f'_{4,1,2} &= a_2 - a_6 \\
f'_{4,1,3} &= a_3 - a_7 \\
f'_{0,2,0} &= (a_0 + a_4) - (a_2 + a_6) \\
f'_{0,2,2} &= (a_1 + a_5) - (a_3 + a_7) \\
f'_{0,4,0} &= (a_0 + a_4) - (a_1 + a_5) + (a_2 + a_6) - (a_3 + a_7) \\
f'_{0,0,0} &= (a_0 + a_4) + (a_1 + a_5) + (a_2 + a_6) + (a_3 + a_7).
\end{aligned} \tag{3.71}$$

These calculations correspond to the 8-point discrete paired transform of the vector $\mathbf{a} = (a_0, a_1, a_2, ..., a_7)'$,

$$\mathbf{P}(4, 1) = [\chi'_8]\mathbf{a}.$$

To express the sums $v(t)$ of the discrete image $f_{n,m}$ through the line-integrals $w(t)$, we consider in detail the part of the mask of the paired function with number $(4, 1, 0)$, which is shown in Figure 3.51. The geometrical ray l_1 between rays $l(6)$ and $l(7)$ passes through six image elements of numbers $(0, 7), (0, 6), (0, 5), (1, 4), (1, 3), (1, 2), (1, 1)$, and $(2, 0)$. The length of the intersection $\Delta l_{0,7}$ of the ray with the IE of number $(0, 7)$ equals the length of the intersection of the ray with each of the remaining seven IEs. Since $\Delta l_{0,7} = \sqrt{17}/4\Delta x$, we can write the following:

$$w_1 = w(l_1) = \frac{\Delta l_{0,7}}{(\Delta x)^2} [v(7) + v(6) + v(5) + v(8)].$$

For the geometrical ray l_2, which is between rays $l(7)$ and $l(8)$, we obtain

$$w_2 = w(l_2) = \frac{\Delta l_{0,7}}{(\Delta x)^2} [v(7) + v(6) + v(9) + v(8)].$$

These equations can be used for calculating other integrals $w(t)$, $t = 0 : 35$, and then the sums $v(t)$ can be found. Similar to the $(p, s) = (1, 4)$ case, these equations can be generalized, and we consider two ways to do that.

Case 1: The line-integral w_1 can be denoted by $w(8)$, and w_2 by $w(9)$.

Case 2: The line-integral w_1 can be denoted by $w(5)$, and w_2 by $w(6)$.

In the first case, we consider the set of geometrical parallel rays, which are defined by the shifting, $t \to t - 3/2$,

$$\tilde{l}(t) = l_{4,1}(t - 3/2) = \left\{ (x, y);\ 4x + 1y = \frac{t}{8} + \frac{1}{8} \right\}, \quad t = 0 : 35.$$

These rays are shown in Figure 3.52, and the original set of arithmetical rays is shown in Figure 3.53.

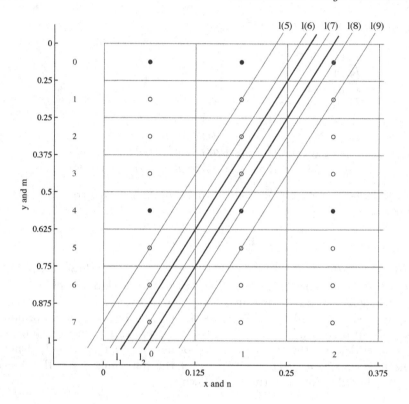

FIGURE 3.51
Set of geometrical rays for calculating line-integrals when $(p, s) = (4, 1)$.

The line-integrals along the set of geometrical rays can be described by the following equations:

$$w(t) = w(\tilde{l}(t)) = 2\sqrt{17}\left[v(t) + v(t-1) + v(t-2) + v(t-3)\right], \quad t = 0 : 35,$$

where $v(-1) = v(-2) = v(-3) = 0$. As shown in (3.66), the solution of this system of linear equations is calculated by $\mathbf{v} = \mathbf{b}/(2\sqrt{17})$, where the vector $\mathbf{b}$ can be calculated by means of the following recursion:

$$
\begin{aligned}
b_0 &= w(0), \\
b_1 &= w(1) - b_0, \\
b_2 &= w(2) - (b_1 + b_0), \\
b_3 &= w(3) - (b_2 + b_1 + b_0), \\
b_4 &= w(4) - (b_3 + b_2 + b_1), \\
b_t &= w(t) - (b_{t-1} + b_{t-2} + b_{t-3}), \quad t = 5, 6, \ldots, 35.
\end{aligned}
\tag{3.72}
$$

For the second case, we consider the set of geometrical parallel rays defined

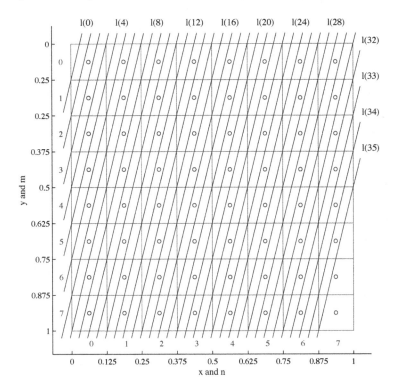

FIGURE 3.52
The set of geometrical rays for the $(4, 1)$-projection.

by the shifting, $t \to t + 3/2$,

$$\tilde{l}(t) = l_{4,1}(t + 3/2) = \left\{ (x, y); \; 4x + 1y = \frac{t}{8} + \frac{1}{2} \right\}, \quad t = 0 : 35.$$

The integrals along the set of geometrical rays can be defined by the following equations:

$$w(t) = w(\tilde{l}(t)) = 2\sqrt{17}\left[v(t) + v(t+1) + v(t+2) + v(t+3)\right], \quad t = 0 : 35,$$

where $v(36) = v(37) = v(38) = 0$. The sums $v(t)$ are defined by $v(t) = b_t/(2\sqrt{17})$, where the components b_t are calculated by the following recurrent procedure:

$$
\begin{aligned}
b_{35} &= w(35), \\
b_{34} &= w(34) - b_{35}, \\
b_{33} &= w(33) - (b_{34} + b_{35}), \\
b_{32} &= w(32) - (b_{33} + b_{34} + b_{35}), \\
b_{31} &= w(31) - (b_{32} + b_{33} + b_{34}), \\
b_{t} &= w(t) - (b_{t+1} + b_{t+2} + b_{t+3}), \quad t = 30, 29, \ldots, 0.
\end{aligned}
\tag{3.73}
$$

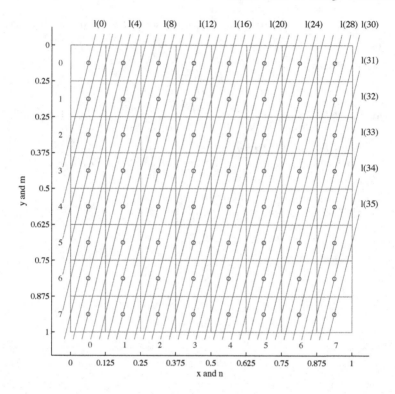

FIGURE 3.53
The set of arithmetical rays for the $(4, 1)$-projection.

The set of 36 geometrical rays $\tilde{l}(t)$ is shown in Figure 3.54.

3.5.7 $(1, 5)$-projection

The set of triplet-numbers generated by the frequency-point $(p, s) = (1, 5)$ is

$$
U(1,5) = \begin{cases} (1,5,0),(1,5,1),(1,5,2),(1,5,3), \\ (2,2,0),(2,2,2), \\ (4,4,0), \\ (0,0,0). \end{cases}
$$

The masks of the paired functions $\chi'_{4,4,0}$ and $\chi'_{0,0,0}$, as well as $\chi'_{2,2,0}$ and $\chi'_{2,2,2}$ were given above. Therefore, we consider only masks with the first four

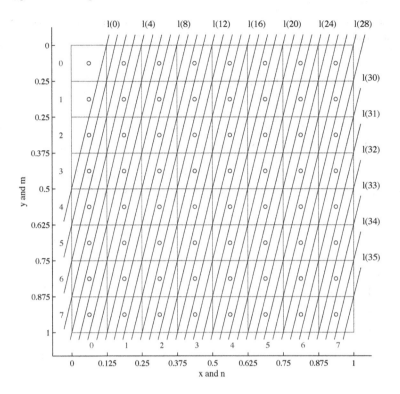

FIGURE 3.54

The 2nd set of geometrical rays for the $(4, 1)$-projection.

triplet-numbers of the subset $U(1, 5)$,

$$[\chi'_{1,5,0}] = \begin{bmatrix} 1 & 0 & 0 & 0 & -1 & 0 & 0 & 0 \\ 0 & 0 & 0 & 1 & 0 & 0 & 0 & -1 \\ 0 & 0 & -1 & 0 & 0 & 0 & 1 & 0 \\ 0 & 1 & 0 & 0 & 0 & -1 & 0 & 0 \\ -1 & 0 & 0 & 0 & 1 & 0 & 0 & 0 \\ 0 & 0 & 0 & -1 & 0 & 0 & 0 & 1 \\ 0 & 0 & 1 & 0 & 0 & 0 & -1 & 0 \\ 0 & -1 & 0 & 0 & 0 & 1 & 0 & 0 \end{bmatrix}, \quad [\chi'_{1,5,1}] = \begin{bmatrix} 0 & 1 & 0 & 0 & 0 & -1 & 0 & 0 \\ -1 & 0 & 0 & 0 & 1 & 0 & 0 & 0 \\ 0 & 0 & 0 & -1 & 0 & 0 & 0 & 1 \\ 0 & 0 & 1 & 0 & 0 & 0 & -1 & 0 \\ 0 & -1 & 0 & 0 & 0 & 1 & 0 & 0 \\ 1 & 0 & 0 & 0 & -1 & 0 & 0 & 0 \\ 0 & 0 & 0 & 1 & 0 & 0 & 0 & -1 \\ 0 & 0 & -1 & 0 & 0 & 0 & 1 & 0 \end{bmatrix}$$

$$[\chi'_{1,5,2}] = \begin{bmatrix} 0 & 0 & 1 & 0 & 0 & 0 & -1 & 0 \\ 0 & -1 & 0 & 0 & 0 & 1 & 0 & 0 \\ 1 & 0 & 0 & 0 & -1 & 0 & 0 & 0 \\ 0 & 0 & 0 & 1 & 0 & 0 & 0 & -1 \\ 0 & 0 & -1 & 0 & 0 & 0 & 1 & 0 \\ 0 & 1 & 0 & 0 & 0 & -1 & 0 & 0 \\ -1 & 0 & 0 & 0 & 1 & 0 & 0 & 0 \\ 0 & 0 & 0 & -1 & 0 & 0 & 0 & 1 \end{bmatrix}, \quad [\chi'_{1,5,3}] = \begin{bmatrix} 0 & 0 & 0 & 1 & 0 & 0 & 0 & -1 \\ 0 & 0 & -1 & 0 & 0 & 0 & 1 & 0 \\ 0 & 1 & 0 & 0 & 0 & -1 & 0 & 0 \\ -1 & 0 & 0 & 0 & 1 & 0 & 0 & 0 \\ 0 & 0 & 0 & -1 & 0 & 0 & 0 & 1 \\ 0 & 0 & 1 & 0 & 0 & 0 & -1 & 0 \\ 0 & -1 & 0 & 0 & 0 & 1 & 0 & 0 \\ 1 & 0 & 0 & 0 & -1 & 0 & 0 & 0 \end{bmatrix}$$

It should be noted that the same set of masks can be referred to as the masks for the subset $U(5,1)$,

$$U(5,1) = \begin{cases} (5,1,0),(5,1,1),(5,1,2),(5,1,3), \\ (2,2,0),(2,2,2), \\ (4,4,0), \\ (0,0,0). \end{cases}$$

Indeed, the following is valid: $\chi'_{5,1,2t} = \chi'_{1,5,2t}$ and $\chi'_{5,1,2t+1} = -\chi'_{1,5,2t+1}$, for $t = 0,1$.

The set of all $(p+s)(N-1)+1 = 6(7)+1 = 43$ arithmetical rays of the projection by the angle $\pi - \tan^{-1}(5)$ is shown in Figure 3.55.

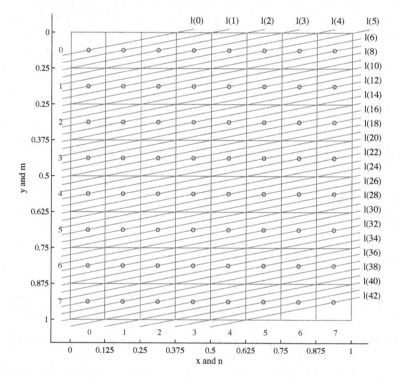

FIGURE 3.55
The set of arithmetical rays for the $(5,1)$-projection.

To reduce this large number of rays, we consider another direction, or projection which defines the same set of masks of the paired functions. The similar substitution of the projections was described in the $N = 4$ example, for the $(p,s) = (1,3)$ case. In the $(p,s) = (1,5)$ case, this alternative projection is determined by the frequency-point $(p, s - N) = (1, -3)$.

As an example, Figures 3.56 and 3.57 show the mask of the paired function

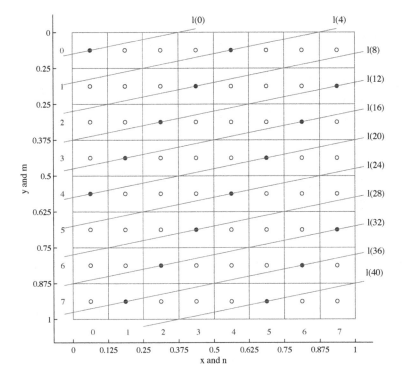

FIGURE 3.56
The mask $[\chi'_{1,5,0}]$ with the set of arithmetical rays of the $(1,5)$-projection.

$\chi'_{1,5,0}$ with the parallel arithmetical rays of two different projections. Eleven rays of the $(1,5)$-projection are shown in the first figure, and seven rays of the $(1,-3)$-projection are shown in the second figure.

The number of parallel rays for the $(1,-3)$-projection equals $(1+3)7+1 = 29$, which is less than 43. The set of all rays for this projection is shown in Figure 3.58. The control points of this set of rays are given below

$$\begin{bmatrix} 0_\bullet & 1_\bullet & 2_\bullet 3_\bullet 4_\bullet 5_\bullet 6_\bullet 7_\bullet \\ -3_\bullet & -2_\bullet & -1_\bullet & \cdot & \cdot & \cdot & \cdot & \cdot \\ -6_\bullet & -5_\bullet & -4_\bullet & \cdot & \cdot & \cdot & \cdot & \cdot \\ -9_\bullet & -8_\bullet & -7_\bullet & \cdot & \cdot & \cdot & \cdot & \cdot \\ -12_\bullet & -11_\bullet & -10_\bullet & \cdot & \cdot & \cdot & \cdot & \cdot \\ -15_\bullet & -14_\bullet & -13_\bullet & \cdot & \cdot & \cdot & \cdot & \cdot \\ -18_\bullet & -17_\bullet & -16_\bullet & \cdot & \cdot & \cdot & \cdot & \cdot \\ -21_\bullet & -20_\bullet & -19_\bullet & \cdot & \cdot & \cdot & \cdot & \cdot \end{bmatrix} \quad \text{(angle of rays is } \tan^{-1}(1/3) = 18.435°\text{)}.$$

The set of the 29 geometrical rays of this projection is described as

$$l(t) = l_{1,-3}(t) = \left\{ (x,y); x - 3y = \frac{t}{8} - \frac{1}{8} \right\}, \quad t = 7:-1:-21, \quad (3.74)$$

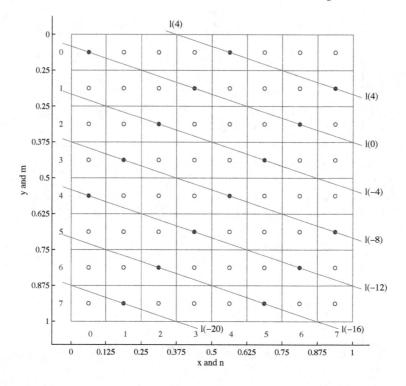

FIGURE 3.57

The mask $[\chi'_{1,5,0}]$ with the set of arithmetical rays of the $(1, -3)$-projection.

where $(x, y) \in [0, 1] \times [0, 1]$. The first six masks for the paired functions are shown in Figure 3.59 and 3.60.

The components of the 2-D paired transform with the triplet-numbers $(1, 5, t)$ are calculated as follows:

$$
\begin{aligned}
f'_{1,5,0} &= -v_4 + v_0 - v_{-4} + v_{-8} - v_{-12} + v_{-16} - v_{-20} \\
&= (v_0 + v_{-8} + v_{-16}) - (v_4 + v_{-4} + v_{-12} + v_{-20}), \\
f'_{1,5,1} &= -v_5 + v_1 - v_{-3} + v_{-7} - v_{-11} + v_{-15} - v_{-19} \\
&= (v_1 + v_{-7} + v_{-15}) - (v_5 + v_{-3} + v_{-11} + v_{-19}), \\
f'_{1,5,2} &= -v_6 + v_2 - v_{-2} + v_{-6} - v_{-10} + v_{-14} - v_{-18} \\
&= (v_2 + v_{-6} + v_{-14}) - (v_6 + v_{-2} + v_{-10} + v_{-18}), \\
f'_{1,5,3} &= -v_7 + v_3 - v_{-1} + v_{-5} - v_{-9} + v_{-13} - v_{-17} + v_{-21} \\
&= (v_3 + v_{-5} + v_{-13} + v_{-21}) - (v_7 + v_{-1} + v_{-9} + v_{-17}),
\end{aligned}
\tag{3.75}
$$

where the sums of the discrete image along the rays are denoted by $v_t = v_{1,-3}(t)$, $t = 7 : -1 : -21$. For the components of the 2-D paired transform

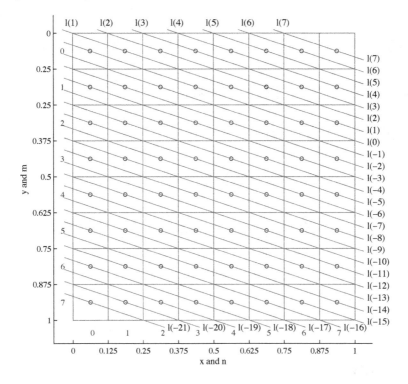

FIGURE 3.58
The set of arithmetical rays for the $(1, -3)$-projection.

with the remaining numbers of the subset $U(1,5)$, we have the following:

$$f'_{2,2,0} = -v_6 + v_4 - v_2 + v_0 - v_{-2} + v_{-4} - v_{-6} + v_{-8} -$$
$$-v_{-10} + v_{-12} - v_{-14} + v_{-16} - v_{-18} + v_{-20}$$
$$= (v_0 + v_{-8} + v_{-16}) + (v_4 + v_{-4} + v_{-12} + v_{-20}) -$$
$$-(v_2 + v_{-6} + v_{-14}) - (v_6 + v_{-2} + v_{-10} + v_{-18}),$$

$$f'_{2,2,2} = -v_7 + v_5 - v_3 + v_1 - v_{-1} + v_{-3} - v_{-5} + v_{-7}$$
$$-v_{-9} + v_{-11} - v_{-13} + v_{-15} - v_{-17} + v_{-19} - v_{-21}$$
$$= (v_1 + v_{-7} + v_{-15}) + (v_5 + v_{-3} + v_{-11} + v_{-19}) -$$
$$-(v_3 + v_{-5} + v_{-13} + v_{-21}) - (v_7 + v_{-1} + v_{-9} + v_{-17}),$$

$$f'_{4,4,0} = -v_7 + v_6 - v_5 + v_4 - v_3 + v_2 - v_1 + v_0 -$$
$$-v_{-1} + v_{-2} - v_{-3} + v_{-4} - v_{-5} + \ldots + v_{-20} - v_{21}$$
$$= (v_0 + v_{-8} + v_{-16}) + (v_4 + v_{-4} + v_{-12} + v_{-20}) -$$
$$-(v_1 + v_{-7} + v_{-15}) - (v_5 + v_{-3} + v_{-11} + v_{-19}) +$$
$$+(v_2 + v_{-6} + v_{-14}) + (v_6 + v_{-2} + v_{-10} + v_{-18}) -$$
$$-(v_3 + v_{-5} + v_{-13} - v_{-21}) - (v_7 + v_{-1} + v_{-9} + v_{-17}),$$

$$f'_{0,0,0} = v_0 + v_1 + v_2 + v_3 + \cdots + v_7 + v_{-1} + v_{-2} + v_{-3} +$$
$$+v_{-4} + v_{-5} + \ldots + v_{-20} + v_{21}.$$

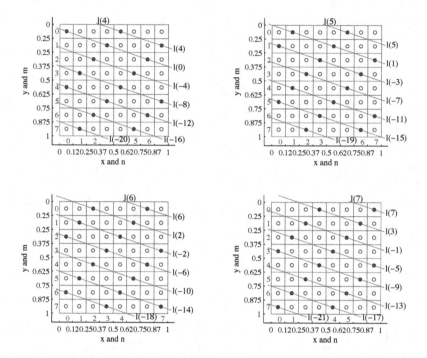

FIGURE 3.59

Four masks $[\chi'_{1,5,k}]$, $k = 0 : 3$, with the arithmetical rays $l_{1,-3}(t)$.

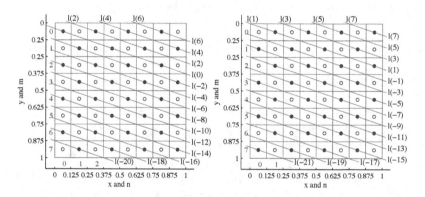

FIGURE 3.60

Two masks $[\chi'_{2,2,2k}]$, $k = 0, 1$, with the arithmetical rays $l_{1,-3}(t)$.

With the new variables

$$a_n = \sum_m v_{n-8m} = \sum_m v_{1,-3,n-8m}, \qquad n = 0 : 7,$$

the complete system of linear equations, for calculating the components of the 2-D paired transform with triplet-numbers of the subset $U(1,5)$, can be written as follows:

$$
\begin{aligned}
f'_{1,5,0} &= a_0 - a_4 \\
f'_{1,5,1} &= a_1 - a_5 \\
f'_{1,5,2} &= a_2 - a_6 \\
f'_{1,5,3} &= a_3 - a_7 \\
f'_{2,2,0} &= (a_0 + a_4) - (a_2 + a_6) \\
f'_{2,2,2} &= (a_1 + a_5) - (a_3 + a_7) \\
f'_{4,4,0} &= (a_0 + a_4) - (a_1 + a_5) + (a_2 + a_6) - (a_3 + a_7) \\
f'_{0,0,0} &= (a_0 + a_4) + (a_1 + a_5) + (a_2 + a_6) + (a_3 + a_7).
\end{aligned}
\tag{3.76}
$$

These calculations correspond to the 8-point discrete paired transform of the vector $\mathbf{a} = (a_0, a_1, a_2, ..., a_7)'$,

$$
\mathbf{P}(1,5) = \mathbf{P}(1,-3) = [\chi'_8]\mathbf{a}.
$$

As in the $(1,3)$ case, we consider the shifted set of the arithmetical rays, $l(t) \to l(t+1)$, as the set of 29 geometrical rays:

$$
\tilde{l}(t) = l_{1,-3}(t+1) = \left\{ (x,y); \; x - 3y = \frac{t}{8} \right\}, \quad t = 7 : -1 : -21,
$$

which are shown in Figure 3.61. We obtain the same system of equations for the line-integrals, as in the $(1,3)$ case considered above (see (3.51)-(3.53)),

$$
w(t) = w(\tilde{l}(t)) = 8\sqrt{10}/3 \, [v(t) + v(t+1) + v(t+2)], \quad t = 7 : -1 : -21,
\tag{3.77}
$$

where $v(8) = v(9) = 0$. The required sums $v(t)$ of the discrete image $f_{n,m}$ are therefore defined as $v(t) = 3b_t/(8\sqrt{10})$, where the components b_t are calculated recursively by

$$
\begin{aligned}
b_7 &= w(7), \\
b_6 &= w(6) - b_7, \\
b_5 &= w(5) - (b_6 + b_7), \\
b_t &= w(t) - (b_{t+1} + b_{t+2}), \quad t = 4,3,2,1,0,-1,\ldots,-20,-21.
\end{aligned}
\tag{3.78}
$$

Remark 3.4 To remove negative numbers in control points, we can consider the change in numbers by $t \to 7 - t$,

$$
\begin{bmatrix}
7 & 6 & 5 & 4 & 3 & 2 & 1 & 0 \\
10 & 9 & 8 & \cdot & \cdot & \cdot & \cdot & \cdot \\
13 & 12 & 11 & \cdot & \cdot & \cdot & \cdot & \cdot \\
16 & 15 & 14 & \cdot & \cdot & \cdot & \cdot & \cdot \\
19 & 18 & 17 & \cdot & \cdot & \cdot & \cdot & \cdot \\
22 & 21 & 20 & \cdot & \cdot & \cdot & \cdot & \cdot \\
25 & 24 & 23 & \cdot & \cdot & \cdot & \cdot & \cdot \\
28 & 27 & 26 & \cdot & \cdot & \cdot & \cdot & \cdot
\end{bmatrix}
\qquad \text{(angle of rays is } \tan^{-1}(1/3) = 18.435°\text{).}
$$

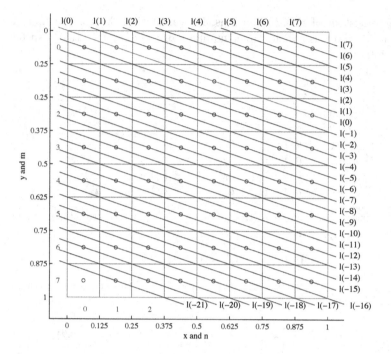

FIGURE 3.61
The set of geometrical rays for the $(1, -3)$-projection.

The set of 29 geometrical rays of this projection in (3.74) can be written as

$$l(t) = l_{1,-3}(7 - t) = \left\{ (x, y); \ x - 3y = -\frac{t}{8} + \frac{3}{4} \right\}, \quad t = 0 : 28.$$

Let $\bar{v}_t = v_{7-t}$ and $\bar{w}_t = w_{7-t}$. Then, the system of equations in (3.77) can be described as

$$\bar{w}(t) = w(\tilde{l}(t)) = 8\sqrt{10}/3 \left[\bar{v}(t) + \bar{v}(t-1) + \bar{v}(t-2) \right], \quad t = 0 : 28, \quad (3.79)$$

and the equations in (3.78) for new coefficients $\bar{b}_t = \bar{v}(t)8\sqrt{10}/3$ are described as

$$\begin{aligned}
\bar{b}_0 &= \bar{w}(0), \\
\bar{b}_1 &= \bar{w}(1) - \bar{b}_0, \\
\bar{b}_2 &= \bar{w}(2) - (\bar{b}_1 + \bar{b}_0), \\
\bar{b}_t &= \bar{w}(t) - (\bar{b}_{t-1} + \bar{b}_{t-2}), \quad t = 3, 4, \ldots, 27, 28.
\end{aligned} \quad (3.80)$$

Remark 3.5 We can consider the shifted set of the arithmetical rays, $l(t) \rightarrow l(t-1)$, as the set of 29 geometrical rays:

$$\tilde{l}(t) = l_{1,-3}(t - 1) = \left\{ (x, y); \ x - 3y = \frac{t}{8} - \frac{1}{4} \right\}, \quad t = 7 : -1 : -21.$$

These parallel rays are shown in Figure 3.62.

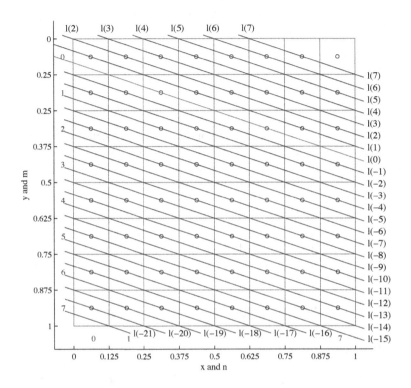

FIGURE 3.62
Another set of geometrical rays for the $(1, -3)$-projection.

The system of equations for the line-integrals is described as

$$w(t) = w(\tilde{l}(t)) = 8\sqrt{10}/3 \left[v(t) + v(t-1) + v(t-2)\right],$$
$$t = 7 : -1 : -21,$$
(3.81)

where $v(-22) = v(-23) = 0$. A similar system of equations was described in (3.57) for the $(3, 1)$-projection with integrals $w(t)$ numbered from 0 to 28. The required sums $v(t)$ of the discrete image are thus calculated by $v(t) = 3b_t/(8\sqrt{10})$, where coefficients b_t are calculated recursively

$$\begin{aligned}
b_{-21} &= w(-21), \\
b_{-20} &= w(-20) - b_{-21}, \\
b_{-19} &= w(-19) - (b_{-20} + b_{-21}), \\
b_t &= w(t) - (b_{t-1} + b_{t-2}), \quad t = -18, -17, \ldots, 6, 7.
\end{aligned}$$
(3.82)

When using the numbering of the control points and rays from 0 to

28, as suggested in Remark 5.1, the solution of the system in (3.57) can be written exactly as in the $(p, s) = (3, 1)$ case (see (3.60)). Then, the system of equations in (3.77) can be described as $\bar{w}(t) = w(\tilde{l}(t)) = 8\sqrt{10}/3 \, [\bar{v}(t) + \bar{v}(t+1) + \bar{v}(t+2)]$, $t = 0 : 28$, when the coefficients $\bar{b}_t$ are calculated by

$$\begin{aligned}
\bar{b}_{28} &= \bar{w}(28), \\
\bar{b}_{27} &= \bar{w}(27) - \bar{b}_{28}, \\
\bar{b}_{26} &= \bar{w}(26) - (\bar{b}_{27} + \bar{b}_{28}), \\
\bar{b}_t &= \bar{w}(t) - (\bar{b}_{t+1} + \bar{b}_{t+2}), \quad t = 25, 24, \ldots, 1, 0.
\end{aligned}$$

Remark 3.6 Since $\chi'_{1,5,t} = (-1)^t \chi'_{5,1,t}$, for $t = 0 : 3$, we can consider the $(-3, 1)$-projection for calculating the components of the 2-D paired transform of the image, whose triplet-numbers are from the subset $U(1, 5)$, or $U(5, 1)$. Four masks with the first four triplet-numbers $(5, 1, t)$ are

$$[\chi'_{5,1,0}] = \begin{bmatrix}
\underline{1} & 0 & 0 & 0 & -1 & 0 & 0 & 0 \\
0 & 0 & 0 & 1 & 0 & 0 & 0 & -1 \\
0 & 0 & -1 & 0 & 0 & 0 & 1 & 0 \\
0 & \underline{1} & 0 & 0 & 0 & -1 & 0 & 0 \\
-1 & 0 & 0 & 0 & 1 & 0 & 0 & 0 \\
0 & 0 & 0 & -1 & 0 & 0 & 0 & 1 \\
0 & 0 & \underline{1} & 0 & 0 & 0 & -1 & 0 \\
0 & -1 & 0 & 0 & 0 & 1 & 0 & 0
\end{bmatrix}, \quad
[\chi'_{5,1,1}] = \begin{bmatrix}
0 & -1 & 0 & 0 & 0 & 1 & 0 & 0 \\
\underline{1} & 0 & 0 & 0 & -1 & 0 & 0 & 0 \\
0 & 0 & 0 & 1 & 0 & 0 & 0 & -1 \\
0 & 0 & -1 & 0 & 0 & 0 & 1 & 0 \\
0 & \underline{1} & 0 & 0 & 0 & -1 & 0 & 0 \\
-1 & 0 & 0 & 0 & 1 & 0 & 0 & 0 \\
0 & 0 & 0 & -1 & 0 & 0 & 0 & 1 \\
0 & 0 & \underline{1} & 0 & 0 & 0 & -1 & 0
\end{bmatrix}$$

$$[\chi'_{5,1,2}] = \begin{bmatrix}
0 & 0 & \underline{1} & 0 & 0 & 0 & -1 & 0 \\
0 & -1 & 0 & 0 & 0 & 1 & 0 & 0 \\
1 & 0 & 0 & 0 & -1 & 0 & 0 & 0 \\
0 & 0 & 0 & \underline{1} & 0 & 0 & 0 & -1 \\
0 & 0 & -1 & 0 & 0 & 0 & 1 & 0 \\
0 & 1 & 0 & 0 & 0 & -1 & 0 & 0 \\
-1 & 0 & 0 & 0 & \underline{1} & 0 & 0 & 0 \\
0 & 0 & -1 & 0 & 0 & 0 & 1 & 0 \, 1
\end{bmatrix}, \quad
[\chi'_{5,1,3}] = \begin{bmatrix}
0 & 0 & 0 & -1 & 0 & 0 & 0 & 1 \\
0 & 0 & \underline{1} & 0 & 0 & 0 & -1 & 0 \\
0 & -1 & 0 & 0 & 0 & 1 & 0 & 0 \\
1 & 0 & 0 & 0 & -1 & 0 & 0 & 0 \\
0 & 0 & 0 & \underline{1} & 0 & 0 & 0 & -1 \\
0 & 0 & -1 & 0 & 0 & 0 & 1 & 0 \\
0 & 1 & 0 & 0 & 0 & -1 & 0 & 0 \\
-1 & 0 & 0 & 0 & \underline{1} & 0 & 0 & 0
\end{bmatrix}$$

In each of these masks, three 1's are underlined, to show the direction of the rays $l_{-3,1}(t)$. The set of 29 geometrical rays for this projection is described as

$$l(t) = l_{-3,1}(t) = \left\{ (x, y); \; -3x + y = \frac{t}{8} - \frac{1}{8} \right\}, \quad t = 7 : -1 : -21.$$

For this set of rays, we consider the following set of control points:

$$\begin{bmatrix}
0 & -3 & -6 & -9 & -12 & -15 & -18 & -21 \\
1 & -2 & -5 & -8 & -11 & -14 & -17 & -20 \\
2 & -1 & -4 & -7 & -10 & -13 & -16 & -19 \\
3 & \cdot & \cdot & \cdot & \cdot & \cdot & \cdot & \cdot \\
4 & \cdot & \cdot & \cdot & \cdot & \cdot & \cdot & \cdot \\
5 & \cdot & \cdot & \cdot & \cdot & \cdot & \cdot & \cdot \\
6 & \cdot & \cdot & \cdot & \cdot & \cdot & \cdot & \cdot \\
7 & \cdot & \cdot & \cdot & \cdot & \cdot & \cdot & \cdot
\end{bmatrix}$$

(angle of rays is $\tan^{-1}(3) = 71.565°$).

It is clear that the diagram of these control points is similar to the diagram of the control points considered above for the $(1, -3)$-projection. Namely, these two diagrams are equivalent up to operation of the matrix transposition. Therefore, our future calculations are similar to the $(1, -3)$ case.

The complete system of linear equations for calculating the components of the 2-D paired transform with triplet-numbers of the subset $U(5, 1)$, or $U(1, 5)$ can be written as follows:

$$
\begin{aligned}
f'_{5,1,0} &= f'_{1,5,0} = a_0 - a_4 \\
f'_{5,1,1} &= -f'_{1,5,1} = a_1 - a_5 \\
f'_{5,1,2} &= f'_{1,5,2} = a_2 - a_6 \\
f'_{5,1,3} &= -f'_{1,5,3} = a_3 - a_7 \\
f'_{2,2,0} &= (a_0 + a_4) - (a_2 + a_6) \\
f'_{2,2,2} &= (a_1 + a_5) - (a_3 + a_7) \\
f'_{4,4,0} &= (a_0 + a_4) - (a_1 + a_5) + (a_2 + a_6) - (a_3 + a_7) \\
f'_{0,0,0} &= (a_0 + a_4) + (a_1 + a_5) + (a_2 + a_6) + (a_3 + a_7)
\end{aligned}
\tag{3.83}
$$

where variables a_n are defined as

$$
a_n = \sum_m v_{n-8m} = \sum_m v_{-3,1,n-8m}, \quad n = 0 : 7.
$$

These calculations correspond to the 8-point discrete paired transform of the vector $\mathbf{a} = (a_0, a_1, a_2, ..., a_7)'$,

$$
\mathbf{P}(5, 1) = \mathbf{P}(-3, 1) = [\chi'_8]\mathbf{a},
$$

which can also be written as $\mathbf{P}(1, 5) = \mathrm{diag}\{1, -1, 1, -1, 1, 1, 1, 1, \}[\chi'_8]\mathbf{a}$.

The set of all rays for the $(-3, 1)$-projection is shown in Figure 3.63.

The components $f'_{5,1,t}$, $t = 0 : 3$, are calculated by the same system of equations as in (3.75), where the sums v_t are defined as $v_{-3,1}(t)$, where $t = 7 : -1 : -21$. For instance, for the mask $[\chi'_{5,1,0}]$ shown in Figure 3.64, we have the following:

$$
\begin{aligned}
f'_{5,1,0} &= -v_4 + v_0 - v_{-4} + v_{-8} - v_{-12} + v_{-16} - v_{-20} \\
&= (v_0 + v_{-8} + v_{-16}) - (v_4 + v_{-4} + v_{-12} + v_{-20}).
\end{aligned}
$$

As in the $(1, 3)$-projection case, the set of 29 geometrical rays can be defined as the shifted set of the arithmetical rays, $l(t) \to l(t + 1)$,

$$
\tilde{l}(t) = l_{-3,1}(t + 1) = \left\{ (x, y); -3x + y = \frac{t}{8} \right\}, \quad t = 7 : -1 : -21,
$$

which are shown in Figure 3.65. We obtain the following system of equations for the line-integrals:

$$
\begin{aligned}
w(t) = w(\tilde{l}(t)) = 8\sqrt{10}/3 \left[v(t) + v(t + 1) + v(t + 2) \right], \\
t = 7 : -1 : -21,
\end{aligned}
\tag{3.84}
$$

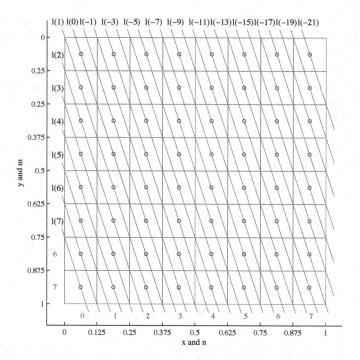

FIGURE 3.63
The set of arithmetical rays for the $(-3, 1)$-projection.

where $v(8) = v(9) = 0$. The required sums $v(t)$ of the discrete image $f_{n,m}$ are therefore defined as $v(t) = 3b_t/(8\sqrt{10})$, where the components b_t are calculated recursively by

$$\begin{aligned}
b_7 &= w(7), \\
b_6 &= w(6) - b_7, \\
b_5 &= w(5) - (b_6 + b_7), \\
b_t &= w(t) - (b_{t+1} + b_{t+2}), \quad t = 4, 3, \ldots, -20, -21.
\end{aligned} \tag{3.85}$$

We now consider another set of control points for the $(-3, 1)$-projection,

$$\begin{bmatrix}
\cdot & \cdot & \cdot & \cdot & \cdot & \cdot & \cdot & 0 \bullet \\
\cdot & \cdot & \cdot & \cdot & \cdot & \cdot & \cdot & 1 \bullet \\
\cdot & \cdot & \cdot & \cdot & \cdot & \cdot & \cdot & 2 \bullet \\
\cdot & \cdot & \cdot & \cdot & \cdot & \cdot & \cdot & 3 \bullet \\
\cdot & \cdot & \cdot & \cdot & \cdot & \cdot & \cdot & 4 \bullet \\
26 \bullet & 23 \bullet & 20 \bullet & 17 \bullet & 14 \bullet & 11 \bullet & 8 \bullet & 5 \bullet \\
27 \bullet & 24 \bullet & 21 \bullet & 18 \bullet & 15 \bullet & 12 \bullet & 9 \bullet & 6 \bullet \\
28 \bullet & 25 \bullet & 22 \bullet & 19 \bullet & 16 \bullet & 13 \bullet & 10 \bullet & 7 \bullet
\end{bmatrix}$$

(angle of rays is $\tan^{-1}(\frac{1}{3} = 18.435°)$).

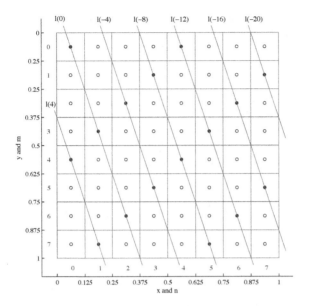

FIGURE 3.64

The mask $[\chi'_{5,1,0}]$ with the rays of the $(-3,1)$-projection.

The set of 29 parallel rays with these control points is shown in Figure 3.66, after shifting them to the right, $t \to t - 1$, to compose the geometrical rays.

The components of the paired transform with the triplet-numbers $(5,1,t)$ are calculated as follows:

$$
\begin{aligned}
f'_{5,1,3} &= v_0 - v_4 + v_8 - v_{12} + v_{16} - v_{20} + v_{24} - v_{28} \\
&= (v_0 + v_8 + v_{16} + v_{24}) - (v_4 + v_{12} + v_{20} + v_{28}), \\
f'_{5,1,0} &= -v_1 + v_5 - v_9 + v_{13} - v_{17} + v_{21} - v_{25} \\
&= -(v_1 + v_9 + v_{17} + v_{25}) + (v_5 + v_{13} + v_{21}), \\
f'_{5,1,1} &= -v_2 + v_6 - v_{10} + v_{14} - v_{18} + v_{22} - v_{26} \\
&= -(v_2 + v_{10} + v_{18} + v_{26}) + (v_6 + v_{14} + v_{22}), \\
f'_{5,1,2} &= -v_3 + v_7 - v_{11} + v_{15} - v_{19} + v_{23} - v_{27} \\
&= -(v_3 + v_{11} + v_{19} + v_{27}) + (v_7 + v_{15} + v_{23}),
\end{aligned}
\tag{3.86}
$$

where the sums of the discrete image along the rays are denoted by v_t, $t = 0 : 28$. Introducing new variables

$$
a_n = \sum_m v_{n+8m} = \sum_m v_{-3,1,n+8m}, \qquad n = 0 : 7,
$$

we can write the complete system of linear equations (3.86) as $f'_{5,1,0} = -a_1 + a_5$, $f'_{5,1,1} = -a_2 + a_6$, $f'_{5,1,2} = -a_3 + a_7$, and $f'_{5,1,3} = a_0 - a_4$.

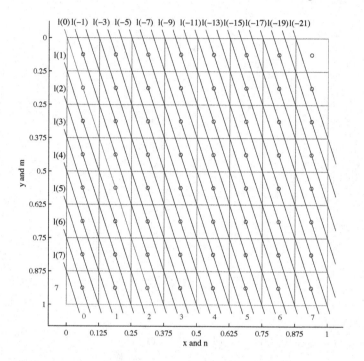

FIGURE 3.65
The set of geometrical rays for the $(-3, 1)$-projection.

The next four components of the 2-D paired transform are

$$f'_{2,2,0} = v_1 + v_5 + v_9 + v_{13} + v_{17} + v_{21} + v_{25}$$
$$- v_3 - v_7 - v_{11} - v_{15} - v_{19} - v_{23} - v_{27}$$
$$= (v_1 + v_9 + v_{17} + v_{25}) + (v_5 + v_{13} + v_{21})$$
$$- (v_3 + v_{11} + v_{19} + v_{27}) - (v_7 + v_{15} + v_{23})$$
$$= (a_1 + a_5) - (a_3 + a_7),$$

$$f'_{2,2,2} = -v_0 - v_4 - v_8 - v_{12} - v_{16} - v_{20} - v_{24} - v_{28}$$
$$+ v_2 + v_6 + v_{10} + v_{14} + v_{18} + v_{22} + v_{26}$$
$$= -(v_0 + v_8 + v_{16} + v_{24}) - (v_4 + v_{12} + v_{20} + v_{28})$$
$$+ (v_2 + v_{10} + v_{18} + v_{26}) + (v_6 + v_{14} + v_{22})$$
$$= -(a_0 + a_4) + (a_2 + a_6),$$

$$f'_{4,4,0} = -v_0 + v_1 - v_2 + v_3 - v_4 + v_5 - \ldots - v_{26} + v_{27} - v_{28}$$
$$= -(a_0 + a_4) + (a_1 + a_5) - (a_2 + a_6) + (a_3 + a_7),$$

$$f'_{0,0,0} = v_0 + v_1 + v_2 + v_3 + v_4 + v_5 + \ldots + v_{26} + v_{27} + v_{28}$$
$$= (a_0 + a_4) + (a_1 + a_5) + (a_2 + a_6) + (a_3 + a_7).$$

$$(3.87)$$

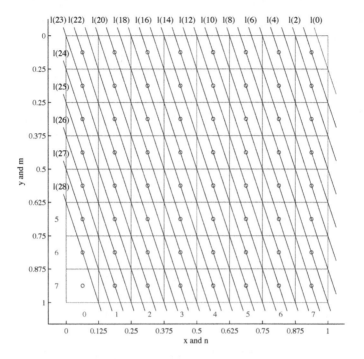

FIGURE 3.66

The set of geometrical rays for the $(-3, 1)$-projection.

Thus, we obtain the following algorithm:

$$
\begin{aligned}
-f'_{5,1,0} &= a_1 - a_5 \\
-f'_{5,1,1} &= a_2 - a_6 \\
-f'_{5,1,2} &= a_3 - a_7 \\
f'_{5,1,3} &= a_0 - a_4 \\
f'_{2,2,0} &= (a_1 + a_5) - (a_3 + a_7) \\
-f'_{2,2,2} &= (a_0 + a_4) - (a_2 + a_6) \\
-f'_{4,4,0} &= (a_0 + a_4) - (a_1 + a_5) + (a_2 + a_6) - (a_3 + a_7) \\
f'_{4,0,0} &= (a_0 + a_4) + (a_1 + a_5) + (a_2 + a_6) + (a_3 + a_7).
\end{aligned}
\tag{3.88}
$$

The calculations correspond to the 8-point discrete paired transform of the vector $\mathbf{a} = (a_0, a_1, a_2, ..., a_7)'$,

$$\mathbf{P}(5, 1) = \operatorname{diag}\{-1, -1, -1, 1, 1, -1, -1, 1\}[\chi'_8]T\mathbf{a},$$

where T is the matrix of the following permutation of the vector:

$$T: \ \{\{a_0, a_1, a_2, a_3\}, \{a_4, a_5, a_6, a_7\}\} \to \{\{a_1, a_2, a_3, a_0\}, \{a_5, a_6, a_7, a_4\}\}.$$

This set of geometrical rays corresponds to the case considered above (see

Figure 3.63), when 29 arithmetical rays $l(t)$, $t = 7 : -1 : -21$, are shifted by $t \to t - 1$, to compose the set of geometrical rays,

$$\tilde{l}(t) = l_{-3,1}(t - 1) = \left\{ (x, y); \ -3x + y = \frac{t}{8} - \frac{1}{4} \right\}, \quad t = 7 : -1 : -21,$$

which are shown in Figure 3.67. For this set of geometrical rays, the system

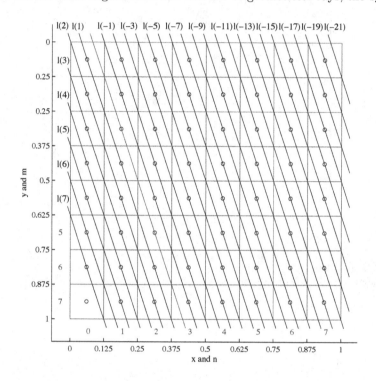

FIGURE 3.67
The set of geometrical rays for the $(-3, 1)$-projection.

of equations for the line-integrals is

$$w(t) = w(\tilde{l}(t)) = 8\sqrt{10}/3 \left[v(t) + v(t - 1) + v(t - 2) \right],$$
$$t = 7 : -1 : -21, \tag{3.89}$$

where $v(-22) = v(-23) = 0$. The required sums $v(t)$ of the discrete image are calculated by $v(t) = 3b_t/(8\sqrt{10})$, where coefficients b_t are calculated recursively

$$
\begin{aligned}
b_{-21} &= w(-21), \\
b_{-20} &= w(-20) - b_{-21}, \\
b_{-19} &= w(-19) - (b_{-20} + b_{-21}), \\
b_t &= w(t) - (b_{t-1} + b_{t-2}), \quad t = -18, -17, \ldots, 6, 7.
\end{aligned} \tag{3.90}
$$

3.5.8 $(1,6)$-projection

The set of triplet-numbers generated by the frequency-point $(p, s) = (1, 6)$ is

$$U(1,6) = \begin{cases} (1,6,0), (1,6,1), (1,6,2), (1,6,3), \\ (2,4,0), (2,4,2), \\ (4,0,0), \\ (0,0,0). \end{cases}$$

The masks of the 2-D paired functions with the first six triplet-numbers of the subset $U(1,6)$ are

$$[\chi'_{1,6,0}] = \begin{bmatrix} 10 & 00 & -10 & 00 \\ 00 & 10 & 00 & -10 \\ -10 & 00 & 10 & 00 \\ -10 & 00 & 10 & 00 \\ 00 & -10 & 00 & 10 \\ 10 & 00 & -10 & 00 \\ 00 & 10 & 00 & -10 \\ -10 & 00 & 10 & 00 \\ 00 & -10 & 00 & 10 \end{bmatrix}, \quad [\chi'_{1,6,1}] = \begin{bmatrix} 0 & 10 & 00 & -10 & 0 \\ 0 & 00 & 10 & 00 & -1 \\ 0 & -10 & 00 & 10 & 0 \\ 0 & 00 & -10 & 00 & 1 \\ 0 & 10 & 00 & -10 & 0 \\ 0 & 00 & 10 & 00 & -1 \\ 0 & -10 & 00 & 10 & 0 \\ 0 & 00 & -10 & 00 & 1 \end{bmatrix},$$

$$[\chi'_{1,6,2}] = \begin{bmatrix} 00 & 10 & 00 & -10 \\ -10 & 00 & 10 & 00 \\ 00 & -10 & 00 & 10 \\ 10 & 00 & -10 & 00 \\ 00 & 10 & 00 & -10 \\ -10 & 00 & 10 & 00 \\ 00 & -10 & 00 & 10 \\ 10 & 00 & -10 & 00 \end{bmatrix}, \quad [\chi'_{1,6,3}] = \begin{bmatrix} 0 & 00 & 10 & 00 & -1 \\ 0 & -10 & 00 & 10 & 0 \\ 0 & 00 & -10 & 00 & 1 \\ 0 & 10 & 00 & -10 & 0 \\ 0 & 00 & 10 & 00 & -1 \\ 0 & -10 & 00 & 10 & 0 \\ 0 & 00 & -10 & 00 & 1 \\ 0 & 10 & 00 & -10 & 0 \end{bmatrix},$$

$$[\chi'_{2,4,0}] = \begin{bmatrix} 10 & -10 & 10 & -10 \\ -10 & 10 & -10 & 10 \\ 10 & -10 & 10 & -10 \\ -10 & 10 & -10 & 10 \\ 10 & -10 & 10 & -10 \\ -10 & 10 & -10 & 10 \\ 10 & -10 & 10 & -10 \\ -10 & 10 & -10 & 10 \end{bmatrix}, \quad [\chi'_{2,4,2}] = \begin{bmatrix} 0 & 10 & -10 & 10 & -1 \\ 0 & -10 & 10 & -10 & 1 \\ 0 & 10 & -10 & 10 & -1 \\ 0 & -10 & 10 & -10 & 1 \\ 0 & 10 & -10 & 10 & -1 \\ 0 & -10 & 10 & -10 & 1 \\ 0 & 10 & -10 & 10 & -1 \\ 0 & -10 & 10 & -10 & 1 \end{bmatrix}.$$

Instead of the parallel rays corresponding to the $(1,6)$-projection, we consider the $(1, -2)$-projection which defines the same set of masks of the paired functions χ'_{p_1, s_1, t_1}, $(p_1, s_1, t_1) \in U(1, 6)$. The change in direction of the rays allows for reducing the total number of rays from $(1 + 6)7 + 1 = 50$ to $(1 + 2)7 + 1 = 22$. As an example, Figures 3.68 and 3.69 show the mask of the paired function $\chi'_{1,6,0}$ with the parallel arithmetical rays of two different projections. Thirteen rays of the $(1, 6)$-projection are shown in the first

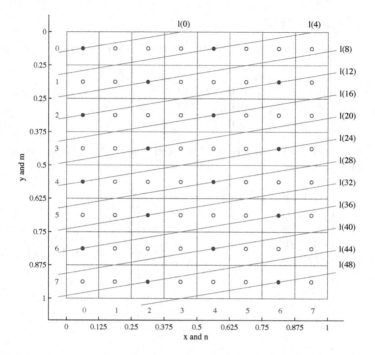

FIGURE 3.68
The mask $[\chi'_{1,6,0}]$ with the set of 13 parallel arithmetical rays.

figure, and 5 rays of the $(1,-2)$-projection are shown in the second figure.

The set of all 22 rays for this projection is shown in Figure 3.70, and it is described by the following system of equations:

$$l(t) = l_{1,-2}(t) = \left\{ (x,y); \; x - 2y = \frac{t}{8} - \frac{1}{16} \right\}, \quad t = 7 : -1 : -14, \quad (3.91)$$

where $(x,y) \in [0,1] \times [0,1]$,

The control points of this set of parallel rays are given below

$$
\begin{bmatrix}
0 & 1 & 2 & 3 & 4 & 5 & 6 & 7 \\
-2 & -1 & \cdot & \cdot & \cdot & \cdot & \cdot & \cdot \\
-4 & -3 & \cdot & \cdot & \cdot & \cdot & \cdot & \cdot \\
-6 & -5 & \cdot & \cdot & \cdot & \cdot & \cdot & \cdot \\
-8 & -7 & \cdot & \cdot & \cdot & \cdot & \cdot & \cdot \\
-10 & -9 & \cdot & \cdot & \cdot & \cdot & \cdot & \cdot \\
-12 & -11 & \cdot & \cdot & \cdot & \cdot & \cdot & \cdot \\
-14 & -13 & \cdot & \cdot & \cdot & \cdot & \cdot & \cdot
\end{bmatrix}
\qquad
\text{(angle of rays is } \tan^{-1}(\tfrac{1}{2}) = 26.5651°).
$$

The first six masks for the paired functions which are defined by the $(1,6)$-projection are shown in Figures 3.71 and 3.72.

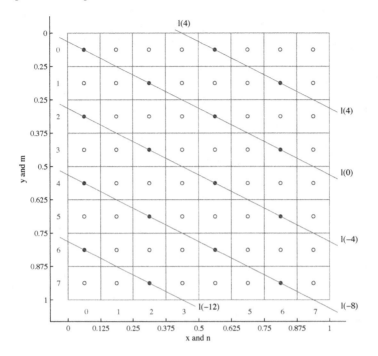

FIGURE 3.69

The mask $[\chi'_{1,6,0}]$ with the set of 5 parallel arithmetical rays.

The components of the 2-D paired transform with the triplet-numbers $(1, 6, t)$ and $(2, 4, 2t)$ are calculated by

$$
\begin{aligned}
f'_{1,6,0} &= (v_0 + v_{-8}) - (v_4 + v_{-4} + v_{-12}), \\
f'_{1,6,1} &= (v_1 + v_{-7}) - (v_5 + v_{-3} + v_{-11}), \\
f'_{1,6,2} &= (v_2 + v_{-6} + v_{-14}) - (v_6 + v_{-2} + v_{-10}), \\
f'_{1,6,3} &= (v_3 + v_{-5} + v_{-13}) - (v_7 + v_{-1} + v_{-9}), \\
f'_{2,4,0} &= (v_0 + v_{-8}) + (v_4 + v_{-4} + v_{-12}) - \\
& \quad - (v_2 + v_{-6} + v_{-14}) - (v_6 + v_{-2} + v_{-10}), \\
f'_{2,4,2} &= (v_1 + v_{-7}) + (v_5 Chapters/chapter3/ + v_{-3} + v_{-11}) - \\
& \quad - (v_3 + v_{-5} + v_{-13}) - (v_7 + v_{-1} + v_{-9}),
\end{aligned}
\tag{3.92}
$$

where the sums of the discrete image along the rays are denoted by $v_t = v_{1,-2}(t)$, $t = 7 : -1 : -14$. For the remaining two components with the triplet-

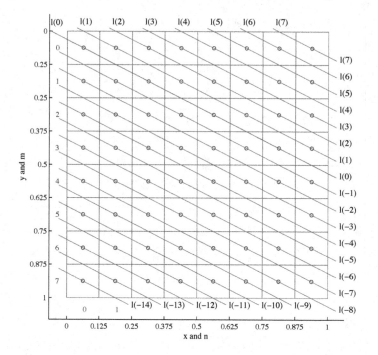

FIGURE 3.70
The set of arithmetical rays for the $(1, -2)$-projection.

numbers $(4, 0, 0)$ and $(0, 0, 0)$ we have the following:

$$f'_{4,0,0} = (v_0 + v_{-8}) + (v_4 + v_{-4} + v_{-12}) - (v_1 + v_{-7}) - (v_5 + v_{-3} + v_{-11})$$
$$+ (v_2 + v_{-6} + v_{-14}) + (v_6 + v_{-2} + v_{-10}) -$$
$$- (v_3 + v_{-5} + v_{-13}) - (v_7 + v_{-1} + v_{-9}),$$
$$f'_{0,0,0} = v_0 + v_1 + v_2 + v_3 + \cdots +$$
$$+ v_7 + v_{-1} + v_{-2} + v_{-3} + v_{-4} + v_{-5} + \dots + v_{-13} + v_{-14}.$$

With the new variables

$$a_n = \sum_m v_{n-8m} = \sum_m v_{1,-2,n-8m}, \qquad n = 0 : 7,$$

the complete system of linear equations, for calculating the eight components of the 2-D paired transform with triplet-numbers of the subset $U(1, 6)$, can be

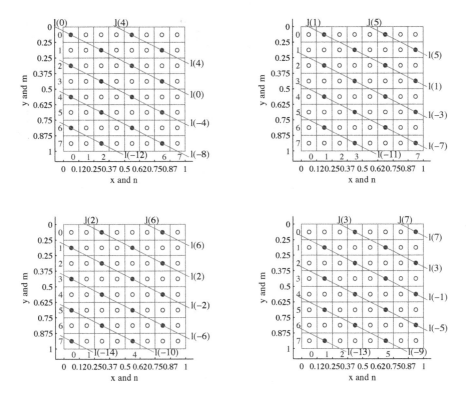

FIGURE 3.71
Masks $[\chi'_{1,6,k}]$, $k = 0 : 3$, with the arithmetical rays $l_{1,-2}(t)$, $t = 7 : -1 : 14$.

written as follows:

$$
\begin{aligned}
f'_{1,6,0} &= a_0 - a_4 \\
f'_{1,6,1} &= a_1 - a_5 \\
f'_{1,6,2} &= a_2 - a_6 \\
f'_{1,6,3} &= a_3 - a_7 \\
f'_{2,4,0} &= (a_0 + a_4) - (a_2 + a_6) \\
f'_{2,4,2} &= (a_1 + a_5) - (a_3 + a_7) \\
f'_{4,0,0} &= (a_0 + a_4) - (a_1 + a_5) + (a_2 + a_6) - (a_3 + a_7) \\
f'_{0,0,0} &= (a_0 + a_4) + (a_1 + a_5) + (a_2 + a_6) + (a_3 + a_7).
\end{aligned}
\tag{3.93}
$$

These calculations correspond to the 8-point discrete paired transform of the vector $\mathbf{a} = (a_0, a_1, a_2, ..., a_7)'$,

$$
\mathbf{P}(1, 6) = \mathbf{P}(1, -2) = [\chi'_8]\mathbf{a}.
$$

As in the case of the $(1, 2)$-projection, two sets of geometrical rays can be

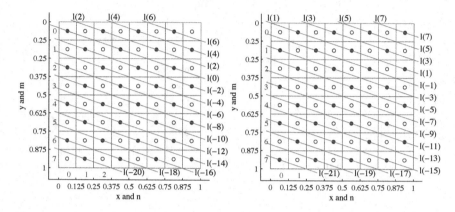

FIGURE 3.72
Two masks $[\chi'_{2,4,2k}]$, $k = 0, 1$, with the arithmetical rays $l_{1,-2}(t)$.

considered to derive a solvable system of linear equations between geometrical and arithmetical rays. The first set is defined by shifting the set of the arithmetical rays by $l(t) \rightarrow l(t + 0.5)$. This set of 22 rays is

$$\tilde{l}(t) = l_{1,-2}(t + 0.5) = \left\{ (x, y); \ x - 2y = \frac{t}{8} \right\}, \quad t = 7 : -1 : -14,$$

and is shown in Figure 3.73. For this set of geometrical rays, we obtain the following system of equations for the line-integrals:

$$w(t) = w(\tilde{l}(t)) = 4\sqrt{5}\left[v(t) + v(t + 1)\right], \quad t = 7 : -1 : -14, \qquad (3.94)$$

where $v(8) = 0$. The required sums $v(t)$ of the discrete image $f_{n,m}$ are therefore defined as $v(t) = b_t/(4\sqrt{5})$, where values of b_t are calculated recursively by

$$\begin{aligned} b_7 &= w(7), \\ b_6 &= w(6) - b_7, \\ b_t &= w(t) - b_{t+1}, \quad t = 5, 4, \ldots, 0, 1, -1, -2, \ldots, -14. \end{aligned} \qquad (3.95)$$

The second set is defined by shifting the set of the arithmetical rays by $l(t) \rightarrow l(t - 0.5)$,

$$\tilde{l}(t) = l_{1,-2}(t - 0.5) = \left\{ (x, y); \ x - 2y = \frac{t}{8} - \frac{1}{8} \right\}, \quad t = 7 : -1 : -14,$$

which is given in Figure 3.74. The system of equations for the line-integrals is defined as

$$w(t) = w(\tilde{l}(t)) = 4\sqrt{5}\left[v(t) + v(t - 1)\right], \quad t = 7 : -1 : -14, \qquad (3.96)$$

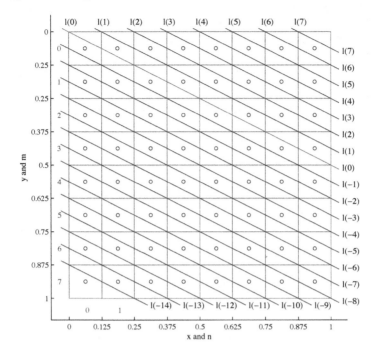

FIGURE 3.73
The set of 22 geometrical rays for the $(1, -2)$-projection.

where $v(-15) = 0$. The required sums $v(t)$ of the discrete image $f_{n,m}$ are therefore defined as $v(t) = b_t/(4\sqrt{5})$, where the components b_t are calculated recursively by

$$
\begin{aligned}
b_{-14} &= w(-14), \\
b_{-13} &= w(-13) - b_{-14}, \\
b_t &= w(t) - b_{t-1}, \quad t = -12, -11, \ldots, 0, 1, \ldots, 7.
\end{aligned}
$$
(3.97)

It should also be noted that the numbering of the control points can be changed, as well as the numbering of all line-integrals $w(t)$ and sums $v_t = v(t)$ by the transformation $t \to 7 - t$, where $t = 7 : -1 : -21$,

$$
\begin{bmatrix}
7_\bullet & 6_\bullet & 5_\bullet & 4_\bullet & 3_\bullet & 2_\bullet & 1_\bullet & 0_\bullet \\
9_\bullet & 8_\bullet & \cdot & \cdot & \cdot & \cdot & \cdot & \cdot \\
11_\bullet & 10_\bullet & \cdot & \cdot & \cdot & \cdot & \cdot & \cdot \\
13_\bullet & 12_\bullet & \cdot & \cdot & \cdot & \cdot & \cdot & \cdot \\
15_\bullet & 14_\bullet & \cdot & \cdot & \cdot & \cdot & \cdot & \cdot \\
17_\bullet & 16_\bullet & \cdot & \cdot & \cdot & \cdot & \cdot & \cdot \\
19_\bullet & 18_\bullet & \cdot & \cdot & \cdot & \cdot & \cdot & \cdot \\
21_\bullet & 20_\bullet & \cdot & \cdot & \cdot & \cdot & \cdot & \cdot
\end{bmatrix} .
$$

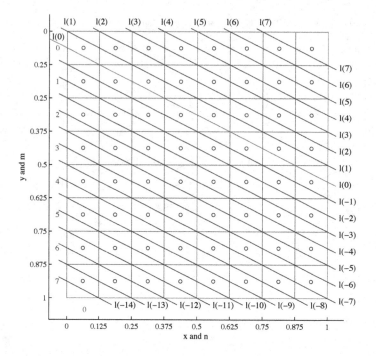

FIGURE 3.74
The second set of 22 geometrical rays for the $(1, -2)$-projection.

3.5.9 $(6, 1)$-projection

The set of triplet-numbers generated by the frequency-point $(p, s) = (6, 1)$ is

$$U(6,1) = \begin{cases} (6,1,0), (6,1,1), (6,1,2), (6,1,3), \\ (4,2,0), (4,2,2), \\ (0,4,0), \\ (0,0,0). \end{cases}$$

The masks of the 2-D paired functions with first six triplet-numbers of the subset $U(6,1)$ are

$$[\chi'_{6,1,0}] = \begin{bmatrix} 1 & 0 & -1 & 0 & 1 & 0 & -1 & 0 \\ 0 & 0 & 0 & 0 & 0 & 0 & 0 & 0 \\ 0 & 1 & 0 & -1 & 0 & 1 & 0 & -1 \\ 0 & 0 & 0 & 0 & 0 & 0 & 0 & 0 \\ -1 & 0 & 1 & 0 & -1 & 0 & 1 & 0 \\ 0 & 0 & 0 & 0 & 0 & 0 & 0 & 0 \\ 0 & -1 & 0 & 1 & 0 & -1 & 0 & 1 \\ 0 & 0 & 0 & 0 & 0 & 0 & 0 & 0 \end{bmatrix}, \; [\chi'_{6,1,1}] = \begin{bmatrix} 0 & 0 & 0 & 0 & 0 & 0 & 0 & 0 \\ 1 & 0 & -1 & 0 & 1 & 0 & -1 & 0 \\ 0 & 0 & 0 & 0 & 0 & 0 & 0 & 0 \\ 0 & 1 & 0 & -1 & 0 & 1 & 0 & -1 \\ 0 & 0 & 0 & 0 & 0 & 0 & 0 & 0 \\ -1 & 0 & 1 & 0 & -1 & 0 & 1 & 0 \\ 0 & 0 & 0 & 0 & 0 & 0 & 0 & 0 \\ 0 & -1 & 0 & 1 & 0 & -1 & 0 & 1 \end{bmatrix}$$

$$[\chi'_{6,1,2}] = \begin{bmatrix} 0 & -1 & 0 & 1 & 0 & -1 & 0 & 1 \\ 0 & 0 & 0 & 0 & 0 & 0 & 0 & 0 \\ 1 & 0 & -1 & 0 & 1 & 0 & -1 & 0 \\ 0 & 0 & 0 & 0 & 0 & 0 & 0 & 0 \\ 0 & 1 & 0 & -1 & 0 & 1 & 0 & -1 \\ 0 & 0 & 0 & 0 & 0 & 0 & 0 & 0 \\ -1 & 0 & 1 & 0 & -1 & 0 & 1 & 0 \\ 0 & 0 & 0 & 0 & 0 & 0 & 0 & 0 \end{bmatrix}, \quad [\chi'_{6,1,3}] = \begin{bmatrix} 0 & 0 & 0 & 0 & 0 & 0 & 0 & 0 \\ 0 & -1 & 0 & 1 & 0 & -1 & 0 & 1 \\ 0 & 0 & 0 & 0 & 0 & 0 & 0 & 0 \\ 1 & 0 & -1 & 0 & 1 & 0 & -1 & 0 \\ 0 & 0 & 0 & 0 & 0 & 0 & 0 & 0 \\ 0 & 1 & 0 & -1 & 0 & 1 & 0 & -1 \\ 0 & 0 & 0 & 0 & 0 & 0 & 0 & 0 \\ -1 & 0 & 1 & 0 & -1 & 0 & 1 & 0 \end{bmatrix}$$

$$[\chi'_{4,2,0}] = \begin{bmatrix} 1 & -1 & 1 & -1 & 1 & -1 & 1 & -1 \\ 0 & 0 & 0 & 0 & 0 & 0 & 0 & 0 \\ -1 & 1 & -1 & 1 & -1 & 1 & -1 & 1 \\ 0 & 0 & 0 & 0 & 0 & 0 & 0 & 0 \\ 1 & -1 & 1 & -1 & 1 & -1 & 1 & -1 \\ 0 & 0 & 0 & 0 & 0 & 0 & 0 & 0 \\ -1 & 1 & -1 & 1 & -1 & 1 & -1 & 1 \\ 0 & 0 & 0 & 0 & 0 & 0 & 0 & 0 \end{bmatrix}, \quad [\chi'_{4,2,2}] = \begin{bmatrix} 0 & 0 & 0 & 0 & 0 & 0 & 0 & 0 \\ 1 & -1 & 1 & -1 & 1 & -1 & 1 & -1 \\ 0 & 0 & 0 & 0 & 0 & 0 & 0 & 0 \\ -1 & 1 & -1 & 1 & -1 & 1 & -1 & 1 \\ 0 & 0 & 0 & 0 & 0 & 0 & 0 & 0 \\ 1 & -1 & 1 & -1 & 1 & -1 & 1 & -1 \\ 0 & 0 & 0 & 0 & 0 & 0 & 0 & 0 \\ -1 & 1 & -1 & 1 & -1 & 1 & -1 & 1 \end{bmatrix}.$$

Instead of parallel rays of the $(6,1)$-projection, we consider the $(-2,1)$-projection which defines the same set of masks of the paired functions. Indeed, the equation $6n + m = t \bmod 8$ can be written as $(8-2)n + m = -2n + m = t \bmod 8$. The change of the projection allows for reducing the number of rays from 50 to 22. The set of all parallel rays for this projection is shown in Figure 3.75.

The control points of this set of rays are given below

$$\begin{bmatrix} 0 & -2 & -4 & -6 & -8 & -10 & -12 & -14 \\ 1 & -1 & -3 & -5 & -7 & -9 & -11 & -13 \\ 2 & \cdot & \cdot & \cdot & \cdot & \cdot & \cdot \\ 3 & \cdot & \cdot & \cdot & \cdot & \cdot & \cdot \\ 4 & \cdot & \cdot & \cdot & \cdot & \cdot & \cdot \\ 5 & \cdot & \cdot & \cdot & \cdot & \cdot & \cdot \\ 6 & \cdot & \cdot & \cdot & \cdot & \cdot & \cdot \\ 7 & \cdot & \cdot & \cdot & \cdot & \cdot & \cdot \end{bmatrix} \quad \text{(angle of rays is } \tan^{-1}(2) = 63.435°\text{)}$$

The set of the 29 geometrical rays of this projection is described as

$$l(t) = l_{-2,1}(t) = \left\{ (x, y); \; -2x + y = \frac{t}{8} - \frac{1}{16} \right\}, \quad t = 7 : -1 : -14, \quad (3.98)$$

where $(x, y) \in [0, 1] \times [0, 1]$. The masks with parallel rays for the first six paired functions are shown in Figures 3.76 and 3.77.

The components of the 2-D paired transform with the triplet-numbers of

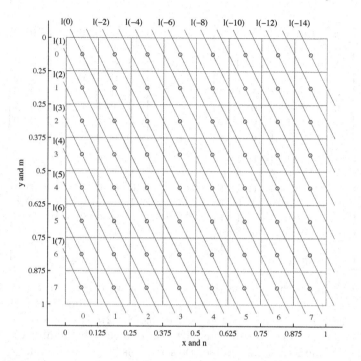

FIGURE 3.75
The set of arithmetical rays for the $(-2, 1)$-projection.

$U(6,1)$ are calculated as follows:

$$f'_{6,1,0} = (v_0 + v_{-8}) - (v_4 + v_{-4} + v_{-12}),$$
$$f'_{6,1,1} = (v_1 + v_{-7}) - (v_5 + v_{-3} + v_{-11}),$$
$$f'_{6,1,2} = (v_2 + v_{-6} + v_{-14}) - (v_6 + v_{-2} + v_{-10}),$$
$$f'_{6,1,3} = (v_3 + v_{-5} + v_{-13}) - (v_7 + v_{-1} + v_{-9}),$$
$$f'_{4,2,0} = (v_0 + v_{-8}) + (v_4 + v_{-4} + v_{-12}) -$$
$$- (v_2 + v_{-6} + v_{-14}) - (v_6 + v_{-2} + v_{-10}),$$
$$f'_{4,2,2} = (v_1 + v_{-7}) + (v_5 + v_{-3} + v_{-11}) -$$
$$- (v_3 + v_{-5} + v_{-13}) - (v_7 + v_{-1} + v_{-9}),$$

$$(3.99)$$

where the sums of the discrete image along the rays are denoted by $v_t = v_{-2,1}(t)$, $t = 7 : -1 : -14$. For the components with two remaining numbers $(0, 4, 0)$ and $(0, 0, 0)$, we have the following:

$$f'_{0,4,0} = (v_0 + v_{-8}) + (v_4 + v_{-4} + v_{-12}) - (v_1 + v_{-7}) -$$
$$- (v_5 + v_{-3} + v_{-11}) + (v_2 + v_{-6} + v_{-14}) + (v_6 + v_{-2} + v_{-10})$$
$$- (v_3 + v_{-5} + v_{-13}) - (v_7 + v_{-1} + v_{-9}),$$
$$f'_{0,0,0} = v_0 + v_1 + v_2 + v_3 + \cdots + v_7 +$$
$$+ v_{-1} + v_{-2} + v_{-3} + v_{-4} + v_{-5} + \ldots + v_{-13} + v_{-14}.$$

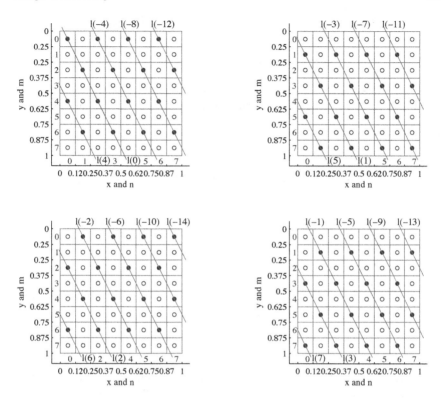

FIGURE 3.76

Four masks $[\chi'_{6,1,k}]$, $k = 0 : 3$, with the arithmetical rays $l_{-2,1}(t)$.

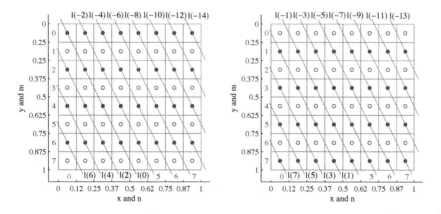

FIGURE 3.77

Two masks $[\chi'_{4,2,2k}]$, $k = 0, 1$, with the arithmetical rays $l_{-2,1}(t)$.

With the new variables

$$a_n = \sum_m v_{n-8m} = \sum_m v_{-2,1,n-8m}, \quad n = 0:7,$$

the complete system of linear equations, for calculating the eight components of the 2-D paired transform with triplets of the subset $U(6,1)$, can be written as follows:

$$\begin{aligned}
f'_{6,1,0} &= a_0 - a_4 \\
f'_{6,1,1} &= a_1 - a_5 \\
f'_{6,1,2} &= a_2 - a_6 \\
f'_{6,1,3} &= a_3 - a_7 \\
f'_{4,2,0} &= (a_0 + a_4) - (a_2 + a_6) \\
f'_{4,2,2} &= (a_1 + a_5) - (a_3 + a_7) \\
f'_{0,4,0} &= (a_0 + a_4) - (a_1 + a_5) + (a_2 + a_6) - (a_3 + a_7) \\
f'_{0,0,0} &= (a_0 + a_4) + (a_1 + a_5) + (a_2 + a_6) + (a_3 + a_7).
\end{aligned} \qquad (3.100)$$

These calculations correspond to the 8-point discrete paired transform of the vector $\mathbf{a} = (a_0, a_1, a_2, ..., a_7)'$,

$$\mathbf{P}(6,1) = \mathbf{P}(-2,1) = [\chi'_8]\mathbf{a}.$$

Two sets of geometrical rays can be used to determine the sums of the image along the arithmetical rays. The first set is defined by shifting the set of arithmetical rays by $l(t) \to l(t+0.5)$. This set of 22 rays is described by

$$\tilde{l}(t) = l_{-2,1}(t+0.5) = \left\{ (x,y); \; -2x + y = \frac{t}{8} \right\}, \quad t = 7:-1:-14,$$

and is shown in Figure 3.78. The line-integrals along this set of geometrical rays can be written as

$$w(t) = w(\tilde{l}(t)) = 4\sqrt{5}\,[v(t) + v(t+1)], \quad t = 7:-1:-14, \qquad (3.101)$$

where $v(8) = 0$. The required sums $v(t)$ of the image $f_{n,m}$ are therefore defined as $v(t) = b_t/(4\sqrt{5})$, where the components b_t are calculated recursively by

$$\begin{aligned}
b_7 &= w(7), \\
b_6 &= w(6) - b_7, \\
b_t &= w(t) - b_{t+1}, \quad t = 5,4,\ldots,0,1,-1,-2,\ldots,-14.
\end{aligned} \qquad (3.102)$$

The second set of geometrical rays is defined by shifting the set of the arithmetical rays by $l(t) \to l(t-0.5)$,

$$\tilde{l}(t) = l_{-2,1}(t-0.5) = \left\{ (x,y); \; -2x + y = \frac{t}{8} - \frac{1}{8} \right\}, \quad t = 7:-1:-14,$$

which is given in Figure 3.79. In this case, the system of equations for the

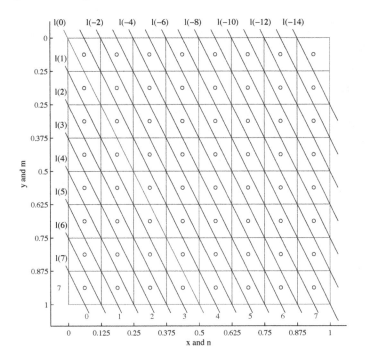

FIGURE 3.78

The set of geometrical rays for the $(-2, 1)$-projection.

line-integrals is described by

$$w(t) = w(\tilde{l}(t)) = 4\sqrt{5}\,[v(t) + v(t-1)], \quad t = 7 : -1 : -14, \qquad (3.103)$$

where $v(-15) = 0$. The required sums $v(t)$ of the discrete image $f_{n,m}$ are therefore defined as $v(t) = b_t/(4\sqrt{5})$, where the components b_t are calculated recursively by

$$\begin{aligned}
b_{-14} &= w(-14), \\
b_{-13} &= w(-13) - b_{-14}, \\
b_t &= w(t) - b_{t-1}, \quad t = -12, -11, \ldots, 0, 1, \ldots, 7.
\end{aligned} \qquad (3.104)$$

It should be noted that the numbering of the control points can be changed, as well as the numbering of all line-integrals $w(t)$ and sums $v_t = v(t)$ by the

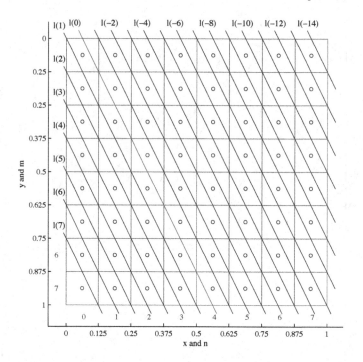

FIGURE 3.79
The second set of geometrical rays for the $(-2, 1)$-projection.

transform $t \to 7 - t$, where $t = 7 : -1 : -21$,

$$\begin{bmatrix} 7 \bullet 9 \bullet 11 \bullet 13 \bullet 15 \bullet 17 \bullet 19 \bullet 21 \bullet \\ 6 \bullet 8 \bullet 10 \bullet 12 \bullet 14 \bullet 16 \bullet 18 \bullet 20 \bullet \\ 5 \bullet \quad \cdot \quad \cdot \quad \cdot \quad \cdot \quad \cdot \quad \cdot \quad \cdot \\ 4 \bullet \quad \cdot \quad \cdot \quad \cdot \quad \cdot \quad \cdot \quad \cdot \quad \cdot \\ 3 \bullet \quad \cdot \quad \cdot \quad \cdot \quad \cdot \quad \cdot \quad \cdot \quad \cdot \\ 2 \bullet \quad \cdot \quad \cdot \quad \cdot \quad \cdot \quad \cdot \quad \cdot \quad \cdot \\ 1 \bullet \quad \cdot \quad \cdot \quad \cdot \quad \cdot \quad \cdot \quad \cdot \quad \cdot \\ 0 \bullet \quad \cdot \quad \cdot \quad \cdot \quad \cdot \quad \cdot \quad \cdot \quad \cdot \end{bmatrix} \qquad \text{(angle of rays is } \tan^{-1}(2) = 63.435°).$$

3.5.10 $(1, 7)$-projection

This is the last set of triplet-numbers in the 8×8 example. The set is generated by the frequency-point $(p, s) = (1, 7)$,

$$U(1, 7) = \begin{cases} (1, 7, 0), (1, 7, 1), (1, 7, 2), (1, 7, 3), \\ (2, 6, 0), (2, 6, 2), \\ (4, 4, 0), \\ (0, 0, 0). \end{cases}$$

We consider masks of the 2-D paired functions with the first six triplet-numbers of the subset $U(1,7)$,

$$[\chi'_{1,7,0}] = \begin{bmatrix} 1 & 0 & 0 & 0 & -1 & 0 & 0 & 0 \\ 0 & 1 & 0 & 0 & 0 & -1 & 0 & 0 \\ 0 & 0 & 1 & 0 & 0 & 0 & -1 & 0 \\ 0 & 0 & 0 & 1 & 0 & 0 & 0 & -1 \\ -1 & 0 & 0 & 0 & 1 & 0 & 0 & 0 \\ 0 & -1 & 0 & 0 & 0 & 1 & 0 & 0 \\ 0 & 0 & -1 & 0 & 0 & 0 & 1 & 0 \\ 0 & 0 & 0 & -1 & 0 & 0 & 0 & 1 \end{bmatrix}, \quad [\chi'_{1,7,1}] = \begin{bmatrix} 0 & 1 & 0 & 0 & 0 & -1 & 0 & 0 \\ 0 & 0 & 1 & 0 & 0 & 0 & -1 & 0 \\ 0 & 0 & 0 & 1 & 0 & 0 & 0 & -1 \\ -1 & 0 & 0 & 0 & 1 & 0 & 0 & 0 \\ 0 & -1 & 0 & 0 & 0 & 1 & 0 & 0 \\ 0 & 0 & -1 & 0 & 0 & 0 & 1 & 0 \\ 0 & 0 & 0 & -1 & 0 & 0 & 0 & 1 \\ 1 & 0 & 0 & 0 & -1 & 0 & 0 & 0 \end{bmatrix}$$

$$[\chi'_{1,7,2}] = \begin{bmatrix} 0 & 0 & 1 & 0 & 0 & 0 & -1 & 0 \\ 0 & 0 & 0 & 1 & 0 & 0 & 0 & -1 \\ -1 & 0 & 0 & 0 & 1 & 0 & 0 & 0 \\ 0 & -1 & 0 & 0 & 0 & 1 & 0 & 0 \\ 0 & 0 & -1 & 0 & 0 & 0 & 1 & 0 \\ 0 & 0 & 0 & -1 & 0 & 0 & 0 & 1 \\ 1 & 0 & 0 & 0 & -1 & 0 & 0 & 0 \\ 0 & 1 & 0 & 0 & 0 & -1 & 0 & 0 \end{bmatrix}, \quad [\chi'_{1,7,3}] = \begin{bmatrix} 0 & 0 & 0 & 1 & 0 & 0 & 0 & -1 \\ -1 & 0 & 0 & 0 & 1 & 0 & 0 & 0 \\ 0 & -1 & 0 & 0 & 0 & 1 & 0 & 0 \\ 0 & 0 & -1 & 0 & 0 & 0 & 1 & 0 \\ 0 & 0 & 0 & -1 & 0 & 0 & 0 & 1 \\ 1 & 0 & 0 & 0 & -1 & 0 & 0 & 0 \\ 0 & 1 & 0 & 0 & 0 & -1 & 0 & 0 \\ 0 & 0 & 1 & 0 & 0 & 0 & -1 & 0 \end{bmatrix}$$

$$[\chi'_{2,6,0}] = \begin{bmatrix} 1 & 0 & -1 & 0 & 1 & 0 & -1 & 0 \\ 0 & 1 & 0 & -1 & 0 & 1 & 0 & -1 \\ -1 & 0 & 1 & 0 & -1 & 0 & 1 & 0 \\ 0 & -1 & 0 & 1 & 0 & -1 & 0 & 1 \\ 1 & 0 & -1 & 0 & 1 & 0 & -1 & 0 \\ 0 & 1 & 0 & -1 & 0 & 1 & 0 & -1 \\ -1 & 0 & 1 & 0 & -1 & 0 & 1 & 0 \\ 0 & -1 & 0 & 1 & 0 & -1 & 0 & 1 \end{bmatrix}, \quad [\chi'_{2,6,2}] = \begin{bmatrix} 0 & 1 & 0 & -1 & 0 & 1 & 0 & -1 \\ -1 & 0 & 1 & 0 & -1 & 0 & 1 & 0 \\ 0 & -1 & 0 & 1 & 0 & -1 & 0 & 1 \\ 1 & 0 & -1 & 0 & 1 & 0 & -1 & 0 \\ 0 & 1 & 0 & -1 & 0 & 1 & 0 & -1 \\ -1 & 0 & 1 & 0 & -1 & 0 & 1 & 0 \\ 0 & -1 & 0 & 1 & 0 & -1 & 0 & 1 \\ 1 & 0 & -1 & 0 & 1 & 0 & -1 & 0 \end{bmatrix}$$

Instead of the parallel rays of the $(1,7)$-projection, we consider the $(1,-1)$-projection which defines the same set of masks of paired functions. This change allows for reducing the number of rays from $(1+7)7+1 = 57$ to $(1+1)7+1 = 15$. As an example, Figures 3.80 and 3.81 show the same mask of the paired function $\chi'_{1,7,0}$ with the parallel arithmetical rays of two different projections. Fifteen rays of the $(1,7)$-projection are shown in the first figure, and three rays of the $(1,-1)$-projection are shown in the second figure.

The set of the 15 geometrical rays of this projection is described by

$$l(t) = l_{1,-1}(t) = \left\{ (x,y); \ x - y = \frac{t}{8} \right\}, \quad t = 7 : -1 : -7, \qquad (3.105)$$

where $(x,y) \in [0,1] \times [0,1]$. This set is shown in Figure 3.82. The control

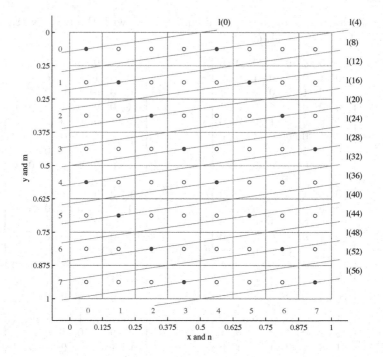

FIGURE 3.80
The mask $[\chi'_{1,7,0}]$ with 15 arithmetical rays of the $(1,7)$-projection.

points of this set of rays are shown below:

$$
\begin{bmatrix}
0_\bullet{}^1_\bullet{}^2_\bullet{}^3_\bullet{}^4_\bullet{}^5_\bullet{}^6_\bullet{}^7_\bullet \\
-1_\bullet \;\; \cdot \;\; \cdot \;\; \cdot \;\; \cdot \;\; \cdot \;\; \cdot \;\; \cdot \\
-2_\bullet \;\; \cdot \;\; \cdot \;\; \cdot \;\; \cdot \;\; \cdot \;\; \cdot \;\; \cdot \\
-3_\bullet \;\; \cdot \;\; \cdot \;\; \cdot \;\; \cdot \;\; \cdot \;\; \cdot \;\; \cdot \\
-4_\bullet \;\; \cdot \;\; \cdot \;\; \cdot \;\; \cdot \;\; \cdot \;\; \cdot \;\; \cdot \\
-5_\bullet \;\; \cdot \;\; \cdot \;\; \cdot \;\; \cdot \;\; \cdot \;\; \cdot \;\; \cdot \\
-6_\bullet \;\; \cdot \;\; \cdot \;\; \cdot \;\; \cdot \;\; \cdot \;\; \cdot \;\; \cdot \\
-7_\bullet \;\; \cdot \;\; \cdot \;\; \cdot \;\; \cdot \;\; \cdot \;\; \cdot \;\; \cdot
\end{bmatrix}
\qquad \text{(angle of rays is } \tan^{-1}(1) = 45°\text{).}
$$

The first six masks for the paired functions with the triplet-numbers of $U(1,7)$ are shown in Figures 3.83 and 3.84.

The components of the 2-D paired transform with the triplet-numbers

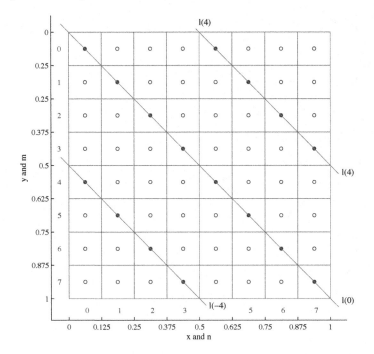

FIGURE 3.81
The mask $[\chi'_{1,7,0}]$ with 3 arithmetical rays of the $(1,-1)$-projection.

$(1,7,t)$ and $(2,6,2t)$ are calculated by

$$
\begin{aligned}
f'_{1,7,0} &= (v_0) - (v_4 + v_{-4}), \\
f'_{1,7,1} &= (v_1 + v_{-7}) - (v_5 + v_{-3}), \\
f'_{1,7,2} &= (v_2 + v_{-6}) - (v_6 + v_{-2}), \\
f'_{1,7,3} &= (v_3 + v_{-5}) - (v_7 + v_{-1}), \\
f'_{2,6,0} &= (v_0) + (v_4 + v_{-4}) - (v_2 + v_{-6}) - (v_6 + v_{-2}), \\
f'_{2,6,2} &= (v_1 + v_{-7}) + (v_5 + v_{-3}) - (v_3 + v_{-5}) - (v_7 + v_{-1}),
\end{aligned}
\tag{3.106}
$$

where the sums of the discrete image along the rays are denoted by $v_t = v_{1,-1}(t)$, $t = 7 : -1 : -7$. For the remaining two components of the 2-D paired transform, which have the triplet-numbers $(4,4,0)$ and $(0,0,0)$, we have the following:

$$
\begin{aligned}
f'_{4,4,0} &= (v_0) + (v_4 + v_{-4}) - (v_1 + v_{-7}) - (v_5 + v_{-3}) + \\
&\quad + (v_2 + v_{-6}) + (v_6 + v_{-2}) - (v_3 + v_{-5}) - (v_7 + v_{-1}), \\
f'_{0,0,0} &= v_0 + v_1 + v_2 + v_3 + \cdots + v_7 + v_{-1} + v_{-2} + \ldots + v_{-6} + v_{-7}.
\end{aligned}
$$

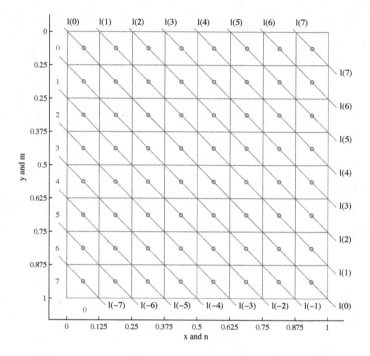

FIGURE 3.82
The set of arithmetical rays for the $(1, -1)$-projection.

Introducing the new variables

$$a_n = \sum_m v_{n-8m} = \sum_m v_{1,-1,n-8m}, \qquad n = 0:7,$$

the complete system of linear equations for calculating the eight components of the 2-D paired transform with triplet-numbers of the subset $U(1,7)$ can be written as follows:

$$
\begin{aligned}
f'_{1,7,0} &= a_0 - a_4 \\
f'_{1,7,1} &= a_1 - a_5 \\
f'_{1,7,2} &= a_2 - a_6 \\
f'_{1,7,3} &= a_3 - a_7 \\
f'_{2,6,0} &= (a_0 + a_4) - (a_2 + a_6) \\
f'_{2,6,2} &= (a_1 + a_5) - (a_3 + a_7) \\
f'_{4,4,0} &= (a_0 + a_4) - (a_1 + a_5) + (a_2 + a_6) - (a_3 + a_7) \\
f'_{0,0,0} &= (a_0 + a_4) + (a_1 + a_5) + (a_2 + a_6) + (a_3 + a_7).
\end{aligned}
\tag{3.107}
$$

These calculations correspond to the 8-point discrete paired transform of the vector $\mathbf{a} = (a_0, a_1, a_2, ..., a_7)'$,

$$\mathbf{P}(1,7) = \mathbf{P}(1,-1) = [\chi'_8]\mathbf{a}.$$

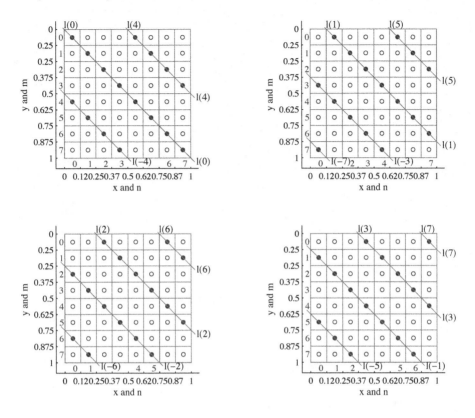

FIGURE 3.83

Four masks $[\chi'_{1,7,k}]$, $k = 0 : 3$, with the arithmetical rays $l_{1,-1}(t)$.

As in the $(p,s) = (1,1)$ case of the diagonal projection, the sums $v(t)$ of the discrete image $f_{n,m}$ can be calculated directly from the line-integrals $w(t)$. The set of geometrical rays coincides with the set of arithmetical rays. Each ray intersects the image element along the path of length $\Delta l = \sqrt{2}\Delta x = \sqrt{2}/8$. Therefore, $w(t) = w_{1,-1}(t) = 8\sqrt{2}v(t)$, or $v(t) = w(t)/(8\sqrt{2})$, $t = 7 : -1 : -7$. We obtain the following: $\mathbf{P}(1,7) = 1/(8\sqrt{2})[\chi'_8]\mathbf{w}$, where the vector $\mathbf{w} = (w_7, w_6, ..., w_0, w_{-1}, ..., w_{-7})'$.

The numbering of the control points can be changed, as well as the numbering of all line-integrals $w(t)$ and sums $v_t = v(t)$, by using the transformation

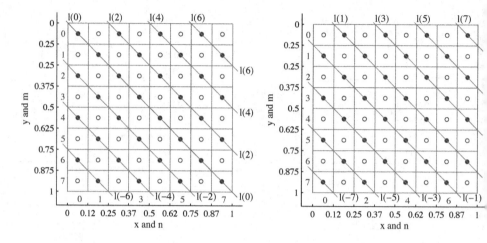

FIGURE 3.84

Two masks $[\chi'_{2,6,2k}]$, $k = 0, 1$, with the arithmetical rays $l_{1,-1}(t)$.

$t \to 7 - t$, where $t = 7 : -1 : -7$,

$$
\begin{bmatrix}
7 & 6 & 5 & 4 & 3 & 2 & 1 & 0 \\
8 & \cdot & \cdot & \cdot & \cdot & \cdot & \cdot & \cdot \\
9 & \cdot & \cdot & \cdot & \cdot & \cdot & \cdot & \cdot \\
10 & \cdot & \cdot & \cdot & \cdot & \cdot & \cdot & \cdot \\
11 & \cdot & \cdot & \cdot & \cdot & \cdot & \cdot & \cdot \\
12 & \cdot & \cdot & \cdot & \cdot & \cdot & \cdot & \cdot \\
13 & \cdot & \cdot & \cdot & \cdot & \cdot & \cdot & \cdot \\
14 & \cdot & \cdot & \cdot & \cdot & \cdot & \cdot & \cdot
\end{bmatrix}
\qquad \text{(angle of rays is } \tan^{-1}(1) = 45°).
$$

3.6   Summary of results

For the $N = 8$ case, all $3N/2 = 12$ projections that are required for reconstructing the image $N \times N$ were considered in detail, and geometrical rays for each of these projections were described. The number of parallel geometrical rays used in the projections is defined as follows:

$$n(1, s) = (1 + s)(N - 1) + 1 = N + s(N - 1), \quad s = 0, 1, ..., N/2,$$
$$n(1, N - s) = N^2 - s(N - 1), \quad s = N/2 + 1, N/2 + 2, ..., N - 1,$$
$$n(N - p, 1) = N^2 - p(N - 1), \quad p = N/2 + 2, N/2 + 4, ..., N - 2,$$
$$n(p, 1) = n(1, p) = N + p(N - 1), \quad p = 0, 2, 4, ..., N/2.$$

(3.108)

Values of these numbers $n = n(p, s)$, for all 12 generators (p, s), are given in Table 3.3. The total number of geometrical rays is 190. The subsets $U(p, s)$ and their intersections (which are underlined) are also given in the table. It shows that the total number of intersections between these subsets equals $96 - 64 = 32$.

TABLE 3.3
Number of geometrical rays and subsets $U(p, s)$ for the 8×8 case

(p, s)	n	triplet-numbers in $U(p, s)$	
$(1, 0)$	8	$(1, 0, 0), (1, 0, 1), (1, 0, 2), (1, 0, 3), (2, 0, 0), (2, 0, 2), (4, 0, 0), (0, 0, 0)$	8
$(1, 1)$	15	$(1, 1, 0), (1, 1, 1), (1, 1, 2), (1, 1, 3), (2, 2, 0), (2, 2, 2), (4, 4, 0), \underline{(0, 0, 0)}$	7
$(1, 2)$	22	$(1, 2, 0), (1, 2, 1), (1, 2, 2), (1, 2, 3), (2, 4, 0), (2, 4, 2), \underline{(4, 0, 0)}, \underline{(0, 0, 0)}$	6
$(1, 3)$	29	$(1, 3, 0), (1, 3, 1), (1, 3, 2), (1, 3, 3), (2, 6, 0), (2, 6, 2), \underline{(4, 4, 0)}, \underline{(0, 0, 0)}$	6
$(1, 4)$	36	$(1, 4, 0), (1, 4, 1), (1, 4, 2), (1, 4, 3), (2, 0, 0), (2, 0, 2), \underline{(4, 0, 0)}, \underline{(0, 0, 0)}$	4
$(1, 5)$	29	$(1, 5, 0), (1, 5, 1), (1, 5, 2), (1, 5, 3), \underline{(2, 2, 0)}, \underline{(2, 2, 2)}, \underline{(4, 4, 0)}, \underline{(0, 0, 0)}$	4
$(1, 6)$	22	$(1, 6, 0), (1, 6, 1), (1, 6, 2), (1, 6, 3), \underline{(2, 4, 0)}, \underline{(2, 4, 2)}, \underline{(4, 0, 0)}, \underline{(0, 0, 0)}$	4
$(1, 7)$	15	$(1, 7, 0), (1, 7, 1), (1, 7, 2), (1, 7, 3), \underline{(2, 6, 0)}, \underline{(2, 6, 2)}, \underline{(4, 4, 0)}, \underline{(0, 0, 0)}$	4
$(0, 1)$	8	$(0, 1, 0), (0, 1, 1), (0, 1, 2), (0, 1, 3), \underline{(0, 2, 0)}, \underline{(0, 2, 2)}, \underline{(0, 4, 0)}, \underline{(0, 0, 0)}$	7
$(2, 1)$	22	$(2, 1, 0), (2, 1, 1), (2, 1, 2), (2, 1, 3), (4, 2, 0), (4, 2, 2), \underline{(0, 4, 0)}, \underline{(0, 0, 0)}$	6
$(4, 1)$	36	$(4, 1, 0), (4, 1, 1), (4, 1, 2), (4, 1, 3), \underline{(0, 2, 0)}, \underline{(0, 2, 2)}, \underline{(0, 4, 0)}, \underline{(0, 0, 0)}$	4
$(6, 1)$	22	$(6, 1, 0), (6, 1, 1), (6, 1, 2), (6, 1, 3), \underline{(4, 2, 0)}, \underline{(4, 2, 2)}, \underline{(0, 4, 0)}, \underline{(0, 0, 0)}$	4
	190		64

Each triplet-number (p, s, t) of masks of the paired functions in this table has the form $2^k(p_1, s_1, t_1)$, where 2^k=g.c.d.(p, s), and $t_1 = 0 : N/2^{k+1} - 1$, and its multiplicity (i.e., the number of subsets U covering this triplet-number) equals 2^k. Therefore, there are two ways to use this property in image reconstruction.

1. One can use the complete 1-D paired transforms over all vectors **a**, whose components are defined as

$$a_n = \sum_m v_{n \pm Nm} = \sum_m v_{p, s, n \pm Nm}, \quad n = 0 : N - 1.$$

Then, the components $f'_{p, s, t}$ of the 2-D paired transform, which have been calculated by 1-D paired transform $\mathbf{P}(p, s) = [\chi'_N]\mathbf{a}$, are normalized by the factor of $2^k = $ g.c.d.(p, s), when $(p, s) \neq (0, 0)$, and N, when $(p, s) = (0, 0)$.

2. One can use the incomplete complete 1-D paired transforms over vectors **a**, to avoid the repetition of calculation of components of the 2-D paired transform of the image. The last column in the table shows the number of outputs to be calculated when performing the incomplete 1-D paired transforms. For instance, the complete 1-D paired transform can be used for the first $(1,0)$-projection, but only six outputs for the transform when calculating the transform $\mathbf{P}(1,2) = [\chi'_8]\mathbf{a}$, for the $(1,2)$-projection, and half of calculations when calculating the transform $\mathbf{P}(1,4) = [\chi'_8]\mathbf{a}$, for the $(1,4)$-projection, and so on. The use of incomplete 1-D paired transforms reduces the time necessary to calculate the reconstructed image, when compared with the first method.

3.6.1 Equations of rays

The systems of equations for both arithmetical and geometrical rays (or simply, A-rays and G-rays) are given in Table 3.4, for the 8×8 case.

TABLE 3.4
Equations of G-rays on the grid 8×8. (The coordinate system is original.)

(p,s)	A-rays	$t \pm t_0$	G-rays $l(t \mp t_0)$	$t_0 = \frac{p+s}{2} - 1$	t
$(1,0)$	$l_{1,0}(t)$	t	$x = t/8 + 1/16$	0	$0:7$
$(1,1)$	$l_{1,1}(t)$	t	$x + y = t/8 + 1/8$	0	$0:14$
$(1,2)$	$l_{1,2}(t)$	$t - 1/2$	$x + 2y = t/8 + 1/8$	$1/2$	$0:21$
	$l_{1,2}(t)$	$t + 1/2$	$x + 2y = t/8 + 1/4$	$1/2$	$21:-1:0$
$(1,3)$	$l_{1,3}(t)$	$t - 1$	$x + 3y = t/8 + 1/8$	1	$0:28$
	$l_{1,3}(t)$	$t + 1$	$x + 3y = t/8 + 3/8$	1	$28:-1:0$
$(1,4)$	$l_{1,4}(t)$	$t - 3/2$	$x + 4y = t/8 + 1/8$	$3/2$	$0:35$
	$l_{1,4}(t)$	$t + 3/2$	$x + 4y = t/8 + 1/2$	$3/2$	$35:-1:0$
(p,s)	A-rays	$t \pm t_0$	G-rays $l(t \mp t_0)$	$t_0 = \frac{p+N-s}{2} - 1$	t
$(1,5)$	$l_{1,-3}(t)$	$t - 1$	$x - 3y = t/8 - 1/4$	1	$7:-1:-21$
	$l_{1,-3}(t)$	$t + 1$	$x - 3y = t/8$	1	$-21:1:7$
$(1,6)$	$l_{1,-2}(t)$	$t - 1/2$	$x - 2y = t/8 - 1/8$	$1/2$	$7:-1:-14$
	$l_{1,-2}(t)$	$t + 1/2$	$x - 2y = t/8$	$1/2$	$-14:1:7$
$(1,7)$	$l_{1,-1}(t)$	t	$x - y = t/8$	0	$7:-1:-7$
(p,s)	A-rays	$t \pm t_0$	G-rays $l(t \mp t_0)$	$t_0 = \frac{p+s}{2} - 1$	t
$(0,1)$	$l_{0,1}(t)$	t	$y = t/8 + 1/16$	0	$0:7$
$(2,1)$	$l_{2,1}(t)$	$t - 1/2$	$2x + y = t/8 + 1/8$	$1/2$	$0:21$
	$l_{2,1}(t)$	$t + 1/2$	$2x + y = t/8 + 1/4$	$1/2$	$21:0$
$(4,1)$	$l_{4,1}(t)$	$t - 3/2$	$4x + y = t/8 + 1/8$	$3/2$	$0:35$
	$l_{4,1}(t)$	$t + 3/2$	$4x + y = t/8 + 1/2$	$3/2$	$35:0$
(p,s)	A-rays	$t \pm t_0$	G-rays $l(t \mp t_0)$	$t_0 = \frac{N-p+s}{2} - 1$	t
$(6,1)$	$l_{-2,1}(t)$	$t - 1/2$	$-2x + y = t/8 - 1/8$	$1/2$	$7:-1:-14$
	$l_{-2,1}(t)$	$t + 1/2$	$-2x + y = t/8$	$1/2$	$-14:1:7$

It should be noted that in this case, as well as in the general $N = 2^r$ case where $r \geq 2$, the system of equations for the G-rays is defined in the following way. We first consider the A-rays $l(t) = l_{p,s}(t)$ which are described by the equation

$$L_{p,s,t}(x,y) = xp + ys = \frac{t}{N} + \frac{p+s}{2N}.$$

The set of all generators of $J_{N,N}$ is divided into four subsets, and we consider each of them separately. The first and second subsets contain all generators $(1, s)$, where $s = 0 : N/2$ and $s = N/2+1 : N-1$, respectively. The third and fourth subsets contain all generators $(p, 1)$, where $p = 2p_1$ and $p_1 = 0 : N/4$ and $p_1 = N/4+1 : N/2-1$, respectively. For each of these subsets we consider case A which is referred to as the case when the arithmetical rays are shifted to the right, to define the corresponding geometrical rays. Case B is referred to as the case when the arithmetical rays are shifted to the left, to define the corresponding geometrical rays.

Subsets I and III: $p = 1$ and $s \leq N/2$, and $s = 1$ and $p \leq N/2$.
G-rays $\tilde{l}(t)$ are defined by the shift

$$t \to t \mp t_0, \quad t_0 = \frac{p+s}{2} - 1,$$

of A-rays, i.e.

$$\tilde{L}_{p,s,t}(x, y) = L_{p,s,t\mp t_0}(x, y) = \frac{t \mp t_0}{N} + \frac{p+s}{2N}.$$

Case A:

$$\tilde{L}_{p,s,t}(x, y) = L_{p,s,t-t_0}(x, y) = \frac{t - t_0}{N} + \frac{p+s}{2N}$$

$$= \frac{t}{N} - \left[\frac{p+s}{2N} - \frac{1}{N}\right] + \frac{p+s}{2N} = \frac{t}{N} + \frac{1}{N}.$$

Case B:

$$\tilde{L}_{p,s,t}(x, y) = L_{p,s,t+t_0}(x, y) = \frac{t + t_0}{N} + \frac{p+s}{2N}$$

$$= \frac{t}{N} + \left[\frac{p+s}{2N} - \frac{1}{N}\right] + \frac{p+s}{2N} = \frac{t}{N} + \frac{p+s-1}{N}.$$

Thus, if $p = 1$,

$$\tilde{L}_{1,s,t}(x, y) = L_{1,s,t+t_0}(x, y) = \frac{t}{N} + \frac{s}{N},$$

and if $s = 1$,

$$\tilde{L}_{p,1,t}(x, y) = L_{p,1,t+t_0}(x, y) = \frac{t}{N} + \frac{p}{N}.$$

The equation for case A is simple, when compared with case B.
Subset II: $p = 1$ and $s > N/2$.
A-rays are considered to be $l(t) = l_{1,s-N}(t)$, which are defined by

$$L_{1,s-N,t}(x, y) = x + y(s - N) = \frac{t}{N} + \frac{1 + (s - N)}{2N}.$$

The G-rays $\tilde{l}(t)$ are defined by the shift of A-rays:

$$t \to t \mp t_0, \quad t_0 = \frac{1 + (N - s)}{2} - 1 = \frac{N - s - 1}{2},$$

$$\tilde{L}_{1,s-N,t}(x, y) = L_{1,s-N,t \mp t_0}(x, y) = \frac{t \mp t_0}{N} + \frac{1 + (s - N)}{2N}.$$

Case A:

$$\tilde{L}_{1,s-N,t}(x, y) = L_{1,s-N,t-t_0}(x, y) = \frac{t - t_0}{N} + \frac{1 + s - N}{2N}$$

$$= \frac{t}{N} - \left[\frac{N - s - 1}{2N} \right] + \frac{s - N + 1}{2N} = \frac{t}{N} - \left[1 - \frac{s + 1}{N} \right].$$

Case B:

$$\tilde{L}_{1,s-N,t}(x, y) = L_{1,s-N,t+t_0}(x, y) = \frac{t + t_0}{N} + \frac{1 + s - N}{2N}$$

$$= \frac{t}{N} + \left[\frac{N - s - 1}{2N} \right] + \frac{s - N + 1}{2N} = \frac{t}{N}.$$

Subset IV: $p > N/2$ and $s = 1$.
A-rays are considered to be $l(t) = l_{p-N,1}(t)$, which are defined by

$$L_{p-N,1,t}(x, y) = x(p - N) + y = \frac{t}{N} + \frac{(p - N) + 1}{2N}.$$

The G-rays $\tilde{l}(t)$ are defined by the shift

$$t \to t \mp t_0, \quad t_0 = \frac{(N - p) + 1}{2} - 1 = \frac{N - p - 1}{2}$$

of A-rays, i.e.

$$\tilde{L}_{p-N,1,t}(x, y) = L_{p-N,1,t \mp t_0}(x, y) = \frac{t \mp t_0}{N} + \frac{(p - N) + 1}{2N}.$$

Case A:

$$\tilde{L}_{p-N,1,t}(x, y) = L_{p-N,1,t-t_0}(x, y) = \frac{t - t_0}{N} + \frac{p - N + 1}{2N}$$

$$= \frac{t}{N} - \left[\frac{N - p - 1}{2N} \right] + \frac{p - N + 1}{2N} = \frac{t}{N} - \left[1 - \frac{p + 1}{N} \right].$$

Case B:

$$\tilde{L}_{p-N,1,t}(x, y) = L_{p-N,1,t+t_0}(x, y) = \frac{t + t_0}{N} + \frac{p - N + 1}{2N}$$

$$= \frac{t}{N} + \left[\frac{N - p - 1}{2N} \right] + \frac{p - N + 1}{2N} = \frac{t}{N}.$$

Thus, for subsets II and IV, equations for case B have a simple form and can therefore be selected for image reconstruction. For subsets I and III, equations for case A are simple. In both cases A and B, geometrical rays are defined from the arithmetical rays by shifting $\tilde{l}(t) = l(t - t_0)$ and $\tilde{l}(t) = l(t + t_0)$, respectively. Here the shift t_0 is calculated by

$$
t_0 = \begin{cases} \dfrac{p+s}{2} - 1, \text{when } 0 < p, s \leq N/2, \\ \dfrac{p+(N-s)}{2} - 1, \text{when } p = 1,\ s > N/2, \\ \dfrac{(N-p)+s}{2} - 1, \text{when } p > N/2,\ s = 1. \end{cases} \tag{3.109}
$$

It should be noted that for each of the subsets of generators (p, s), the use of only case A, or case B leads to similar systems of equations that describe the procedure of calculating the sums $v(t)$ from integrals $w(t)$. Indeed, these equations are described respectively by the upper or lower triangular Toeplitz matrices. One can use only, for instance, the upper triangular Toeplitz matrices, if only case A will be used for all subsets of generators. It is also not difficult to notice that the solution of the system described by the low upper triangular Toeplitz matrix, $\mathbf{b} = A^{-1}\mathbf{w}$, is reduced to the solution of the system with the upper triangular Toeplitz matrix, $\tilde{\mathbf{b}} = \tilde{A}^{-1}\tilde{\mathbf{w}}$, after permutation $t \to M - t$ of rows of the vectors $\mathbf{b}, \mathbf{w}$, and the matrix A^{-1}. Here $M \times M$ is the size of the Toeplitz matrix.

3.6.2 Equations for line-integrals

For each (p, s)-projection, except the horizontal, vertical, and diagonal projections, the set of arithmetical rays is used for image reconstruction, after shifting the rays either to the left, $t \to t - t_0$, or to the right $t \to t + t_0$, where the positive parameter $t_0 = t_0(p, s)$ is calculated by (3.109). For the described $N = 8$ example, the systems of equations which describe the solutions $\mathbf{b} = A^{-1}\mathbf{w}$ of the convolution equations $\mathbf{w} = KA\mathbf{v}$, for all (p, s)-projections, are given together in Table 3.5 for case A, i.e., when all shifts are by $t \to t - t_0$. Constants K are defined by the lengths of intersection of the ray with the image element. For instance, $K = 8\sqrt{2}, 8\sqrt{5}/2$, and $8\sqrt{10}/3$ for the $(1, 1), (1, 2)$, and $(1, 3)$-projections, respectively.

Another set of systems of equations, which describe the solutions of the convolution equations $\mathbf{w} = KA\mathbf{v}$, is given together in Table 3.6, for case B, i.e., when all shifting of arithmetical rays are performed by $t \to t + t_0$.

TABLE 3.5
Model I (case A): Convolution equations for the 8×8 case

G-rays	$\mathbf{w} = KA\mathbf{v}, \quad \mathbf{b} = A^{-1}\mathbf{w}$	$\mathbf{v} = B\mathbf{w} \ (\mathbf{v} = B\mathbf{b})$
$l_{1,0}(t)$	$w(t) = 8v(t) \quad t = 0:7, \quad (K = 8)$	$v(t) = w(t)/8$
$l_{1,1}(t)$	$w(t) = 8\sqrt{2}v(t) \quad t = 0:14, \quad (K = 8\sqrt{2})$	$v(t) = w(t)/(8\sqrt{2})$
$l_{1,2}(t - 0.5)$	$w(t) = 4\sqrt{5}[v(t-1) + v(t)]$ $b_0 = w(0), \ b_t = w(t) - b_{t-1}$	$v(t) = b_t/(4\sqrt{5})$ $t = 0:21$
$l_{1,3}(t - 1)$	$w(t) = 8\sqrt{10}/3[v(t) + v(t-1) + v(t-2)]$ $b_0 = w(0), \ b_t = w(t) - (b_{t-1} + b_{t-2})$	$v(t) = 3b_t/(8\sqrt{10})$ $t = 0:28$
$l_{1,4}(t - 1.5)$	$w(t) = 2\sqrt{17}[v(t) + v(t-1) + v(t-2) + v(t-3)]$ $b_0 = w(0), \ b_t = w(t) - (b_{t-1} + b_{t-2} + b_{t-3})$	$v(t) = b_t/(2\sqrt{17})$ $t = 0:35$
$l_{1,-3}(t - 1)$	$w(t) = 8\sqrt{10}/3[v(t) + v(t-1) + v(t-2)]$ $b_{-21} = w(-21), \ b_t = w(t) - (b_{t-1} + b_{t-2})$	$v(t) = 3b_t/(8\sqrt{10})$ $t = -21:7$
$l_{1,-2}(t - 0.5)$	$w(t) = 4\sqrt{5}[v(t) + v(t-1))]$ $b_{-14} = w(-14), \ b_t = w(t) - b_{t-1}$	$v(t) = b_t/(4\sqrt{5})$ $t = -14:7$
$l_{1,-1}(t)$	$w(t) = 8\sqrt{2}v(t)$ $b_t = w(t)$	$v(t) = w(t)/(8\sqrt{2})$ $t = -7:7$
$l_{0,1}(t)$	$w(t) = 8v(t), \quad t = 0:7$	$v(t) = w(t)/8$
$l_{2,1}(t - 0.5)$	$w(t) = 4\sqrt{5}[v(t) + v(t-1)]$ $b_0 = w(0), \ b_t = w(t) - b_{t-1}$	$v(t) = b_t/(4\sqrt{5})$ $t = 0:21$
$l_{4,1}(t - 1.5)$	$w(t) = 2\sqrt{17}[v(t) + v(t-1) + v(t-2) + v(t-3)]$ $b_0 = w(0), \ b_t = w(t) - (b_{t-1} + b_{t-2} + b_{t-3})$	$v(t) = b_t/(2\sqrt{17})$ $t = 0:35$
$l_{-2,1}(t - 0.5)$	$w(t) = 4\sqrt{5}[v(t) + v(t-1))]$ $b_{-14} = w(-14), \ b_t = w(t) - b_{t-1}$	$v(t) = b_t/(4\sqrt{5})$ $t = -14:7$

3.7 Equations in the coordinate system $(X, 1 - Y)$

It is common to consider the coordinate system as shown in Figure 3.2. In the first realization of the proposed method of image reconstruction in the MATLAB®-based program, this system was used for calculating all rays, arithmetical and geometrical. Thus, the realization was implemented in the system of coordinates $(x, 1 - y)$ instead of the original system (x, y).

In the new coordinate system, A-rays $l(t) = l_{p,s}(t)$ within the unit square $[0, 1] \times [0, 1]$ can be written as

$$L_{p,s,t}(x, 1 - y) = xp + (1 - y)s = \frac{t}{N} + \frac{p + s}{2N}.$$

Therefore, we obtain another set of equations, as shown in Table 3.7, for the $N = 8$ case. We now consider separately each subset of generators and write the corresponding equations.

Subsets I and III: $p = 1$ and $s \leq N/2$, and $s = 1$ and $p \leq N/2$.

The G-rays $\tilde{l}(t)$ are defined by the shift

$$t \to t \mp t_0, \quad t_0 = \frac{p + s}{2} - 1,$$

TABLE 3.6
Model I (case B): Convolution equations for the 8×8 case

G-rays	$\mathbf{w} = KA\mathbf{v}, \quad \mathbf{b} = A^{-1}\mathbf{w}$	$\mathbf{v} = B\mathbf{w} \ (\mathbf{v} = B\mathbf{b})$
$l_{1,0}(t)$	$w(t) = 8v(t), \quad t = 0:7, \quad (K=8)$	$v(t) = w(t)/8$
$l_{1,1}(t)$	$w(t) = 8\sqrt{2}v(t), \quad t = 0:14 \quad (K = 8\sqrt{2})$	$v(t) = w(t)/(8\sqrt{2})$
$l_{1,2}(t+0.5)$	$w(t) = 4\sqrt{5}[v(t) + v(t+1)]$ $b_{21} = w(21), \ b_t = w(t) - b_{t+1}$	$v(t) = b_t/(4\sqrt{5})$ $t = 21 : -1 : 0$
$l_{1,3}(t+1)$	$w(t) = 8\sqrt{10}/3[v(t) + v(t+1) + v(t+2)]$ $b_{28} = w(28), \ b_t = w(t) - (b_{t+1} + b_{t+2})$	$v(t) = b_t/(8/3\sqrt{10})$ $t = 28 : -1 : 0$
$l_{1,4}(t+1.5)$	$w(t) = 2\sqrt{17}[v(t) + v(t+1) + v(t+2) + v(t+3)]$ $b_{35} = w(35), \ b_t = w(t) - (b_{t+1} + b_{t+2} + b_{t+3})$	$v(t) = b_t/(2\sqrt{17})$ $t = 35 : -1 : 0$
$l_{1,-3}(t+1)$	$w(t) = 8\sqrt{10}/3[v(t) + v(t+1) + v(t+2)]$ $b_7 = w(7), \ b_t = w(t) - (b_{t+1} + b_{t+2})$	$v(t) = b_t/(8/3\sqrt{10})$ $t = 7 : -1 : -21$
$l_{1,-2}(t+0.5)$	$w(t) = 4\sqrt{5}[v(t) + v(t+1))]$ $b_7 = w(7), \ b_t = w(t) - b_{t+1}$	$v(t) = b_t/(4\sqrt{5})$ $t = 7 : -1 : -14$
$l_{1,-1}(t)$	$w(t) = \sqrt{2}/8v(t)$ $b_t = w(t)$	$v(t) = w(t)/(8\sqrt{2})$ $t = 7 : -1 : -7$
$l_{0,1}(t)$	$w(t) = 8v(t), \quad t = 0:7$	$v(t) = w(t)/8$
$l_{2,1}(t+0.5)$	$w(t) = 4\sqrt{5}[v(t) + v(t+1)]$ $b_{21} = w(21), \ b_t = w(t) - b_{t+1}$	$v(t) = b_t/(4\sqrt{5})$ $t = 21 : -1 : 0$
$l_{4,1}(t+1.5)$	$w(t) = 2\sqrt{17}[v(t) + v(t+1) + v(t+2) + v(t+3)]$ $b_{35} = w(35), \ b_t = w(t) - (b_{t+1} + b_{t+2} + b_{t+3})$	$v(t) = b_t/(2\sqrt{17})$ $t = 35 : -1 : 0$
$l_{-2,1}(t+0.5)$	$w(t) = 4\sqrt{5}[v(t) + v(t+1))]$ $b_7 = w(7), \ b_t = w(t) - b_{t+1}$	$v(t) = b_t/(4\sqrt{5})$ $t = 7 : -1 : -14$

of A-rays,

$$\tilde{L}_{p,s,t}(x, 1-y) = L_{p,s,t \mp t_0}(x, 1-y) = \frac{t \mp t_0}{N} + \frac{p+s}{2N}.$$

Case A:

$$\tilde{L}_{p,s,t}(x, 1-y) = L_{p,s,t-t_0}(x, 1-y) = \frac{t - t_0}{N} + \frac{p+s}{2N}$$

$$= \frac{t}{N} - \left[\frac{p+s}{2N} - \frac{1}{N}\right] + \frac{p+s}{2N} = \frac{t}{N} + \frac{1}{N}.$$

Thus we have the equation:

$$xp + (1-y)s = \frac{t}{N} + \frac{1}{N}, \quad t = 0 : (p+s)(N-1). \tag{3.110}$$

If $p = 1$,

$$x + (1-y)s = \frac{t}{N} + \frac{1}{N}, \quad t = 0 : (s+1)(N-1).$$

If $s = 1$,

$$xp + (1-y) = \frac{t}{N} + \frac{1}{N}, \quad t = 0 : (p+1)(N-1).$$

TABLE 3.7

Model II: Equations of G-rays on the grid 8×8

(p,s)	A-rays	G-rays $\tilde{l}(t)$	t
$(1,0)$	$l_{1,0}(t)$	$x = t/8 + 1/16$	$0:7$
$(1,1)$	$l_{1,1}(t)$	$x + (1-y) = t/8 + 1/8$	$0:14$
$(1,2)$	$l_{1,2}(t)$	$x + 2(1-y) = t/8 + 1/8$	$0:21$
	$l_{1,2}(t)$	$x + 2(1-y) = t/8 + 1/4$	$21:-1:0$
$(1,3)$	$l_{1,3}(t)$	$x + 3(1-y) = t/8 + 1/8$	$0:28$
	$l_{1,3}(t)$	$x + 3(1-y) = t/8 + 3/8$	$28:-1:0$
$(1,4)$	$l_{1,4}(t)$	$x + 4(1-y) = t/8 + 1/8$	$0:35$
	$l_{1,4}(t)$	$x + 4(1-y) = t/8 + 1/2$	$35:-1:0$
(p,s)	A-rays	G-rays $\tilde{l}(t)$	t
$(1,5)$	$l_{1,-3}(t)$	$x - 3(1-y) = t/8 - 1/4$	$7:-1:-21$
	$l_{1,-3}(t)$	$x - 3(1-y) = t/8$	$-21:1:7$
$(1,6)$	$l_{1,-2}(t)$	$x - 2(1-y) = t/8 - 1/8$	$7:-1:-14$
	$l_{1,-2}(t)$	$x - 2(1-y) = t/8$	$-14:1:7$
$(1,7)$	$l_{1,-1}(t)$	$x - (1-y) = t/8$	$7:-1:-7$
(p,s)	A-rays	G-rays $\tilde{l}(t)$	t
$(0,1)$	$l_{0,1}(t)$	$(1-y) = t/8 + 1/16$	$0:7$
$(2,1)$	$l_{2,1}(t)$	$2x + (1-y) = t/8 + 1/8$	$0:21$
	$l_{2,1}(t)$	$2x + (1-y) = t/8 + 1/4$	$21:-1:0$
$(4,1)$	$l_{4,1}(t)$	$4x + (1-y) = t/8 + 1/8$	$0:35$
	$l_{4,1}(t)$	$4x + (1-y) = t/8 + 1/2$	$35:-1:0$
(p,s)	A-rays	G-rays $\tilde{l}(t)$	t
$(6,1)$	$l_{-2,1}(t)$	$-2x + (1-y) = t/8 - 1/8$	$7:-1:-14$
	$l_{-2,1}(t)$	$-2x + (1-y) = t/8$	$-14:1:7$

Case B:

$$\tilde{L}_{p,s,t}(x, 1-y) = L_{p,s,t+t_0}(x, 1-y) = \frac{t+t_0}{N} + \frac{p+s}{2N}$$

$$= \frac{t}{N} + \left[\frac{p+s}{2N} - \frac{1}{N}\right] + \frac{p+s}{2N} = \frac{t}{N} + \frac{p+s-1}{N},$$

where $t = (p+s)(N-1) : -1 : 0$.

If $p = 1$,

$$\tilde{L}_{1,s,t}(x, 1-y) = L_{1,s,t+t_0}(x, 1-y) = \frac{t}{N} + \frac{s}{N},$$

and, if $s = 1$,

$$\tilde{L}_{p,1,t}(x, 1-y) = L_{p,1,t+t_0}(x, 1-y) = \frac{t}{N} + \frac{p}{N}.$$

Thus, if $p = 1$,

$$x + (1-y)s = \frac{t}{N} + \frac{s}{N}, \quad t = (s+1)(N-1) : -1 : 0,$$

if $s = 1$, the G-rays are described by

$$xp + (1 - y) = \frac{t}{N} + \frac{p}{N}, \quad t = (p+1)(N-1) : -1 : 0.$$

Subset II: $p = 1$ and $s > N/2$.
A-rays are considered to be $l(t) = l_{1,s-N}(t)$, which are defined by

$$L_{1,s-N,t}(x, 1 - y) = x + (1 - y)(s - N) = \frac{t}{N} + \frac{1 + (s - N)}{2N}$$

The G-rays $\tilde{l}(t)$ are defined by the shift of A-rays:

$$t \to t \mp t_0, \quad t_0 = \frac{1 + (N - s)}{2} - 1 = \frac{N - s - 1}{2},$$

$$\tilde{L}_{1,s-N,t}(x, 1 - y) = L_{1,s-N,t\mp t_0}(x, 1 - y) = \frac{t \mp t_0}{N} + \frac{1 + (s - N)}{2N}.$$

Case A:

$$\tilde{L}_{1,s-N,t}(x, 1 - y) = L_{1,s-N,t-t_0}(x, 1 - y) = \frac{t - t_0}{N} + \frac{1 + s - N}{2N}$$

$$= \frac{t}{N} - \left[\frac{N - s - 1}{2N}\right] + \frac{s - N + 1}{2N} = \frac{t}{N} - \left[1 - \frac{s + 1}{N}\right].$$

Thus

$$x + (1 - y)(s - N) = \frac{t}{N} - \left[1 - \frac{s + 1}{N}\right], \quad t = (N - 1) : -1 : -(N - s)(N - 1).$$

Case B:

$$\tilde{L}_{1,s-N,t}(x, 1 - y) = L_{1,s-N,t+t_0}(x, 1 - y) = \frac{t + t_0}{N} + \frac{1 + s - N}{2N}$$

$$= \frac{t}{N} + \left[\frac{N - s - 1}{2N}\right] + \frac{s - N + 1}{2N} = \frac{t}{N}.$$

Thus

$$x + (1 - y)(s - N) = \frac{t}{N}, \quad t = -(N - s)(N - 1) : 1 : (N - 1).$$

Subset IV: $p > N/2$ and $s = 1$.
A-rays are considered to be $l(t) = l_{p-N,1}(t)$, which are defined by

$$L_{p-N,1,t}(x, 1 - y) = x(p - N) + (1 - y) = \frac{t}{N} + \frac{(p - N) + 1}{2N}.$$

The G-rays $\tilde{l}(t)$ are defined by the shift

$$t \to t \mp t_0, \quad t_0 = \frac{(N - p) + 1}{2} - 1 = \frac{N - p - 1}{2}$$

of A-rays

$$\tilde{L}_{p-N,1,t}(x,1-y) = L_{p-N,1,t\mp t_0}(x,1-y) = \frac{t\mp t_0}{N} + \frac{(p-N)+1}{2N}.$$

Case A:

$$\tilde{L}_{p-N,1,t}(x,1-y) = L_{p-N,1,t-t_0}(x,1-y) = \frac{t-t_0}{N} + \frac{p-N+1}{2N}$$

$$= \frac{t}{N} - \left[\frac{N-p-1}{2N}\right] + \frac{p-N+1}{2N} = \frac{t}{N} - \left[1 - \frac{p+1}{N}\right].$$

Thus

$$x(p-N) + (1-y) = \frac{t}{N} - \left[1 - \frac{p+1}{N}\right], \quad t = (N-1) : -1 : -(N-p)(N-1).$$

Case B:

$$\tilde{L}_{p-N,1,t}(x,y) = L_{p-N,1,t+t_0}(x,y) = \frac{t+t_0}{N} + \frac{p-N+1}{2N}$$

$$= \frac{t}{N} + \left[\frac{N-p-1}{2N}\right] + \frac{p-N+1}{2N} = \frac{t}{N}.$$

Thus

$$x(p-N) + (1-y) = \frac{t}{N}, \quad t = -(N-p)(N-1) : 1 : (N-1).$$

For each subset of generators, there are two cases, A or B, and one of them can be selected to accomplish the method of image reconstruction. For two scenarios, when only case A or B is considered for all four subsets of generators of the set $J_{N,N}$, the equations for G-rays are given in Tables 3.8 and 3.9, respectively. In the first column, the number I, II, III, or IV is assigned to the subset containing generators (p,s).

Here we recall that for the generators $(1,s)$, and $(p,1)$, such that $p,s > N/2$, all equations for rays are written and the components of the 2-D paired transform, or splitting-signals $f_{T'_{p,s}}$ that are generated by these frequency-points, are calculated by considering the change in direction of projections. The $(1,s)$-projections are changed by the corresponding $(1,s-N)$-projections, and the $(p,1)$-projections by the $(p-N,1)$-projections.

For the $N=8$ case, Figure 3.85 shows two sets of angles of 12 projections before and after changing the directions, $\Phi_{8,8} = \{\arctan(s/p); (p,s) \in J_{8,8}\}$ and $\Phi^*_{8,8} = \{\arctan(s/p); (p,s) \in J^*_{8,8}\}$. The sets of generators are defined as

$$J_{8,8} = \{\{(p,1); p = 0 : 7\}, \{(1,2s); s = 0 : 3\}\},$$
$$J^*_{8,8} = \{\{(p,1); p = 0 : 4\}, \{(\bar{p},1); \bar{p} = -3 : -1\}, \{(1,2s); s = 0 : 2\}, (1,-2)\}.$$

Herewith, all these projections together with the remaining ones, are used to calculate the 2-D discrete paired transform of the image, with generators (p,s) from the original set $J_{N,N}$. That allows for using the existing program for the inverse 2-D discrete paired transform (which is available at http://www.fasttransforms.com).

TABLE 3.8

Model II (case A): Equations of G-rays for image $N \times N$

	(p,s)	G-rays	
I	$(1,0)$	$x - \dfrac{t}{N} - \dfrac{1}{2N} = 0$ $t = 0 : (N-1)$	$s = 0$
I	$(1,s)$	$x - ys + s - \dfrac{t}{N} - \dfrac{1}{N} = 0$ $t = 0 : (s+1)(N-1)$	$s \leq N/2$
II	$(1,s)$	$x + y(N-s) + (s-N) - \dfrac{t}{N} + \left(1 - \dfrac{s+1}{N}\right) = 0$ $t = (N-1) : -1 : -(N-s)(N-1)$	$s > N/2$
III	$(0,1)$	$-y + 1 - \dfrac{t}{N} - \dfrac{1}{2N} = 0,$ $t = 0 : (N-1)$	$p = 0$
III	$(p,1)$	$xp - y + 1 - \dfrac{t}{N} - \dfrac{1}{N} = 0$ $t = 0 : (p+1)(N-1)$	$p \leq N/2$
IV	$(p,1)$	$x(p-N) - y + 1 - \dfrac{t}{N} + \left(1 - \dfrac{p+1}{N}\right) = 0$ $t = (N-1) : -1 : -(N-p)(N-1)$	$p > N/2$

TABLE 3.9

Model II (case B): Equations of G-rays for image $N \times N$

	(p,s)	G-rays	
I	$(1,0)$	$x - \dfrac{t}{N} - \dfrac{1}{2N} = 0$ $t = 0 : (N-1)$	$s = 0$
I	$(1,s)$	$x - ys + s - \dfrac{t}{N} - \dfrac{s}{N} = 0$ $t = (s+1)(N-1) : -1 : 0$	$s \leq N/2$
II	$(1,s)$	$x - y(s-N) + (s-N) - \dfrac{t}{N} = 0$ $t = -(N-s)(N-1) : 1 : (N-1)$	$s > N/2$
III	$(0,1)$	$-y + 1 - \dfrac{t}{N} - \dfrac{1}{2N} = 0,$ $t = 0 : (N-1)$	$p = 0$
III	$(p,1)$	$xp - y + 1 - \dfrac{t}{N} - \dfrac{p}{N} = 0$ $t = (p+1)(N-1) : -1 : 0$	$p \leq N/2$
IV	$(p,1)$	$x(p-N) - y + 1 - \dfrac{t}{N} = 0$ $t = -(N-p)(N-1) : 1 : (N-1)$	$p > N/2$

3.7.1 Convolution equations

For all generators $(p,s) \in J_{N,N}$, the relations between the sums $v(t) = v_{p,s}(t)$ of the discrete image $f_{n,m}$ and line-integrals $w(t) = w_{p,s}(t)$ of the original

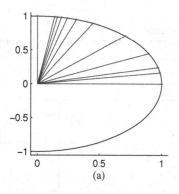
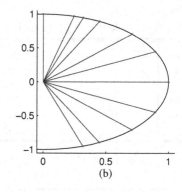

(a)   (b)

FIGURE 3.85
Central angles of (a) original and (b) new 12 projections.

unknown image $f(x, y)$ are described in matrix form by the Toeplitz matrices, which are similar to the described $N = 8$ and $N = 4$ examples. We consider separately the convolution equations of line-integrals for cases A and B.

1. Equations in case A

Subset I: $p = 1$ and $s \leq N/2$.

When $s \neq 0$, the convolution equation $\mathbf{w} = \mathbf{A}\mathbf{v}$ is described as

$$w(t) = N\frac{\sqrt{1+s^2}}{s}[v(t) + v(t-1) + v(t-2) + \ldots + v(t-s+1)], \quad (3.111)$$
$$t = 0 : (1+s)(N-1),$$

where it is assumed that $v(-1) = v(-2) = \ldots = v(-s+1) = 0$. The inverse transform $\mathbf{b} = \mathbf{A}^{-1}\mathbf{w}$ is calculated by the following recurrent form:

$$
\begin{aligned}
b_0 &= w(0), \\
b_1 &= w(1) - b_0, \\
b_2 &= w(2) - (b_1 + b_0), \\
b_3 &= w(3) - (b_2 + b_1 + b_0), \\
\ldots \quad &\ldots \quad \ldots \quad \ldots \quad \ldots \quad \ldots \\
b_{s-1} &= w(s-1) - (b_{s-2} + b_{s-3} + \ldots + b_1 + b_0), \\
b_t &= w(t) - (b_{t-1} + b_{t-2} + \ldots + b_{t-s+2} + b_{t-s+1)}), \\
&\quad t = s, s+1, \ldots, (1+s)(N-1).
\end{aligned}
\quad (3.112)
$$

The required sums $v(t)$ of the discrete image are calculated by

$$v(t) = b_t \frac{s}{N\sqrt{1+s^2}}, \quad t = 0 : (1+s)(N-1). \quad (3.113)$$

In the $s = 0$ case, we have the following simple solution: $v(t) = w(t)/N$, $t = 0 : (N-1)$.

Subset II: $p = 1$ and $s > N/2$.

The $(1, s)$-projection is described as the $(1, s - N)$-projection, i.e., s is considered as $s - N$. Therefore, the following convolution equation $\mathbf{w} = \mathbf{A}\mathbf{v}$ is used:

$$w(t) = N\frac{\sqrt{1 + \bar{s}^2}}{\bar{s}}[v(t) + v(t-1) + v(t-2) + \ldots + v(t - \bar{s} + 1)], \quad (3.114)$$
$$t = (N-1) : -1 : -\bar{s}(N-1),$$

where $\bar{s} = N - s$, and $v(-k) = 0$, when $k > \bar{s}(N-1))$.

We denote by M the negative number $-\bar{s}(N-1)$. The inverse transform $\mathbf{b} = \mathbf{A}^{-1}\mathbf{w}$ can be calculated by the following recurrent form:

$$
\begin{aligned}
b_M &= w(M), \\
b_{M+1} &= w(M+1) - b_M, \\
b_{M+2} &= w(M+2) - (b_{M+1} + b_M), \\
b_{M+3} &= w(M+3) - (b_{M+2} + b_{M+1} + b_M), \\
\ldots &\quad \ldots \quad \ldots \quad \ldots \quad \ldots \quad \ldots \quad \ldots \quad \ldots \\
b_{M+\bar{s}-1} &= w(M + \bar{s} - 1) - (b_{M+\bar{s}-2} + b_{M+\bar{s}-3} + \ldots + b_{M+1} + b_M), \\
b_t &= w(t) - (b_{t-1} + b_{t-2} + \ldots + b_{t-\bar{s}+2} + b_{t-\bar{s}+1)}), \\
&\quad t = M + \bar{s}, M + \bar{s} + 1, \ldots, (N-1).
\end{aligned}
\quad (3.115)
$$

The required sums $v(t)$ of the discrete image are calculated by

$$v(t) = b_t\frac{\bar{s}}{N\sqrt{1 + \bar{s}^2}}, \quad t = (N-1) : -1 : -\bar{s}(N-1). \quad (3.116)$$

Subset III: $p \leq N/2$ and $s = 1$.

This subset is described similarly to subset I. When $p \neq 0$, the convolution equation $\mathbf{w} = \mathbf{A}\mathbf{v}$ is described as

$$w(t) = N\frac{\sqrt{1 + p^2}}{p}[v(t) + v(t-1) + v(t-2) + \ldots + v(t - p + 1)], \quad (3.117)$$
$$t = 0 : (p+1)(N-1).$$

The inverse transform $\mathbf{b} = \mathbf{A}^{-1}\mathbf{w}$ can be calculated in the recurrent form

$$
\begin{aligned}
b_0 &= w(0), \\
b_1 &= w(1) - b_0, \\
b_2 &= w(2) - (b_1 + b_0), \\
b_3 &= w(3) - (b_2 + b_1 + b_0), \\
\ldots &\quad \ldots \quad \ldots \quad \ldots \quad \ldots \quad \ldots \\
b_{p-1} &= w(p-1) - (b_{p-2} + b_{p-3} + \ldots + b_1 + b_0), \\
b_t &= w(t) - (b_{t-1} + b_{t-2} + \ldots + b_{t-p+2} + b_{t-p+1)}), \\
&\quad t = p, p+1, \ldots, (p+1)(N-1).
\end{aligned}
\quad (3.118)
$$

The required sums $v(t)$ of the discrete image are calculated by

$$v(t) = b_t\frac{p}{N\sqrt{1 + p^2}}, \quad t = 0 : M. \quad (3.119)$$

In the $p = 0$ case, we have the following simple solution: $v(t) = w(t)/N$, $t = 0 : (N-1)$.

Subset IV: $p > N/2$ and $s = 1$.

The projection $(p, 1)$ is considered as the $(p - N, 1)$-projection, i.e., p is considered as $p - N$. The convolution equation $\mathbf{w} = \mathbf{A}\mathbf{v}$ is described as

$$w(t) = N\frac{\sqrt{1 + \bar{p}^2}}{\bar{p}}[v(t) + v(t-1) + v(t-2) + \ldots + v(t - \bar{p} + 1)], \tag{3.120}$$
$$t = (N-1) : -1 : -\bar{p}(N-1),$$

where $\bar{p} = N - p$, and $v(k) = 0$, when $k < -\bar{p}(N-1))$. The inverse transform $\mathbf{b} = \mathbf{A}^{-1}\mathbf{w}$ can be calculated by the recurrent form

$$\begin{aligned}
b_M &= w(M), \\
b_{M+1} &= w(M+1) - b_M, \\
b_{M+2} &= w(M+2) - (b_{M+1} + b_M), \\
b_{M+3} &= w(M+3) - (b_{M+2} + b_{M+1} + b_M), \\
&\ldots \quad \ldots \quad \ldots \quad \ldots \quad \ldots \quad \ldots \quad \ldots \quad \ldots \\
b_{M+\bar{p}-1} &= w(M+\bar{p}-1) - (b_{M+\bar{p}-2} + b_{M+\bar{p}-3} + \ldots + b_{M+1} + b_M), \\
b_t &= w(t) - (b_{t-1} + b_{t-2} + \ldots + b_{t-\bar{p}+2} + b_{t-\bar{p}+1})), \\
&\quad t = M + \bar{p}, M + \bar{p} + 1, \ldots, (N-1),
\end{aligned} \tag{3.121}$$

where $M = -\bar{p}(N-1)$. The required sums $v(t)$ of the discrete image are calculated by

$$v(t) = b_t\frac{\bar{p}}{N\sqrt{1 + \bar{p}^2}}, \quad t = (N-1) : -1 : -\bar{p}(N-1). \tag{3.122}$$

We now write all convolution equations for case B.

2. Equations in case B

Subset I: $p = 1$ and $s \leq N/2$.

When $s \neq 0$, the convolution equation $\mathbf{w} = \mathbf{A}\mathbf{v}$ is described as

$$w(t) = N\frac{\sqrt{1 + s^2}}{s}[v(t) + v(t+1) + v(t+2) + \ldots + v(t+s-1)], \tag{3.123}$$
$$t = (1+s)(N-1) : -1 : 0,$$

where it is assumed that $v(k) = 0$, for $k > (1+s)(N-1)$. The inverse transform $\mathbf{b} = \mathbf{A}^{-1}\mathbf{w}$ is calculated by the recurrent form

$$\begin{aligned}
b_M &= w(M), \\
b_{M-1} &= w(M-1) - b_M, \\
b_{M-2} &= w(M-2) - (b_{M-1} + b_M), \\
b_{M-3} &= w(M-3) - (b_{M-2} + b_{M-1} + b_M), \\
&\ldots \quad \ldots \quad \ldots \quad \ldots \quad \ldots \quad \ldots \\
b_{M-s+1} &= w(M-s+1) - (b_{M-s+2} + b_{M-s+3} + \ldots + b_{M-1} + b_M), \\
b_t &= w(t) - (b_{t+1} + b_{t+2} + \ldots + b_{t+s-2} + b_{t+s-1})), \\
&\quad t = M - s, M - s - 1, \ldots, 0,
\end{aligned} \tag{3.124}$$

where $(M = (1+s)(N-1)$. The required sums $v(t)$ of the discrete image are calculated by

$$v(t) = b_t \frac{s}{N\sqrt{1+s^2}}, \quad t = 0 : (1+s)(N-1). \tag{3.125}$$

In the $s = 0$ case, we have the following: $v(t) = w(t)/N, \ t = 0 : (N-1)$.

Subset II: $p = 1$ and $s > N/2$.

The projection $(1,s)$ is described as the $(1, s-N)$-projection. The following convolution equation $\mathbf{w} = \mathbf{Av}$ is considered:

$$w(t) = N\frac{\sqrt{1+\bar{s}^2}}{\bar{s}}[v(t) + v(t+1) + v(t+2) + \ldots + v(t+\bar{s}-1)], \tag{3.126}$$
$$t = (N-1) : -1 : -\bar{s}(N-1),$$

where $\bar{s} = N - s$, and $v(N) = v(N+1) = \ldots = v(N+\bar{s}-2) = 0$. The inverse transform $\mathbf{b} = \mathbf{A}^{-1}\mathbf{w}$ is calculated by the recurrent form

$$\begin{aligned}
b_{N-1} &= w(N-1), \\
b_{N-2} &= w(N-2) - b_{N-1}, \\
b_{N-3} &= w(N-3) - (b_{N-2} + b_{N-1}), \\
b_{N-4} &= w(N-4) - (b_{N-3} + b_{N-2} + b_{N-1}), \\
\ldots \quad & \ldots \quad \ldots \quad \ldots \quad \ldots \quad \ldots \quad \ldots \quad \ldots \\
b_{N-\bar{s}} &= w(N-\bar{s}) - (b_{N-\bar{s}+1} + b_{N-\bar{s}+2} + \ldots + b_{N-2} + b_{N-1}), \\
b_t &= w(t) - (b_{t+1} + b_{t+2} + \ldots + b_{t+\bar{s}+2} + b_{t+\bar{s}-1)}), \\
& t = N - \bar{s} - 1, N - \bar{s} - 2, \ldots, -\bar{s}(N-1).
\end{aligned} \tag{3.127}$$

where $M = -\bar{s}(N-1)$. The required sums $v(t)$ of the discrete image are calculated by

$$v(t) = b_t \frac{\bar{s}}{N\sqrt{1+\bar{s}^2}}, \quad t = (N-1) : -1 : -\bar{s}(N-1). \tag{3.128}$$

Subset III: $p \leq N/2$ and $s = 1$.

When $p \neq 0$, the convolution equation $\mathbf{w} = \mathbf{Av}$ is described as

$$w(t) = N\frac{\sqrt{1+p^2}}{p}[v(t) + v(t+1) + v(t+2) + \ldots + v(t+p-1)], \tag{3.129}$$
$$t = (p+1)(N-1) : -1 : 0,$$

where it is assumed that $v(k) = 0$, for $k > (p+1)(N-1)$. The inverse transform $\mathbf{b} = \mathbf{A}^{-1}\mathbf{w}$ is calculated by the following recurrent form: $(M = (p+1)(N-1))$:

$$\begin{aligned}
b_M &= w(M), \\
b_{M-1} &= w(M-1) - b_M, \\
b_{M-2} &= w(M-2) - (b_{M-1} + b_M), \\
b_{M-3} &= w(M-3) - (b_{M-2} + b_{M-1} + b_M), \\
\ldots \quad & \ldots \quad \ldots \quad \ldots \quad \ldots \quad \ldots \\
b_{M-p+1} &= w(M-p+1) - (b_{M-p+2} + b_{M-p+3} + \ldots + b_{M-1} + b_M), \\
b_t &= w(t) - (b_{t+1} + b_{t+2} + \ldots + b_{t+p-2} + b_{t+p-1)}), \\
& t = M - p, M - p - 1, \ldots, 0.
\end{aligned} \tag{3.130}$$

The required sums $v(t)$ of the discrete image are calculated by

$$v(t) = b_t \frac{p}{N\sqrt{1+p^2}}, \quad t = 0 : (p+1)(N-1). \tag{3.131}$$

In the $p = 0$ case, $v(t) = w(t)/N$, $t = 0 : (N-1)$.

Subset IV: $p > N/2$ and $s = 1$.

The projection $(p, 1)$ is described as the $(p-N, 1)$-projection. The following convolution equation $\mathbf{w} = \mathbf{Av}$ is considered:

$$w(t) = N\frac{\sqrt{1+\bar{p}^2}}{\bar{p}}[v(t) + v(t+1) + v(t+2) + ... + v(t+\bar{p}-1)], \tag{3.132}$$
$$t = (N-1) : -1 : -\bar{p}(N-1),$$

where $\bar{p} = N - p$, and $v(N) = v(N+1) = ... = v(N+\bar{p}-1) = 0$. The inverse transform $\mathbf{b} = \mathbf{A}^{-1}\mathbf{w}$ is calculated by the recurrent form

$$\begin{aligned}
b_{N-1} &= w(N-1), \\
b_{N-2} &= w(N-2) - b_{N-1}, \\
b_{N-3} &= w(N-3) - (b_{N-2} + b_{N-1}), \\
b_{N-4} &= w(N-4) - (b_{N-3} + b_{N-2} + b_{N-1}), \\
... \quad & ... \quad ... \quad ... \quad ... \quad ... \quad ... \quad ... \\
b_{N-\bar{p}} &= w(N-\bar{p}) - (b_{N-\bar{p}+1} + b_{N-\bar{p}+2} + ... + b_{N-2} + b_{N-1}), \\
b_t \quad &= w(t) - (b_{t+1} + b_{t+2} + ... + b_{t+\bar{p}+2} + b_{t+\bar{p}-1})), \\
& \quad t = N - \bar{p} - 1, N - \bar{p} - 2, ..., -\bar{p}(N-1).
\end{aligned} \tag{3.133}$$

where $M = -\bar{p}(N-1)$. The required sums $v(t)$ of the discrete image are calculated by

$$v(t) = b_t \frac{\bar{p}}{N\sqrt{1+\bar{p}^2}}, \quad t = (N-1) : -1 : -\bar{p}(N-1). \tag{3.134}$$

Problems

Problem 3.1 Different sets of generators (p, s) can be used to cover the Cartesian lattice 8×8 by sets $T_{p,s}$, but each such set will have a cardinality greater than or equal to 12. For each (p, s)-projection, there are different ways to select the set of G-rays which will be used to calculate the line-integrals $w_{p,s}(t)$. The number of these integrals depends on (p, s). How would you choose the sets of generators and G-rays, to obtain the minimum total number of G-rays for reconstructing the image composed of rectangles on the lattice?

Problem 3.2 Consider the following image on the square $[0, 1] \times [0, 1]$ to be reconstructed as the discrete image 8×8 :

$$[f(x,y);\ 0 \le x, y \le 1] = \begin{array}{|c|c|c|c|c|c|c|c|}
\hline
0 & 0 & 4 & 4 & 0 & 0 & 2 & 2 \\
\hline
0 & 0 & 4 & 4 & 0 & 0 & 2 & 2 \\
\hline
3 & 3 & 0 & 0 & 1 & 1 & 0 & 0 \\
\hline
3 & 3 & 0 & 0 & 1 & 1 & 0 & 0 \\
\hline
0 & 0 & 2 & 2 & 0 & 0 & 4 & 4 \\
\hline
0 & 0 & 2 & 2 & 0 & 0 & 4 & 4 \\
\hline
1 & 1 & 0 & 0 & 3 & 3 & 0 & 0 \\
\hline
1 & 1 & 0 & 0 & 3 & 3 & 0 & 0 \\
\hline
\end{array}.$$

Each square is referred to as an image element of size $1/8 \times 1/8$. Sketch the graphs of line-integrals $w_{p,s}(t)$ of the $(1,2)$- and $(1,5)$-projections, i.e., when $(p,s) = (1,2)$ and $(1,5)$. Use these integrals to calculate the line-sums $v_{p,s}(t)$ and splitting-signals $\{f'_{p,s,t};\ t = 0:3\}$. Verify if these splitting-signals can be used to calculate the 2-D DFT of the image in the sets of frequency-points $T'_{1,2}$ and $T'_{1,5}$.

Problem 3.3 Suppose that you are given an image in the square $[0,1] \times [0,1]$, which is composed of image elements of size $1/16 \times 1/16$ each with a constant intensity. In this case, we say that the image is on the Cartesian lattice 16×16. Sketch line-integrals along G-rays of the $(1,4)$-projection and prove that these integrals can be used to calculate the paired splitting-signal $\{f'_{1,4,t};\ t = 0:7\}$ of the reconstructed image 16×16. This means you can calculate the 2-D DFT of this image at frequency-points of the set $T'_{1,4}$.

Problem 3.4 Suppose that you are given an image on the square $[0,1] \times [0,1]$ and the Cartesian lattice 10×10.

A. Determine the number of projections required to reconstruct the image.

B. Show that for each (p,s)-projection in A, a set of G-rays exists such that the line-integrals along these rays can be used to calculate the splitting-signals $\{f_{p,s,t};\ t = 0:9\}$.

C. Develop the algorithm for full reconstruction of the given image from its projections and show your results. As an example, you can use the image in Problem 3.2 after extending the image to the size 10×10, by adding a zero column or row to each side of the image.

Problem 3.5* In the described model 8×8, the G-rays are considered as lines. Assume that the projections were obtained from the rays of width $\neq 0$. It means the integrals $w_{p,s}(t)$ represent the integrals of the image along the wide rays. Assume that the width of G-rays is a small number, for instance $d < 1/64$. Is it possible to generalize the method of reconstruction described in this chapter to the case with wide rays? Propose the method of transferring the geometry of wide G-rays to line A-rays, to solve the problem of image reconstruction from its ray-projections.

Problem 3.6* For $N = 2^r$, where $r > 1$, the basic functions of the 2-D discrete paired transformation are described by the masks, or matrices with coefficients $-1, 1$, and 0. It was shown in the $N = 8$ example, that all coefficients 1 and -1 are located on parallel lines. Propose the concept of 2-D paired functions with the width $\neq 0$. In other words, propose such 2-D binary functions on the square $[0,1] \times [0,1]$, whose masks contain coefficients 1 and -1, which are located on different sets of parallel wide rays.

A. Give examples of such wide-ray paired functions.

B. Discuss the property of orthogonality for the functions in A.

For instance, we studied the following functions. Given $(p, s) \neq (0, 0)$, we consider the set of parallel rays $px + sy = t/N$ at the angle $\vartheta = -\tan^{-1}(p/s)$ to the horizontal line, where t runs a set of integers. We consider rays with width $\Delta = \Delta(p, s)$ and the central line $px + sy = t/N$. Let a be the maximum width of the rays, which is assumed to be the width of the disjoint next rays for the vertical or horizontal projections, when $(p, s) = (0, 1)$ and $(1, 0)$, respectively. For wide rays with central lines $y = t/N$ or $x = t/N$, the maximum width is $a = 1/N$. This is the distance between two neighbor points of the lattice $N \times N$ places in the square $[0, 1] \times [0, 1]$.

Definition 1: On the square $[0, 1] \times [0, 1]$ with the lattice $X_{N,N}$ on it, the wide ray is defined by

$$r_{p,s,t}(x, y; a) = \begin{cases} 1, & \text{if } -\frac{a}{2} < px + sy - \frac{t}{N} \leq \frac{a}{2}, \\ 0, & \text{otherwise}, \end{cases} \tag{3.135}$$

where $x, y \in [0, 1]$.

The width $\Delta(p, s)$ of the ray $r_{p,s,t}$ is calculated by $\Delta(p, s) = a/s \cos(\vartheta) = a/\sqrt{p^2 + s^2}$, $(s \neq 0)$. In the $(1, s) = (1, 0)$ case, we consider $\Delta = a$.

Definition 2: Given triplet (p, s, t), the paired function of wide rays, or the wide ray paired function on the square $[0, 1] \times [0, 1]$ is defined as

$$\chi'_{p,s,t}(x, y; a) = \sum_{k=-\infty}^{\infty} (-1)^k r_{p,s,t+k2^{r-n-1}}(x, y; a), \quad x, y \in [0, 1], \tag{3.136}$$

where $t \in [0, N/2 - 1]$. For $(p, s) = (0, 0)$, the paired function $\chi'_{0,0,0}(x, y; a) \equiv 1$.

As an example for the $N = 8$ case, Figure 3.86 shows two gray-scale images of the paired wide ray functions for $(p, s, t) = (1, 2, 1)$ and $(4, 1, 3)$ in parts a and b, respectively. White, black, and gray colors on the image correspond to the values of 1, -1, and 0, respectively.

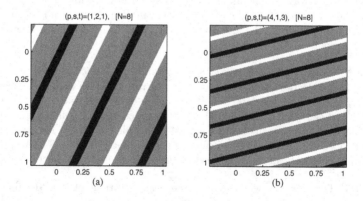

FIGURE 3.86

Images of functions (a) $\chi'_{1,2,1}(x, y; a)$ and (b) $\chi'_{4,1,3}(x, y; a)$, when $a = 1/8$.

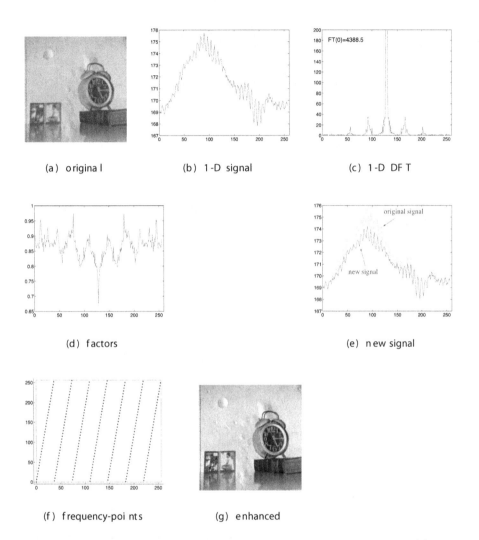

(a) original (b) 1-D signal (c) 1-D DFT

(d) factors (e) new signal

(f) frequency-points (g) enhanced

FIGURE 1.4: Fast transform-based method of image enhancement. (a) The original image, (b) the image-signal, (c) the magnitude spectrum of the signal, (d) factors, (e) the processed signal, (f) marked locations of 256 frequency-points, at which the 2-D DFT was changed, and (g) the image enhanced by one signal.

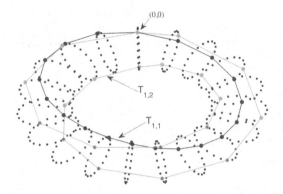

FIGURE 1.6: Torus of the lattice 20×20 with two spirals corresponding to the groups $T_{1,1}$ and $T_{1,2}$.

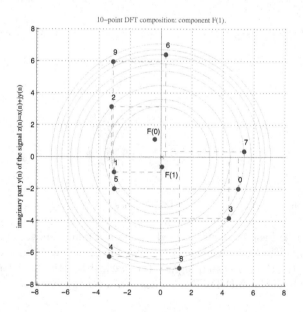

FIGURE 1.15: Calculation of the second component of the DFT, F_1.

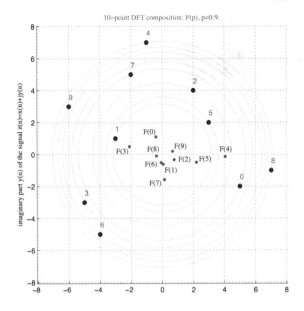

FIGURE 1.16: Result of ten rotations of the signal-data.

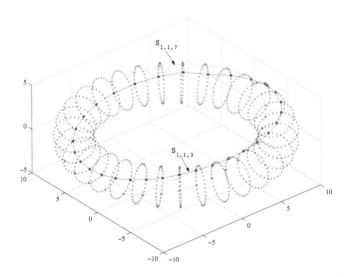

FIGURE 2.4: The net with knots of the grid 32×32 in the 3-D space with locus of two spirals $S_{1,1,3}$ and $S_{1,1,7}$.

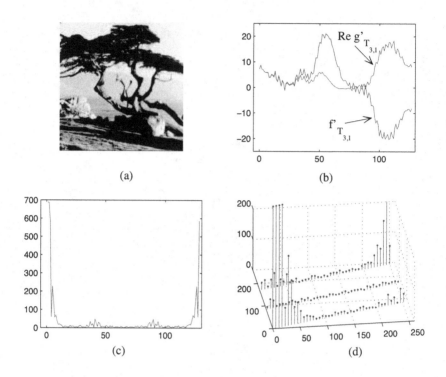

FIGURE 2.18: (a) Tree image 256×256, (b) the paired splitting-signal $f_{T'_{3,1}}$ and the real part of the modified signal, (c) the 1-D DFT of the modified signal, and (d) the arrangement of values of the 1-D DFT at frequency-points of the subset $T'_{3,1}$. (The 1-D DFT is shown in the absolute scale.)

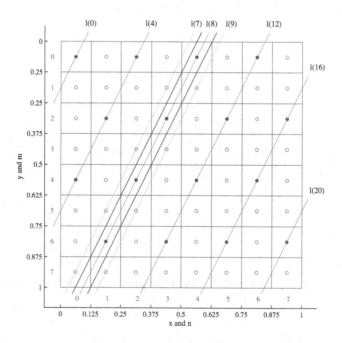

FIGURE 3.22: Mask of the paired function $\chi'_{2,1,0}(n,m)$ and rays.

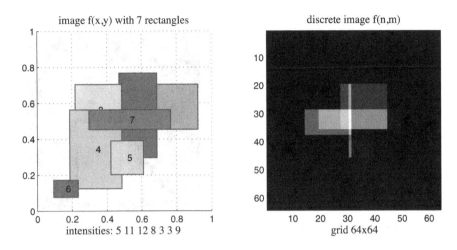

FIGURE 4.2: Seven color rectangles and the discrete gray-scale image composed by these rectangles.

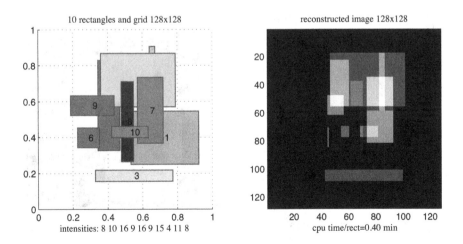

FIGURE 4.6: Ten random rectangles (left) and the reconstructed image (right).

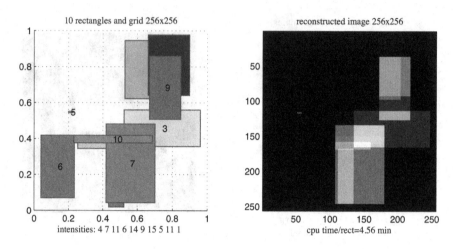

FIGURE 4.7: Ten random rectangles (left) and the reconstructed image (right).

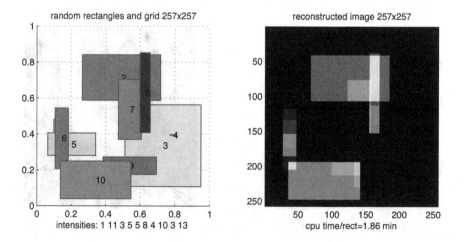

FIGURE 5.37: Ten colored rectangles and the discrete gray-scale image of size 257×257, which is composed by these rectangles.

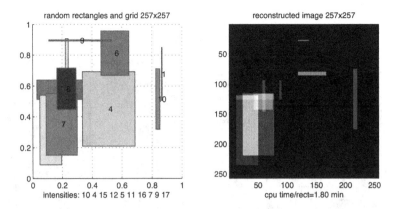

FIGURE 5.41: Image with ten rectangles and its reconstruction 257×257.

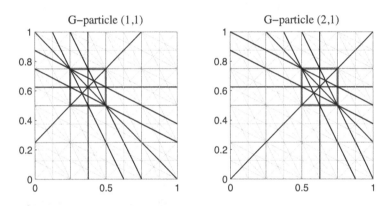

FIGURE 6.13: The field functions of two G-particles $(1, 1)$ and $(2, 1)$.

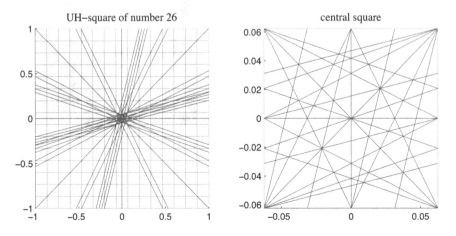

FIGURE 6.14: Map of G-rays in the UH-square of number 26 and its central component.

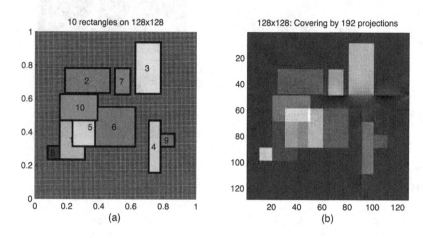

FIGURE 6.18: (a) Ten rectangles and the lattice 128×128 and (b) the reconstructed image.

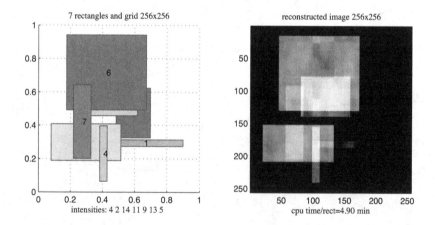

FIGURE 7.1: Image with seven random rectangles and the reconstructed discrete image 256×256.

4

Main Program of Image Reconstruction

The method of transformation of the geometry of projection data, i.e., line-integrals $w_{p,s}(t)$ along the parallel rays $l_{p,s}(t)$, into the sums $v_{p,s}(t)$ of the discrete image to be reconstructed can be used for writing an uncomplicated code for exact reconstructing the images on the Cartesian lattice $N \times N$, with N being a power of two. Different schemes for changing the geometry of the arithmetic rays with following solutions of the convolution equations were considered in Chapter 3, and the examples for the reconstruction on lattices 4×4 and 8×8 were described in detail. In this chapter, we give the description of one of our codes, which was first written in MATLAB® and then in C+ and used for reconstruction of an image up to the size 2048×2048. The image $f(x, y)$ is considered on the unit square with the Cartesian lattice, i.e. the image is the composition of image elements (squares) of size $1/N \times 1/N$ each. Each image element has a constant intensity at all its points.

4.1 The main diagram of the reconstruction

The image is generated from the random rectangles in our program. The calculation of the reconstructed image from its projections by the proposed method of image reconstruction can be separated into the following parts, which are shown in the block-diagram in Figure 4.1.

1. The image $N \times N$ is composed by random rectangles on the square $[0, 1] \times [0, 1]$.

2. For the given frequency-point (p, s), the (p, s)-projection is calculated, i.e., all line-integrals $w(t)$ along the set of geometrical rays $\tilde{l}_{p,s}(t)$.

3. The value of (p, s) is taken from the set $J = J_{N,N}$ of $3N/2$ generators, which is calculated for the given N.

4. The geometry of the G-rays of the (p, s)-projection is transformed to A-rays, i.e., the set of sums $v(t)$ of the discrete image $f_{n,m}$ is calculated from the line-integrals $w(t)$, by solving the system of linear equations with a Toeplitz matrix $(M \times M)$, where $M = (\bar{p} + \bar{s} - 1)(N - 1) + 1$. Here, $\bar{p} = N - p$ and $\bar{s} = N - s$ when p and s are greater than $N/2$, and $\bar{p} = p$ and $\bar{s} = s$ when $p, s \leq N/2$. The Toeplitz matrix is triangular, therefore the transformation of geometry of G-rays to A-rays is fast.

image $f(x, y)$

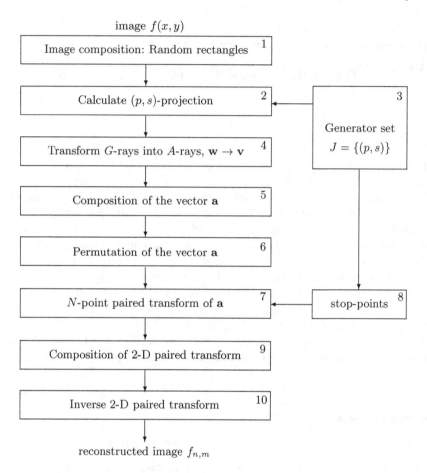

FIGURE 4.1
Block-diagram of image reconstruction (case $N = 2^r$, $r > 1$), when the image
is composed of rectangles on the grid $N \times N$ placed in the square $[0, 1] \times [0, 1]$.

5. The N-dimensional vector **a** is calculated from the sums $v(t)$.

6. The correction in indexing of this vector is performed, when the generator (p, s) has a coordinate which greater than $N/2$.

7. The 1-D N-point fast discrete paired transform (DPT) is calculated over the vector **a**. For many projections, the incomplete paired transform can be used to reduce the calculations.

8. Stop-points are calculated to accomplish the incomplete 1-D DPT, when the case with incomplete 1-D DPT is selected in the program.

9. The 2-D DPT of the discrete image is filled by the 1-D paired transform.

10. After processing all $3N/2$ projections, the inverse 2-D DPT is calculated to obtain the reconstructed discrete image $f_{n,m}$.

Now we describe each part in this block-diagram in more detail.

4.2 Part 1: Image model

The image $f(x, y)$ to be reconstructed is considered on the unit square $[0, 1] \times [0, 1]$ and in the form of N^2 image elements (IE) with intensities $f_{n,m}$,

$$ f(x, y) = \frac{1}{(\Delta x)^2} \sum_{m=0}^{N-1} \sum_{n=0}^{N-1} f_{n,m} \text{rect}_2 \left(Nx - n - \frac{1}{2}, Ny - m - \frac{1}{2} \right), \quad (4.1) $$

where the function $\text{rect}_2(x, y) = \text{rect}(x)\text{rect}(y)$, and $\text{rect}(t) = 1$, if $|t| < 1/2$, and 0, otherwise. The image $f(x, y)$ is generated from a few rectangles $r_k(x, y)$ defined by positions $(x_k, y_k, \Delta x_k, \Delta y_k)$, which are randomly placed in the square on the grid $N \times N$,

$$ f(x, y) = \frac{1}{(\Delta x)^2} \sum_{k=1}^{K} R_k \text{rect} \left(\frac{x - x_k}{\Delta x_k} - \frac{1}{2}, \frac{y - y_k}{\Delta y_k} - \frac{1}{2} \right). \quad (4.2) $$

Here, R_k is the intensity of the k-th rectangle, and integer $K \geq 1$. It is assumed that the sizes of the IEs are equal, $\Delta x = \Delta y = 1/N$, for all k. As an example, for the $N = 64$ case, Figure 4.2 shows the gray-scale image of size 64×64, which is composed of seven rectangles (on the left), and its discrete representation (on the right), after the image reconstruction. The image is the sum of these rectangles.

The coordinates of all rectangles and their intensities are given in Table 4.1.

TABLE 4.1
Data of rectangles

k	x_k	y_k	Δx_k	Δy_k	R_k
1	0.4219	0.4531	0.5000	0.2500	5
2	0.4688	0.2969	0.2188	0.4688	11
3	0.2188	0.4219	0.2656	0.2813	12
4	0.1875	0.1250	0.2969	0.4375	8
5	0.4219	0.2031	0.1875	0.1875	3
6	0.0938	0.0781	0.1406	0.0938	3
7	0.2969	0.4531	0.4688	0.1094	9

All data of random rectangles are generated first in the integer form as shown in Table 4.2, and then they are used for the discrete grid $N \times N$ by

$$ x_k = x'_k/N, \quad y_k = y'_k/N, \quad \Delta x_k = \Delta x'_k/N, \quad \Delta y_k = \Delta y'_k/N, \quad k = 1:7. $$

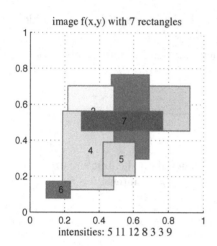
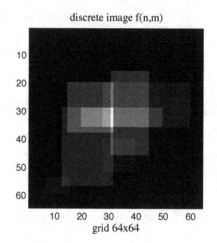

FIGURE 4.2 (See color insert)
Seven color rectangles and the discrete gray-scale image composed of these
rectangles.

TABLE 4.2
Integer data of rectangles

k	x'_k	y'_k	$\Delta x'_k$	$\Delta y'_k$	R_k
1	27	29	32	16	5
2	30	19	14	30	11
3	14	27	17	18	12
4	12	8	19	28	8
5	27	13	12	12	3
6	6	5	9	6	3
7	19	29	30	7	9

Below is the beginning of the script `image_reconstionA3A4B1A2.m`, where
the random rectangles are generated and saved as data `data_allrect` in integer
format [x,y,dx,dy,d_intensity].

```
% ------------------------------------------------------------------
% call: image_reconstionA3A4B1A2.m
% Art-Meruzhan Grigoryan, January 04, 2011
   . . .
fprintf('\n   Start the code \n');
N=input('  input the value of N: ');
n=input('  input the number of rectangles (1,2,...,16): ');
many_rectangles=n;
data_allrect=zeros(many_rectangles,5);
for k=1:many_rectangles
    kk=1;
    while kk==1
        data_1rect=round(rand(1,4)/2*N);
```

```
        if (min(data_1rect)>1)
            kk=0;
        end
    end
    % data_1rect is in integer format [x,y,dx,dy].
    % intensity (<=16) of the image is added:
    data_1rect(5)=round(rand*16)+1;
    data_allrect(k,:)=data_1rect;
end
draw_allfigures;  % draw all rectangles on the square [0,1]x[0,1].
    . . .
```

4.3 The coordinate system and rays

When writing the first version of our program of image reconstruction, the discrete image $f_{n,m}$ and all masks of the paired functions $\chi'_{p,s,t}(n,m)$ were considered in the original coordinate system (n,m), as shown in Figure 4.2 in part b. The parallel arithmetical rays $l_{p,s}(t)$ along only the units $+1$ and -1 in each such mask were described with respect to the coordinate system $(x, 1-y)$, instead of the original system with coordinates (x,y). In this system, all rectangles and lines are drawn using MATLAB commands. This means the top left corner of the square $[0,1] \times [0,1]$ has coordinates $(0,1)$, not $(0,0)$ as in the original coordinate system (x,y), as shown in Figure 4.2 in part a. In these two systems of coordinates, the arithmetical rays and geometrical rays are described by $np + ms = t$ on discrete grid $X_{N,N}$ and $xp + (1-y)s = t_1$ on the square $[0,1] \times [0,1]$, respectively.

In addition, all diaphanous equations for the A-rays and G-rays are considered in the form of $ns + mp = t$ and $xs + (1-y)p = t$, instead of $np + ms = t$ and $xp + (1-y)s = t$, respectively. In other words, the parameter s is referred to as p, and p as s. The generators (p,s) of the set $J_{N,N}$ composed of subsets I, II, III, and IV are taken in the order III, IV, I, and II. We denote by M1, M2, M3, and M4 these four cases with corresponding families of equations for the G-rays and convolutions. Equations for M1, M2, M3, and M4 are selected of types A, A, B, and A, respectively. In other words, different types of equations for geometrical rays are considered. Therefore the program of image reconstruction is called A3-A4-B1-A2 code.

It should be said that in the program all control points of the rays are numbered from 0 to $(p+s)(N-1)$, for the (p,s)-projections, in order to avoid negative indexing. Therefore, all equations which were given above for the A-rays and G-rays have been changed, as described below.

Case M1 (A3): Generators are $(1, s)$, $0 \le s \le N/2$, and equations are

$$\begin{cases} xs - y + 1 - \dfrac{t}{N} - \dfrac{1}{N} = 0, \text{when } s \ne 0 \\ -y + 1 - \dfrac{t}{N} - \dfrac{1}{2N} = 0, \quad \text{when } s = 0, \end{cases}$$

where $t = 0 : (s + 1)(N - 1)$. This is the case of A3 (see Table 3.8), when changing the parameter p to s, and $(p, 1)$ to $(1, s)$.

Case M2 (A4): Generators are $(1, s)$, $s > N/2$, and equations are

$$x(s - N) - y - \frac{t}{N} + (N - s) + \left(1 - \frac{1}{N}\right) = 0,$$

where $t = 0 : (N - s + 1)(N - 1)$. These equations correspond to case A4

$$x(p - N) + (1 - y) - \frac{t}{N} + \left[1 - \frac{p + 1}{N}\right] = 0,$$
$$t = (N - 1) : -1 : -(N - p)(N - 1),$$

after permutation $t \to t - (N - p)(N - 1)$ and the change $p \to s$.

Case M3 (B1): Generators are $(p, 1)$, $0 < p \le N/2$, and equations are

$$x - yp + p - \frac{t}{N} - \frac{p}{N} = 0.$$

This is the case of B1 after changing the parameter p to s.

Case M4 (A2): Generators are $(p, 1)$, $p > N/2$, and equations are

$$x + y(N - p) - \frac{t}{N} - \frac{1}{N} = 0, \quad t = 0 : (N - p + 1)(N - 1).$$

These equations correspond to equations in case A2,

$$x + y(N - s) + (s - N) - \frac{t}{N} + \left(1 - \frac{s + 1}{N}\right) = 0$$
$$t = (N - 1) : -1 : -(N - s)(N - 1),$$

after permutation $t \to t - (N - s)(N - 1)$ and replacement of s by p.

All equations above for G-rays are given in Tables 4.3.

4.4 Part 2: Projection data

The set of projections is defined by the following set of generators (p, s):

$$J_{N,N} = \{\{(1, s); \ s = 0 : (N - 1)\}, \{(2p, 1); \ p = 0 : (N/2 - 1)\}\}.$$

TABLE 4.3

Model II (case M): Equations of G-rays for image $N \times N$

	(p, s)	G-rays	
I	$(1, 0)$	$-y + 1 - \dfrac{t}{N} - \dfrac{1}{2N} = 0$ $t = 0 : (N - 1)$	$s = 0$
I	$(1, s)$	$xs - y + 1 - \dfrac{t}{N} - \dfrac{1}{N} = 0,$ $t = 0 : (s + 1)(N - 1)$	$s \leq N/2$
II	$(1, s)$	$x(s - N) - y - \dfrac{t}{N} + (N - s) + (1 - \dfrac{1}{N}) = 0$ $t = 0 : (N - s + 1)(N - 1)$	$s > N/2$
III	$(0, 1)$	$x - \dfrac{t}{N} - \dfrac{1}{2N} = 0$ $t = 0 : (N - 1)$	$p = 0$
III	$(p, 1)$	$x - yp + p - \dfrac{t}{N} - \dfrac{p}{N} = 0$ $t = 0 : (p + 1)(N - 1)$	$p \leq N/2$
IV	$(p, 1)$	$x + y(N - p) - \dfrac{t}{N} - \dfrac{1}{N} = 0$ $t = 0 : (N - p + 1)(N - 1)$	$p > N/2$

For each (p, s) of this set, the corresponding (p, s)-projection is calculated. The measurements of the projections, i.e., the line-integrals $w_l(t)$ of the image $f(x, y)$ along the rays $l(t) = l_{p,s}(t)$,

$$w_l(t) = \int_{l(t)} f_d(x, y) dl, \qquad (4.3)$$

are calculated as the sum of measurements $w_{l;k}(t)$ of each rectangle $r_k(x, y)$. In other words,

$$w_l(t) = \sum_{k=1}^{K} w_{l;k}(t) = \sum_{k=1}^{K} R_k \int_{l(t)} r_k(x, y) dl. \qquad (4.4)$$

Each such integral $w_{l;k}(t)$ is calculated by finding two points of intersection of the ray $l(t)$ with the k-th rectangle. We denote these points by $A(t) = (a_1, b_1)$ and $B(t) = (a_2, b_2)$, and the distance between them as $\Delta l_k(t) = \sqrt{(a_2 - a_1)^2 + (b_2 - b_1)^2}$. Then, the value of the line-integral $w_{l;k}(t)$ equals

$$w_{l;k}(t) = R_k \Delta l_k(t).$$

To find the points of intersection of the line $l_{p,s}(t)$ with the rectangle $r_k(x, y)$, one can perform a few verifications, namely, the maximum four following verifications:

1. $\qquad x = \dfrac{t - s(y_k + \Delta y_k)}{p} \in [x_k, x_k + \Delta x_k],$

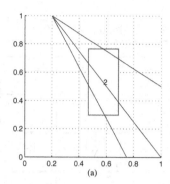
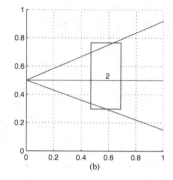

FIGURE 4.3
The rectangle and rays passing through it.

if the line $px + sy = t$ intersects the horizontal line $y = y_k + \Delta y_k$,

$$2. \qquad y = \frac{t - p(x_k + \Delta x_k)}{s} \in [y_k, y_k + \Delta y_k],$$

if the line $px + sy = t$ intersects the vertical line $x = x_k + \Delta x_k$,

$$3. \qquad x = \frac{t - sy_k}{p} \in [x_k, x_k + \Delta x_k],$$

if the line $px + sy = t$ intersects the horizontal line $y = y_k$,

$$4. \qquad y = \frac{t - px_k}{s} \in [y_k, y_k + \Delta y_k],$$

if the line $px + sy = t$ intersects the vertical line $x = x_k$. Verification #1 includes the $(1,0)$ case, when $s = 0$, and verification #4 includes the $(0,1)$ case, when $p = 0$.

The positive answer in verification #1 results in the intersection point $(x, y_k + \Delta y_k)$. The positive answers in verifications #2, 3, and 4 result in the intersection points $(x_k + \Delta x_k, y)$, (x, y_k), and (x_k, y), respectively. These different examples can be seen in Figure 4.3 for the rectangle $r_2(x, y)$.

Each of these verifications requires a maximum of one multiplication or division, because all generators $(p, s) \in J_{N,N}$ have the form $(1, s)$ or $(p, 1)$. Two multiplications and one square root are required for computing $\Delta l_k(t)$. Thus, a total of six multiplications and one square root are used to find the intersection of the line with the rectangle and calculate the length of the intersection. One additional multiplication is used in the product $w_{l;k}(t) = R_k \Delta l_k(t)$.

It should be noted that three positive answers can be obtained in the case when the line $l_{p,s}(t)$ passes along a side of the rectangle. These cases are $(p, s) = (1, 0)$ and $(0, 1)$, because a rotation of rectangles in the image $f(x, y)$ is not considered. Four positive answers can also be obtained in the case where the line $l_{p,s}(t)$ passes along a diagonal of the rectangle. In both such cases, only two different points of intersection are considered. The case where the line $l_{p,s}(t)$ intersects the rectangle in one point, i.e., the intersection is in a corner of the rectangle, can be omitted.

The calculation of points of intersection of the line $px + sy = t$ with one of the sides of the rectangle can also be written on the unique form. Indeed, consider the general form of the vertical and horizontal lines, $\hat{p}x + \hat{s}y = \hat{t}$, where $\hat{p}, \hat{s} = 0$ or 1, and $\hat{t}$ is one value from $\{x_k, y_k, x_k + \Delta x_k, y_k + \Delta y_k\}$. The the solution of the system

$$\begin{cases} px + sy = t \\ \hat{p}x + \hat{s}y = \hat{t} \end{cases}$$

is

$$\begin{bmatrix} x \\ y \end{bmatrix} = \begin{bmatrix} p & s \\ \hat{p} & \hat{s} \end{bmatrix}^{-1} \begin{bmatrix} t \\ \hat{t} \end{bmatrix} = \frac{1}{p\hat{s} - s\hat{p}} \begin{bmatrix} \hat{s} & -s \\ -\hat{p} & p \end{bmatrix} \begin{bmatrix} t \\ \hat{t} \end{bmatrix}.$$

The determinant does not require a multiplication. It is probably the simplest method of calculating the points of intersection of the line with the rectangle, which uses two multiplications.

Below is the script of function intersection_withrectangleA, which is used in the main program for computing the intersection of the line with one rectangle.

```
% ------------------------------------------------------------
% call: intersection_withrectangleA.m  / A.-M. Grigoryan, 2011
% The calculation of coordinates (x,y) of intersection
% of line "p_line=[p,s,-t]" which defines px+sy-t=0
% with the rectangle "p_rect=[x0,y0,x1,y1]"
% This code can be optimized, because p or s equals +1,-1.
% ------------------------------------------------------------
function p=intersection_withrectangleA(p_rect,p_line)
  pp=[];
  x0=p_rect(1,1);   y0=p_rect(1,2);
  x1=p_rect(2,1);   y1=p_rect(2,2);
  p=p_line(1); s=p_line(2); t=-p_line(3);
  k=0;            % number of intersections
  if s==0        % (1,0) case, i.e. p=1 -> line is x=t
      if (t>=x0) & (t<=x1)
          r1=[t,y0]; r2=[t,y1]; pp=[pp;r1;r2]; k=k+2;
      end
  elseif p==0    % (0,1) case, i.e. s=1 -> y=t
      if (t>=y0) & (t<=y1)
          r1=[x0,t]; r2=[x1,t]; pp=[pp;r1;r2]; k=k+2;
    end
  else           % case when p,s>0
      x=(t-s*y1)/p;              % 1. for line y=y1
      if (x>=x0) & (x<=x1)
          r1=[x,y1]; pp=[pp;r1]; k=k+1;
      end
```

```
    y=(t-p*x1)/s;
    if (y>=y0) & (y<=y1)        % 2. for line x=x1
        r1=[x1,y]; pp=[pp;r1]; k=k+1;
    end
    x=(t-s*y0)/p;               % 3. for line y=y0
    if (x>=x0) & (x<=x1)
        r1=[x,y0]; pp=[pp;r1]; k=k+1;
    end
    y=(t-p*x0)/s;
    if (y>=y0) & (y<=y1)        % 4. for line x=x0
        r1=[x0,y]; pp=[pp;r1]; k=k+1;
    end
end
% Calculate the points of intersection (all cases):
p=[];
if k==0 return end
if k==1 p=[pp;pp]; return end
if k==2
    p=pp; return
elseif k>2
    if pp(1,:)~=pp(2,:)
        p(1:2,:)=pp(1:2,:); return
    else
        p(1,:)=pp(1,:); p(2,:)=pp(3,:);
    end
end
% --------------------------------------------------------------
```

Another Method: We now describe another method of calculating the points of intersection of the line with a rectangle. Consider the four parallel rays passing the corners of the rectangle $r_k(x, y)$, as shown in Figure 4.4. We

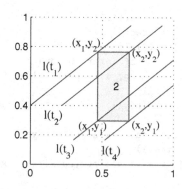

FIGURE 4.4

The rectangle and four parallel rays passing its corners.

denote the coordinates of the corners by (x_1, y_2), (x_2, y_2), (x_1, y_1), and (x_2, y_1), which are on the parallel lines $l(t_1), l(t_2), l(t_3)$, and $l(t_4)$, respectively. Given numbers p and s, we consider equations of these parallel lines $px + sy = t_k$, $k = 1 : 4$, where

$$t_1 = px_1 + sy_2, \quad t_2 = px_2 + sy_2, \quad t_3 = px_1 + sy_1, \quad t_4 = px_2 + sy_1,$$

by assuming $t_1 > t_2 > t_3 > t_4$.

Let $l(t)$ be the line $px + sy = t$. The intersection of this line with the rectangle can be defined by looking the location of the point t in the partition of the real line $(-\infty, t_4, t_3, t_2, t_1, \infty)$. Indeed, it is not difficult to verify that

$$\Delta l(t) = \begin{cases} 0, & \text{if } t \geq t_1, \\ \Delta_1(t), & \text{if } t \in [t_2, t_1), \\ \Delta_2(t), & \text{if } t \in [t_3, t_2), \\ \Delta_3(t), & \text{if } t \in (t_4, t_3), \\ 0, & \text{if } t \leq t_4. \end{cases} \tag{4.5}$$

The length Δ_2 is calculated by

$$\Delta_2(t) = \sqrt{(x_2 - x_1)^2 + (y_0 - y_1)^2} = \sqrt{(x_2 - x_1)^2 + ((x_1 - x_2)p/s)^2}$$
$$= |x_2 - x_1|\sqrt{1 + (p/s)^2},$$

where y_0 is the coordinate of the point of intersection of the line with the side $x = x_2$, i.e., $y_0 = y_1 + (x_1 - x_2)p/s = x_1 + (t_1 - t_2)/s$. The values of $\Delta_1(t)$ and $\Delta_3(t)$ are calculated by

$$\Delta_1(t) = \frac{t - t_1}{t_2 - t_1}\Delta_2(t), \qquad \Delta_3(t) = \frac{t - t_4}{t_3 - t_4}\Delta_2(t).$$

To calculate each value of t_k, one operation of multiplication is required, and one multiplication for $\Delta_2(t)$, assuming that the coefficient $k(p, s) = \sqrt{1 + (s/p)^2}$ can be calculated in advance. Calculation of $\Delta_1(t)$ or $\Delta_3(t)$ required two multiplications. Thus, seven operations of multiplication and one square root are used in this method.

4.5 Part 3: Transformation of geometry

Given generator (p, s), the set $\mathbf{w}$ of line-integrals $w_{l;k}(t)$ is used for calculating the line-sums $v(t)$, by multiplying the inverse Toeplitz matrix by the integrals, as $\mathbf{b} = \mathbf{A}^{-1}\mathbf{w}$, with the following calculations

$$\mathbf{v} = \frac{1}{\Delta l(p, s)}\mathbf{b},$$

where $\Delta l(p, s)$ denotes the length of the geometrical ray in the image element. The calculation of the vector **b** from the set of integrals **w** is performed as follows. We consider separately each of four subsets of generators (p, s).

Subset M1: $(p = 1$ and $s \leq N/2)$.

The number of line-integrals $w(t)$ calculated along the geometrical rays $\bar{l}_{1,s}(t)$ equals $(1 + s)(N - 1) + 1$. In the $s = 0$ case, **b** = **w**, and

$$v(t) = b(t)/N, \quad t = 0 : (N - 1).$$

When $s \neq 0$, the convolution equation **w** = **Av** is described by

$$w(t) = N\frac{\sqrt{1 + s^2}}{s}[v(t) + v(t - 1) + v(t - 2) + \dots + v(t - s + 1)], \quad (4.6)$$
$$t = 0 : (s + 1)(N - 1).$$

Therefore, the inverse transform **b** is calculated in the recurrent form,

$$
\begin{aligned}
b_0 &= w(0), \\
b_1 &= w(1) - b_0, \\
b_2 &= w(2) - (b_1 + b_0), \\
b_3 &= w(3) - (b_2 + b_1 + b_0), \\
&\dots \quad \dots \quad \dots \quad \dots \quad \dots \quad \dots \\
b_{s-1} &= w(s - 1) - (b_{s-2} + b_{s-3} + \dots + b_1 + b_0), \\
b_t &= w(t) - (b_{t-1} + b_{t-2} + \dots + b_{t-s+2} + b_{t-s+1)}), \\
&\quad t = s, s + 1, \dots, (s + 1)(N - 1).
\end{aligned}
\quad (4.7)
$$

The required sums $v(t)$ of the discrete image are calculated by

$$v(t) = b_t\frac{s}{N\sqrt{1 + s^2}}, \quad t = 0 : (1 + s)(N - 1). \quad (4.8)$$

Below is the script (`solution_of_linearsystemA.m`) of the function that is used for calculating the vector **b**.

```
% ------------------------------------------------------------
% call: solution_of_linearsystemA.m
% solution of the system with the upper triangular T-matrix
%       w(t)= v(t)+v(t-1)+...+v(t-p+1)
%       w=Av
%       b=inv(A)w
% such systems of equations correspond to the case A.
    function b=solution_of_linearsystemA(p,w)
    N=length(w);
    z=0;
    for i1=1:p
        b(i1)=w(i1)-z;   z=z+b(i1);
    end
    for i1=p+1:N
        b(i1)=w(i1)-z;   z=z-b(i1-p)+b(i1);
    end
% ------------------------------------------------------------
```

Subset M2: ($p = 1$ and $s > N/2$).

The number of line-integrals $w(t)$ calculated along the geometrical rays $\bar{l}_{1,s-N}(t)$ equals $(1 + N - s)(N - 1) + 1$. The convolution equation $\mathbf{w} = \mathbf{Av}$ is described as in (4.6), and the inverse transform $\mathbf{b}$ is calculated in the recurrent form as in (4.7), where s is substituted by $N - s$. The required sums $v(t)$ of the discrete image are calculated by

$$v(t) = b_t \frac{N - s}{N\sqrt{1 + (N - s)^2}}, \quad t = 0 : (1 + N - s)(N - 1).$$

The numbering is not negative, as it was in A4 ($t = (N - 1) : -1 : (N - s)(N - 1)$), because of the permutation of rays by $t \to t - (N - s)(N - 1)$.

Subset M3: ($p \le N/2$ and $s = 1$).

The number of line-integrals $w(t)$ calculated along the geometrical rays $\bar{l}_{p,1}(t)$ equals $(p + 1)(N - 1) + 1$. In the $p = 0$ case, $\mathbf{b} = \mathbf{w}$ and

$$v(t) = b(t)/N, \quad t = 0 : (N - 1).$$

When $p \ne 0$, the convolution equation $\mathbf{w} = \mathbf{Av}$ is described by

$$w(t) = N\frac{\sqrt{1 + p^2}}{p}[v(t) + v(t + 1) + v(t + 2) + \ldots + v(t + p - 1)], \tag{4.9}$$
$$t = (1 + p)(N - 1) : -1 : 0,$$

where it is assumed that $v(k) = 0$, for $k > (1+p)(N-1)$. The inverse transform $\mathbf{b} = \mathbf{A}^{-1}\mathbf{w}$ is calculated in the following recurrent form ($M = (p+1)(N-1)$):

$$
\begin{aligned}
b_M &= w(M), \\
b_{M-1} &= w(M - 1) - b_M, \\
b_{M-2} &= w(M - 2) - (b_{M-1} + b_M), \\
b_{M-3} &= w(M - 3) - (b_{M-2} + b_{M-1} + b_M), \\
\ldots \quad &\ldots \quad \ldots \quad \ldots \quad \ldots \quad \ldots \\
b_{M-p+1} &= w(M - p + 1) - (b_{M-p+2} + b_{M-p+3} + \ldots + b_{M-1} + b_M), \\
b_t &= w(t) - (b_{t+1} + b_{t+2} + \ldots + b_{t+p-2} + b_{t+p-1})), \\
& \quad t = M - p, M - p - 1, \ldots, 0.
\end{aligned}
\tag{4.10}
$$

The required sums $v(t)$ of the discrete image are calculated by

$$v(t) = b_t \frac{p}{N\sqrt{1 + p^2}}, \quad t = 0 : (1 + p)(N - 1).$$

By changing the parameter t as $(p+1)(N-1) - t$, Equation (4.9) becomes

$$w(t) = N\frac{\sqrt{1 + p^2}}{p}[v(t) + v(t - 1) + v(t - 2) + \ldots + v(t - p + 1)], \tag{4.11}$$
$$t = 0 : (p + 1)(N - 1),$$

where it is assumed that $v(k) = 0$, for $k < 0$. Therefore, the solution in (4.10) is calculated by

$$
\begin{aligned}
b_0' &= w(0), \\
b_1' &= w(1) - b_0', \\
b_2' &= w(2) - (b_1' + b_0'), \\
b_3' &= w(3) - (b_2' + b_1' + b_0'), \\
&\quad\cdots \quad\cdots \quad\cdots \quad\cdots \quad\cdots \\
b_{p-1}' &= w(p-1) - (b_{p-2}' + b_{p-3}' + \ldots + b_1' + b_0'), \\
b_t' &= w(t) - (b_{t-1}' + b_{t-2}' + \ldots + b_{t-p+2}' + b_{t-p+1}')), \\
&\quad t = p, p+1, \ldots, (p+1)(N-1),
\end{aligned}
\tag{4.12}
$$

where $b_t' = b_{M-t}$ (or $b_t = b_{M-t}'$), for $t = 0 : M$.

This is the same system of equations that was given above in (4.7) for cases M1 and M2. Therefore, the function `solution_of_linearsystemA` can be used with the following commands:

```
w=fliplr(w);
b=solution_of_linearsystemA(s,w);
b=fliplr(b);
```

Subset M4: $(p > N/2$ and $s = 1)$.

The $(p,1)$ case is described by the $(p-N,1)$-projection, i.e., p is considered as $p - N$. The number of line-integrals $w(t)$ calculated along the geometrical rays $\bar{l}_{p-N,1}(t)$ equals $(N-p+1)(N-1)+1$. The convolution equation $\mathbf{w} = \mathbf{A}\mathbf{v}$ is described as

$$
w(t) = N \frac{\sqrt{1 + (N-p)^2}}{N-p} [v(t) + v(t-1) + v(t-2) + \ldots + v(t-N+p+1)],
$$
$$
t = -(N-p)(N-1) : (N-1),
$$

$$
\tag{4.13}
$$

where $v(-k) = 0$, when $k > (N-p)(N-1)$. After permutation $t \to t - (N-p)(N-1)$, the numbering in this convolution equation runs the numbers $t = 0 : (N-p+1)(N-1)$.

The inverse transform $\mathbf{b} = \mathbf{A}^{-1}\mathbf{w}$ is calculated in the recurrent form as

$$
\begin{aligned}
b_0 &= w(0), \\
b_1 &= w(1) - b_0, \\
b_2 &= w(2) - (b_1 + b_0), \\
b_3 &= w(3) - (b_2 + b_1 + b_0), \\
&\quad\cdots \quad\cdots \quad\cdots \quad\cdots \quad\cdots \quad\cdots \\
b_{N-p-1} &= w(N-p-1) - (b_{N-p-2} + b_{N-p-3} + \ldots + b_1 + b_0), \\
b_t &= w(t) - (b_{t-1} + b_{t-2} + \ldots + b_{t-N+p+2} + b_{t-N+p+1})), \\
&\quad t = N-p, N-p+1, \ldots, (N-p+1)(N-1).
\end{aligned}
\tag{4.14}
$$

Therefore the function `solution_of_linearsystemA.m` is used to calculate the vector **b**, and the required sums $v(t)$ of the discrete image are calculated by

$$v(t) = b_t \frac{N - p}{N\sqrt{1 + (N - p)^2}}, \quad t = 0 : (N - p + 1)(N - 1). \quad (4.15)$$

For all generators $(p, s) \in J_{N,N}$, in equations $v(t) = b_t / \Delta l(p, s)$, the corresponding normalized coefficients $\Delta l(p, s)$ equal the lengths of the geometrical rays in the image element. These coefficients are calculated in advance (as `delta_r` in the program) by

$$\Delta l(p, s) = \frac{1}{N} \frac{\sqrt{p^2 + s^2}}{p + s - 1},$$

when $p, s \leq N/2$. Otherwise, if $p > N/2$,

$$\Delta l(p, 1) = \frac{1}{N} \frac{\sqrt{(N - p)^2 + 1}}{N - p},$$

and, if $s > N/2$,

$$\Delta l(1, s) = \frac{1}{N} \frac{\sqrt{(1 + (N - s)^2}}{N - s}.$$

4.6   Part 4: Linear transformation of projections

For the given (p, s)-projection, all line-sums $v(t)$ are used to compose the components $f'_{p,s,t}$ of the 2-D discrete paired transform of the reconstructed image $f_{n,m}$. After solving the convolution equation $\mathbf{w} = \mathbf{Av}$, the vector $\mathbf{v}$ is transferred to the corresponding vector $\mathbf{a} = (a_0, a_1, ..., a_{N-1})'$ with the following transformation into the 2-D discrete paired transform.

This transformation is described in detail in Chapter 3 for the $N = 8$ case. For instance, the first four components of the paired transform with the triplet-numbers of the subset $U(1, 2)$ are calculated as follows (see (3.54)):

$$f'_{1,2,0} = v_0 - v_4 + v_8 - v_{12} + v_{16} - v_{20},$$
$$f'_{1,2,1} = v_1 - v_5 + v_9 - v_{13} + v_{17} - v_{21},$$
$$f'_{1,2,2} = v_2 - v_6 + v_{10} - v_{14} + v_{18},$$
$$f'_{1,2,3} = v_3 - v_7 + v_{11} - v_{15} + v_{19}.$$

From the set of 22 components $v_t = v_{1,2}(t)$, $t = 0 : 21$, the components of the vector $\mathbf{a} = (a_0, a_1, ..., a_7)'$ are composed as $a_n = v_n + v_{n+8} + v_{n+16}$, $n = 0 : 7$. It is assumed that $v_{n+16} = 0$, if $n + 16 > 21$. By means of the vector $\mathbf{a}$, we obtain the following system of equations:

$$f'_{1,2,0} = a_0 - a_4$$
$$f'_{1,2,1} = a_1 - a_5$$
$$f'_{1,2,2} = a_2 - a_6$$
$$f'_{1,2,3} = a_3 - a_7$$
$$f'_{2,4,0} = (a_0 + a_4) - (a_2 + a_6)$$
$$f'_{2,4,2} = (a_1 + a_5) - (a_3 + a_7)$$
$$f'_{4,0,0} = (a_0 + a_4) - (a_1 + a_5) + (a_2 + a_6) - (a_3 + a_7)$$
$$f'_{0,0,0} = (a_0 + a_4) + (a_1 + a_5) + (a_2 + a_6) + (a_3 + a_7).$$

The 2-D paired transform data at triplet-points of $U(1,2)$ is thus calculated by the 8-point 1-D discrete paired transform, $P(1,2) = [\chi'_8]\mathbf{a}$.

Similar composition of the vector $\mathbf{a}$ holds in the general case, when $N = 2^r$, $r > 1$,

$$a_n = v_n + v_{n+N} + v_{n+2N} + \cdots + v_{n+(p+s-1)N}, \quad n = 0 : (N-1).$$

Here, $v_t = v_{p,s}(t)$, and $t = 0 : (p+s-1)(N-1)$, for the first and third subsets of generators, when $(p,s) = (1,s)$, $s \leq N/2$, and $(p,s) = (p,1)$, $p \leq N/2$. It is assumed that $v_t = 0$, if $t > (p+s-1)(N-1)$.

Below is the script of the function called `Avector_composition`, to accomplish the composition of the N-dimensional vector $\mathbf{a}$ from the sums $v(t)$.

```
% ------------------------------------
% call: Avector_composition.m
    function A=Avector_composition(N,E)
    M=length(E);
    A=zeros(1,N);
    for i1=1:N
        A(i1)=sum(E(i1:N:M));
    end
% ------------------------------------
```

When calculating vector $\mathbf{a}$ from the sums $v_{p,s}(t)$ for the second and fourths subsets of generators (i.e., for subsets M2 and M4), we need to take into account the fact that the control points of geometrical rays $\bar{l}_{p,s}(t)$ have negative values as well. For the $N = 8$ case, Table 4.4 shows the original numbering of the rays by t, for these subsets of generators, when $(p,s) = (1,s)$, $s > N/2$, and $(p,s) = (p,1)$, $p > N/2$, respectively.

Because in the program negative numbering is not used, the numbering of control points for rays of (p,s)-projections, for such generators, should be changed. To compose new sets of control points, we first consider the $(1,5)$-, or $(1,-3)$-projection in the $N = 8$ case. This projection is described by the set of 29 geometrical rays which are described by

$$l(t) = l_{1,-3}(t) = \left\{ (x,y); \ x - 3y = \frac{t}{8} - \frac{1}{8} \right\}, \quad t = 7 : -1 : -21,$$

TABLE 4.4

Case M: Composition of the vector **a** for the image 8×8

(p,s)	$(a_0, ..., a_n, ..., a_7)$	$\mathbf{v} = B\mathbf{w}$
$(1,0)$	$a_n = v_n$	$v(t) = b(t)/8,\ t = 0:7$
$(1,1)$	$a_n = \sum_m v_{n+8m}$	$v(t) = b(t)/(8\sqrt{2}),\ t = 0:14$
$(1,2)$	$a_n = \sum_m v_{n+8m}$	$v(t) = b_t/(4\sqrt{5}),\ t = 0:21$
$(1,3)$	$a_n = \sum_m v_{n+8m}$	$v(t) = b_t/(8/3\sqrt{10}),\ t = 0:28$
$(1,4)$	$a_n = \sum_m v_{n+8m}$	$v(t) = b_t/(2\sqrt{17}),\ t = 0:35$
$(1,5)$	$a_n = \sum_m v_{n-8m}$	$v(t) = b_t/(2\sqrt{17}),\ t = 7:-1:-21$
$(1,6)$	$a_n = \sum_m v_{n-8m}$	$v(t) = b_t/(4\sqrt{5}),\ t = 7:-1:-14$
$(1,7)$	$a_n = \sum_m v_{n-8m}$	$v(t) = b(t)/(8\sqrt{2}),\ t = 7:+1:-7$
$(0,1)$	$a_n = v_n$	$v(t) = b(t)/8\ \ t = 0:7$
$(2,1)$	$a_n = \sum_m v_{n+8m}$	$v(t) = b_t/(4\sqrt{5}),\ t = 0:21$
$(4,1)$	$a_n = \sum_m v_{n+8m}$	$v(t) = b_t/(2\sqrt{17}),\ t = 0:35$
$(6,1)$	$a_n = \sum_m v_{n-8m}$	$v(t) = b_t/(4\sqrt{5}),\ t = 7:-1:-14$

where $(x,y) \in [0,1] \times [0,1]$.

The control points of this set of rays are given below:

$$
\begin{bmatrix}
0 & 1 & 2 & 3 & 4 & 5 & 6 & 7 \\
-3 & -2 & -1 & \cdot & \cdot & \cdot & \cdot & \cdot \\
-6 & -5 & -4 & \cdot & \cdot & \cdot & \cdot & \cdot \\
-9 & -8 & -7 & \cdot & \cdot & \cdot & \cdot & \cdot \\
-12 & -11 & -10 & \cdot & \cdot & \cdot & \cdot & \cdot \\
-15 & -14 & -13 & \cdot & \cdot & \cdot & \cdot & \cdot \\
-18 & -17 & -16 & \cdot & \cdot & \cdot & \cdot & \cdot \\
-21 & -20 & -19 & \cdot & \cdot & \cdot & \cdot & \cdot
\end{bmatrix} .
$$

As shown in (3.75), the four components of the paired transform with the triplet-numbers $(1,5,0), (1,5,1), (1,5,2)$, and $(1,5,3)$ are calculated as follows:

$$
\begin{aligned}
f'_{1,5,0} &= (v_0 + v_{-8} + v_{-16}) - (v_4 + v_{-4} + v_{-12} + v_{-20}), \\
f'_{1,5,1} &= (v_1 + v_{-7} + v_{-15}) - (v_5 + v_{-3} + v_{-11} + v_{-19}), \\
f'_{1,5,2} &= (v_2 + v_{-6} + v_{-14}) - (v_6 + v_{-2} + v_{-10} + v_{-18}), \\
f'_{1,5,3} &= (v_3 + v_{-5} + v_{-13} + v_{-21}) - (v_7 + v_{-1} + v_{-9} + v_{-17}).
\end{aligned}
\tag{4.16}
$$

These components are written in (3.83) by components of the vector **a** as

$$
\begin{aligned}
f'_{1,5,0} &= a_0 - a_4, \quad f'_{1,5,1} = a_1 - a_5, \\
f'_{1,5,2} &= a_2 - a_6, \quad f'_{1,5,3} = a_3 - a_7.
\end{aligned}
\tag{4.17}
$$

The components a_n, are calculated by

$$
a_n = \sum_m v_{n-8m} = v_n + v_{n-8} + v_{n-16} + v_{n-24}, \qquad n = 0:7,
$$

where $v_{-22} = v_{-23} = v_{-24} = 0$. Because subset M2 corresponds to subset A4 with the transformation $t \to t - (N - p)(N - 1)$, where $p = 5$, we consider the change $t \to t - 21$ in numbering of the control points,

$$
\begin{bmatrix}
21 & 22 & 23 & 24 & 25 & 26 & 27 & 28 & \\
18 & 19 & 20 & & \cdot & \cdot & \cdot & \cdot & \\
15 & 16 & 17 & & \cdot & \cdot & \cdot & \cdot & \\
12 & 13 & 14 & & \cdot & \cdot & \cdot & \cdot & \cdot \\
9 & 10 & 11 & & \cdot & \cdot & \cdot & \cdot & \\
6 & 7 & 8 & & \cdot & \cdot & \cdot & \cdot & \\
3 & 4 & 5 & & \cdot & \cdot & \cdot & \cdot & \\
0 & 1 & 2 & & \cdot & \cdot & \cdot & \cdot &
\end{bmatrix}
$$

Denoting the sums $v'_k = v_{k-21}$, equations in (4.16) and (4.17) can be written respectively as

$$
\begin{aligned}
f'_{1,5,0} &= (v'_{21} + v'_{13} + v'_5) - (v'_{25} + v'_{17} + v'_9 + v'_1), \\
f'_{1,5,1} &= (v'_{22} + v'_{14} + v'_6) - (v'_{26} + v'_{18} + v'_{10} + v'_2), \\
f'_{1,5,2} &= (v'_{23} + v'_{15} + v'_7) - (v'_{27} + v'_{19} + v'_{11} + v'_3), \\
f'_{1,5,3} &= (v'_{24} + v'_{16} + v'_8 + v'_0) - (v'_{28} + v'_{20} + v'_{12} + v'_4),
\end{aligned} \tag{4.18}
$$

and

$$
\begin{aligned}
f'_{1,5,0} &= -a'_1 + a'_5, & f'_{1,5,1} &= -a'_2 + a'_6, \\
f'_{1,5,2} &= -a'_3 + a'_7, & f'_{1,5,3} &= -a'_4 + a'_0.
\end{aligned} \tag{4.19}
$$

The new components a'_n are calculated by

$$
a'_n = \sum_m v_{n+8m} = v'_n + v'_{n+8} + v'_{n+16} + v'_{n+24}, \quad n = 0 : 7,
$$

and $v'_{29} = v'_{30} = v'_{31} = 0$. Therefore, to obtain the order of components a'_n as in (4.17), the sign of all components can be changed with the following permutation: $\{1, 2, 3, 4, 5, 6, 7, 0\} \to \{0, 1, 2, 3, 4, 5, 6, 7\}$, or $n \to (n-1) \bmod 8$, $n = 0 : 7$.

It is not difficult to see that for the $(1, 6)$-projection, the similar renumbering, $t \to t - 14$, of the control points results in the following representation of the components with triplet-numbers $(1, 6, 0), (1, 6, 1), (1, 6, 2)$ and $(1, 6, 3)$:

$$
\begin{aligned}
f'_{1,6,0} &= -a'_2 + a'_6, & f'_{1,6,1} &= -a'_3 + a'_7, \\
f'_{1,6,2} &= -a'_4 + a'_0, & f'_{1,6,3} &= -a'_5 + a'_1.
\end{aligned}
$$

Therefore, the permutation $n \to (n-2) \bmod 8$ of components a'_n can be used, with following change of the sign, in order to write the above equations as

$$
\begin{aligned}
f'_{1,6,0} &= a'_0 - a'_4, & f'_{1,6,1} &= a'_1 - a'_5, \\
f'_{1,6,2} &= a'_2 - a'_6, & f'_{1,6,3} &= a'_3 - a'_7.
\end{aligned}
$$

In the general case of N, for $s > N/2$, the permutation of components a'_n for $f'_{1,s,0}, f'_{1,s,1}, ..., f'_{1,s,N/2-1}$ is described by

$$
n \to (n - s + N/2) \bmod N, \quad n = 0 : (N - 1). \tag{4.20}
$$

Therefore, for such generators $(1, s)$, the function called `vectorB_indexcorection` is used after `Avector_composition`, to accomplish the correct composition of the N-dimensional vector **a** from the sums $v(t)$.

```
% ----------------------------------------------
% call: vectorB_indexcorrection.m
   function A=vectorB_indexcorection(p,s,N,E)
   N2=N/2;
   if p==1
      sN2=s-N2;
      for t=0:N-1
         t1=mod(t-sN2,N)+1;
         a=E(t+1); A(t1)=-a;
      end
   end
   if s==1
      pN2=p-N2;
      for t=0:N-1
         t1=mod(t-pN2,N)+1;
         a=E(t+1); A(t1)=-a;
      end
   end
% ----------------------------------------------
```

For the subset M4 (or A2), when the generators are $(p, 1)$ and p is even and greater that $N/2$, the change in the numbering of the control points is calculated in the same way and the function `vectorB_indexcorection` is used after `Avector_composition`, to accomplish the composition of the N-dimensional vector **a** from the sums $v(t) = v_{p,1}(t)$.

4.7   Part 5: Calculation the 2-D paired transform

After calculating the vector $\mathbf{a} = (a_0, a_1, a_2, ..., a_{N-1})'$ for the (p, s)-projection, the N-point 1-D discrete paired transform (DPT) of this vector results in the part of the 2-D DPT of the reconstructed image, which is denoted by $\mathbf{P}(p, s)$,

$$\mathbf{P}(p, s) = [\chi'_N]\mathbf{a}.$$

This part contains N components that belong to $r + 1$, or $\log_2(N) + 1$ paired splitting-signals, which are generated by the frequency-points (p, s), $(\overline{2p}, \overline{2s})$,

$(\overline{4p}, \overline{4s})$, $(\overline{8p}, \overline{8s})$, ..., $(\overline{N/2p}, \overline{N/2s})$, and $(0,0)$. In other words,

$$\mathbf{P}(p,s) = \begin{cases} \{f'_{p,s,0}, f'_{p,s,1}, f'_{p,s,2}, f'_{p,s,3}, f'_{p,s,4}, ..., f'_{p,s,N/2-1}\}, \\ \{f'_{\overline{2p},\overline{2s},0}, f'_{\overline{2p},\overline{2s},2}, f'_{\overline{2p},\overline{2s},4}, ..., f'_{\overline{2p},\overline{2s},N/2-2}\}, \\ \{f'_{\overline{4p},\overline{4s},0}, f'_{\overline{4p},\overline{4s},4}, ..., f'_{\overline{4p},\overline{4s},N/2-4}\}, \\ \quad ... \qquad ... \qquad ... \qquad ... \\ \{f'_{\overline{N/2p},\overline{N/2s},0}\}, \\ \{f'_{0,0,0}\}. \end{cases} \qquad (4.21)$$

The number of (p, s)-projections is $3N/2$, and the number of splitting-signals in the 2-D DPT equals $3N - 2$, from which, $3N/2$ signals are of length $N/2$ each, $3N/4$ signals are of length $N/4$ each, ..., 6 signals are of length 2 each, 3 signals are of length 1 each, and the last one is $\{f'_{0,0,0}\}$.

The part $\mathbf{P}(p, s)$ of the 2-D discrete paired transform contains r splitting-signals plus one-point signal $\{f'_{0,0,0}\}$ which is common for all parts. For different (p, s), parts $\mathbf{P}(p, s)$ may contain the same splitting-signals of length $N/4$, or $N/8, ..., 1$. Indeed, each splitting-signal of length $N/4$ can be seen in two different parts, each splitting-signal of length $N/8$ can be seen in four different parts, and so on. Therefore, when performing the complete calculation of the N-point DPT over the vectors $\mathbf{a}$ calculated for all $3N/2$ projections, the components $f'_{p,s,t}$ of the 2-D DPT of the discrete image $f_{n,m}$ are normalized as

$$f'_{p,s,t} \to \frac{1}{2^k} f'_{p,s,t}, \quad 2^k = \text{g.c.d.}(p,s), \quad (p,s) \in J_{N,N} \setminus (0,0),$$

and $f'_{0,0,0} \to f'_{0,0,0}/N$. In our program, this method is implemented separately in function `reconstruction_ofAllrectanglesOpt2fast`.

4.7.1   Method of incomplete 1-D DPT

Another method, which is based on calculation of incomplete 1-D DPT over vectors $\mathbf{a}$, is implemented in the function `reconstruction_ofAllrectanglesFast`. For this method, the set of $3N/2$ "stop-points" is calculated first and then used for calculating the incomplete 1-D DFT. Each such stop-point indicates how many splitting-signals are calculated.

For instance, in the $N = 8$ case, $J_{8,8} = \{(1,0), (1,1), (1,2), ..., (1,7), (0,1), (2,1), (4,1), (6,1)\}$, and each part $\mathbf{P}(p, s)$ contains four splitting-signals of lengths $4, 2, 1$, and 1. The set of stop-points equals

$$SP(8) = \{4, 3, 2, 2, 1, 1, 1, 1, 3, 2, 1, 1\}.$$

This means the following (see Table 3.3): the 8-point DPT for the $(1,0)$-projection is complete, the 8-point DPT for the $(1,1)$-projection is used to calculate only the first 3 splitting-signals, the 8-point DPT for the $(1,2)$-projection is used to calculate only the first 2 splitting-signals, ..., and the 8-point DPT for the $(6,1)$-projection is used to calculate only the first splitting-signal. The sum of stop-points is 22 and equals the number of splitting-signals in the 8×8-point 2-D DPT.

Below is the script of the function called find_stoppoints, to calculate the set of stop-points for all $3N/2$ projections in image reconstruction.

```
% ------------------------------------------------------------
% call: find_stoppoints.m    (M. Grigoryan, 2011)
function Stop_points=find_stoppoints(N)
    r=log2(N);
    Stop_points=zeros(1,N-1);
    o=zeros(1,N-1);
    N2=N/2; n1=1; n2=N2; M2=N2;
    for k=1:r
        o(n1:n2)=M2;
        N2=N2/2; M2=M2+N2; n1=n2+1; n2=n2+N2;
    end
    o=N-o(N-1:-1:1);
    Stop_points=r-log2(o(1:N-1));
    Stop_points=[Stop_points(1)+1 Stop_points];
    % add the second part for the generators of parts 3,4.
    s1=Stop_points(1)-1;
    Stop_points=[Stop_points s1 Stop_points(3:2:N)];
    % sum(Stop_points) = 3N-2
% ------------------------------------------------------------
```

We now consider briefly the concepts of the 1-D DPT and incomplete 1-D DPT, and then we give the scripts of codes that perform these transforms.

4.7.2   Fast 1-D paired transform

The 1-D N-point discrete paired transformation χ'_N, which was described in §2.4.1, can be defined similarly to the paired transformation in the 2-D case. The complete system of the paired functions for the N-point DPT is composed of the paired functions [46, 51]:

$$
\chi'_{2^k,2^k t}(n) = \begin{cases} 1, & \text{if } 2^k n = (2^k t) \bmod N \\ -1, & \text{if } 2^k n = (2^k t + N/2) \bmod N \\ 0, & \text{otherwise} \end{cases} \tag{4.22}
$$
$$
t = 0 : (2^{r-k-1} - 1), \ \ k = 0 : (r-1),
$$

and $\chi'_{0,0}(n) \equiv 1$, $n = 0 : (N-1)$.

The paired transform is the unitary frequency-time representation of the discrete-time signal f_n in the form of the set of $(r+1)$ splitting-signals,

$$
\chi'_N : f_n \to \left\{ \{ f'_{2^k,0}, f'_{2^k,2^k}, f'_{2^k,2^k 2}, ..., f'_{2^k,N/2^{k+1}-1} \}, k = 0 : (r-1) \right\}, \{ f'_{0,0} \},
$$

which are generated by frequencies $p = 1, 2, 4, 8, ..., N/2$, and 0.

The 1-D DPT is fast and requires $2N - 2$ operations of addition and subtraction. Figure 4.5 shows the signal-flow graph of the 8-point transform with 7 operations of addition and 7 subtractions.

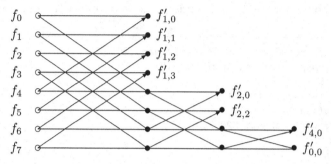

FIGURE 4.5
Signal-flow graph of the fast eight-point DPT.

Below is the script `fst_1d.m` of the function which is used for calculating the N-point DPT, when $N > 1$ is a power of two.

```
% ----------------------------------------
% call: fst_1d.m   (A.M. Grigoryan, 1996)
function A=fst_1d(A,N,M)
    % N=length(A); M=log2(N);
    LK=0; NK=N; I=1;  II=1;
    while I <= M
        NK=NK/2; LK=LK+NK; J=II;
        while J <= LK
            J1=J+NK;
            T=A(J1); T1=A(J);
            A(J1)=T1+T; A(J)=T1-T; J=J+1;
        end
        II=LK+1; I=I+1;
    end
% ----------------------------------------
```

One can also use another function with script `fastpaired_1d.m` for calculating the 1-D DPT, which is written in a simple form, but it does not accomplish the calculation faster than `fst_1d.m`.

```
% ----------------------------------------------
% call: fastpaired_1d.m   (M.M. Grigoryan, 2007)
  function y=fastpaired_1d(x)
     N=length(x);
     if N==1
         y=x;
     else
        N2=N/2;
```

```
            x1=x(1:N2); x2=x(N2+1:N);
            y1=x1+x2;    y2=x1-x2;
            y=[y2 fastpaired_1d(y1)];
        end
% ------------------------------------------------
```

To accomplish the calculation of the incomplete N-point DPT with the stop-point $Ls \in SP(N)$, the function incomplete_dpt can be used. For simplicity of reading, this function is also written with the iteration, as the function fastpaired_1d.

```
% ------------------------------------------------
% call: incomplete_dpt.m   (M.M. Grigoryan, 2011)
function y=incomplete_dpt(x,Ls)
    N=length(x);
    if (length(x)==1) | (Ls==0)
        y=x;
    else
        N2=N/2;
        y(1,1:N2)=x(1,1:N2)-x(1,N2+1:end);
        Ls=Ls-1;
        if Ls~=0
            z(1,1:N2)=x(1,1:N2)+x(1,N2+1:end);
            y=[y incomplete_dpt(z,Ls)];
        end
    end
% ------------------------------------------------
```

For fast implementation of the inverse 1-D DPT, the function incopletefst_1d is used in the main program, which works as the function fst_1d, when the stop-point Ls equals $r + 1$.

```
% ------------------------------------------------
% call: incopletefst_1d.m   (A.M. Grigoryan, 2011)
function B=incompletefst_1d(A,Ls)
    N=length(A); M=log2(N);
    LK=0; NK=N;
    if Ls>M Ls=M+1; end
    I=1;  II=1;
    while I <= Ls
        NK=NK/2; LK=LK+NK; J=II;
        while J <= LK
            J1=J+NK;
            T=A(J1); T1=A(J);
            A(J1)=T1+T; A(J)=T1-T; J=J+1;
        end
        II=LK+1; I=I+1;
    end
    if Ls<M B=A(1:LK);
```

```
    elseif Ls==M B=A(1:N-1);
    else B=A;
    end
% ----------------------------------------------------
```

4.7.3   Inverse 2-D DPT

The 1-D N-point discrete paired transforms of vectors **a**, which are calculated for all $3N/2$ projections, are used to compose the 2-D paired transform of the reconstructed image, which we consider in the form of the square matrix $N \times N$. For example, the components $f'_{p,s,t}$ of the 4×4-point 2-D DPT are located in such a matrix as follows:

'1,0,0'	'1,0,1'	'0,1,0'	'0,1,1'
'1,1,0'	'1,1,1'	'2,1,0'	'2,1,1'
'1,2,0'	'1,2,1'	'2,0,0'	'0,2,0'
'1,3,0'	'1,3,1'	'2,2,0'	'0,0,0'

For the 8×8 case, the 2-D DPT is transformed into the square matrix as

'1,0,0'	'1,0,1'	'1,0,2'	'1,0,3'	'0,1,0'	'0,1,1'	'0,1,2'	'0,1,3'
'1,1,0'	'1,1,1'	'1,1,2'	'1,1,3'	'2,1,0'	'2,1,1'	'2,1,2'	'2,1,3'
'1,2,0'	'1,2,1'	'1,2,2'	'1,2,3'	'4,1,0'	'4,1,1'	'4,1,2'	'4,1,3'
'1,3,0'	'1,3,1'	'1,3,2'	'1,3,3'	'6,1,0'	'6,1,1'	'6,1,2'	'6,1,3'
'1,4,0'	'1,4,1'	'1,4,2'	'1,4,3'	'2,0,0'	'2,0,2'	'0,2,0'	'0,2,2'
'1,5,0'	'1,5,1'	'1,5,2'	'1,5,3'	'2,2,0'	'2,2,2'	'4,2,0'	'4,2,2'
'1,6,0'	'1,6,1'	'1,6,2'	'1,6,3'	'2,4,0'	'2,4,2'	'4,0,0'	'0,4,0'
'1,7,0'	'1,7,1'	'1,7,2'	'1,7,3'	'2,6,0'	'2,6,2'	'4,4,0'	'0,0,0'

This representation of the 2-D DPT was illustrated for the "tomo" image in Figures 2.16 and 2.17.

Therefore, we first need to locate each part $\mathbf{P}(p, s)$ in this square matrix. Since the components of the 2-D DPT are numbered by triplets (p, s, t), we use function map_pst2nm to transfer the triplets into the corresponding numbers (n, m), which show the locations of components $f'_{p,s,t}$ in the matrix.

```
% -------------------------------------------
% call: map_pst2nm.m
 function nm=map_pst2nm(N,p,s,t)
    if (p==0) & (s==0) & (t==0)
        nm=[N,N];
        return
    end
    if p==1
        nm=[s+1,t+1];
    elseif s==1
        nm=[p/2+1,N/2+t+1];
    else
        N2=N/2;
        nm=N2+map_pst2nm(N2,p/2,s/2,t/2);
```

```
        end
% ------------------------------------------
```

For example, for the image 256×256, the command `map_pst2nm(256,p,s,t)` gives $nm = (1,2), (4,3)$, and $(130,133)$, when $(p,s,t) = (1,0,1), (1,3,2)$ and $(2,2,8)$, respectively. The function `checking_table` is written to calculate the entire matrix of such triplet-numbers in the general $N = 2^r$ case, when $r \geq 2$.

```
% ----------------------------------------------------
% call: checking_table.m
% To calculate the table NxN of triplets (p,s,t).
    N=8; L=3*N-2-1; N2=N/2-1;
    Jps=setof_ps(N);
    A=cell(N);
    for k=1:L
        p=Jps(k,1); s=Jps(k,2); g=gcd(p,s);
        p_instring=num2str(p);   s_instring=num2str(s);
        ss=strcat(p_instring,',',s_instring,',');
        for t=0:g:N2
            t_instring=num2str(t);   ss2=strcat(ss,t_instring);
            nm=map_pst2nm(N,p,s,t); n=nm(1); m=nm(2);
            A(n,m)={ss2};
        end
    end
    A(N,N)={'0,0,0'};
% ----------------------------------------------------
```

The inverse 2-D DPT is calculated by the formula of reconstruction of the image, which is given in *Statement 2* in (2.49),

$$f_{n,m} = \frac{1}{2N} \sum_{k=0}^{r-1} \frac{1}{2^k} \sum_{(p,s) \in J_{2^{r-k},2^{r-k}}} f'_{p,s,(np+ms) \bmod N} + \frac{1}{N^2} f'_{0,0,0}, \tag{4.23}$$

$$n, m = 0 : (N-1).$$

The inverse 2-D DPT requires only operations of addition/subtraction and division by powers of two. To accomplish this inverse transform, the function `inverse_2DDPT` with the following scripts is used.

```
% ----------------------------------------------------------------
%   Call: inverse_2DDPT.m
%   split_matrix is the image-matrix of the 2-D DPT.
%   Jps=setof_ps(N) is the set of all 3N-2 generators (p,s).
%   A.M. Grigoryan and Nan Du,                    10/27/2009
% ----------------------------------------------------------------
  function image_fromPT=inverse_2DDPT(split_matrix,Jps,N)
        L=length(Jps);
        % The inverse formula of the 2-D paired transform:
        image_fromPT=zeros(N);
```

```
NN=2*N;  N1=N/2;   N2=NN;    Np1=N+1;
for m=0:N-1
    m1=m+1;
    for n=0:N-1
        n1=n+1;
        total=0; con=0; N_new=N; row=1; col=1; M=1;
        r=1; col_add=N; col_r=0; mm=1; Ncol=N1; Nnn=N1;
        for k=1:L
            p=Jps(k,1);  s=Jps(k,2);  t=mod(p*n+m*s,N);
            if t>=N1
                t=t-N1;  t=t/mm;
                total=total-split_matrix(row,col+t);
            else
                t=t/mm;
                total=total+ split_matrix(row,col+t);
            end
            N_new2=N_new/2;  C=N_new+N_new2;
            if k==con+C
                con=k;  N_new=N_new2;  mm=2*mm;
            end
            row=row+1;
            if row==Np1
                row=M;  M=M+Nnn;  r=r+1;  Nnn=Nnn/2;
            end
            if k==col_add
                cc=Ncol;  col=col+cc;
                col_add=col_add+2*cc;
                col_r=col_r+1;  Ncol=Ncol/2;
            end
        end
        image_fromPT(m1,n1)=total;
    end
end
image_fromPT=image_fromPT/NN;
% --------------------------------------------------------------
```

4.7.4 Preliminary results

Main program main_reconstionofRCT8fast.m, which is written in MATLAB, accomplishes the reconstruction of the image $f(x,y)$ of size $N \times N$, $N = 2^r$, from $3N/2$ projections. The image is generated randomly as a set of rectangles with coordinates on the Cartesian lattice $X_{N,N}$ located in the square $[0,1] \times [0,1]$.

As an example, Figure 4.6 shows ten rectangles on the square with different intensities on the left. These rectangles are on the grid 128×128. The image calculated from 192 projections is shown on the right. The reconstruction is exact, and the main program required 24 seconds to calculate all projections, i.e., line-integrals $w(t)$, and reconstruct the image. Incomplete 1-D DPTs were

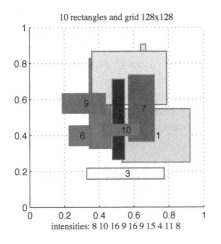
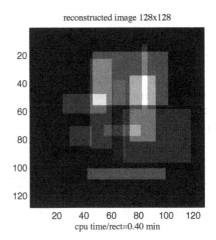

FIGURE 4.6 (See color insert)
Ten random rectangles (left) and the reconstructed image (right).

used in calculations. The use of the complete 1-D DPT for all projections required 24.87 seconds. The CPU time is calculated as the average time for one rectangle. The process of calculation is organized so that for each (p, s)-projection, the integrals are calculated for each rectangle separately, and then the results are added.

Figure 4.7 shows ten rectangles on the square $[0, 1] \times [0, 1]$ on the left. The coordinates of these rectangles are on the grid 256×256. The image calculated from 384 projections is shown on the right. The reconstruction is

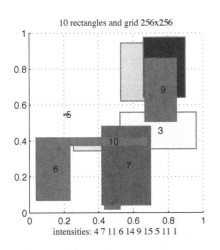
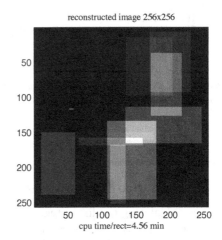

FIGURE 4.7 (See color insert)
Ten random rectangles (left) and the reconstructed image (right).

exact, and the main program required 4.56 minutes to calculate all projections and reconstruct the image. Incomplete 1-D DPTs were used in calculations. The use of complete 1-D DPTs for all projections required 4.68 minutes, which exceeds the time with incomplete DPTs, because of many repetitions in the process of calculating the 2-D DPT of the discrete image.

Time characteristics of calculation, when the program was implemented in MATLAB are given in Table 4.5. The time for calculation of the 2-D paired transform and its inverse transform are also given.

TABLE 4.5
Time for MATLAB-based program

N	incomplete 1-D DPT	complete 1-D DPT	2-D DPT	2-D iDPT
32	0.48s/rec	0.59s/rec	0.0312s	0.0156s
64	2.58s/rec	2.68s/rec	0.0468s	0.0780s
128	24.01s/rec	24.87s/rec	0.2028s	0.5304s
256	4.56m/rec	4.68m/rec	1.8564s	4.4460s

Time data of the program implemented in Ci+ is given in Table 4.6, when the reconstruction is performed by incomplete 1-D DPTs. The time includes the calculation of all line-integrals $w(t)$, line-sums $v(t)$, and the calculation of the 2-D direct and inverse paired transforms. The data in these tables were

TABLE 4.6
Total time for the Ci-based program (14 random rectangles)

$N \times N$	Image scanning & reconstruction with incomplete 1-D DPTs
32×32	$00 : 00 : 00.03$
64×64	$00 : 00 : 00.11$
128×128	$00 : 00 : 00.79$
256×256	$00 : 00 : 06.17$
512×512	$00 : 00 : 49.76$
1024×1024	$00 : 06 : 59.09$

obtained from running programs on a personal computer with Intel Dual CPU Processor at 3.20-GHz speed. This is the first realization of the method on Ci+, which can be improved, in order to achieve a fast reconstruction of images on the large size Cartesian lattices.

4.8   Fast projection integrals by squares

When calculating the line-integrals $w(t)$ for each (p, s)-projection by the program with the block-diagram of Figure 4.1, the intersections of G-rays $\hat{l}_{p,s}(t)$ with all rectangles are calculated, which slows the program. To make this process more effective, we consider the following method of G-ray and A-ray calculation, which can be used for any image $f(x, y)$ on the grid $N \times N$.

All image elements (IE) have the same size $\Delta x \times \Delta y$, which is considered equal to the size of the small square $1/N \times 1/N$ in the program. There is no need to calculate the coordinates of intersections of G-rays with each rectangle and then calculate the lengths of intersections, because the lengths of intersection of rays with each square are equal. It is sufficient, therefore, to determine the set of squares, along which the rays pass. In other words, one can represent each ray by the set of such squares

$$l_{p,s}(t) = \{(n_k, m_k); \; k = 1 : K\}.$$

We number each square by the coordinates (n_k, m_k) of its bottom left corner. The number K of such squares depends on p, s, and t, i.e., $K = K(p, s, t)$. Then, the line-integrals of the image can be calculated by

$$w(t) = \Delta l(p, s) \sum_{(n_k, m_k) \in l_{p,s}(t)} f(n_k - 0.5, m_k - 0.5), \quad t = 0 : (p + s)(N - 1).$$

Now we describe this method in detail, which we call the *method of fast projection integrals by squares (FPIS)*.

For that we first consider the simple script (`control_points_rays.m`) of the function that calculates all control points of rays $l_{p,s}(t)$ for the given (p, s)-projection. The control points show how the rays are numbered. If we consider the process of projection data collection for the parallel X-ray set, as in computerized tomography [29], the control points determine the location and numbering of radiation sources on the transmission side and the detectors on the receiving side.

```
% ================================================================
function nm=control_points_rays(p,s,N) / M.M. Grigoryan, May 2011
%------------------------------- 1st part of (p,s)
  if (p==1) & (s<=N/2)
     b=1;
     if s==0
        b=0;
     end
     for i1=1:N-s+b
         nm(i1,1:2)=[i1,1];
     end
     for i2=1:N
         for s1=s:-1:1
             nm(i1,1:2)=[N-s1+1 i2]; i1=i1+1;
         end
     end
     return
  end
%------------------------------- 2nd part of (p,s)
  if (p==1) & (s>N/2)
     s=N-s;
```

```
    for i1=1:N-s+1
        nm(i1,1:2)=[i1,N];
    end
    for i2=N:-1:1
        for s1=s:-1:1
            nm(i1,1:2)=[N-s1+1 i2]; i1=i1+1;
        end
    end
    return
end
%-------------------------------- 3rd part of (p,s)
 if (s==1) & (p~=1) & (p<=N/2)
    b=0;
    if p~=0 b=1; end
    for i1=1:N-p+b
        nm(i1,1:2)=[1,i1];
    end
    for i2=1:N
        for p1=p:-1:1
            nm(i1,1:2)=[i2 N-p1+1]; i1=i1+1;
        end
    end
    return
end
%-------------------------------- 4th part of (p,s)
 if (s==1) & (p>N/2)
    p=N-p;
    i1=1;
    for i2=N:-1:1
        for p1=1:p
            nm(i1,1:2)=[i2 p1]; i1=i1+1;
        end
    end
    for i2=p+1:N
        nm(i1,1:2)=[1 i2]; i1=i1+1;
    end
    return
end
% ================================================================
```

The number of rays $l_{p,s}(t)$ in the (p,s)-projection equals $(p+s)(N-1)+1$, if $p,s \le N/2$, and $(N-p+s)(N-1)+1$, and $(p+N-s)(N-1)+1$, if $p > N/2$ and $s > N/2$, respectively.

For example, when $N = 8$, the set of 22 control points on the lattice 8×8

for the $(2,1)$-projection equals

$$CP(2,1) = \begin{cases} (1,1),(1,2),(1,3),(1,4), ..., (1,8), \\ (2,7),(2,8), \\ (3,7),(3,8), \\ ... \\ (8,7),(8,8). \end{cases}$$

For the $(1,5)$-projection, the set of 29 control points equals

$$CP(1,5) = \begin{cases} (1,8),(2,8),(3,8),(4,8), ..., (8,8), \\ (6,7),(7,7),(8,7), \\ (6,6),(7,6),(8,6), \\ ... \\ (6,1),(7,1),(8,1). \end{cases}$$

In the program, the numbering of coordinates is performed from 1, not 0. These two sets of control-points are shown in Figure 4.8.

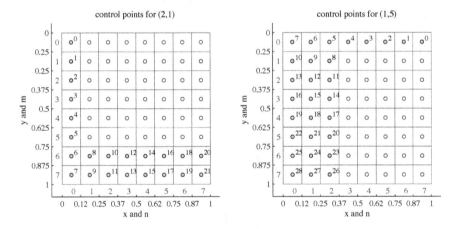

FIGURE 4.8
Control points for the $(2,1)$- and $(1,5)$-projections in the 8×8 case.

Now we can determine and separate all centers of image elements, or squares through which the t-th rays $l_{p,s}(t)$ pass. These squares and their centers are numbered by the same integer coordinates (n_k, m_k) of the lattice $N \times N$. For example, when $N = 8$ and $(p,s) = (1,1)$, the A-rays are represented by the following pairs of coordinates in the lattice 8×8:

$$
\{l_{1,1}(t)\} = \left\{
\begin{array}{l}
l(0) \\
l(1) \\
l(2) \\
l(3) \\
l(4) \\
l(5) \\
l(6) \\
l(7) \\
l(8) \\
l(9) \\
l(10) \\
l(11) \\
l(12) \\
l(13) \\
l(14)
\end{array}
\right\}
\rightarrow
\left\{
\begin{array}{l}
(1,1) \\
(2,1)\,(1,2) \\
(3,1)\,(2,2)\,(1,3) \\
(4,1)\,(3,2)\,(2,3)\,(1,4) \\
(5,1)\,(4,2)\,(3,3)\,(2,4)\,(1,5) \\
(6,1)\,(5,2)\,(4,3)\,(3,4)\,(2,5)\,(1,6) \\
(7,1)\,(6,2)\,(5,3)\,(4,4)\,(3,5)\,(2,6)\,(1,7) \\
(8,1)\,(7,2)\,(6,3)\,(5,4)\,(4,5)\,(3,6)\,(2,7)\,(1,8) \\
(8,2)\,(7,3)\,(6,4)\,(5,5)\,(4,6)\,(3,7)\,(2,8) \\
(8,3)\,(7,4)\,(6,5)\,(5,6)\,(4,7)\,(3,8) \\
(8,4)\,(7,5)\,(6,6)\,(5,7)\,(4,8) \\
(8,5)\,(7,6)\,(6,7)\,(5,8) \\
(8,6)\,(7,7)\,(6,8) \\
(8,7)\,(7,8) \\
(8,8)
\end{array}
\right\}
$$

This projection is defined by the corresponding splitting of the grid 8×8 by 15 subsets of points.

For the $(1,2)$-projection, 22 A-rays are represented by the coordinates in following way:

$$
\{l_{1,2}(t)\} = \left\{
\begin{array}{l}
l(0) \\
l(1) \\
l(2) \\
l(3) \\
l(4) \\
l(5) \\
l(6) \\
l(7) \\
l(8) \\
l(9) \\
l(10) \\
l(11) \\
l(12) \\
l(13) \\
l(14) \\
l(15) \\
l(16) \\
l(17) \\
l(18) \\
l(19) \\
l(20) \\
l(21)
\end{array}
\right\}
\rightarrow
\left\{
\begin{array}{l}
(1,1) \\
(2,1) \\
(3,1)\,(1,2) \\
(4,1)\,(2,2) \\
(5,1)\,(3,2)\,(1,3) \\
(6,1)\,(4,2)\,(2,3) \\
(7,1)\,(5,2)\,(3,3)\,(1,4) \\
(8,1)\,(6,2)\,(4,3)\,(2,4) \\
(7,2)\,(5,3)\,(3,4)\,(1,5) \\
(8,2)\,(6,3)\,(4,4)\,(2,5) \\
(7,3)\,(5,4)\,(3,5)\,(1,6) \\
(8,3)\,(6,4)\,(4,5)\,(2,6) \\
(7,4)\,(5,5)\,(3,6)\,(1,7) \\
(8,4)\,(6,5)\,(4,6)\,(2,7) \\
(7,5)\,(5,6)\,(3,7)\,(1,8) \\
(8,5)\,(6,6)\,(4,7)\,(2,8) \\
(7,6)\,(5,7)\,(3,8) \\
(8,6)\,(6,7)\,(4,8) \\
(7,7)\,(5,8) \\
(8,7)\,(6,8) \\
(7,8) \\
(8,8)
\end{array}
\right\}
$$

In general, for the given (p, s)-projection, the corresponding set of points can be calculated by the program `ps_Arays_points.m` with the following script.

```
% ================================================================
function S=ps_Arays_points(p,s,N) / M.M. Grigoryan, May 2011
% ----------------------------------------------------------------
S=1;
nm=control_points_rays(p,s,N); sz=size(nm);
%--------  1th set of generators ---------
if (p==1) & (s<=N/2)
  for i1=1:sz(1)
    iy=nm(i1,1);  ix=nm(i1,2);
    six=s*ix; t=iy+six;
    while(iy>0) & (ix<=N)
        fprintf(' (%d,%d) ' ,iy,ix);     % (*)
        ix=ix+1;  six=six+s; iy=t-six;
    end
    fprintf('\n');
  end
  return
end
%--------  2nd set of generators  ---------
if (p==1) & (s>N/2)
  s=N-s;
  for i1=1:sz(1)
    iy=nm(i1,1);    ix=nm(i1,2);
    six=s*ix; t=iy-six;
    while(ix>0) & (iy>0)
        fprintf(' (%d,%d) ' ,iy,ix);     % (*)
        ix=ix-1;  six=six+s; iy=t+six;
    end
    fprintf('\n');
  end
  return
end
%--------  3rd set of generators  ---------
if (s==1) & (p<=N/2)
  for i1=1:sz(1)
    iy=nm(i1,1);  ix=nm(i1,2);
    piy=p*ix; t=piy+ix;
    while(ix>0) & (iy<=N)
        fprintf(' (%d,%d)' ,iy,ix);      % (*)
        iy=iy+1;  piy=piy+p; ix=t-piy;
    end
    fprintf('\n');
  end
  return
end
%-------  4th set of generators  ---------
if (s==1) & (p>N/2)
```

```
    p=N-p;
    for i1=1:sz(1)
      iy=nm(i1,1); ix=nm(i1,2);
      piy=p*iy; t=ix-piy;
      while(iy<=N & ix<=N)
           fprintf(' (%d,%d)' ,iy,ix);        % (*)
           iy=iy+1;  piy=piy+p; ix=t+piy;
      end
      fprintf('\n');
    end
    return
end
% ========================================================
```

The above two sets of coordinates for rays of the $(1,1)$- and $(1,2)$-projections are calculated by using the following commands:

```
N=8;
p=1; s=1;  S=ps_Arays_points(p,s,N);
p=1; s=2;  S=ps_Arays_points(p,s,N);
```

To obtain values of the sums $v(t)$ along each A-ray of the (p, s)-projection, the sums of the image $f(x, y)$ in the corresponding set of points is calculated. To accomplish this calculation, we can make a few changes in program ps_Arays_points.m. Namely, we substitute the first command S=1 by S=zeros(1,size(sz)), the printing commands which are marked as (*) by S(1,i1)=S(1,i1)+A(iy,ix), and remove commands fprintf('\n'). Here "A" stands for the image $f(x, y)$. We add this new function with the syntax definition function S=ps_Arays(A,p,s,N).

Example 4.1 We consider the following image on the lattice 4×4 :

$$[f_d(x, y)] = \begin{bmatrix} 1 & 5 & 9 & 13 \\ 2 & 6 & 10 & 14 \\ 3 & 7 & 11 & 15 \\ 4 & 8 & 12 & 16 \end{bmatrix} .$$

The call of the program

```
N=4; A=1:N*N; A=reshape(A,4,4);
p=1;
for s=0:3
     Vps=ps_Arays(A,p,s,N)
end
```

results in the following values of the $(1,0)$-, $(1,1)$-, $(1,2)$-, and $(1,3)$-projections.

$$\begin{aligned}
\{v_{1,0}(t); \ t = 0 : 3\} &= \{28, 32, 36, 40\}, \\
\{v_{1,1}(t); \ t = 0 : 6\} &= \{1, 7, 18, 34, 33, 27, 16\}, \\
\{v_{1,2}(t); \ t = 0 : 9\} &= \{1, 2, 8, 10, 16, 18, 24, 26, 15, 16\}, \\
\{v_{1,-1}(t); \ t = 0 : 6\} &= \{13, 23, 30, 34, 21, 11, 4\}.
\end{aligned}$$

It is not difficult to modify the script of the function ps_Arays_points.m and save all coordinates (ix,iy) of points which represent the A-rays in one array S of size $L \times N \times 2$. The first dimension $L = (p + s)(N - 1) + 1$ is for the number of A-rays in the (p, s)-projection, the second one is for the maximal number of coordinates of points, and the last dimension is for the coordinates. This function in our library is called Arays2grid.m with syntax definition S=Array2grid(p,s,N). The part of the script of this function with commands for the first set of generators $(1, s)$, when $s \leq N/2$, is given below.

```
% ===========================================================
function S=Arays2grid(p,s,N) / M.M. Grigoryan, June 2011
% -----------------------------------------------------------
nm=control_points_rays(p,s,N); sz=size(nm);
S=zeros(sz(1),N,2);
%-------- 1th set of generators ---------
if (p==1) & (s<=N/2)
  for i1=1:sz(1)
    iy=nm(i1,1);   ix=nm(i1,2);
    six=s*ix; t=iy+six;
    n_pst=0;
    while(iy>0) & (ix<=N)
          n_pst=n_pst+1; S(i1,n_pst,:)=[iy,ix];
          ix=ix+1;   six=six+s; iy=t-six;
    end
  end
  return
end
... ... ...
```

For fast calculation of line-integrals $w(t) = w_{p,s}(t)$ of the (p, s)-projection, we can use the system of linear equations that is described in Section 4.5. For instance, for the generator $(p, s) = (1, s)$, when $0 < s \leq N/2$, we consider the convolution equation

$$w(t) = \Delta l_{1,s}[v(t) + v(t - 1) + v(t - 2) + \ldots + v(t - s + 1)],$$
$$t = 0 : (s + 1)(N - 1), \tag{4.24}$$

where the length $\Delta l_{1,s} = N\sqrt{1 + s^2}/s$. This system of equations also describes two other cases, when generators (p, s) are from the subsets $M2$ and $M4$.

Below is the script of the function Aray2Gray_case1.m which accomplishes the summation along the G-rays from A-rays. The multiplication of the integrals $w(t)$ by the factor of $\Delta l_{1,s}$ can be omitted and not used when reconstructing the sums $v(t)$ from the line-integrals $w(t)$, to simplify the calculations.

```
% ==================================================
% call: Aray2Gray_case1.m   / A.-M. Grigoryan 2011
% Vps is the set of sums v(t) of (p,s)-projections:
% Vps={Vps(t),t=1:L}, L=(1+s)(N-1)+1.
```

```
% The generators (p,s)=(1,s), 0<s<=N/2.
% ----------------------------------------------------
  function Wps=Aray2Gray_case1(Vps,s,L)
    Wps=zeros(1,L);
    s1=s-1;
    for t=1:s1
        Wps(t)=sum(Vps(1:t));
    end
    for t=s:L
        Wps(t)=sum(Vps(t-s1:t));
    end
    dl=N*sqrt(1+s^2)/s;  Wps=Wps*dl;
% ===================================================
```

For generators of the subset $M3$, when $(p,s) = (p,1)$ and $0 < p \leq N/2$, we have another system of equations

$$
\begin{aligned}
w(t) = \Delta l_{p,1}[v(t) + v(t+1) + v(t+2) + \ldots + v(t+p-1)], \\
t = (1+p)(N-1) : -1 : 0,
\end{aligned}
\tag{4.25}
$$

where the length $\Delta l_{p,1} = N\sqrt{p^2 + 1}/p$. The function `Aray2Gray_case2.m` accomplishes the calculation of G-rays from A-rays in this case.

```
% ===================================================
% call: Aray2Gray_case2.m    / A.-M. Grigoryan 2011
% Vps is the set of sums v(t) of (p,s)-projections:
% Vps={Vps(t),t=1:L}, L=(p+1)(N-1)+1.
% The generators (p,s)=(p,1), 0<p<=N/2.
% ----------------------------------------------------
  function Wps=Aray2Gray_case2(Vps,p,L)
    Wps=zeros(1,L);
    p1=p-1;
    for t=1:L-p1
        Wps(t)=sum(Vps(t:t+p1));
    end
    for t=L-p1+1:L
        Wps(t)=sum(Vps(t:L));
    end
    dl=N*sqrt(1+p^2)/p;  Wps=Wps*dl;
% ===================================================
```

As described in Section 4.5, the program `solution_of_linearsystemA.m` calculates the inverse transformation, i.e., the sums $v(t)$ from the integrals $w(t)$.

Example 4.2 To illustrate the line-integrals $w(t)$ and sums $v(t)$, we consider the image composed of seven rectangles shown in Figure 4.9 in part a. The data of these rectangles are given in integer form in Table 4.7, where

$$
x'_k = 64x_k, \quad y'_k = 64y_k, \quad \Delta x'_k = 64\Delta x_k, \quad \Delta y'_k = 64\Delta y_k, \quad k = 1:7.
$$

TABLE 4.7
Integer data of seven
rectangles

k	x'_k	y'_k	$\Delta x'_k$	$\Delta y'_k$	R_k
1	18	19	6	15	9
2	29	29	14	25	6
3	39	39	18	18	13
4	41	14	17	17	3
5	14	38	2	10	10
6	36	11	10	8	15
7	24	9	7	25	4

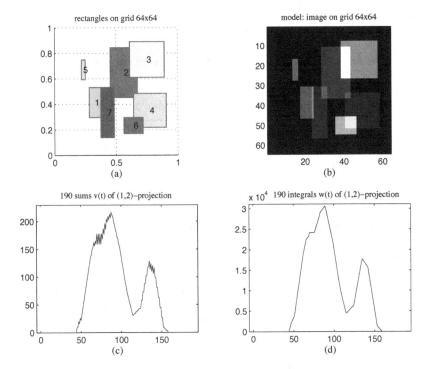

FIGURE 4.9
(a) Seven rectangles on the square $[0,1] \times [0,1]$, (b) image composed from
these rectangles, (c) sums $v(t)$, and (d) integrals $w(t)$ of the $(1,2)$-projection.

The gray-scale image $f(x,y)$ composed of these seven rectangles on the
grid 64×64 is shown in part b. The geometries of A-rays and G-rays, i.e.,
the sums $v(t)$ and integrals $w(t)$ of the $(1,2)$-projection, are shown in parts c
and d, respectively. The number of rays in the $(1,2)$-projection equals $190 =
(1+2)63+1$. This projection is at angle $90° - \arctan(1/2) = 63.4349°$ to the
horizontal axis.

Below is the script of the program for calculating the image $f(x, y)$ from the random rectangles. Here, `imageSQ_model` stands for the discrete image $f(x, y)$ and `data_allrect` is used for the data of seven rectangles, as in code `image_reconstructionA3A4B1A2.m`, which is given in Section 4.2.

```
% ------------------------------------------------------------------
% composition of the gray-scale image fd(x,y) from rectangles
N=64;
many_rectangles=7;
imageSQ_model=zeros(N,N);
for k=1:many_rectangles
    data_1rect=data_allrect(k,:);
    x=data_1rect(1); dx=data_1rect(3);
    y=data_1rect(2); dy=data_1rect(4);   x1=x+dx; y1=y+dy;
    d_intensity=data_1rect(5);
    imageSQ_model(x:x1,y:y1)=imageSQ_model(x:x1,y:y1)+d_intensity;
end
imageSQ_model=imageSQ_model(:,N:-1:1);
% to display the image use:  imagesc(imageSQ_model');
% ------------------------------------------------------------------
```

For the same image, Figure 4.10 shows sums $v(t)$ and integrals $w(t)$ of the $(1, 3)$- and $(2, 1)$-projections in parts (a), (b), and (c), (d), respectively. These projections are at angles $90° - \arctan(1/3) = 71.5651°$ and $90° - \arctan(2) = 26.5651°$ to the horizontal axis. From these graphs, one can notice the difference between A-ray and G-ray integrals, namely, the straightening out of oscillations in the line-integrals $w(t)$.

Example 4.3 We consider the gray-scale discrete image $f_{n,m}$, which is the modified Shepp-Logan phantom from the MATLAB library. We can assume that this image represents the image $f(x, y)$ on the grid 256×256. Each value of the discrete image corresponds to the value of $f(x, y)$ on the (n, m)-th square of size $1/256 \times 1/256$. To call this image and put it in the gray-scale $[0, 255]$, we use the following commands:

```
N=256; X=phantom('Modified Shepp-Logan',N);
imageSQ_model=round(double(X)*255);
```

Figure 4.10 shows the phantom image in part a, along with the sums $v(t)$ and integrals $w(t)$ of the $(1, 2)$-projection in parts c and d, respectively. The number of rays in the $(1, 2)$-projection equals $766 = (1 + 2)255 + 1$.

The FIPS method is fast. For comparison we consider the time data given in Table 4.5 for the main program A3-A4-B1-A2. The total time of processing images of sizes 64×64, 128×128, and 256×256 equals 0.9048 s, 8.4865 s, and 3.48 min, respectively.

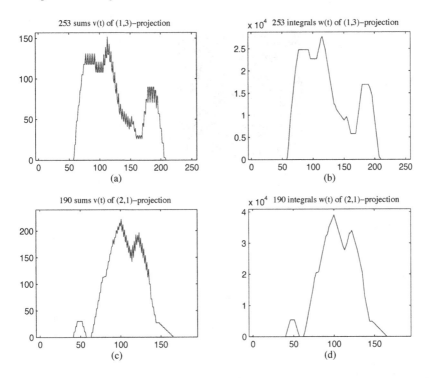

FIGURE 4.10
(a,b) Sums $v(t)$ and integrals $w(t)$ of the $(1,3)$-projection, and (c,d) sums $v(t)$ and integrals $w(t)$ of the $(2,1)$-projection.

4.9 Selection of projections

In the described case $N = 2^r$, $r > 1$, we consider the $(1, s - N)$-projections instead of $(1,s)$-projections when $s > N/2$, and the $(p - N, 1)$-projections instead of $(p, 1)$-projections when $p > N/2$. This change allows for reducing the number of projections for these projections, as well as moving the half of the angles to the second part of the interval of $[0, \pi]$. As an example, Figure 4.12 shows these angles before and after the changing projections in the $N = 8$ case.

Figure 4.13 shows new angles of projections in the interval $[0, \pi]$ in the $N = 8$ case, as well as for the $N = 16$ and 32 cases. One can observe that a few angles are located very close to each other, which may lead to difficulties in obtaining projections for such angles in practical applications. We therefore consider the problem of changing this set of angles by another more uniform distributed set of angles in the interval $[0, \pi]$, which can also be used in our proposed method of image reconstruction for the set of generators $J_{N,N}$.

phantom on the grid 256x256

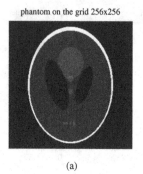

(a)

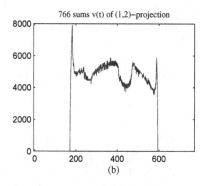

766 sums v(t) of (1,2)–projection

(b)

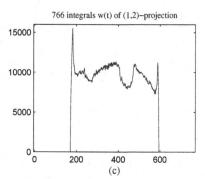

766 integrals w(t) of (1,2)–projection

(c)

FIGURE 4.11

(a) Phantom image, (b) sums $v(t)$, and (c) integrals $w(t)$ of the $(1,2)$-projection.

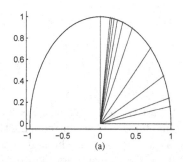

(a)

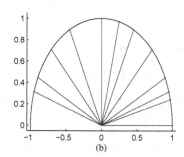

(b)

FIGURE 4.12

Sets of angles for 12 projections for model 8×8.

For that, we first consider the $N = 16$ case. The set of angles $\Psi(16)$ is

$$
\Psi = \begin{cases}
1.5708 \ (\pi/2), \\
0.7854,\ 0.4636,\ 0.3218,\ 0.2450,\ 0.1974,\ 0.1651,\ \underline{0.1419},\ 0.1244, \\
2.9997,\ 2.9764,\ 2.9442,\ 2.8966,\ 2.8198,\ 2.6779,\ 2.3562, \\
0, \\
1.1071,\ 1.3258,\ 1.4056,\ 1.4464,\ 1.7359,\ 1.8158,\ 2.0344.
\end{cases}
$$

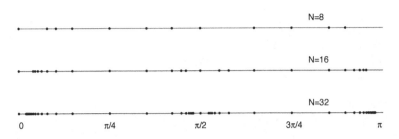

FIGURE 4.13
Sets of angles for $3N/2$ projections for the cases when $N = 8, 16$, and 32.

Below is the script of the commands to calculate these angles.

```
N=16; L=N+N/2;
Jps=ps_generators(N);
A1=zeros(1,L);   A1(1)=pi/2;
for k=2:L
     p=Jps(k,1); if p>N/2 p=p-N; end
     s=Jps(k,2); if s>N/2 s=s-N; end
     A1(k)=-atan(s/p);   % in radians
end
A1=pi/2+A1;   A1(1)=pi/2+0;
```

Let us try to substitute one of these angles, let it be $\vartheta(1,7) = 0.1419$, by another angle. The projection number 8 in the set $\Psi(16)$ is the $(1,7)$-projection, which is used to calculate a few splitting-signals, including $f_{T'_{1,7}} = \{f_{1,7,0}, f_{1,7,1}, f_{1,7,2}, \ldots, f_{1,7,7}\}$. This signals determines the 2-D DFT of the image at frequency-points of $T'_{1,7} = \{(1,7), (3,5), (5,3), (7,1), (9,15), \ldots \}$.

```
n=8;   a=A1(n);   % 0.1419
p=Jps(n,1);   s=Jps(n,2);   % (p,s)=(1,7)
Tps=zeros(N/2,2);
k=1;
for m=1:2:N/2
     Tps(k,1)=mod(m*p,N); Tps(k,2)=mod(m*s,N); k=k+1;
end; Tps'
%    1   3   5   7
%    7   5   3   1
```

The generators $(1,7), (3,5), (5,3)$, and $(7,1)$ are equivalent in the sense that $T'_{1,7} = T'_{3,5} = T'_{5,3} = T'_{7,1}$, and the number of parallel rays for the $(1,7)$-, $(3,5)$-, $(5,3)$-, and $(7,1)$-projections is the same, $(1+7)15+1 = (3+5)15+1 = \ldots = 121$. The angles of three new projections are $0.5404, 1.0304$, and 1.4289, respectively (in radians).

```
An=zeros(1,N/2);
for k=1:N/2
    p1=Tps(k,1); if p1>N/2 p1=p1-N; end
    s1=Tps(k,2); if s1>N/2 s1=s1-N; end
    An(k)=-atan(s1/p1);
end
An=pi/2+An
%   0.1419  0.5404  1.0304  1.4289   1.4289  1.0304  0.5404  0.1419
```

The first three angles $0.1419, 0.5404$, and 1.0304 are shown beneath the first line in Figure 4.14. The splitting-signal $f_{T'_{1,7}}$, as well as two other splitting-

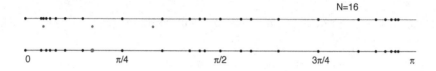

FIGURE 4.14
Two sets of angles for 24 projections for the $N = 16$ case.

signals $f_{T'_{3,5}}$ and $f_{T'_{5,3}}$ determine the 2-D DFT of the image at the same set of frequency-points, $T'_{1,7}$. Thus, one of these projections can be used instead of the $(1, 7)$-projection. For example, we consider the $(3, 5)$-projection. That means the angle 0.1419 of the projection is changed by 0.5404, as shown in the figure in the second line of angles. In the obtained new set of generators $J_{16,16}$, we can also substitute other angles and make all angles more uniformly distributed in the interval $[0, 1]$. The question is, can one write an optimal code to obtain such distribution of the angles in the interval $[0, \pi]$, for the general N case?

Problems

Problem 4.1 Develop the program for image reconstruction in the square $[0, 1] \times [0, 1]$, by using the method of fast projection integrals by squares, which is given in Section 4.8. Discuss the performance of your program. How fast is your program in comparison with the main program with script image_reconstionA3A4B1A2.m? Show by example that this method is effective. Assume that the reconstruction is on the lattice 256×256 and the image $f(x, y)$ is composed of a few random rectangles.

Problem 4.2 In the 2nd and 3rd columns of Table 4.5, the time data describe the total time of program performance. It includes the time required for calculating the line-integrals $w(t)$ for all projections, the time for calculating the sums $v(t)$ for these integrals, and the time for calculating the 2-D paired and its inverse transformations. Assume that the measurement data, or line-integrals are given. Determine how much of the time is used for reconstructing the image directly from the line-integrals. Give your answer in the form of a table for images of sizes 64×64, 128×128, and 256×256.

Problem 4.3 Different sets $J_{N,N}$ of generators (p, s) can be used when implementing the proposed method of image reconstruction by transferring the geometry of G-rays to A-rays. The angles can be spread approximately uniformly in the interval $[0, \pi)$, or they can be moved into a small subinterval of $[0, \pi)$. How you would select the angles for projections for image reconstruction? Discuss the $N = 32$ example.

Problem 4.4 As mentioned in Section 4.3, the main program of image reconstruction is referred to as A3-A4-B1-A2 code. Different types of equations for geometrical rays and convolution are considered. Propose the A3-A4-A1-A2 code, when the same type, A, equations are selected for all four subsets M1, M2, M3, and M4 of generators (p, s) of the set $J_{N,N}$.

Problem 4.5 Discuss how would you improve the main program for image reconstruction, which is described in this chapter. Propose your program for image reconstruction and give an example of the reconstruction.

Problem 4.6* Assume that you are given the image with a rectangle whose one or a few vertices are not on the Cartesian lattice. For instance, let $N = 8$ and let the image be the following rectangle of length $1.5(1/8)$:

$$f(x, y) = \begin{cases} 1; \text{if } x \in [1/4, 3/8 + 1/16], \ y \in [1/4, 3/8], \\ 0; \text{otherwise.} \end{cases}$$

The image consists of one square (which we call *particle*) and one half-square (which we call *semi-particle*) and it is shown in Figure 4.15 in part a. In the ideal case, the image we would like to obtain is

$$\mathbf{f}_{1.5} = \begin{bmatrix} 0 & 0 & 0 & 0 & 0 & 0 & 0 & 0 \\ 0 & 0 & 0 & 0 & 0 & 0 & 0 & 0 \\ 0 & 0 & 0 & 0 & 0 & 0 & 0 & 0 \\ 0 & 0 & 0 & 0 & 0 & 0 & 0 & 0 \\ 0 & 0 & 0 & 0 & 0 & 0 & 0 & 0 \\ 0 & 0 & 1 & 0.5 & 0 & 0 & 0 & 0 \\ 0 & 0 & 0 & 0 & 0 & 0 & 0 & 0 \\ 0 & 0 & 0 & 0 & 0 & 0 & 0 & 0 \end{bmatrix}.$$

Discuss the result of reconstruction of this image. Would you receive a good reconstruction after thresholding the reconstruction? If yes, propose an example of such thresholding and show your result.

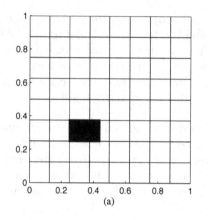 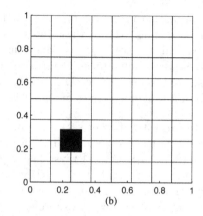

(a)   (b)

FIGURE 4.15
Image with (a) one rectangle and (b) shifted particle on the lattice 8×8.

Problem 4.7* Consider an image that is composed of the following particle shifted in the lattice 8×8 :

$$f_s(x, y) = \begin{cases} 1; \text{if } x, y \in [1/4 - 1/8 : 1/4 + 1/8], \\ 0; \text{otherwise.} \end{cases}$$

This image is shown in Figure 4.15 in part b. The image we would like to obtain is probably the following:

$$\mathbf{f}_s = \begin{bmatrix} 0 & 0 & 0 & 0 & 0 & 0 & 0 & 0 \\ 0 & 0 & 0 & 0 & 0 & 0 & 0 & 0 \\ 0 & 0 & 0 & 0 & 0 & 0 & 0 & 0 \\ 0 & 0 & 0 & 0 & 0 & 0 & 0 & 0 \\ 0 & 0 & 0 & 0 & 0 & 0 & 0 & 0 \\ 0 & 1 & 1 & 0 & 0 & 0 & 0 & 0 \\ 0 & 1 & 1 & 0 & 0 & 0 & 0 & 0 \\ 0 & 0 & 0 & 0 & 0 & 0 & 0 & 0 \end{bmatrix}.$$

Discuss the result of reconstruction of this image. Would you receive a good reconstruction after thresholding the reconstruction? If yes, propose an example of such thresholding and show your result.

5

RECONSTRUCTION FOR PRIME SIZE IMAGE

In this chapter, we describe the method of transferring the geometry of line-integrals of the image along the rays from the image plane to the Cartesian lattice $N \times N$, for the case when N is prime. The tensor representation of the image is considered and the calculation of the discrete image is accomplished by the inverse tensor transform. The proposed method of image reconstruction is described in detail for the $N = 7$ case. A general algorithm and programs are also given. The tensor representation is very effective for discrete images of size $N \times N$. The tensor transformation is not orthogonal, but not redundant [48, 55]. The transform and its inverse transform are fast.

5.1 Image reconstruction: Model II

In tensor representation, the discrete image $f_{n,m}$ of size $N \times N$ is defined as a set of splitting-signals of length N each [39],

$$\chi : \{f_{n,m}\} \to \left\{ f_{T_{p,s}} = \{f_{p,s,t}; t = 0 : (N-1)\} \right\}_{(p,s) \in J}. \qquad (5.1)$$

For the N prime case, the set of generators of splitting-signals is defined as

$$J = J_{N,N} = \{(1,0), (1,1), (1,2), \ldots, (1, N-1), (0,1)\}.$$

The components of the splitting-signals are calculated as the sum of the discrete image along the parallel lines

$$f_{p,s,t} = \sum_{n,m} \{f_{n,m}; np + ms = t \bmod N\}.$$

According to the principle of superposition by direction images, the image $f_{n,m}$ of size $N \times N$ can be composed of $(N+1)$ splitting-signals as follows:

$$f_{n,m} = \frac{1}{N} \left[\sum_{s=0}^{N-1} f_{1,s,(n+sm) \bmod N} + f_{0,1,m} \right] - NE[f], \qquad (5.2)$$

$$n, m = 0 : (N-1).$$

For image reconstruction, we will use this simple formula of the inverse tensor transformation, which is the transformation from the 2-D frequency and 1-D time space to the 2-D image space.

5.2   Example with image 7×7

The $N = 7$ case is described. $N(N + 1) = 56$ triplet-numbers (p, s, t) of components of the 7×7-point tensor transform can be defined by the set

$$
U_{7,7} = \begin{cases}
(1,0,0),(1,0,1),(1,0,2),(1,0,3),(1,0,4),(1,0,5),(1,0,6) \\
(1,1,0),(1,1,1),(1,1,2),(1,1,3),(1,1,4),(1,1,5),(1,1,6) \\
(1,2,0),(1,2,1),(1,2,2),(1,2,3),(1,2,4),(1,2,5),(1,2,6) \\
(1,3,0),(1,3,1),(1,3,2),(1,3,3),(1,3,4),(1,3,5),(1,3,6) \\
(1,4,0),(1,4,1),(1,4,2),(1,4,3),(1,4,4),(1,4,5),(1,4,6) \\
(1,5,0),(1,5,1),(1,5,2),(1,5,3),(1,5,4),(1,5,5),(1,5,6) \\
(1,6,0),(1,6,1),(1,6,2),(1,6,3),(1,6,4),(1,6,5),(1,6,6) \\
(0,1,0),(0,1,1),(0,1,2),(0,1,3),(0,1,4),(0,1,5),(0,1,6)
\end{cases}
$$

In this set, the number of directions is $N + 1 = 8$. These directions are calculated by the angles $\vartheta = \pi - \text{arctg}(p/s)$ to the horizontal, where $(p, s) \in J_{7,7}$. Given the generator (p, s), the corresponding subset of triplet-numbers $\{(p, s, t); t = 0 : 6\}$ is called $U(p, s)$.

We consider the image $f(x, y)$ on the square $[0, 1] \times [0, 1]$ with the Cartesian lattice 7×7. The image elements (IE) are considered to have the following values in the square $[0, 1] \times [0, 1]$:

$$
[f(x, y)] =
\begin{array}{|c|c|c|c|c|c|c|}
\hline
1 & 1 & 3 & 1 & 2 & 2 & 5 \\
\hline
1 & 1 & 1 & 2 & 1 & 1 & 4 \\
\hline
0 & 2 & 2 & 1 & 1 & 1 & 2 \\
\hline
2 & 3 & 4 & 2 & 1 & 1 & 3 \\
\hline
0 & 3 & 4 & 8 & 9 & 2 & 0 \\
\hline
2 & 3 & 1 & 8 & 7 & 6 & 6 \\
\hline
1 & 9 & 2 & 4 & 3 & 1 & 2 \\
\hline
\end{array}
\times 49 \Rightarrow \{f_{n,m}\} =
\begin{bmatrix}
1131225 \\
1112114 \\
0221112 \\
2342113 \\
0348920 \\
2318766 \\
1924312
\end{bmatrix} .
$$

Here the image $f(x, y)$ is given together with the discrete image $f_{n,m}$ to be reconstructed from eight projections of $f(x, y)$. For simplicity of calculations, all values of the image $f(x, y)$ are given with the factor of 49, which is N^2. Each IE has the size $\Delta x \Delta y = (1/7) \cdot (1/7) = 1/49$. Therefore, for instance, the value of $f_{2,0}$ is calculated as the integral of the image $f(x, y)$ in the $(2, 0)$-th IE, which equals $(3 \times 49)\Delta x \Delta y = 3$.

5.2.1 Horizontal projection

We consider the first subset of triplet-numbers $(1, 0, t)$,

$$U(1,0) = \big\{(1,0,0),(1,0,1),(1,0,2),(1,0,3),(1,0,4),(1,0,5),(1,0,6)\big\}.$$

The masks of the corresponding seven tensor functions $\chi_{1,0,t}$, $t = 0 : 6$, equal

$$[\chi_{1,0,0}] = \begin{bmatrix} 1000000 \\ 1000000 \\ 1000000 \\ 1000000 \\ 1000000 \\ 1000000 \\ 1000000 \end{bmatrix}, \ [\chi_{1,0,1}] = \begin{bmatrix} 0100000 \\ 0100000 \\ 0100000 \\ 0100000 \\ 0100000 \\ 0100000 \\ 0100000 \end{bmatrix}, \ \dots, \ [\chi_{1,0,6}] = \begin{bmatrix} 0000001 \\ 0000001 \\ 0000001 \\ 0000001 \\ 0000001 \\ 0000001 \\ 0000001 \end{bmatrix}.$$

Figure 5.1 shows the masks of the first two basic functions. Seven rays

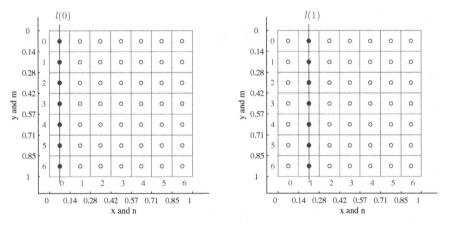

FIGURE 5.1
Masks $[\chi_{1,0,0}]$ and $[\chi_{1,0,1}]$ of the tensor functions for the $(1,0)$-projection.

of the horizontal projection are used to define all components of the tensor transforms with triplet-numbers of $U(1,0)$. We define the control points of these rays by the bullets through which the rays pass. These control points are numbered as

$$\begin{bmatrix} \cdot & \cdot & \cdot & \cdot & \cdot & \cdot & \cdot \\ \vdots & \vdots & \vdots & \vdots & \vdots & \vdots & \vdots \\ 0_{\bullet} & 1_{\bullet} & 2_{\bullet} & 3_{\bullet} & 4_{\bullet} & 5_{\bullet} & 6_{\bullet} \end{bmatrix} \qquad \text{(angle of rays is } 90°\text{).}$$

The sums of the image along the rays $l(t) = \{x = t + 1/(2N)\}$ are denoted by $v_t = v_{1,0}(t)$, where $t = 0 : 6$. The components of the tensor transform with the triplet-numbers $(1, 0, t)$ are equal to these sums:

$$f_{1,0,0} = v_0, \quad f_{1,0,1} = v_1, \quad f_{1,0,2} = v_2, \quad \dots, \quad f_{1,0,6} = v_6. \tag{5.3}$$

The length of intersection of the geometrical ray $l(t)$ with the image element equals $\Delta x = 1/7$. Therefore, the line-integrals equal

$$w(t) = w_{1,0}(t) = \frac{1}{\Delta x}v(t) = 7v(t), \quad t = 0:6,$$

and $v(t) = w_{1,0}(t)/7$, $t = 0:6$. The values of these line-integrals for the given image $f(x,y)$ can be calculated by

$$\begin{aligned}
w(0) &= 7(1+1+0+2+0+2+1) = 7 \times 7 = 49 \\
w(1) &= 7(1+1+2+3+3+3+9) = 7 \times 22 = 154 \\
w(2) &= 7(3+1+2+4+4+1+2) = 7 \times 17 = 119 \\
w(3) &= 7(1+2+1+2+8+8+4) = 7 \times 26 = 182 \\
w(4) &= 7(2+1+1+1+9+7+3) = 7 \times 24 = 168 \\
w(5) &= 7(2+1+1+1+2+6+1) = 7 \times 14 = 98 \\
w(6) &= 7(5+4+2+3+0+6+2) = 7 \times 22 = 154.
\end{aligned}$$

The splitting-signal $\{f_{1,0,t}; t = 0:6\} = \mathbf{v} = \mathbf{w}/7 = \{7, 22, 17, 26, 24, 14, 22\}$. Here, we denote by $\mathbf{v}$ and $\mathbf{w}$ the vectors of data $v(t)$ and $w(t)$, respectively.

5.2.2   Vertical projection

Consider the vertical projection, or the $(0,1)$-projection. The subset of triplet-numbers (p, s, t) for this projection is

$$U(0,1) = \{(0,1,0), (0,1,1), (0,1,2), (0,1,3), (0,1,4), (0,1,5), (0,1,6)\},$$

and the masks of the corresponding seven tensor functions are

$$[\chi_{0,1,0}] = \begin{bmatrix} 1111111 \\ 0000000 \\ 0000000 \\ 0000000 \\ 0000000 \\ 0000000 \\ 0000000 \end{bmatrix}, \ [\chi_{0,1,1}] = \begin{bmatrix} 0000000 \\ 1111111 \\ 0000000 \\ 0000000 \\ 0000000 \\ 0000000 \\ 0000000 \end{bmatrix}, \ ..., \ [\chi_{0,1,6}] = \begin{bmatrix} 0000000 \\ 0000000 \\ 0000000 \\ 0000000 \\ 0000000 \\ 0000000 \\ 1111111 \end{bmatrix}.$$

The numbered control points of seven rays for this projection are denoted by the bullets, as shown below

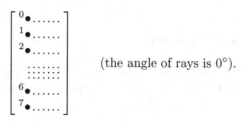

(the angle of rays is $0°$).

From the projection data $w(t) = w_{0,1}(t)$, $t = 0 : 6$, the vector $\mathbf{v}$, or the splitting-signal $\{f_{0,1,t}; t = 0 : 6\}$ is calculated by $f_{0,1,t} = v_{0,1}(t) = w_{0,1}(t)/7$, $t = 0 : 6$. For the given image $f(x,y)$, seven line-integrals $w(t)$ along the horizontal lines $l(t) = \{y = t + 1/(2N)\}$ are calculated by

$$\begin{aligned}
w(0) &= 7(1+1+3+1+2+2+5) = 7 \times 15 = 105 \\
w(1) &= 7(1+1+1+2+1+1+4) = 7 \times 11 = 77 \\
w(2) &= 7(0+2+2+1+1+1+2) = 7 \times 9 = 63 \\
w(3) &= 7(2+3+4+2+1+1+3) = 7 \times 16 = 112 \\
w(4) &= 7(0+3+4+8+9+2+0) = 7 \times 26 = 182 \\
w(5) &= 7(2+3+1+8+7+6+6) = 7 \times 33 = 231 \\
w(6) &= 7(1+9+2+4+3+1+2) = 7 \times 22 = 154.
\end{aligned}$$

Therefore, the splitting-signal $\{f_{0,1,t}\} = \mathbf{v} = \mathbf{w}/7 = \{15, 11, 9, 16, 26, 33, 22\}$. Figure 5.2 shows the plots of line-integrals $w(t)$ and splitting-signals for the $(1,0)$- and $(0,1)$-projections.

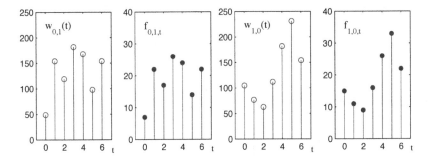

FIGURE 5.2
Line-integrals $w(t)$ and splitting-signals for the $(0,1)$- and $(1,0)$-projections.

5.2.3   Diagonal projection

Now we consider the subset of triplet-numbers

$$U(1,1) = \big\{(1,1,0),(1,1,1),(1,1,2),(1,1,3),(1,1,4),(1,1,5),(1,1,6)\big\}.$$

The masks of the corresponding seven tensor functions equal

$$[\chi_{1,1,0}] = \begin{bmatrix} 1000000 \\ 0000001 \\ 0000010 \\ 0000100 \\ 0001000 \\ 0010000 \\ 0100000 \end{bmatrix}, \quad [\chi_{1,1,1}] = \begin{bmatrix} 0100000 \\ 1000000 \\ 0000001 \\ 0000010 \\ 0000100 \\ 0001000 \\ 0010000 \end{bmatrix}, \quad [\chi_{1,1,2}] = \begin{bmatrix} 0010000 \\ 0100000 \\ 1000000 \\ 0000001 \\ 0000010 \\ 0000100 \\ 0001000 \end{bmatrix}$$

$$[\chi_{1,1,3}] = \begin{bmatrix} 0001000 \\ 0010000 \\ 0100000 \\ 1000000 \\ 0000001 \\ 0000010 \\ 0000100 \end{bmatrix}, \ [\chi_{1,1,4}] = \begin{bmatrix} 0000100 \\ 0001000 \\ 0010000 \\ 0100000 \\ 1000000 \\ 0000001 \\ 0000010 \end{bmatrix}, \ [\chi_{1,1,5}] = \begin{bmatrix} 0000010 \\ 0000100 \\ 0001000 \\ 0010000 \\ 0100000 \\ 1000000 \\ 0000001 \end{bmatrix}$$

$$[\chi_{1,1,6}] = \begin{bmatrix} 0000001 \\ 0000010 \\ 0000100 \\ 0001000 \\ 0010000 \\ 0100000 \\ 1000000 \end{bmatrix} = \begin{bmatrix} 1111111 \\ 1111111 \\ 1111111 \\ 1111111 \\ 1111111 \\ 1111111 \\ 1111111 \end{bmatrix} - \sum_{t=0}^{5} [\chi_{1,1,t}]. \tag{5.4}$$

The composition of the last matrix by the first six matrices shows the redundancy of the 2-D tensor transform, when the component $f_{1,1,6}$ can be calculated as

$$f_{1,1,6} = \sum_{n=0}^{5} \sum_{m=0}^{5} f_{n,m} - \sum_{t=0}^{5} f_{1,1,t} = \sum_{t=0}^{6} f_{1,0,t} - \sum_{t=0}^{5} f_{1,1,t}.$$

We define and number the control points of the parallel rays for this projection by the bullets as

$$\begin{bmatrix} 0 \bullet & \cdot & \cdot & & \cdot & & \cdot \\ 1 \bullet & \cdot & \cdot & \cdot & \cdot & \cdot \\ 2 \bullet & \cdot & \cdot & \cdot & \cdot & \cdot \\ \vdots & \vdots & \vdots & \vdots & \vdots & \vdots \\ 5 \bullet & \cdot & \cdot & \cdot & \cdot & \cdot \\ 6 \bullet 7 \bullet 8 \bullet 9 \bullet 10 \bullet 11 \bullet 12 \bullet \end{bmatrix}$$ (the angle of rays is $-45°$).

Seven masks of tensor functions $\chi_{1,1,t}(n, m)$ with the parallel rays on which the coefficients 1 are situated are shown in Figures 5.3–5.5. The set of all 14 rays is also shown in Figure 5.5.

The sums of the discrete image along these rays are denoted by $v_t = v_{1,1}(t)$, where $t = 0 : 12$. One can notice that the components of the tensor transform with the triplet-numbers $(p, s, t) \in U(1, 1)$ can be calculated as follows:

$$\begin{array}{ll} f_{1,1,0} = v_0 + v_7, & f_{1,1,1} = v_1 + v_8, \\ f_{1,1,2} = v_2 + v_9, & f_{1,1,3} = v_3 + v_{10}, \\ f_{1,1,4} = v_4 + v_{11}, & f_{1,1,5} = v_5 + v_{12}, \quad f_{1,1,6} = v_6. \end{array} \tag{5.5}$$

Thus, the above system of linear equations can be written as

$$f_{1,1,t} = v_t + v_{t+7} = v_{1,1,t} + v_{1,1,t+7}, \quad t = 0 : 6,$$

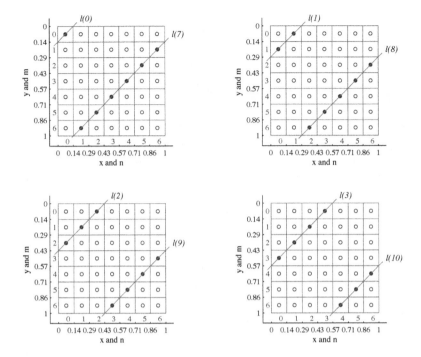

FIGURE 5.3
Parallel rays in the masks $[\chi_{1,1,t}]$, $t = 0 : 3$, for the $(1,1)$-projection.

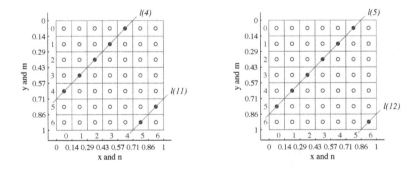

FIGURE 5.4
Parallel rays in the masks $[\chi_{1,1,t}]$, $t = 4, 5$ for the $(1,1)$-projection.

where $v_{13} = 0$. The sums of the image $v(t) = v_{1,1,t}$ can be calculated from the integrals of the image $f(x, y)$ along the corresponding geometrical rays,

$$l(t) = l_{1,1}(t) = \left\{(x,y); \ x + y = \frac{t}{7} + \frac{1}{7}\right\}, \quad t = 0 : 12,$$

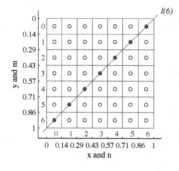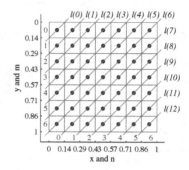

FIGURE 5.5
Mask $[\chi_{1,1,6}]$ and the set of arithmetical rays of the $(1,1)$-projection.

where $(x,y) \in [0,1] \times [0,1]$. Indeed, the length of the intersection of the geometrical rays in the image element equals $\Delta l = \sqrt{2}\Delta x = \sqrt{2}/7$. Each integral along the geometrical ray is proportional to the sum of the image along the corresponding arithmetical ray,

$$w(t) = w_{1,1}(t) = \Delta l \frac{v(t)}{(\Delta x)^2} = 7\sqrt{2}v(t), \quad t = 0:12.$$

Therefore $v(t) = w(t)/(7\sqrt{2})$, $t = 0:12$. The number of rays is calculated by $(1+1)(7-1)+1 = 13$.

For the given image $f(x,y)$, the line-integrals equal

$$
\begin{aligned}
w(0) &= 7\sqrt{2}(1) & &= 7\sqrt{2} \\
w(1) &= 7\sqrt{2}(1+1) & &= 7\sqrt{2}\cdot 2 \\
w(2) &= 7\sqrt{2}(3+1+0) & &= 7\sqrt{2}\cdot 4 \\
w(3) &= 7\sqrt{2}(1+1+2+2) & &= 7\sqrt{2}\cdot 6 \\
w(4) &= 7\sqrt{2}(2+2+2+3+0) & &= 7\sqrt{2}\cdot 9 \\
w(5) &= 7\sqrt{2}(2+1+1+4+3+2) & &= 7\sqrt{2}\cdot 13 \\
w(6) &= 7\sqrt{2}(5+1+1+2+4+3+1) & &= 7\sqrt{2}\cdot 17 \\
w(7) &= 7\sqrt{2}(4+1+1+8+1+9) & &= 7\sqrt{2}\cdot 24 \\
w(8) &= 7\sqrt{2}(2+1+9+8+2) & &= 7\sqrt{2}\cdot 22 \\
w(9) &= 7\sqrt{2}(3+2+7+4) & &= 7\sqrt{2}\cdot 16 \\
w(10) &= 7\sqrt{2}(0+6+3) & &= 7\sqrt{2}\cdot 9 \\
w(11) &= 7\sqrt{2}(6+1) & &= 7\sqrt{2}\cdot 7 \\
w(12) &= 7\sqrt{2}(2) & &= 7\sqrt{2}\cdot 2
\end{aligned}
$$

The sums of the image along the corresponding arithmetical rays can be written as the vector $\mathbf{v} = (1,2,4,6,9,13,17,24,22,16,9,7,2)'$. The components of the splitting-signal are calculated by

$$
\begin{aligned}
f_{1,1,0} &= v(0) + v(7) = 1 + 24 = 25 \\
f_{1,1,1} &= v(1) + v(8) = 2 + 22 = 24 \\
f_{1,1,2} &= v(2) + v(9) = 4 + 16 = 20 \\
f_{1,1,3} &= v(3) + v(10) = 6 + 9 = 15 \\
f_{1,1,4} &= v(4) + v(11) = 9 + 7 = 16 \\
f_{1,1,5} &= v(5) + v(12) = 13 + 2 = 15 \\
f_{1,1,6} &= v(6) = 17.
\end{aligned}
$$

Figure 5.6 shows the graphs of the line-integrals $w(t)$, sums $v(t)$, and splitting-signal for the $(1,1)$-projection.

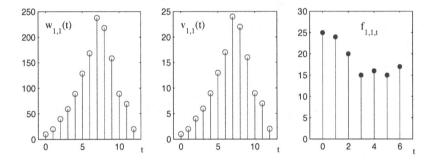

FIGURE 5.6
Line-integrals $w(t)$, sums $v(t)$, and splitting-signal of the $(1,1)$-projection.

5.2.4 $(1,2)$-Projection

Consider the next subset of triplet-numbers,

$$
U(1,2) = \big\{(1,2,0),(1,2,1),(1,2,2),(1,2,3),(1,2,4),(1,2,5),(1,2,6)\big\}.
$$

The masks of the tensor functions with seven triplet-numbers of $U(1,2)$ are

$$
[\chi_{1,2,0}] = \begin{bmatrix} 1000000 \\ 0000010 \\ 0001000 \\ 0100000 \\ 0000001 \\ 0000100 \\ 0010000 \end{bmatrix}, \;
[\chi_{1,2,1}] = \begin{bmatrix} 0100000 \\ 0000001 \\ 0000100 \\ 0010000 \\ 1000000 \\ 0000010 \\ 0001000 \end{bmatrix}, \;
[\chi_{1,2,2}] = \begin{bmatrix} 0010000 \\ 1000000 \\ 0000010 \\ 0001000 \\ 0100000 \\ 0000001 \\ 0000100 \end{bmatrix},
$$

$$[\chi_{1,2,3}] = \begin{bmatrix} 0001000 \\ 0100000 \\ 0000010 \\ 0001000 \\ 0100000 \\ 0000001 \\ 0000100 \end{bmatrix}, \ [\chi_{1,2,4}] = \begin{bmatrix} 0000100 \\ 0010000 \\ 1000000 \\ 0000010 \\ 0001000 \\ 0100000 \\ 0000001 \end{bmatrix}, \ [\chi_{1,2,5}] = \begin{bmatrix} 0000010 \\ 0001000 \\ 0100000 \\ 0000001 \\ 0000100 \\ 0010000 \\ 1000000 \end{bmatrix},$$

$$[\chi_{1,2,6}] = \begin{bmatrix} 0000001 \\ 0000100 \\ 0010000 \\ 1000000 \\ 0000010 \\ 0001000 \\ 0100000 \end{bmatrix} = \begin{bmatrix} 1111111 \\ 1111111 \\ 1111111 \\ 1111111 \\ 1111111 \\ 1111111 \\ 1111111 \end{bmatrix} - \sum_{t=0}^{5} [\chi_{1,2,t}].$$

We define the control points of these rays by the bullets and number them as shown below:

$$\begin{bmatrix} {}^0\bullet{}^1\bullet{}^2\bullet{}^3\bullet{}^4\bullet{}^5\bullet{}^6\bullet \\ \cdot\ \cdot\ \cdot\ \cdot\ \cdot\ {}^7\bullet{}^8\bullet \\ \cdot\ \cdot\ \cdot\ \cdot\ {}^9\bullet{}^{10}\bullet \\ \cdot\ \cdot\ \cdot\ \cdot\ {}^{11}\bullet{}^{12}\bullet \\ \cdot\ \cdot\ \cdot\ \cdot\ {}^{13}\bullet{}^{14}\bullet \\ \cdot\ \cdot\ \cdot\ \cdot\ {}^{15}\bullet{}^{16}\bullet \\ \cdot\ \cdot\ \cdot\ \cdot\ {}^{17}\bullet{}^{18}\bullet \end{bmatrix} \quad (\text{angle of rays is } -\arctan 1/2 = -26.565°).$$

The number of parallel rays is calculated by $(p+s)(N-1)+1 = 3\cdot 6 + 1 = 19$. The angle of this projection is $\arctan(2)$. These masks with the rays, on which the coefficients 1 are situated, are shown in Figures 5.7 and 5.8. Nineteen arithmetical rays of this projection

$$l(t) = l_{1,2}(t) = \left\{ (x,y); x + 2y = \frac{t}{7} + \frac{3}{14} \right\}, \quad t = 0:18,$$

are also shown in Figure 5.8. The components of the tensor transform with the triplet-numbers of the subset $U(1,2)$ can be calculated by

$$\begin{aligned} f_{1,2,0} &= v_0 + v_7 + v_{14}, & f_{1,2,1} &= v_1 + v_8 + v_{15}, \\ f_{1,2,2} &= v_2 + v_9 + v_{16}, & f_{1,2,3} &= v_3 + v_{10} + v_{17}, & & (5.6) \\ f_{1,2,4} &= v_4 + v_{11} + v_{18}, & f_{1,2,5} &= v_5 + v_{12}, & f_{1,2,6} &= v_6 + v_{13}, \end{aligned}$$

where variables $v_t = v_{1,2}(t)$ denote the sums of the discrete image along the rays $l(t), t = 0:18$. Therefore, the system of equations can be written as

$$f_{1,2,t} = v_t + v_{t+7} + v_{t+14} = v_{1,2,t} + v_{1,2,t+7} + v_{1,2,t+14}, \quad t = 0:6,$$

where $v_{19} = v_{20} = 0$.

To calculate the signal $\{f_{1,2,t}\}$ from the projection data, we will simplify the equation describing the linear relation between the line-integral $w(t)$ and

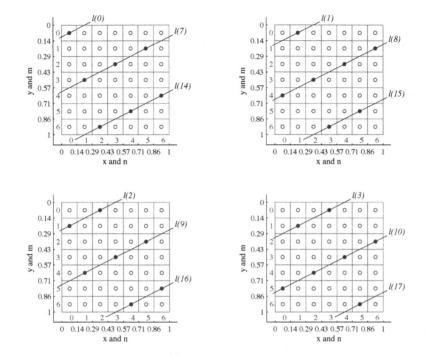

FIGURE 5.7
Four masks $[\chi_{1,2,t}]$, $t = 0 : 3$, for the $(1,2)$-projection.

sums $v(t)$, $t = 0 : 18$, by considering another set of 19 geometric rays. For that, we first consider the part of the mask of the tensor function $\chi_{1,2,0}(n,m)$ with coefficients 1 on the rays $l(0)$ and $l(7)$, which is shown in Figure 5.9. The first ray intersects the IE by the length $\sqrt{5}/2\Delta x = 7\sqrt{5}/2$. The ray $l(7)$ intersects the $(1,3)$-th IE by the same length, but each of the $(0,3)$- and $(2,3)$-th IEs by a length that is twice as small. The integral of the image along the geometrical ray between the rays $l(6)$ and $l(7)$ can be calculated from the sums of the image along two arithmetical rays by $w = 7\sqrt{5}/2[v_{1,2}(6) + v_{1,2}(7)]$. This geometrical ray is considered as ray number 7 in the new set of parallel rays, which are shown in Figure 5.10. These rays are defined by

$$\tilde{l}(t) = l_{1,2}(t - 0.5) = \{(x,y); x + 2y = \frac{t}{7} + \frac{1}{7}\}, \quad t = 0 : 18.$$

The relation between the line-integrals along the geometrical rays $\tilde{l}(t)$ and the line-sums along the arithmetical rays $l(t)$ is described by the following system of linear equations:

$$w(t) = w(\tilde{l}(t)) = 7\sqrt{5}/2[v_{1,2}(t - 1) + v_{1,2}(t)], \quad t = 0 : 18,$$

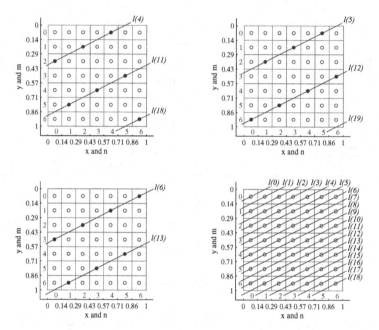

FIGURE 5.8

Masks $[\chi_{1,2,t}]$, $t = 4 : 6$, and the set of 19 arithmetical rays.

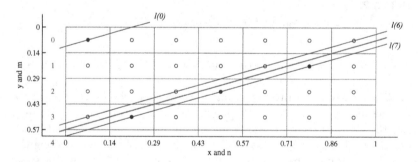

FIGURE 5.9

Four parallel rays of the $(1,2)$-projection.

where $v_{1,2}(-1) = 0$. These equations can be written in matrix form as

$$
\mathbf{w} =
\begin{bmatrix}
w_{1,2}(0) \\
w_{1,2}(1) \\
w_{1,2}(2) \\
w_{1,2}(3) \\
w_{1,2}(4) \\
\vdots \\
w_{1,2}(17) \\
w_{1,2}(18)
\end{bmatrix}
= \frac{7\sqrt{5}}{2}
\begin{bmatrix}
10000\cdots 000 \\
11000\cdots 000 \\
01100\cdots 000 \\
00110\cdots 000 \\
00011\cdots 000 \\
\vdots\ \vdots\ \vdots\ \vdots\ \vdots\ \ddots\ \vdots\ \vdots\ \vdots \\
00000\cdots 110 \\
00000\cdots 011
\end{bmatrix}
\begin{bmatrix}
v_{1,2}(0) \\
v_{1,2}(1) \\
v_{1,2}(2) \\
v_{1,2}(3) \\
v_{1,2}(4) \\
\vdots \\
v_{1,2}(17) \\
v_{1,2}(18)
\end{bmatrix} .
$$

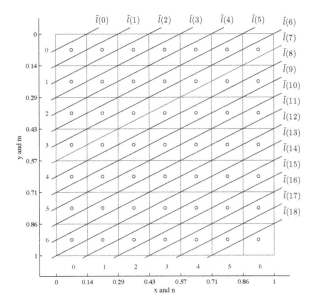

FIGURE 5.10
Geometrical rays for calculating line-integrals of the $(1, 2)$-projection.

The above Toeplitz matrix 19×19 has the triangle inverse matrix, and all sums along the arithmetical rays are calculated by

$$
\begin{bmatrix}
v_{1,2}(0) \\
v_{1,2}(1) \\
v_{1,2}(2) \\
v_{1,2}(3) \\
v_{1,2}(4) \\
\vdots \\
v_{1,2}(17) \\
v_{1,2}(18)
\end{bmatrix}
=
\frac{2}{7\sqrt{5}}
\begin{bmatrix}
1 & 0 & 0 & 0 & 0 & \cdots & 0 & 0 \\
-1 & 1 & 0 & 0 & 0 & \cdots & 0 & 0 \\
1 & -1 & 1 & 0 & 0 & \cdots & 0 & 0 \\
-1 & 1 & -1 & 1 & 0 & \cdots & 0 & 0 \\
1 & -1 & 1 & -1 & 1 & \cdots & 0 & 0 \\
\vdots & \vdots & \vdots & \vdots & \vdots & \ddots & \vdots & \vdots \\
-1 & 1 & -1 & 1 & -1 & \cdots & 1 & 0 \\
1 & -1 & 1 & -1 & 1 & \cdots & -1 & 1
\end{bmatrix}
\begin{bmatrix}
w_{1,2}(0) \\
w_{1,2}(1) \\
w_{1,2}(2) \\
w_{1,2}(3) \\
w_{1,2}(4) \\
\vdots \\
w_{1,2}(17) \\
w_{1,2}(18)
\end{bmatrix}.
$$

The required sums $v(t)$ of the discrete image $f_{n,m}$ are calculated by $v(t) = b_t/(7\sqrt{5}/2)$, where the coefficients b_t are calculated as follows:

$$
\begin{aligned}
b_0 &= w(0), \\
b_1 &= w(1) - b_0, \\
b_t &= w(t) - b_{t-1}, \quad t = 2, 3, \ldots, 18.
\end{aligned}
\tag{5.7}
$$

For the given image $f(x, y)$, the line-integrals are calculated by

$$
\begin{aligned}
w(0) &= K(1) & &= K \cdot 1 \\
w(1) &= K(1 + 1) & &= K \cdot 2 \\
w(2) &= K(3 + 1 + 1) & &= K \cdot 5 \\
w(3) &= K(1 + 3 + 1 + 1) &= K \cdot 6
\end{aligned}
$$

$$\begin{aligned}
w(4) &= K(2+1+1+1+0) &&= K \cdot 5 \\
w(5) &= K(2+2+2+1+2+0) &&= K \cdot 9 \\
w(6) &= K(5+2+1+2+2+2+2) &&= K \cdot 16 \\
w(7) &= K(5+1+1+1+2+3+2) &&= K \cdot 15 \\
w(8) &= K(4+1+1+1+4+3+0) &&= K \cdot 14 \\
w(9) &= K(4+1+1+2+4+3+0) &&= K \cdot 15 \\
w(10) &= K(2+1+1+2+4+3+2) &&= K \cdot 15 \\
w(11) &= K(2+1+1+8+4+3+2) &&= K \cdot 21 \\
w(12) &= K(3+1+9+8+1+3+1) &&= K \cdot 26 \\
w(13) &= K(3+2+9+8+1+9+1) &&= K \cdot 33 \\
w(14) &= K(0+2+7+8+2+9) &&= K \cdot 28 \\
w(15) &= K(0+6+7+4+2) &&= K \cdot 19 \\
w(16) &= K(6+6+3+4) &&= K \cdot 19 \\
w(17) &= K(6+1+3) &&= K \cdot 10 \\
w(18) &= K(2+1) &&= K \cdot 3
\end{aligned}$$

where $K = 7\sqrt{5}/2$. Therefore, the sums $v(t) = K^{-1}w(t) - v(t-1)$ of the discrete image $f_{n,m}$ can be calculated as follows:

$$\begin{aligned}
v(0) &= K^{-1}w(0) &&= 1 &&= 1 \\
v(1) &= K^{-1}w(1) - v(0) &&= 2-1 &&= 1 \\
v(2) &= K^{-1}w(2) - v(1) &&= 5-1 &&= 4 \\
v(3) &= K^{-1}w(3) - v(2) &&= 6-4 &&= 2 \\
v(4) &= K^{-1}w(4) - v(3) &&= 5-2 &&= 3 \\
v(5) &= K^{-1}w(5) - v(4) &&= 9-3 &&= 6 \\
v(6) &= K^{-1}w(6) - v(5) &&= 16-6 &&= 10 \\
v(7) &= K^{-1}w(7) - v(6) &&= 15-10 &&= 5 \\
v(8) &= K^{-1}w(8) - v(7) &&= 14-5 &&= 9 \\
v(9) &= K^{-1}w(9) - v(8) &&= 15-9 &&= 6 \\
v(10) &= K^{-1}w(10) - v(9) &&= 15-6 &&= 9 \\
v(11) &= K^{-1}w(11) - v(10) &&= 21-9 &&= 12 \\
v(12) &= K^{-1}w(12) - v(11) &&= 26-12 &&= 14 \\
v(13) &= K^{-1}w(13) - v(12) &&= 33-14 &&= 19 \\
v(14) &= K^{-1}w(14) - v(13) &&= 28-19 &&= 9 \\
v(15) &= K^{-1}w(15) - v(14) &&= 19-9 &&= 10 \\
v(16) &= K^{-1}w(16) - v(15) &&= 19-10 &&= 9 \\
v(17) &= K^{-1}w(17) - v(16) &&= 10-9 &&= 1 \\
v(18) &= K^{-1}w(18) - v(17) &&= 3-1 &&= 2.
\end{aligned}$$

The components of the splitting-signal $\{f_{1,2,t}; \ t = 0:6\}$ are calculated by

$$\begin{aligned}
f_{1,2,0} &= v(0) + v(7) + v(14) = 1 + 5 + 9 = 15 \\
f_{1,2,1} &= v(1) + v(8) + v(15) = 1 + 9 + 10 = 20 \\
f_{1,2,2} &= v(2) + v(9) + v(16) = 4 + 6 + 9 = 19 \\
f_{1,2,3} &= v(3) + v(10) + v(17) = 2 + 9 + 1 = 12 \\
f_{1,2,4} &= v(4) + v(11) + v(18) = 3 + 12 + 2 = 17 \\
f_{1,2,5} &= v(5) + v(12) = 6 + 14 = 20 \\
f_{1,2,6} &= v(6) + v(13) = 10 + 19 = 29.
\end{aligned}$$

Figure 5.11 shows the graphs of the integrals $w(t)$, sums $v(t)$, and splitting-signal for the $(1,2)$-projection.

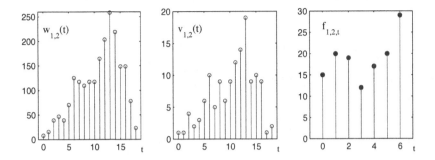

FIGURE 5.11
Line-integrals $w(t)$, sums $v(t)$, and splitting-signal for the $(1,2)$-projection.

5.2.5 $(1,3)$-projection

We consider the subset of triplets

$$U(1,3) = \big\{(1,3,0),(1,3,1),(1,3,2),(1,3,3),(1,3,4),(1,3,5),(1,3,6)\big\}.$$

Masks of the seven tensor functions with triplet-numbers of this subset are

$$[\chi_{1,3,0}] = \begin{bmatrix} 1000000 \\ 0000100 \\ 0100000 \\ 0000010 \\ 0010000 \\ 0000001 \\ 0001000 \end{bmatrix}, \ [\chi_{1,3,1}] = \begin{bmatrix} 0100000 \\ 0000010 \\ 0010000 \\ 0000001 \\ 0001000 \\ 1000000 \\ 0000100 \end{bmatrix}, \ [\chi_{1,3,2}] = \begin{bmatrix} 0010000 \\ 0000001 \\ 0001000 \\ 1000000 \\ 0000100 \\ 0100000 \\ 0000010 \end{bmatrix},$$

$$[\chi_{1,3,3}] = \begin{bmatrix} 0001000 \\ 1000000 \\ 0000100 \\ 0100000 \\ 0000010 \\ 0010000 \\ 0000001 \end{bmatrix}, \ [\chi_{1,3,4}] = \begin{bmatrix} 0000100 \\ 0100000 \\ 0000010 \\ 0010000 \\ 0000001 \\ 0001000 \\ 1000000 \end{bmatrix}, \ [\chi_{1,3,5}] = \begin{bmatrix} 0000010 \\ 0010000 \\ 0000001 \\ 0001000 \\ 1000000 \\ 0000100 \\ 0100000 \end{bmatrix},$$

$$[\chi_{1,3,6}] = \begin{bmatrix} 0000001 \\ 0001000 \\ 1000000 \\ 0000100 \\ 0100000 \\ 0000010 \\ 0010000 \end{bmatrix} = \begin{bmatrix} 1111111 \\ 1111111 \\ 1111111 \\ 1111111 \\ 1111111 \\ 1111111 \\ 1111111 \end{bmatrix} - \sum_{t=0}^{5} [\chi_{1,3,t}].$$

The masks of these tensor functions are shown in Figures 5.12 and 5.13, together with the rays passing through coefficients 1 of the masks.

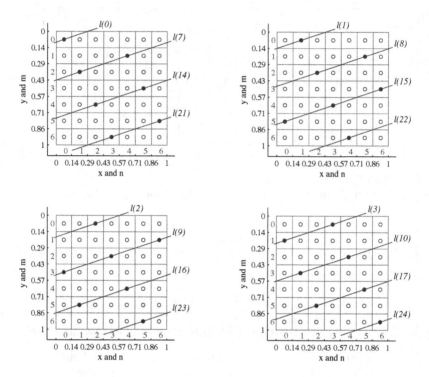

FIGURE 5.12
Four masks $[\chi_{1,3,t}]$, $t = 0 : 3$, and parallel rays for the $(1,3)$-projection.

The number of parallel rays of this projection equals $(1 + 3)6 + 1 = 25$, and they are defined as

$$l(t) = l_{1,3}(t) = \left\{(x,y); x + 3y = \frac{t}{7} + \frac{2}{7}\right\}, \quad t = 0 : 24,$$

where $(x,y) \in [0,1] \times [0,1]$. Figure 5.13 shows these rays, which are numbered in accordance with the following set of control points:

$$\begin{bmatrix} 0 & 1 & 2 & 3 & 4 & 5 & 6 \\ & & & & 7 & 8 & 9 \\ & & & 10 & 11 & 12 \\ & & & 13 & 14 & 15 \\ & & & 16 & 17 & 18 \\ & & & 19 & 20 & 21 \\ & & & 22 & 23 & 24 \end{bmatrix} \quad \text{(angle of rays is } -\tan^{-1}(1/3) = -18.435°\text{).}$$

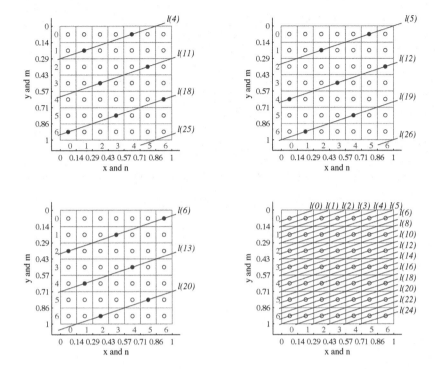

FIGURE 5.13
Three masks $[\chi_{1,3,t}]$, $t = 4 : 6$, and the set of 25 rays for the $(1,3)$-projection.

The components of the tensor transform with the triplet-numbers of the set $U(1,3)$ are calculated as

$$
\begin{aligned}
f_{1,3,0} &= v_0 + v_7 + v_{14} + v_{21}, & f_{1,3,1} &= v_1 + v_8 + v_{15} + v_{22}, \\
f_{1,3,2} &= v_2 + v_9 + v_{16} + v_{23}, & f_{1,3,3} &= v_3 + v_{10} + v_{17} + v_{24}, \\
f_{1,3,4} &= v_4 + v_{11} + v_{18}, & f_{1,3,5} &= v_5 + v_{12} + v_{19}, \\
f_{1,3,6} &= v_6 + v_{13} + v_{20},
\end{aligned} \tag{5.8}
$$

where variables $v_t = v_{1,3}(t)$ denote the sums of the discrete image along the corresponding rays $l(t)$, $t = 0 : 24$. The system of equations can also be written as

$$
f_{1,3,t} = \sum_{m=0}^{3} v_{t+7m} = \sum_{m=0}^{3} v_{1,3,t+7m}, \quad t = 0 : 6,
$$

where $v_t = 0$, if $t > 24$.

We now define a set of geometrical rays that allows for expressing the line-integrals $w(t)$ by means of the sums $v(t)$, $t = 0 : 24$. For that, we consider a few arithmetical rays as the geometrical rays in the part of the square $[0, 1] \times [0, 1]$, which are shown in Figure 5.14. Each of these rays equally intersects image elements, that is, the length of intersection of the ray with each IE equals $\Delta l_{0,6} = \sqrt{10}/3\Delta x$. The integrals along the geometrical rays 15, 16, and 17 can be calculated by the line-sums of the discrete image as

$$w(15) = w_{1,3}(15) = \frac{7\sqrt{10}}{3}[v(15) + v(14) + v(16)],$$
$$w(16) = w_{1,3}(16) = \frac{7\sqrt{10}}{3}[v(15) + v(16) + v(17)],$$
$$w(17) = w_{1,3}(17) = \frac{7\sqrt{10}}{3}[v(18) + v(16) + v(17)].$$

Similar equations hold for all integrals $w(t)$ of the considered projections:

$$w(t) = w_{1,3}(t) = 7\sqrt{10}/3[v(t-1) + v(t) + v(t+1)], \quad t = 0 : 24.$$

In matrix form, this system of linear equations is determined by the Toeplitz

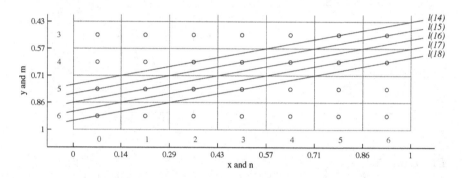

FIGURE 5.14
Five parallel geometrical rays of the $(1, 3)$-projection.

matrix 25×25 with right bandwidth of 2. This matrix has determinant 0 and the given set of geometrical rays cannot be used directly to define all required sums $v(t), t = 0 : 24$. We consider other sets of geometrical rays. For instance, this set of 25 geometrical rays can be defined as (see Figure 5.15)

$$\tilde{l}(t) = l_{1,3}(t-1) = \left\{(x,y); x + 3y = \frac{t}{7} + \frac{1}{7}\right\}, \quad t = 0 : 24.$$

The integrals of the image along these rays are denoted by $w(t) = w(\tilde{l}(t))$.
The corresponding system of equations for integrals $w(t)$ is described by

$$w(t) = \frac{7\sqrt{10}}{3}[v(t) + v(t-1) + v(t-2)], \quad t = 0 : 24,$$

where it is assumed that $v(-1)$ and $v(-2)$ are 0. This system of equations can

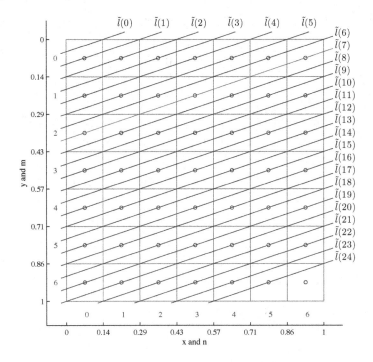

FIGURE 5.15
A set of geometrical rays for the $(1,3)$-projection.

be written in matrix form

$$\mathbf{w} = \frac{7\sqrt{10}}{3}\mathbf{A}\mathbf{v} = \frac{7\sqrt{10}}{3}\begin{bmatrix} 10000\cdots000 \\ 11000\cdots000 \\ 11100\cdots000 \\ 01110\cdots000 \\ \vdots\vdots\vdots\vdots\vdots\vdots\ddots\vdots\vdots\vdots \\ 00000\cdots110 \\ 00000\cdots111 \end{bmatrix}\mathbf{v}. \qquad (5.9)$$

The above Toeplitz matrix 25×25 has the lower triangle inverse matrix

$$
\mathbf{A}^{-1} = \begin{bmatrix}
1 & 0 & 0 & 0 & 0 & 0 & 00\cdots & 0 & 00 \\
-1 & 1 & 0 & 0 & 0 & 0 & 00\cdots & 0 & 00 \\
0 & -1 & 1 & 0 & 0 & 0 & 00\cdots & 0 & 00 \\
1 & 0 & -1 & 1 & 0 & 0 & 00\cdots & 0 & 00 \\
-1 & 1 & 0 & -1 & 1 & 0 & 00\cdots & 0 & 00 \\
0 & -1 & 1 & 0 & -1 & 1 & 00\cdots & 0 & 00 \\
1 & 0 & -1 & 1 & 0 & -1 & 10\cdots & 0 & 00 \\
\vdots & \vdots & \vdots & \vdots & \vdots & \vdots & \vdots\vdots\ddots & \vdots & \vdots\vdots \\
1 & 0 & -1 & 1 & 0 & -1 & 10\cdots & -1 & 10 \\
-1 & 1 & 0 & -1 & 1 & 0 & -11\cdots & 0 & -11
\end{bmatrix} . \qquad (5.10)
$$

The inverse transform $\mathbf{b} = \mathbf{A}^{-1}\mathbf{w}$ can be calculated in the recurrent form

$$
\begin{aligned}
b_0 &= w(0), \quad b_1 = w(1) - b_0, \\
b_2 &= w(2) - (b_1 + b_0), \\
b_t &= w(t) - (b_{t-1} + b_{t-2}), \quad t = 3, 4, 5, \cdots, 24,
\end{aligned} \qquad (5.11)
$$

and $v(t) = b_t/(7\sqrt{10}/3)$, where $t = 0 : 24$.

For the considered image $f(x, y)$, the integrals $w(t)$ are calculated by

$$
\begin{aligned}
w(0) &= K(1) & &= K \\
w(1) &= K(1 + 1) & &= K \cdot 2 \\
w(2) &= K(3 + 1 + 1) & &= K \cdot 5 \\
w(3) &= K(1 + 3 + 1 + 1) & &= K \cdot 6 \\
w(4) &= K(2 + 1 + 3 + 1 + 1 + 0) & &= K \cdot 8 \\
w(5) &= K(2 + 2 + 1 + 1 + 1 + 1) & &= K \cdot 8 \\
w(6) &= K(5 + 2 + 2 + 2 + 1 + 1 + 0) &= K \cdot 13 \\
w(7) &= K(5 + 2 + 1 + 2 + 1 + 2 + 0) &= K \cdot 13 \\
w(8) &= K(5 + 1 + 1 + 2 + 2 + 2 + 0) &= K \cdot 13 \\
w(9) &= K(4 + 1 + 1 + 1 + 2 + 2 + 2) &= K \cdot 13 \\
w(10) &= K(4 + 1 + 1 + 1 + 2 + 3 + 2) &= K \cdot 14 \\
w(11) &= K(4 + 1 + 1 + 1 + 4 + 3 + 2) &= K \cdot 16 \\
w(12) &= K(2 + 1 + 1 + 2 + 4 + 3 + 0) &= K \cdot 13 \\
w(13) &= K(2 + 1 + 1 + 2 + 4 + 3 + 0) &= K \cdot 13 \\
w(14) &= K(2 + 1 + 1 + 2 + 4 + 3 + 0) &= K \cdot 13 \\
w(15) &= K(3 + 1 + 1 + 8 + 4 + 3 + 2) &= K \cdot 22 \\
w(16) &= K(3 + 1 + 9 + 8 + 4 + 3 + 2) &= K \cdot 30 \\
w(17) &= K(3 + 2 + 9 + 8 + 1 + 3 + 2) &= K \cdot 28 \\
w(18) &= K(0 + 2 + 9 + 8 + 1 + 3 + 1) &= K \cdot 24 \\
w(19) &= K(0 + 2 + 7 + 8 + 1 + 9 + 1) &= K \cdot 28 \\
w(20) &= K(0 + 6 + 7 + 8 + 2 + 9 + 1) &= K \cdot 33 \\
w(21) &= K(6 + 6 + 7 + 4 + 2 + 9) & &= K \cdot 34 \\
w(22) &= K(6 + 6 + 3 + 4 + 2) & &= K \cdot 21 \\
w(23) &= K(6 + 1 + 3 + 4) & &= K \cdot 14 \\
w(24) &= K(2 + 1 + 3) & &= K \cdot 6,
\end{aligned}
$$

where $K = 7\sqrt{10}/3$.

Therefore, the sums $v(t)$ of the discrete image $f_{n,m}$ are calculated by

$$
\begin{aligned}
v(0) &= K^{-1}w(0) & &= 1 & &= 1 \\
v(1) &= K^{-1}w(1) - v(0) & &= 2 - 1 & &= 1 \\
v(2) &= K^{-1}w(2) - \big(v(1) + v(0)\big) & &= 5 - (1 + 1) & &= 3 \\
v(3) &= K^{-1}w(3) - \big(v(2) + v(1)\big) & &= 6 - (3 + 1) & &= 2 \\
v(4) &= K^{-1}w(4) - \big(v(3) + v(2)\big) & &= 8 - (2 + 3) & &= 3 \\
v(5) &= K^{-1}w(5) - \big(v(4) + v(3)\big) & &= 8 - (3 + 2) & &= 3 \\
v(6) &= K^{-1}w(6) - \big(v(5) + v(4)\big) & &= 13 - (3 + 3) & &= 7 \\
v(7) &= K^{-1}w(7) - \big(v(6) + v(5)\big) & &= 13 - (7 + 3) & &= 3 \\
v(8) &= K^{-1}w(8) - \big(v(7) + v(6)\big) & &= 13 - (3 + 7) & &= 3 \\
v(9) &= K^{-1}w(9) - \big(v(8) + v(7)\big) & &= 13 - (3 + 3) & &= 7 \\
v(10) &= K^{-1}w(10) - \big(v(9) + v(8)\big) & &= 14 - (7 + 3) & &= 4 \\
v(11) &= K^{-1}w(11) - \big(v(10) + v(9)\big) & &= 16 - (4 + 7) & &= 5 \\
v(12) &= K^{-1}w(12) - \big(v(11) + v(10)\big) & &= 13 - (5 + 4) & &= 4 \\
v(13) &= K^{-1}w(13) - \big(v(12) + v(11)\big) & &= 13 - (4 + 5) & &= 4 \\
v(14) &= K^{-1}w(14) - \big(v(13) + v(12)\big) & &= 13 - (4 + 4) & &= 5 \\
v(15) &= K^{-1}w(15) - \big(v(14) + v(13)\big) & &= 22 - (5 + 4) & &= 13 \\
v(16) &= K^{-1}w(16) - \big(v(15) + v(14)\big) & &= 30 - (13 + 5) & &= 12 \\
v(17) &= K^{-1}w(17) - \big(v(16) + v(15)\big) & &= 28 - (12 + 13) & &= 3 \\
v(18) &= K^{-1}w(18) - \big(v(17) + v(16)\big) & &= 24 - (3 + 12) & &= 9 \\
v(19) &= K^{-1}w(15) - \big(v(14) + v(17)\big) & &= 28 - (9 + 3) & &= 16 \\
v(20) &= K^{-1}w(16) - \big(v(15) + v(18)\big) & &= 33 - (16 + 9) & &= 8 \\
v(21) &= K^{-1}w(17) - \big(v(16) + v(19)\big) & &= 34 - (8 + 16) & &= 10 \\
v(22) &= K^{-1}w(18) - \big(v(17) + v(20)\big) & &= 21 - (10 + 8) & &= 3 \\
v(23) &= K^{-1}w(17) - \big(v(16) + v(21)\big) & &= 14 - (3 + 10) & &= 1 \\
v(24) &= K^{-1}w(18) - \big(v(17) + v(22)\big) & &= 6 - (1 + 3) & &= 2.
\end{aligned}
$$

The components of the splitting-signal $\{f_{1,3,t}\}$ are calculated by

$$
\begin{aligned}
f_{1,3,0} &= v(0) + v(7) + v(14) + v(21) = 1 + 3 + 5 + 10 = 19 \\
f_{1,3,1} &= v(1) + v(8) + v(15) + v(22) = 1 + 3 + 13 + 3 = 20 \\
f_{1,3,2} &= v(2) + v(9) + v(16) + v(23) = 3 + 7 + 12 + 1 = 23 \\
f_{1,3,3} &= v(3) + v(10) + v(17) + v(24) = 2 + 4 + 3 + 2 = 11 \\
f_{1,3,4} &= v(4) + v(11) + v(18) = 3 + 5 + 9 = 17 \\
f_{1,3,5} &= v(5) + v(12) = 3 + 4 + 16 = 23 \\
f_{1,3,6} &= v(6) + v(13) = 7 + 4 + 8 = 19.
\end{aligned}
$$

Figure 5.16 shows the graphs of the line-integrals $w(t)$, sums $v(t)$, and the splitting-signal for the $(1,3)$-projection.

5.2.6 $(1,4)$-projection

We now consider the subset of triplet-numbers

$$
U(1,4) = \big\{(1,4,0), (1,4,1), (1,4,2), (1,4,3), (1,4,4), (1,4,5), (1,4,6)\big\}.
$$

Masks of the tensor functions which are numbered by these triplets are:

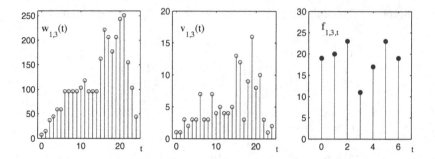

FIGURE 5.16
Line-integrals $w(t)$, sums $v(t)$, and splitting-signal for the $(1,3)$-projection.

$$[\chi_{1,4,0}] = \begin{bmatrix} 1000000 \\ 0001000 \\ 0000001 \\ 0010000 \\ 0000010 \\ 0100000 \\ 0000100 \end{bmatrix}, \ [\chi_{1,4,1}] = \begin{bmatrix} 0100000 \\ 0000100 \\ 1000000 \\ 0001000 \\ 0000001 \\ 0010000 \\ 0000010 \end{bmatrix}, \ [\chi_{1,4,2}] = \begin{bmatrix} 0010000 \\ 0000010 \\ 0100000 \\ 0000100 \\ 1000000 \\ 0001000 \\ 0000001 \end{bmatrix},$$

$$[\chi_{1,4,3}] = \begin{bmatrix} 0001000 \\ 0000001 \\ 0010000 \\ 0000010 \\ 0100000 \\ 0000100 \\ 1000000 \end{bmatrix}, \ [\chi_{1,4,4}] = \begin{bmatrix} 0000100 \\ 1000000 \\ 0001000 \\ 0000001 \\ 0010000 \\ 0000010 \\ 0100000 \end{bmatrix}, \ [\chi_{1,4,5}] = \begin{bmatrix} 0000010 \\ 0100000 \\ 0000100 \\ 1000000 \\ 0001000 \\ 0000001 \\ 0010000 \end{bmatrix},$$

$$[\chi_{1,4,6}] = \begin{bmatrix} 0000001 \\ 0010000 \\ 0000010 \\ 0100000 \\ 0000100 \\ 1000000 \\ 0001000 \end{bmatrix} = \begin{bmatrix} 1111111 \\ 1111111 \\ 1111111 \\ 1111111 \\ 1111111 \\ 1111111 \\ 1111111 \end{bmatrix} - \sum_{t=0}^{5} [\chi_{1,4,t}].$$

The set of all $(p + s)(N - 1) + 1 = 31$ arithmetical rays of the projection by the angle $\operatorname{arctg}(4)$ are shown in Figure 5.17.

To reduce the large number of rays, we consider another direction, namely, the $(1, -3)$-projection, which defines the same set of masks of tensor functions. This projection is defined as $(1, s - N) = (1, 4 - 7)$. Masks of the tensor

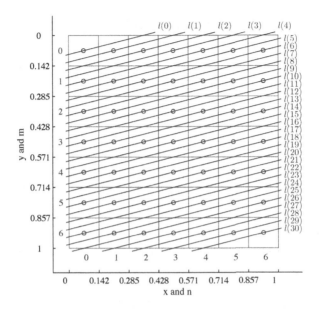

FIGURE 5.17
The set of 25 parallel rays for the $(1,4)$-projection.

functions with the first three triplet-numbers of the subset $U(1,-3)$ are

$$[\chi_{1,-3,0}] = \begin{bmatrix} 1000000 \\ 0001000 \\ 0000001 \\ 0010000 \\ 0000010 \\ 0100000 \\ 0000100 \end{bmatrix}, \quad [\chi_{1,-3,1}] = \begin{bmatrix} 0100000 \\ 0000100 \\ 1000000 \\ 0001000 \\ 0000001 \\ 0010000 \\ 0000010 \end{bmatrix}, \quad [\chi_{1,-3,2}] = \begin{bmatrix} 0010000 \\ 0000010 \\ 0100000 \\ 0000100 \\ 1000000 \\ 0001000 \\ 0000001 \end{bmatrix}.$$

As an example, Figure 5.18 shows the mask of the tensor function $\chi_{1,4,0}$ with the parallel rays of two different projections. Five rays of the $(1,4)$-projection are shown on the left, and three rays of the $(1,-3)$-projection are shown on the right.

For the remaining four triplet-numbers, we also have the equality of masks, $[\chi_{1,-3,t}] = [\chi_{1,4,t}]$, $t = 3, 4, 5, 6$.

We can see that for the subsets $U(1,4)$ and $U(1,-3)$, we have the same masks of the tensor functions. The number of parallel rays for $(1,-3)$-projection equals $(1+3)6 + 1 = 25$, which is less than 31. The set of all rays for this projection is shown in Figure 5.19.

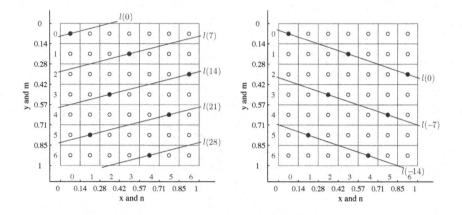

FIGURE 5.18

The mask $[\chi_{1,4,0}]$ with arithmetical rays of the $(1,4)$- and $(1,-3)$-projections.

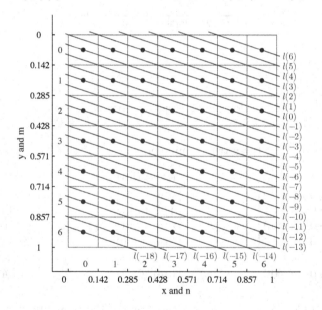

FIGURE 5.19

The set of parallel rays for the $(1,-3)$-projection.

The control points of this set of rays are defined as

$$
\begin{bmatrix}
0 & 1 & 2 & 3 & 4 & 5 & 6 \\
-3 & -2 & -1 & & & & \\
-6 & -5 & -4 & & & & \\
-9 & -8 & -7 & & & & \\
-12 & -11 & -10 & & & & \\
-15 & -14 & -13 & & & & \\
-18 & -17 & -16 & & & &
\end{bmatrix}
$$

(angle of rays is $\arctan(1/3) = 18.435°$).

The set of the 25 geometrical rays of this projection is described by

$$l(t) = l_{1,-3}(t) = \left\{ (x,y); x - 3y = \frac{t}{7} - \frac{1}{7} \right\}, \quad t = 6 : -1 : -18, \qquad (5.12)$$

where $(x,y) \in [0,1] \times [0,1]$. Six masks of the tensor functions $\chi_{1,4,t}$, $t = 1 : 6$, are shown in Figures 5.20 and 5.21. The components of the tensor transform

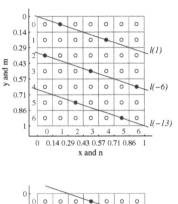

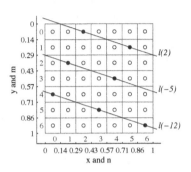

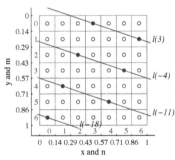

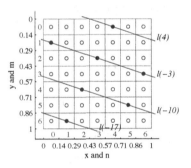

FIGURE 5.20
Masks $[\chi_{1,4,t}]$, $t = 1 : 4$, and the rays of the $(1, -3)$-projection.

with the triplet-numbers $(1, 4, t)$ can be calculated by

$$
\begin{aligned}
f_{1,4,0} &= v_0 + v_{-7} + v_{-14}, \quad f_{1,4,1} = v_1 + v_{-6} + v_{-13}, \\
f_{1,4,2} &= v_2 + v_{-5} + v_{-12}, \quad f_{1,4,3} = v_3 + v_{-4} + v_{-11} + v_{-18}, \\
f_{1,4,4} &= v_4 + v_{-3} + v_{-10} + v_{-17}, \quad f_{1,4,5} = v_5 + v_{-2} + v_{-9} + v_{-16}, \\
f_{1,4,6} &= v_6 + v_{-1} + v_{-8} + v_{-15},
\end{aligned}
\qquad (5.13)
$$

where the sums of the discrete image along the rays are denoted by $v_t = v(t) = v_{1,-3}(t)$, $t = 6 : -1 : -18$. This system of linear equations can also be written as follows:

$$f_{1,4,t} = \sum_m v_{t-7m} = \sum_m v_{1,-3,t-7m}, \quad t = 0 : 6.$$

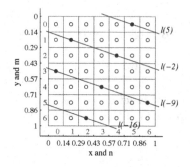
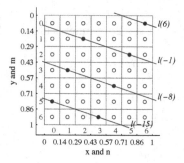

FIGURE 5.21
Two masks $[\chi_{1,4,t}]$, $t = 5, 6$, and the rays of the $(1, -3)$-projection.

As in the $(1, 3)$ case, we consider the shifted set of the rays, $l(t) \to l(t-1)$, as the set of 25 geometrical rays

$$\tilde{l}(t) = l_{1,-3}(t-1) = \left\{(x,y); x - 3y = \frac{t}{7} - \frac{2}{7}\right\}, \quad t = 6 : -1 : -18,$$

which are shown in Figure 5.22.

For the line-integrals, we obtain the system of equations, which is similar to the system described for the $(1, 3)$-projection,

$$w(t) = w(\tilde{l}(t)) = \frac{7\sqrt{10}}{3}\Big[v(t) + v(t-1) + v(t-2)\Big], \quad t = 6 : -1 : -18, \quad (5.14)$$

where $v(-19) = v(-20) = 0$. The required sums $v(t)$ of the discrete image $f_{n,m}$ can therefore be defined as $v(t) = 3b_t/(7\sqrt{10})$, where the components b_t are calculated recursively by

$$
\begin{aligned}
b_{-18} &= w(-18), \quad b_{-17} = w(-17) - b_{-18}, \\
b_{-16} &= w(-16) - (b_{-17} + b_{-18}), \\
b_t &= w(t) - (b_{t-1} + b_{t-2}), \quad t = -15, -14, \cdots, 6.
\end{aligned}
\quad (5.15)
$$

The line-integrals $w(t)$ for the given image $f(x, y)$ can be calculated by

$$
\begin{aligned}
w(6) &= K(5 + 2 + 2) & &= K \cdot 9 \\
w(5) &= K(4 + 2 + 2 + 1) & &= K \cdot 9 \\
w(4) &= K(4 + 1 + 2 + 1 + 3) & &= K \cdot 11 \\
w(3) &= K(4 + 1 + 1 + 1 + 3 + 1) & &= K \cdot 11 \\
w(2) &= K(2 + 1 + 1 + 2 + 3 + 1 + 1) &= K \cdot 11 \\
w(1) &= K(2 + 1 + 1 + 2 + 1 + 1 + 1) &= K \cdot 9 \\
w(0) &= K(2 + 1 + 1 + 2 + 1 + 1 + 1) &= K \cdot 9 \\
w(-1) &= K(3 + 1 + 1 + 1 + 1 + 1 + 1) &= K \cdot 9 \\
w(-2) &= K(3 + 1 + 1 + 1 + 2 + 1 + 1) &= K \cdot 10 \\
w(-3) &= K(3 + 1 + 1 + 1 + 2 + 2 + 1) &= K \cdot 11 \\
w(-4) &= K(0 + 1 + 1 + 2 + 2 + 2 + 0) &= K \cdot 8 \\
w(-5) &= K(0 + 2 + 1 + 2 + 4 + 2 + 0) &= K \cdot 11 \\
w(-6) &= K(0 + 2 + 9 + 2 + 4 + 3 + 0) &= K \cdot 20
\end{aligned}
$$

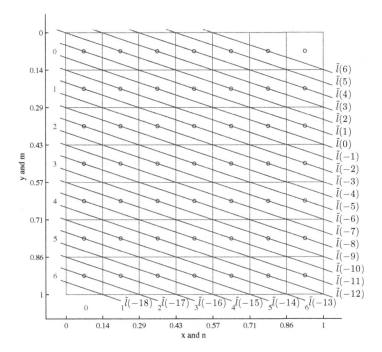

FIGURE 5.22

The set of geometrical rays for the $(1, -3)$-projection.

$$w(-7) = K(6 + 2 + 9 + 8 + 4 + 3 + 2) = K \cdot 34$$
$$w(-8) = K(6 + 6 + 9 + 8 + 4 + 3 + 2) = K \cdot 38$$
$$w(-9) = K(6 + 6 + 7 + 8 + 4 + 3 + 2) = K \cdot 36$$
$$w(-10) = K(2 + 6 + 7 + 8 + 4 + 3 + 0) = K \cdot 30$$
$$w(-11) = K(2 + 1 + 7 + 8 + 1 + 3 + 0) = K \cdot 22$$
$$w(-12) = K(2 + 1 + 3 + 8 + 1 + 3 + 0) = K \cdot 18$$
$$w(-13) = K(1 + 3 + 4 + 1 + 3 + 2) \quad = K \cdot 14$$
$$w(-14) = K(3 + 4 + 2 + 3 + 2) \quad = K \cdot 14$$
$$w(-15) = K(4 + 2 + 9 + 2) \quad = K \cdot 17$$
$$w(-16) = K(2 + 9 + 1) \quad = K \cdot 12$$
$$w(-17) = K(9 + 1) \quad = K \cdot 10$$
$$w(-18) = K(1) \quad = K \cdot 1$$

where $K = 7\sqrt{10}/3$.

The sums $v(t)$ of the discrete image $f_{n,m}$ are calculated as follows:

$$v(-18) = K^{-1}w(-18) \qquad\qquad = 1 \qquad = 1$$
$$v(-17) = K^{-1}w(-17) - v(-18) \qquad = 10 - 1 \qquad = 9$$
$$v(-16) = K^{-1}w(-16) - [v(-17) + v(-18)] = 12 - [9 + 1] = 2$$
$$v(-15) = K^{-1}w(-15) - [v(-16) + v(-17)] = 17 - [2 + 9] = 6$$
$$v(-14) = K^{-1}w(-14) - [v(-15) + v(-16)] = 14 - [6 + 2] = 6$$
$$v(-13) = K^{-1}w(-13) - [v(-14) + v(-15)] = 14 - [6 + 6] = 2$$

$$
\begin{aligned}
v(-12) &= K^{-1}w(-12) - [v(-13) + v(-14)] = 18 - [2 + 6] &= 10 \\
v(-11) &= K^{-1}w(-11) - [v(-12) + v(-13)] = 22 - [10 + 2] &= 10 \\
v(-10) &= K^{-1}w(-10) - [v(-11) + v(-12)] = 30 - [10 + 10] &= 10 \\
v(-9) &= K^{-1}w(-9) - [v(-10) + v(-11)] &= 36 - [10 + 10] = 16 \\
v(-8) &= K^{-1}w(-8) - [v(-9) + v(-10)] &= 38 - [16 + 10] = 12 \\
v(-7) &= K^{-1}w(-7) - [v(-8) + v(-9)] &= 34 - [12 + 16] = 6 \\
v(-6) &= K^{-1}w(-6) - [v(-7) + v(-8)] &= 20 - [6 + 12] = 2 \\
v(-5) &= K^{-1}w(-5) - [v(-6) + v(-7)] &= 11 - [2 + 6] = 3 \\
v(-4) &= K^{-1}w(-4) - [v(-5) + v(-6)] &= 8 - [3 + 2] = 3 \\
v(-3) &= K^{-1}w(-3) - [v(-4) + v(-5)] &= 11 - [3 + 3] = 5 \\
v(-2) &= K^{-1}w(-2) - [v(-3) + v(-4)] &= 10 - [5 + 3] = 2 \\
v(-1) &= K^{-1}w(-1) - [v(-2) + v(-3)] &= 9 - [2 + 5] = 2 \\
v(0) &= K^{-1}w(0) - [v(-1) + v(-2)] &= 9 - [2 + 2] = 5 \\
v(1) &= K^{-1}w(1) - [v(0) + v(-1)] &= 9 - [5 + 2] = 2 \\
v(2) &= K^{-1}w(2) - [v(1) + v(0)] &= 11 - [2 + 5] = 4 \\
v(3) &= K^{-1}w(3) - [v(2) + v(1)] &= 11 - [4 + 2] = 5 \\
v(4) &= K^{-1}w(4) - [v(3) + v(2)] &= 11 - [5 + 4] = 2 \\
v(5) &= K^{-1}w(5) - [v(4) + v(3)] &= 9 - [2 + 5] = 2 \\
v(6) &= K^{-1}w(6) - [v(5) + v(4)] &= 9 - [2 + 2] = 5.
\end{aligned}
$$

The splitting-signal $\{f_{1,4,t}\}$ is calculated by

$$
\begin{aligned}
f_{1,4,0} &= v(0) + v(-7) + v(-14) &&= 5 + 6 + 6 = 17 \\
f_{1,4,1} &= v(1) + v(-6) + v(-13) &&= 2 + 2 + 2 = 6 \\
f_{1,4,2} &= v(2) + v(-5) + v(-12) &&= 4 + 3 + 10 = 17 \\
f_{1,4,3} &= v(3) + v(-4) + v(-11) + v(-18) &&= 5 + 3 + 10 + 1 = 19 \\
f_{1,4,4} &= v(4) + v(-3) + v(-10) + v(-17) &&= 2 + 5 + 10 + 9 = 26 \\
f_{1,4,5} &= v(5) + v(-2) + v(-9) + v(-16) &&= 2 + 2 + 16 + 2 = 22 \\
f_{1,4,6} &= v(6) + v(-1) + v(-8) + v(-15) &&= 5 + 2 + 12 + 6 = 25.
\end{aligned}
$$

Figure 5.23 shows the graphs of the integrals $w(t)$, sums $v(t)$, and splitting-signal for the $(1, -3)$-projection.

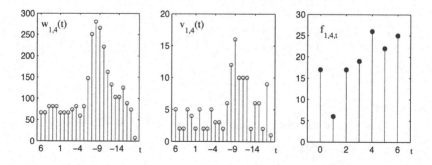

FIGURE 5.23
Line-integrals $w(t)$, sums $v(t)$, and splitting-signal of the $(1, -3)$-projection.

Remark 5.1 To work with only nonnegative numbers for control points, we consider the transformation $t \rightarrow 6 - t$,

$$
\begin{bmatrix}
6 & 5 & 4 & 3 & 2 & 1 & 0 \\
9 & 8 & 7 & \cdot & \cdot & \cdot & \cdot \\
12 & 11 & 10 & \cdot & \cdot & \cdot & \cdot \\
15 & 14 & 13 & \cdot & \cdot & \cdot & \cdot \\
18 & 17 & 16 & \cdot & \cdot & \cdot & \cdot \\
21 & 20 & 19 & \cdot & \cdot & \cdot & \cdot \\
24 & 23 & 22 & \cdot & \cdot & \cdot & \cdot
\end{bmatrix}
\qquad \text{(angle of rays is } \tan^{-1}(1/3) = 18.435°).
$$

With such numbering of the control points, the set of 25 geometrical rays of the $(1, -3)$-projection in (5.12) can be written as

$$
l(t) = l_{1,-3}(6 - t) = \left\{ (x, y); x - 3y = -\frac{t}{7} + \frac{5}{7} \right\}.
$$

Therefore, the system of equations in (5.14) can be written as

$$
w(t) = w(\tilde{l}(t)) = K \left[v(t) + v(t + 1) + v(t + 2) \right], \quad t = 0 : 24, \tag{5.16}
$$

where $v(25) = v(26) = 0$. The system of equations in (5.15) for the coefficients $b_t = Kv(t)$ is described by

$$
\begin{aligned}
b_{24} &= w(24), \\
b_{23} &= w(23) - b_{24}, \\
b_{22} &= w(22) - (b_{23} + b_{24}), \\
b_t &= w(t) - (b_{t+1} + b_{t+2}), \quad t = 21, 20, \cdots, 1, 0.
\end{aligned} \tag{5.17}
$$

5.2.7 $(1, 5)$-projection

Consider the subset of triplets

$$
U(1, 5) = \left\{ (1, 5, 0), (1, 5, 1), (1, 5, 2), (1, 5, 3), (1, 5, 4), (1, 5, 5), (1, 5, 6) \right\}.
$$

Seven masks of the tensor functions with triplet-numbers of $U(1, 5)$ are

$$
[\chi_{1,5,0}] = \begin{bmatrix}
1000000 \\
0010000 \\
0000100 \\
0000001 \\
0100010 \\
0001000 \\
0000010
\end{bmatrix}, \;
[\chi_{1,5,1}] = \begin{bmatrix}
0100000 \\
0001000 \\
0000010 \\
1000000 \\
0010000 \\
0000100 \\
0000001
\end{bmatrix}, \;
[\chi_{1,5,2}] = \begin{bmatrix}
0010000 \\
0000100 \\
0000001 \\
0100000 \\
0001000 \\
0000010 \\
1000000
\end{bmatrix},
$$

$$[\chi_{1,5,3}] = \begin{bmatrix} 0001000 \\ 0000010 \\ 1000000 \\ 0010010 \\ 0000100 \\ 0000001 \\ 0100000 \end{bmatrix}, \; [\chi_{1,5,4}] = \begin{bmatrix} 0000100 \\ 0000001 \\ 0100000 \\ 0001000 \\ 0000010 \\ 1000000 \\ 0010000 \end{bmatrix}, \; [\chi_{1,5,5}] = \begin{bmatrix} 0000010 \\ 1000000 \\ 0010000 \\ 0000100 \\ 0000001 \\ 0100000 \\ 0001000 \end{bmatrix},$$

$$[\chi_{1,5,6}] = \begin{bmatrix} 0000001 \\ 0100000 \\ 0001000 \\ 0000010 \\ 1000000 \\ 0010000 \\ 0000100 \end{bmatrix} = \begin{bmatrix} 1111111 \\ 1111111 \\ 1111111 \\ 1111111 \\ 1111111 \\ 1111111 \\ 1111111 \end{bmatrix} - \sum_{t=0}^{5} [\chi_{1,5,t}].$$

Because $(1, 5 - 7) = (1, -2)$, the functions $\chi_{1,5,t}$ equal $\chi_{1,-2,t}$, for $t = 0 : 6$. Therefore, to reduce the number of rays from $(1 + 5)6 + 1 = 37$ to $(1 + |-2|)6 + 1 = 19$, we consider the direction of the rays, which is determined by the $(1, -2)$-projection. As an example, Figure 5.24 shows the mask of the tensor function $\chi_{1,5,0}$ with the parallel rays of two different projections. Six

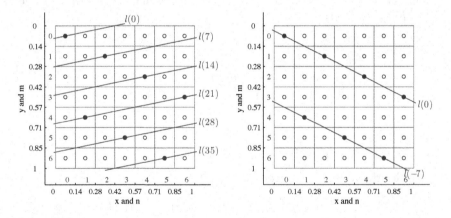

FIGURE 5.24
The mask $[\chi_{1,5,0}]$ with two different sets of parallel rays.

rays of the $(1, 5)$-projection are shown on the left mask, and two rays of the $(1, -2)$-projection are shown on the right mask. The set of parallel rays for the $(1, -2)$-projection is shown in Figure 5.25.

The control points of parallel rays of the $(1, 5)$-projection are numbered

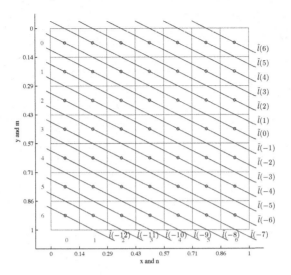

FIGURE 5.25
The set of parallel rays for the $(1, -2)$-projection.

and located as

$$
\begin{bmatrix}
0 & 1 & 2 & 3 & 4 & 5 & 6 \\
-2 & -1 & & & & & \\
-4 & -3 & \cdot & \cdot & \cdot & \cdot & \cdot \\
-6 & -5 & \cdot & \cdot & \cdot & \cdot & \cdot \\
-8 & -7 & \cdot & \cdot & \cdot & \cdot & \cdot \\
-10 & -9 & \cdot & \cdot & \cdot & \cdot & \cdot \\
-12 & -11 & \cdot & \cdot & \cdot & \cdot & \cdot
\end{bmatrix}
\qquad \text{(angle of rays is } \arctan(1/2) = 26.5651°).
$$

The set of 19 geometrical rays of this projection is described by

$$
l(t) = l_{1,-2}(t) = \left\{ (x, y); x - 2y = \frac{t}{7} - \frac{1}{14} \right\}, \quad t = 6 : -1 : -12, \qquad (5.18)
$$

where $(x, y) \in [0, 1] \times [0, 1]$. Six masks $[\chi_{1,5,t}]$, $t = 1 : 6$, of the tensor functions
are shown in Figures 5.26 and 5.27.

The components of the tensor transform with the triplet-numbers of
$U(1, 5)$ can be calculated by

$$
\begin{aligned}
f_{1,5,0} &= v_0 + v_{-7}, \quad f_{1,5,1} = v_1 + v_{-6}, \quad f_{1,5,2} = v_2 + v_{-5} + v_{-12}, \\
f_{1,5,3} &= v_3 + v_{-4} + v_{-11}, \quad f_{1,5,4} = v_4 + v_{-3} + v_{-10}, \\
f_{1,5,5} &= v_5 + v_{-2} + v_{-9}, \quad f_{1,5,6} = v_6 + v_{-1} + v_{-8},
\end{aligned} \qquad (5.19)
$$

where the sums of the discrete image along the rays are denoted by $v_t = v(t) = v_{1,-2}(t)$, $t = 6 : -1 : -12$. The above system of linear equations can be

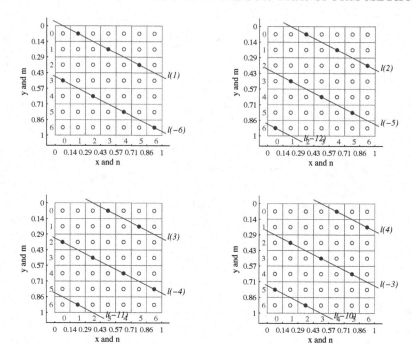

FIGURE 5.26
Masks $[\chi_{1,5,t}]$, $t = 1 : 4$, and parallel rays of the $(1, -2)$-projection.

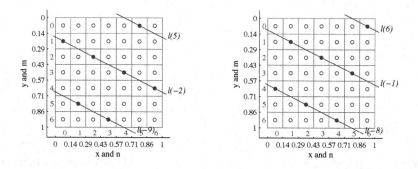

FIGURE 5.27
Masks $[\chi_{1,5,t}]$, $t = 5, 6$, and parallel rays of the $(1, -2)$-projection.

written as

$$f_{1,5,t} = \sum_{m=0}^{3} v_{t-7m} = \sum_{m=0}^{3} v_{1,-2,t-7m}, \quad t = 0 : 6,$$

where $v_{-13} = v_{-14} = 0$. As in the $(1, 2)$-projection case, we consider the shifted set of the rays, $l(t) \to l(t - 0.5)$, to be the set of 19 geometrical rays

$$\tilde{l}(t) = l_{1,-2}(t - 0.5) = \left\{ (x, y); x - 2y = \frac{t}{7} - \frac{1}{7} \right\}, \quad t = 6 : -1 : -12.$$

These rays are shown in Figure 5.28. For this set of geometrical rays, the

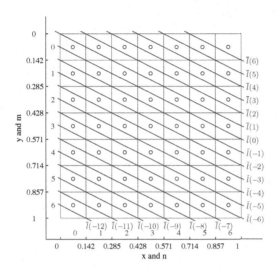

FIGURE 5.28
The set of geometrical rays for the $(1, -2)$-projection.

system of equations for the line-integrals is described by

$$w(t) = w(\tilde{l}(t)) = \frac{7\sqrt{5}}{2} \Big[v(t) + v(t - 1) \Big], \quad t = 6 : -1 : -12, \tag{5.20}$$

where $v(-13) = 0$. The required sums $v(t)$ of the discrete image $f_{n,m}$ are
therefore defined as $v(t) = b_t/(7\sqrt{5}/2)$, where components b_t are calculated
recursively by

$$\begin{aligned}
b_{-12} &= w(-12), \quad b_{-11} = w(-11) - b_{-12}, \quad b_{-10} = w(-10) - b_{-11} \\
b_t &= w(t) - b_{t-1}, \quad t = -9, -8, -7, -6, \cdots, 6.
\end{aligned} \tag{5.21}$$

For the considered image $f(x, y)$, the line-integrals $w(t)$ of the $(1, -2)$-
projection are calculated by

$$\begin{aligned}
w(6) &= K(5 + 2) & &= K \cdot 7 \\
w(5) &= K(4 + 2 + 2) & &= K \cdot 8 \\
w(4) &= K(4 + 1 + 2 + 1) & &= K \cdot 8 \\
w(3) &= K(2 + 1 + 1 + 1 + 3) & &= K \cdot 8 \\
w(2) &= K(2 + 1 + 1 + 2 + 3 + 1) & &= K \cdot 10 \\
w(1) &= K(3 + 1 + 1 + 2 + 1 + 1 + 1) &= K \cdot 10 \\
w(0) &= K(3 + 1 + 1 + 1 + 1 + 1 + 1) &= K \cdot 9
\end{aligned}$$

$$w(-1) = K(0+1+1+1+2+1+1) = K \cdot 7$$
$$w(-2) = K(0+2+1+2+2+2+1) = K \cdot 10$$
$$w(-3) = K(6+2+9+2+4+2+0) = K \cdot 25$$
$$w(-4) = K(6+6+9+8+4+3+0) = K \cdot 36$$
$$w(-5) = K(2+6+7+8+4+3+2) = K \cdot 32$$
$$w(-6) = K(2+1+7+8+4+3+2) = K \cdot 27$$
$$w(-7) = K(1+3+8+1+3+0) \qquad = K \cdot 16$$
$$w(-8) = K(3+4+1+3+0) \qquad = K \cdot 11$$
$$w(-9) = K(4+2+3+2) \qquad = K \cdot 11$$
$$w(-10) = K(2+9+2) \qquad = K \cdot 13$$
$$w(-11) = K(9+1) \qquad = K \cdot 10$$
$$w(-12) = K(1) \qquad = K \cdot 1$$

where $K = 7\sqrt{5}/2$. The sums $v(t)$ of the image $f_{n,m}$ are calculated by

$$v(-12) = K^{-1}w(-12) \qquad = 1 \qquad = 1$$
$$v(-11) = K^{-1}w(-11) - v(-12) = 10 - 1 \ = 9$$
$$v(-10) = K^{-1}w(-10) - v(-11) = 13 - 9 \ = 4$$
$$v(-9) = K^{-1}w(-9) - v(-10) = 11 - 4 \ = 7$$
$$v(-8) = K^{-1}w(-8) - v(-9) = 11 - 7 \ = 4$$
$$v(-7) = K^{-1}w(-7) - v(-8) = 16 - 4 \ = 12$$
$$v(-6) = K^{-1}w(-6) - v(-7) = 27 - 12 = 15$$
$$v(-5) = K^{-1}w(-5) - v(-6) = 32 - 15 = 17$$
$$v(-4) = K^{-1}w(-4) - v(-5) = 36 - 17 = 19$$
$$v(-3) = K^{-1}w(-3) - v(-4) = 25 - 19 = 6$$
$$v(-2) = K^{-1}w(-2) - v(-3) = 10 - 6 \ = 4$$
$$v(-1) = K^{-1}w(-1) - v(-2) = 7 - 4 \ = 3$$
$$v(0) = K^{-1}w(0) - v(-1) = 9 - 3 \ = 6$$
$$v(1) = K^{-1}w(1) - v(0) = 10 - 6 \ = 4$$
$$v(2) = K^{-1}w(2) - v(1) = 10 - 4 \ = 6$$
$$v(3) = K^{-1}w(3) - v(2) = 8 - 6 \ = 2$$
$$v(4) = K^{-1}w(4) - v(3) = 8 - 2 \ = 6$$
$$v(5) = K^{-1}w(5) - v(4) = 8 - 6 \ = 2$$
$$v(6) = K^{-1}w(6) - v(5) = 7 - 2 \ = 5.$$

The components of the splitting-signal $\{f_{1,5,t}\}$ are calculated by

$$f_{1,5,0} = v(0) + v(-7) \qquad = 6 + 12 = 18$$
$$f_{1,5,1} = v(1) + v(-6) \qquad = 4 + 15 = 19$$
$$f_{1,5,2} = v(2) + v(-5) + v(-12) = 6 + 17 + 1 = 24$$
$$f_{1,5,3} = v(3) + v(-4) + v(-11) = 2 + 19 + 9 = 30$$
$$f_{1,5,4} = v(4) + v(-3) + v(-10) = 6 + 6 + 4 = 16$$
$$f_{1,5,5} = v(5) + v(-2) + v(-9) \ = 2 + 4 + 7 = 13$$
$$f_{1,5,6} = v(6) + v(-1) + v(-8) \ = 5 + 3 + 4 = 12.$$

Figure 5.29 shows the graphs of the line-integrals $w(t)$, sums $v(t)$, and the splitting-signal for the $(1,-2)$-projection.

To avoid the negative numbering of rays, or control points, we consider

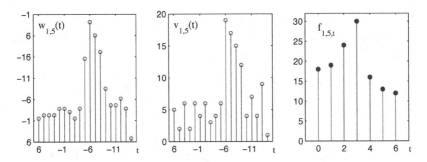

FIGURE 5.29
Line-integrals $w(t)$, sums $v(t)$, and splitting-signal of the $(1,5)$-projection.

the transformation $t \to 6 - t$,

$$
\begin{bmatrix}
6 & 5 & 4 & 3 & 2 & 1 & 0 \\
8 & 7 & & & & & \\
10 & 9 & & & & & \\
12 & 11 & & \cdot & \cdot & \cdot & \cdot \\
14 & 13 & & & & & \\
16 & 15 & & & & & \\
18 & 17 & & \cdot & \cdot & \cdot & \cdot
\end{bmatrix}
\qquad \text{(angle of rays is } \tan^{-1}(1/2) = 26.5651°\text{)}.
$$

The set of 19 geometrical rays of the projection in (5.18) can be written as

$$
l(t) = l_{1,-2}(6 - t) = \left\{ (x, y); x - 2y = -\frac{1}{7} + \frac{11}{14} \right\}.
$$

Then, the system of equations in (5.20) is written in the form of

$$
w(t) = w(\tilde{l}(t)) = K\Big[v(t) + v(t + 1)\Big], \quad t = 0 : 18, \tag{5.22}
$$

where $v(19) = 0$. The system of equations (5.21) for calculating the coefficients $b_t = Kv(t)$ is described by

$$
\begin{aligned}
b_{18} &= w(18), \\
b_{17} &= w(17) - b_{18}, \\
b_{16} &= w(16) - b_{17}, \\
b_t &= w(t) - b_{t+1}, \quad t = 15, 14, \cdots, 0,
\end{aligned} \tag{5.23}
$$

and the equations for sums $v(t)$ are written as

$$
\begin{aligned}
v(18) &= K^{-1}w(18), \\
v(t) &= K^{-1}w(t) - v(t + 1), \quad t = 17 : -1 : 0.
\end{aligned}
$$

This numbering was considered when plotting the integrals $w(t)$ and sums $v(t)$ in Figure 5.29. Data $\{w(6), w(5), w(4), ..., w(-11), w(-12)\}$ and $\{v(6), v(5), v(4), ..., v(-11), v(-12)\}$ were plotted versus time-points $0, 1, 2, ..., 17, 18$.

5.2.8 $(1,6)$-projection

We consider the subset of triplet-numbers corresponding to the generator $(1,6)$,

$$U(1,6) = \{(1,6,0),(1,6,1),(1,6,2),(1,6,3),(1,6,4),(1,6,5),(1,6,6)\}.$$

Masks of the tensor functions $\chi_{1,6,t}$ with triplet-numbers of this subset are

$$[\chi_{1,6,0}] = \begin{bmatrix} 1000000 \\ 0100000 \\ 0010000 \\ 0001000 \\ 0000100 \\ 0000010 \\ 0000001 \end{bmatrix}, \quad [\chi_{1,6,1}] = \begin{bmatrix} 0100000 \\ 0010000 \\ 0001000 \\ 0000100 \\ 0000010 \\ 0000001 \\ 1000000 \end{bmatrix}, \quad [\chi_{1,6,2}] = \begin{bmatrix} 0010000 \\ 0001000 \\ 0000100 \\ 0000010 \\ 0000001 \\ 1000000 \\ 0100000 \end{bmatrix},$$

$$[\chi_{1,6,3}] = \begin{bmatrix} 0001000 \\ 0000100 \\ 0000010 \\ 0000001 \\ 1000000 \\ 0100000 \\ 0010000 \end{bmatrix}, \quad [\chi_{1,6,4}] = \begin{bmatrix} 0000100 \\ 0000010 \\ 0000001 \\ 1000000 \\ 0100000 \\ 0010000 \\ 0001000 \end{bmatrix}, \quad [\chi_{1,6,5}] = \begin{bmatrix} 0000010 \\ 0000001 \\ 1000000 \\ 0100000 \\ 0010000 \\ 0001000 \\ 0000100 \end{bmatrix},$$

$$[\chi_{1,6,6}] = \begin{bmatrix} 0000001 \\ 1000000 \\ 0100000 \\ 0010000 \\ 0001000 \\ 0000100 \\ 0000010 \end{bmatrix} = \begin{bmatrix} 1111111 \\ 1111111 \\ 1111111 \\ 1111111 \\ 1111111 \\ 1111111 \\ 1111111 \end{bmatrix} - \sum_{t=0}^{5} [\chi_{1,6,t}].$$

To reduce the number of parallel rays, along which the corresponding components $f_{1,6,t}$ of the tensor transform are defined (see Figure 5.30), we consider the parallel rays of the $(1,-1)$-projection. The masks of the tensor functions for these two projections are the same, i.e., $[\chi_{1,6,t}] = [\chi_{1,-1,t}]$, for $t = 0:6$. The use of the $(1,-1)$-projection allows for reducing the number of rays from $(1+6)6+1 = 43$ to $(1+|-1|)6+1 = 13$.

As an example, Figure 5.31 shows the mask of the tensor function $\chi_{1,5,0}$ with the parallel arithmetical rays of two different projections. Seven rays of the $(1,6)$-projection are shown in the mask on the left, and one ray of the $(1,-1)$-projection is shown in the same mask on the right. The set of 13 rays for the $(1,-1)$-projection is shown in Figure 5.32.

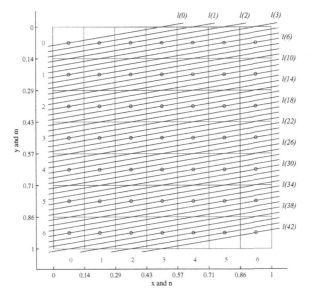

FIGURE 5.30
The set of parallel rays for the $(1, 6)$-projection.

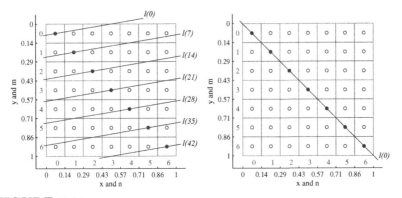

FIGURE 5.31
The mask $|\chi'_{1,6,0}|$ with two different sets of arithmetical rays.

The control points of this set of rays are defined as

$$
\begin{bmatrix}
0 \bullet^1 \bullet^2 \bullet^3 \bullet^4 \bullet^5 \bullet^6 \bullet \\
-1 \bullet \quad \cdot \quad \cdot \quad \cdot \quad \cdot \quad \cdot \quad \cdot \\
-2 \bullet \quad \cdot \quad \cdot \quad \cdot \quad \cdot \quad \cdot \quad \cdot \\
-3 \bullet \quad \cdot \quad \cdot \quad \cdot \quad \cdot \quad \cdot \quad \cdot \\
-4 \bullet \quad \cdot \quad \cdot \quad \cdot \quad \cdot \quad \cdot \quad \cdot \\
-5 \bullet \quad \cdot \quad \cdot \quad \cdot \quad \cdot \quad \cdot \quad \cdot \\
-6 \bullet \quad \cdot \quad \cdot \quad \cdot \quad \cdot \quad \cdot \quad \cdot
\end{bmatrix}
\qquad (\text{angle of rays is } \tan^{-1}(1) = 45°).
$$

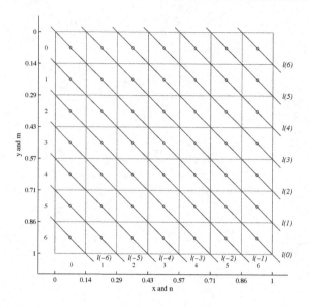

FIGURE 5.32
The set of parallel rays for the $(1, -1)$-projection.

The set of 13 parallel rays of this projection is described by

$$l(t) = l_{1,-1}(t) = \left\{ (x, y); x - y = \frac{t}{7} \right\}, \quad t = 6 : -1 : -6, \quad (5.24)$$

where $(x, y) \in [0, 1] \times [0, 1]$. Six masks for the tensor functions $\chi_{1,6,t}$, $t = 1 : 6$, are shown in Figures 5.33 and 5.34.

The components of the tensor transform with the triplet-numbers $(1, 6, t)$ can be calculated by

$$\begin{aligned}
f_{1,6,0} &= v_0, \\
f_{1,6,1} &= v_1 + v_{-6}, \quad f_{1,6,2} = v_2 + v_{-5}, \\
f_{1,6,3} &= v_3 + v_{-4}, \quad f_{1,6,4} = v_4 + v_{-3}, \\
f_{1,6,5} &= v_5 + v_{-2}, \quad f_{1,6,6} = v_6 + v_{-1},
\end{aligned} \quad (5.25)$$

where the sums of the discrete image along the rays are denoted by $v_t = v(t) = v_{1,-1}(t)$, $t = 6 : -1 : -6$. This complete system of linear equations for calculating the components of the 2-D tensor transform with triplet-numbers of the subset $U(1, 6)$ can be written as

$$f_{1,6,t} = v_t + v_{t-7} = v_{1,-1,t} + v_{1,-1,t-7}, \quad t = 0 : 6,$$

where $v_{-7} = 0$. As in the $(1, 1)$-projection case, the line-sums $v(t)$ of the discrete image $f_{n,m}$ can be calculated from the line-integrals $w(t)$. The set of geometrical rays coincides with the set of arithmetical rays. Each ray intersects

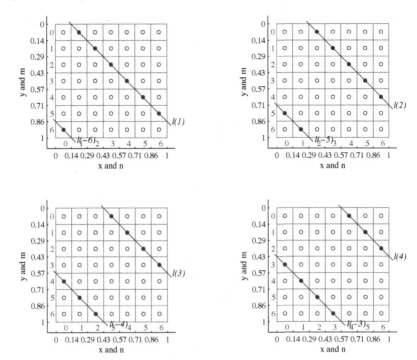

FIGURE 5.33
Masks $[\chi_{1,6,t}]$, $t = 1 : 4$, and rays of the $(1, -1)$-projection.

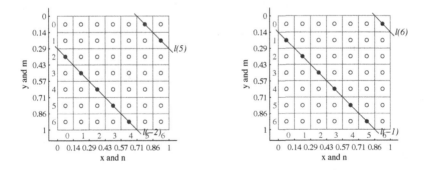

FIGURE 5.34
Masks $[\chi_{1,6,t}]$, $t = 5, 6$ and rays of the $(1, -1)$-projection.

the image element along the path of length $\Delta l = \sqrt{2}\Delta x = \sqrt{2}/7$. Therefore, $w(t) = w_{1,-1}(t) = 7\sqrt{2}v(t)$, or

$$v(t) = \frac{1}{7\sqrt{2}}w(t), \quad t = 6 : -1 : -6. \tag{5.26}$$

For the given image $f(x, y)$, the line-integrals $w(t)$ are calculated by

$$
\begin{aligned}
w(6) &= 7\sqrt{2}(5) & &= 7\sqrt{2} \cdot 5 \\
w(5) &= 7\sqrt{2}(2 + 4) & &= 7\sqrt{2} \cdot 6 \\
w(4) &= 7\sqrt{2}(2 + 1 + 2) & &= 7\sqrt{2} \cdot 5 \\
w(3) &= 7\sqrt{2}(1 + 1 + 1 + 3) & &= 7\sqrt{2} \cdot 6 \\
w(2) &= 7\sqrt{2}(3 + 2 + 1 + 1 + 0) & &= 7\sqrt{2} \cdot 7 \\
w(1) &= 7\sqrt{2}(1 + 1 + 1 + 1 + 2 + 6) & &= 7\sqrt{2} \cdot 12 \\
w(0) &= 7\sqrt{2}(1 + 1 + 2 + 2 + 9 + 6 + 2) & &= 7\sqrt{2} \cdot 23 \\
w(-1) &= 7\sqrt{2}(1 + 2 + 4 + 8 + 7 + 1) & &= 7\sqrt{2} \cdot 23 \\
w(-2) &= 7\sqrt{2}(0 + 3 + 4 + 8 + 3) & &= 7\sqrt{2} \cdot 18 \\
w(-3) &= 7\sqrt{2}(2 + 3 + 1 + 4) & &= 7\sqrt{2} \cdot 10 \\
w(-4) &= 7\sqrt{2}(0 + 3 + 2) & &= 7\sqrt{2} \cdot 5 \\
w(-5) &= 7\sqrt{2}(2 + 9) & &= 7\sqrt{2} \cdot 11 \\
w(-6) &= 7\sqrt{2}(1) & &= 7\sqrt{2} \cdot 1.
\end{aligned}
$$

The sums $v(t)$, $t = 6 : -1 : -6$, of the image along the corresponding arithmetical rays compose the vector $\mathbf{v} = (5, 6, 5, 6, 7, 12, 23, 23, 18, 10, 5, 11, 1)'$, and the components of the splitting-signal $\{f_{1,6,t}; \ t = 0 : 6\}$ are calculated by

$$
\begin{aligned}
f_{1,6,0} &= v(0) = 23 \\
f_{1,6,1} &= v(1) + v(-6) = 12 + 1 = 13 \\
f_{1,6,2} &= v(2) + v(-5) = 7 + 11 = 18 \\
f_{1,6,3} &= v(3) + v(-4) = 6 + 5 = 11 \\
f_{1,6,4} &= v(4) + v(-3) = 5 + 10 = 15 \\
f_{1,6,5} &= v(5) + v(-2) = 6 + 18 = 24 \\
f_{1,6,6} &= v(6) + v(-1) = 5 + 23 = 28.
\end{aligned}
$$

Figure 5.35 shows the graphs of the line-integrals $w(t)$, sums $v(t)$, and splitting-signal for the $(1, -1)$-projection. The control points can be numbered

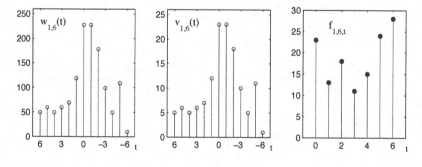

FIGURE 5.35
Line-integrals $w(t)$, sums $v(t)$, and splitting-signal for the $(1, -1)$-projection.

from $0:12$, by using the transformation $t \to 6 - t$,

$$
\begin{bmatrix}
6_\bullet \, 5_\bullet \, 4_\bullet \, 3_\bullet \, 2_\bullet \, 1_\bullet \, 0_\bullet \\
7_\bullet \quad \cdot \quad \cdot \quad \cdot \quad \cdot \quad \cdot \quad \cdot \\
8_\bullet \quad \cdot \quad \cdot \quad \cdot \quad \cdot \quad \cdot \quad \cdot \\
9_\bullet \quad \cdot \quad \cdot \quad \cdot \quad \cdot \quad \cdot \quad \cdot \\
10_\bullet \quad \cdot \quad \cdot \quad \cdot \quad \cdot \quad \cdot \quad \cdot \\
1_\bullet \quad \cdot \quad \cdot \quad \cdot \quad \cdot \quad \cdot \quad \cdot \\
12_\bullet \quad \cdot \quad \cdot \quad \cdot \quad \cdot \quad \cdot \quad \cdot
\end{bmatrix}
\qquad (\text{angle of rays is } \tan^{-1}(1) = 45°).
$$

The set of 13 geometrical rays of this projection, which is given in (5.24), can be written as

$$
l(t) = l_{1,-1}(6 - t) = \left\{ (x, y); x - y = -\frac{t}{7} + \frac{6}{7} \right\},
$$

and the system of equations in (5.26) as

$$
v(t) = w(t)/(7\sqrt{2}) = w(\tilde{l}(t))/(7\sqrt{2}), \quad t = 0 : 12. \tag{5.27}
$$

5.2.9 Reconstructed image 7×7

To summarize all calculations accomplished above for obtaining eight splitting-signals from the given image $f(x, y)$, we pack these signals into the matrix 8×7 of the 2-D discrete tensor transform (DTT) as shown below

$$
\begin{bmatrix}
f_{0,1,0} \, f_{0,1,1} \, f_{0,1,2} \, f_{0,1,3} \, f_{0,1,4} \, f_{0,1,5} \, f_{0,1,6} \\
f_{1,0,0} \, f_{1,0,1} \, f_{1,0,2} \, f_{1,0,3} \, f_{1,0,4} \, f_{1,0,5} \, f_{1,0,6} \\
f_{1,1,0} \, f_{1,1,1} \, f_{1,1,2} \, f_{1,1,3} \, f_{1,1,4} \, f_{1,1,5} \, f_{1,1,6} \\
f_{1,2,0} \, f_{1,2,1} \, f_{1,2,2} \, f_{1,2,3} \, f_{1,2,4} \, f_{1,2,5} \, f_{1,2,6} \\
f_{1,3,0} \, f_{1,3,1} \, f_{1,3,2} \, f_{1,3,3} \, f_{1,3,4} \, f_{1,3,5} \, f_{1,3,6} \\
f_{1,4,0} \, f_{1,4,1} \, f_{1,4,2} \, f_{1,4,3} \, f_{1,4,4} \, f_{1,4,5} \, f_{1,4,6} \\
f_{1,5,0} \, f_{1,5,1} \, f_{1,5,2} \, f_{1,5,3} \, f_{1,5,4} \, f_{1,5,5} \, f_{1,5,6} \\
f_{1,6,0} \, f_{1,6,1} \, f_{1,6,2} \, f_{1,6,3} \, f_{1,6,4} \, f_{1,6,5} \, f_{1,6,6}
\end{bmatrix}
=
\begin{bmatrix}
15 & 11 & 9 & 16 & 26 & 33 & 22 \\
7 & 22 & 17 & 26 & 24 & 14 & 22 \\
25 & 24 & 20 & 15 & 16 & 15 & 17 \\
15 & 20 & 19 & 12 & 17 & 20 & 29 \\
19 & 20 & 23 & 11 & 17 & 23 & 19 \\
17 & 6 & 17 & 19 & 26 & 22 & 25 \\
18 & 19 & 24 & 30 & 16 & 13 & 12 \\
23 & 13 & 18 & 11 & 15 & 24 & 28
\end{bmatrix}.
$$

By using the inverse tensor transformation, which is given in (5.2), we obtain the following reconstruction of the image $f(x, y)$:

$$
[f_{n,m}] =
\begin{bmatrix}
1 & 1 & 3 & 1 & 2 & 2 & 5 \\
1 & 1 & 1 & 2 & 1 & 1 & 4 \\
0 & 2 & 2 & 1 & 1 & 1 & 2 \\
2 & 3 & 4 & 2 & 1 & 1 & 3 \\
0 & 3 & 4 & 8 & 9 & 2 & 0 \\
2 & 3 & 1 & 8 & 7 & 6 & 6 \\
1 & 9 & 2 & 4 & 3 & 1 & 2
\end{bmatrix}.
$$

Below is the script `example7x7_inverseDTT.m` of the program that computes the inverse 7×7-point DTT. The calculations are performed without the matrix of the tensor transform, as is done in the program `example7x7_inverseDTT.m` given in §2.2.2.1.

```
% --------------------------------------------------------
% call: example7x7_inverseDTT.m
% The inverse 7x7-point 2-D DTT, for Example 5.1
% A.M. Grigoryan,                    November 2, 2011
% --------------------------------------------------------
% Set of generators is considered to be calculated as
% J77=ones(2,8); J77(:,1)=[0 1]; J77(2,2:8)=0:6;
  DTT_ofimage=[15   11    9   16   26   33   22
                7   22   17   26   24   14   22
               25   24   20   15   16   15   17
               15   20   19   12   17   20   29
               19   20   23   11   17   23   19
               17    6   17   19   26   22   25
               18   19   24   30   16   13   12
               23   13   18   11   15   24   28];
% Eq 5.2 of the inverse 2-D discrete tensor transform:
  image_fromTT=zeros(7,7);
  for n1=0:6
      n=n1+1;
      for m1=0:6
          m=m1+1;
          image_fromTT(n,m)=DTT_ofimage(1,n);
          col=m1-n1;
          for p=2:8
              col=col+n1;
              if col>=7 col=col-7; end
              image_fromTT(n,m)=image_fromTT(n,m)+...
                  DTT_ofimage(p,col+1);
          end
      end
  end
  image_power=sum(DTT_ofimage(1,:));
  image_fromTT=(image_fromTT-image_power)/7
%    1   1   3   1   2   2   5
%    1   1   1   2   1   1   4
%    0   2   2   1   1   1   2
%    2   3   4   2   1   1   3
%    0   3   4   8   9   2   0
%    2   3   1   8   7   6   6
%    1   9   2   4   3   1   2
% --------------------------------------------------------
```

The number of parallel geometrical rays used in each projection is defined as follows (see Table 5.1):

$$n(1,0) = n(0,1) = N,$$
$$n(1,s) = (1+s)(N-1) + 1 = N - 1 + s(N-1), \quad s = 1 : (N-1)/2,$$
$$n(1, N-s) = N^2 - s(N-1), \quad s = (N+1)/2 : N - 1.$$

The total number of rays is 138. The sufficient number of rays is less than

138, if we consider the redundancy of the tensor transform. The calculation of the transform requires one full splitting-signal, let it be $\{f_{0,1,t}; t = 0 : 6\}$, and seven incomplete splitting-signals. The mask of the tensor function $\chi_{1,0,6}$ is determined by one ray. The masks of the tensor functions $\chi_{1,s,6}$, when $s = 1 : 6$, are defined by $1, 2, 3, 4, 3, 2, 1$ rays, respectively. Therefore the sufficient number of rays to reconstruct the image on the Cartesian grid 7×7, equals $138 - (1 + 2 + 3 + 4 + 3 + 2 + 1) = 122$. Thus, we need 122 measurements,

TABLE 5.1
Projection data for reconstruction of the discrete image 7×7

(p, s)	$n(p, s)$	masks in $U(p, s)$
$(1, 0)$	7	$(1, 0, 0), (1, 0, 1), (1, 0, 2), (1, 0, 3), (1, 0, 4), (1, 0, 5), (1, 0, 6)$
$(1, 1)$	13	$(1, 1, 0), (1, 1, 1), (1, 1, 2), (1, 1, 3), (1, 1, 4), (1, 1, 5), (1, 1, 6)$
$(1, 2)$	19	$(1, 2, 0), (1, 2, 1), (1, 2, 2), (1, 2, 3), (1, 2, 4), (1, 2, 5), (1, 2, 6)$
$(1, 3)$	25	$(1, 3, 0), (1, 3, 1), (1, 3, 2), (1, 3, 3), (1, 3, 4), (1, 3, 5), (1, 3, 6)$
$(1, 4)$	25	$(1, 4, 0), (1, 4, 1), (1, 4, 2), (1, 4, 3), (1, 4, 4), (1, 4, 5), (1, 4, 6)$
$(1, 5)$	19	$(1, 5, 0), (1, 5, 1), (1, 5, 2), (1, 5, 3), (1, 5, 4), (1, 5, 5), (1, 5, 6)$
$(1, 6)$	13	$(1, 6, 0), (1, 6, 1), (1, 6, 2), (1, 6, 3), (1, 6, 4), (1, 6, 5), (1, 6, 6)$
$(0, 1)$	7	$(0, 1, 0), (0, 1, 1), (0, 1, 2), (0, 1, 3), (0, 1, 4), (0, 1, 5), (0, 1, 6)$
	138 [122]	

or line-integrals $w(t) = w_{p,s}(t)$ of the image $f(x, y)$, to reconstruct 49 values of the discrete image $f_{n,m}$. Equations of the geometrical rays for the image 7×7 are given in Table 5.2.

TABLE 5.2
Equations of geometrical rays for the image 7×7 (case A)

(p, s)	A-rays	$t \to t - t_0$	G-rays $l(t) = l(t - t_0)$	t
$(1, 0)$	$l_{1,0}(t)$	$t \to t$	$x = t/7 + 1/14$	$0 : 6$
$(1, 1)$	$l_{1,1}(t)$	$t \to t$	$x + y = t/7 + 1/7$	$0 : 12$
$(1, 2)$	$l_{1,2}(t)$	$t \to t - 1/2$	$x + 2y = t/7 + 1/7$	$0 : 18$
$(1, 3)$	$l_{1,3}(t)$	$t \to t - 1$	$x + 3y = t/7 + 1/7$	$0 : 24$
$(1, 4)$	$l_{1,-3}(t)$	$t \to t - 1$	$x - 3y = t/7 - 2/7$	$6 : -1 : -18$
$(1, 5)$	$l_{1,-2}(t)$	$t \to t - 1/2$	$x - 2y = t/7 - 1/7$	$6 : -1 : -12$
$(1, 6)$	$l_{1,-1}(t)$	$t \to t$	$x - y = t/7$	$6 : -1 : -6$
$(0, 1)$	$l_{0,1}(t)$	$t \to t$	$y = t/7 + 1/14$	$0 : 6$

Equations for solving the convolution equations are given in Table 5.3.

5.3 General algorithm of image reconstruction

For N prime, the block-diagram of the transform-based method of reconstruction of the image on the Cartesian grid $N \times N$ from $N + 1$ projections is shown in Figure 5.36.

TABLE 5.3
Solutions of equations: Case 7×7

(p,s)	$\mathbf{a}, a_n$	$\mathbf{w} = KA\mathbf{v}, \ \mathbf{b} = A^{-1}\mathbf{w}$	$\mathbf{v} = B\mathbf{w}(B\mathbf{b})$
$(1,0)$	v_n	$w_t = 7v_t \quad (K=7)$	$v(t) = \frac{1}{7}w(t)$
$(1,1)$	$\sum_m v_{n+7m}$	$w(t) = 7\sqrt{2}v(t) \quad (K = 7\sqrt{2})$	$v(t) = \frac{1}{7\sqrt{2}}w(t)$
$(1,2)$	$\sum_m v_{n+7m}$	$w(t) = \frac{7\sqrt{5}}{2}[v(t-1) + v(t)]$ $b_0 = w(0), b_t = w(t) - b_{t-1}$	$v(t) = \frac{b_t}{7\sqrt{5}/2}$ $t = 0:18$
$(1,3)$	$\sum_m v_{n+7m}$	$w(t) = \frac{7\sqrt{10}}{3}[v(t) + v(t-1) + v(t-2)]$ $b_0 = w(0), b_t = w(t) - (b_{t-1} + b_{t-2})$	$v(t) = \frac{b_t}{7\sqrt{10}/3}$ $t = 0:24$
$(1,4)$	$\sum_m v_{n-7m}$	$w(t) = \frac{7\sqrt{10}}{3}[v(t) + v(t-1) + v(t-2)]$ $b_{-18} = w(-18), b_t = w(t) - (b_{t-1} + b_{t-2})$	$v(t) = \frac{b_t}{7\sqrt{10}/3}$ $t = -18:6$
$(1,5)$	$\sum_m v_{n-7m}$	$w(t) = \frac{7\sqrt{5}}{2}[v(t) + v(t-1)]$ $b_{-12} = w(-12), b_t = w(t) - b_{t-1}$	$v(t) = \frac{b_t}{7\sqrt{5}/2}$ $t = -12:-1:6$
$(1,6)$	$\sum_m v_{n-7m}$	$w(t) = 7\sqrt{2}v(t), \quad t = -6:6$	$v(t) = \frac{1}{7\sqrt{2}}w(t)$
$(0,1)$	v_n	$w_t = 7v_t$	$v(t) = \frac{1}{7}w(t)$

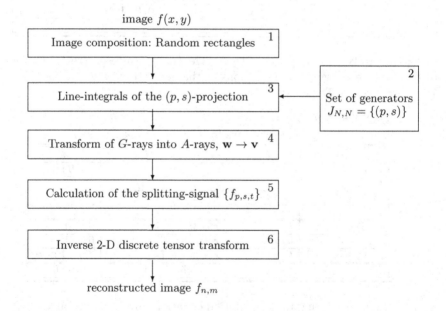

FIGURE 5.36
Block-diagram of the reconstruction of the image $f(x,y)$ composed of rectangles on the Cartesian grid $N \times N$ placed in the square $[0,1] \times [0,1]$.

The following steps describe the proposed method of image reconstruction.

1. The image $N \times N$ is composed of a few random rectangles on the square $[0,1] \times [0,1]$.

2. The set $J_{N,N}$ of generators (p, s) is calculated to define $N+1$ projections. The considered set is $J_{N,N} = \{(1, s); s = 0 : (N - 1)\} \cup \{(0, 1)\}$.

3. Given frequency-point (p, s), the (p, s)-projection is calculated, i.e. all line-integrals $w(t)$ along the set of geometrical rays $\tilde{l}_{p,s}(t)$. For that, the intersections of each ray of the projection with the rectangles are calculated.

4. The geometry of G-rays of the (p, s)-projection is transformed to A-rays, i.e. the set of sums $v(t)$ of the discrete image $f_{n,m}$ is calculated from the integrals $w(t)$, by solving the system of linear equations with the Toeplitz matrix $(M \times M)$, where $M = (p + s - 1)(N - 1) + 1$, if $s \leq (N - 1)/2$, and $M = (p + (N - s) - 1)(N - 1) + 1$, if $s > (N - 1)/2$. The Toeplitz matrix is triangular and the transform of geometry of G-rays to A-rays is fast.

5. The splitting-signal $\{f_{p,s,0}, f_{p,s,1}, ..., f_{p,s,N-1}\}$ is calculated from $v(t)$. This signal is written into the corresponding row of the matrix $N \times (N + 1)$ of the 2-D discrete tensor transform.

6. The inverse 2-D tensor transform is calculated, which is the reconstructed discrete image $f_{n,m}$.

5.4 Program description and image model

The image $f(x, y)$ is considered in the form of N^2 cells, or image elements (IE) with intensities $f_{n,m}$ for the (n, m)-th IE, where $n, m = 0 : (N-1)$. To compose such an image, a few rectangles $r_k(x, y)$, $k = 1 : K$ are randomly placed on the grid $N \times N$ in the square $[0, 1] \times [0, 1]$. We denote by $(x_k, y_k, \Delta x_k, \Delta y_k)$ the rectangle data which include the position and size of the rectangle. The intensity of the k-th rectangle is denoted by R_k. Thus, the image is described as

$$
\begin{aligned}
f_d(x, y) &= \frac{1}{(\Delta x)^2} \sum_{m=0}^{N-1} \sum_{n=0}^{N-1} f_{n,m} \text{rect}_2 \left(Nx - n - \frac{1}{2}, Ny - m - \frac{1}{2} \right) \\
&= \frac{1}{(\Delta x)^2} \sum_{k=1}^{K} R_k \text{rect}_2 \left(\frac{x - x_k}{\Delta x_k} - \frac{1}{2}, \frac{y - y_k}{\Delta y_k} - \frac{1}{2} \right).
\end{aligned}
\tag{5.28}
$$

The value of $f_{n,m}$ is defined as the sum of intensities of rectangles that cover the (n, m)-th IE.

As an example, Figure 5.37 shows the gray-scale image of size 257×257, which is composed of ten rectangles (on the left), and its discrete representation (on the right), after the image reconstruction. The image is the sum of these rectangles.

The program generates data of random rectangles in the integer form as shown in Table 5.4, for the image shown in the figure.

The coordinates of these rectangles are used for the discrete grid $N \times N$, by normalizing them, $x_k = x'_k/N$, $y_k = y'_k/N$, $\Delta x_k = \Delta x'_k/N$, $\Delta y_k = \Delta y'_k/N$,

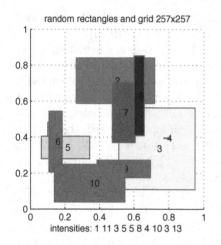
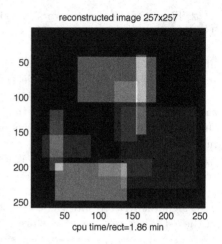

FIGURE 5.37 (See color insert)
Ten colored rectangles and the discrete gray-scale image of size 257×257, which is composed of these rectangles.

TABLE 5.4
Integer data of 10 rectangles

k	x'_k	y'_k	$\Delta x'_k$	$\Delta y'_k$	R_k
1	25	84	22	40	1
2	68	151	116	65	11
3	132	28	112	117	3
4	198	101	10	1	5
5	16	72	72	32	5
6	27	53	20	87	8
7	122	95	34	86	4
8	154	105	15	114	10
9	99	45	79	25	3
10	35	10	106	54	13

$k = 1 : 10$. These coordinates of all rectangles and their intensities are given in Table 5.4.

Below is the beginning of the script (`image_reconstionA3A4.m`), where the random rectangles are generated and saved as data `data_allrect` in integer format [x,y,dx,dy,d_intensity].

```
% ----------------------------------------------------------------
% call: image_reconstructionA3A4m.m
% Modified General code (N is prime)
% The rectangles are assumed on the grid NxN.
% One set of projections is used to scan all rectangles.
% The inverse 2-D tensor transform is used
% Artyom and Meruzhan Grigoryan, June 30 / July 12, 2011
```

TABLE 5.5

Data of 10 rectangles

k	x_k	y_k	Δx_k	Δy_k	R_k
1	0.0973	0.3268	0.0856	0.1556	1
2	0.2646	0.5875	0.4514	0.2529	11
3	0.5136	0.1089	0.4358	0.4553	3
4	0.7704	0.3930	0.0389	0.0039	5
5	0.0623	0.2802	0.2802	0.1245	5
6	0.1051	0.2062	0.0778	0.3385	8
7	0.4747	0.3696	0.1323	0.3346	4
8	0.5992	0.4086	0.0584	0.4436	10
9	0.3852	0.1751	0.3074	0.0973	3
10	0.1362	0.0389	0.4125	0.2101	13

```
% Input data ------------------------------------------------------
fprintf('\n   Start the code (case N is prime) \n');
  N=input('   input the value of N: ');
  if ~isprime(N)
      P=primes(N); mprime=length(P); N=P(mprime);
      fprintf('   N to be the prime as %g \n',N);
  end
% ------------------------------------------------------------------
n=input('   input the number of rectangles (1,2,...,16): ');
many_rectangles=n;
% ------------------------------------------------------------------
data_allrect=zeros(many_rectangles,5);
NM=(N-1)/2; N10=N-10;
for k=1:many_rectangles
    kk=1;
    while kk==1
        for kkk=1:7
            x=rand*N10+4; dx=rand*NM+1; y=rand*N10+4; dy=rand*NM+1;
        end
         if (x+dx<N-4) & (y+dy<N-4)
            data_1rect=round([x,y,dx,dy]); kk=0;
        end
    end
    % data_1rect is in integer format [x,y,dx,dy]
    % plus the intensity (<=16) of the image:
    d_intensity=round(rand*16)+1; data_1rect(5)=d_intensity;
    data_allrect(k,:)=data_1rect;
end
draw_allfigures;  % draw all rectangles on the square [0,1]x[0,1].
  . . .
```

5.5 System of equations

Three different systems of equations are considered for the parallel rays of (p, s)-projections. All $N + 1$ projections can be divided into three parts. The first part is defined by the $(1, s)$-projections, where $0 < s \leq (N-1)/2$, and the second part contains $(1, s)$-projections, where $(N - 1)/2 < s < N$. The first part is referred to as case A3, and the second part of projections is referred to as case A4, similar to the $N = 2^r$ case, which is described in Chapter 4. The $(1, 0)$- and $(0, 1)$-projections, i.e., the horizontal and vertical projections, are considered separately.

Case (A3): For generators $(1, s)$, $0 < s \leq N/2$, the A-rays are defined as

$$l(t) = l_{1,s}(t) = \left\{ x + sy = \frac{t}{N} + \frac{1+s}{2N} \right\},$$

where $t = 0 : (s + 1)(N - 1)$. The set of G-rays are the shifted versions of the A-rays; the shift is $t_0 = (1 + s)/2 - 1$. Therefore, the following equations are used for the G-rays:

$$\tilde{l}(t) = \tilde{l}_{1,s}(t) = l_{1,s}(t - t_0) = \left\{ x + sy = \frac{t}{N} + \frac{1}{N} \right\}. \tag{5.29}$$

Case (A4): For generators $(1, s)$, $(N - 1)/2 < s < N$, the $(1, s - N)$-projections are considered. Therefore, the A-rays are described by equations

$$l(t) = l_{1,s-N}(t) = \left\{ x - (N - s)y = \frac{t}{N} + \frac{1+s-N}{2N} \right\},$$

where $t = (N - 1) : -1 : (s - N)(N - 1)$. The set of G-rays are defined from the A-rays by the shift $t_0 = (1 + N - s)/2 - 1$. The equations for the G-rays are described as

$$\tilde{l}(t) = \tilde{l}_{1,s}(t) = l_{1,s}(t - t_0) = \left\{ x - (N - s)y = \frac{t}{N} - 1 + \frac{1+s}{N} \right\}. \tag{5.30}$$

When $(p, s) = (0, 1)$ or $(1, 0)$, the G-rays are defined as

$$l(t) = l_{p,s}(t) = \left\{ px + sy = \frac{t}{N} + \frac{1}{2N} \right\}, \quad t = 0 : (N - 1).$$

It should be noted that the above equations correspond to the considered (X, Y) coordinate system. In the system $(X, 1 - Y)$, the same equations of parallel rays can be considered, after changing y by $(1 - y)$. In the program, the control points of parallel rays for $(1, s)$-projections, where $s > (N - 1)/2$, are denoted from 0 to $(p+s)(N-1)$, in order to avoid negative indexing. This new numbering of rays is accomplished by the transformation: $t \rightarrow (N-1)-t$. Therefore, the equations given above for the A-rays and G-rays are described in the following way.

Case (A3): Generators are $(1, s)$, when $0 < s \leq N/2$. The equations for A-and G-rays are:

$$l(t) = l_{1,s}(t) = \left\{ x - sy = \frac{t}{N} - s - \frac{1+s}{2N} \right\},$$
$$\tilde{l}(t) = \tilde{l}_{1,s}(t) = \left\{ x - sy = \frac{t}{N} - s + \frac{1}{N} \right\}, \tag{5.31}$$

where $t = 0 : (s+1)(N-1)$.

Case (A4): Generators are $(1, s)$, when $(N-1)/2 < s < N$. The A-rays and G-rays are described by equations

$$l(t) = l_{1,s-N}(t) = \left\{ x - (N-s)y = -\frac{t}{N} + \frac{N+s-1}{2N} \right\},$$
$$\tilde{l}(t) = \tilde{l}_{1,s}(t) = \left\{ x - (N-s)y = -\frac{t}{N} + \frac{s}{N} \right\}, \tag{5.32}$$
$$t = 0 : (1+N-s)(N-1).$$

When $(p, s) = (0, 1)$ and $(1, 0)$, the G-rays are defined as

$$l(t) = l_{p,s}(t) = \left\{ px - sy = \frac{t}{N} + \frac{1}{2N} - s \right\}, \quad t = 0 : (N-1).$$

5.6 Solutions of convolution equations

For generators $(p, s) \in J_{N,N}$, the relations between the sums $v(t) = v_{p,s}(t)$ of the discrete image $f_{n,m}$ and line-integrals $w(t) = w_{p,s}(t)$ of the original image $f(x, y)$ are described in matrix form by Toeplitz matrices, which are similar to the described 7×7 example. The set of generators $J_{N,N}$ is divided by three subsets. Therefore, we consider the corresponding convolution equations for each subset separately.

Subset I: $p = 1$ and $0 < s \leq (N-1)/2$.

The convolution equation $\mathbf{w} = \mathbf{A}\mathbf{v}$ is described as

$$w(t) = N \frac{\sqrt{1+s^2}}{s} [v(t) + v(t-1) + v(t-2) + ... + v(t-s+1)], \tag{5.33}$$
$$t = 0 : (1+s)(N-1).$$

Here, it is assumed that $v(-1) = v(-2) = ... = v(-s+1) = 0$. The inverse transform $\mathbf{b} = \mathbf{A}^{-1}\mathbf{w}$ is calculated in the following recurrent form:

$$
\begin{aligned}
b_0 &= w(0), \\
b_1 &= w(1) - b_0, \\
b_2 &= w(2) - (b_1 + b_0), \\
b_3 &= w(3) - (b_2 + b_1 + b_0), \\
&\dots \quad \dots \quad \dots \quad \dots \quad \dots \quad \dots \\
b_{s-1} &= w(s-1) - (b_{s-2} + b_{s-3} + \dots + b_1 + b_0), \\
b_t &= w(t) - (b_{t-1} + b_{t-2} + \dots + b_{t-s+2} + b_{t-s+1)}), \\
&\quad t = s, s+1, \dots, (1+s)(N-1).
\end{aligned}
\tag{5.34}
$$

The required sums $v(t)$ of the discrete image are calculated by

$$
v(t) = b_t \frac{s}{N\sqrt{1+s^2}}, \quad t = 0 : (1+s)(N-1). \tag{5.35}
$$

In the $s = 0$ case, the line-integrals and sums are equal, up to a normalized factor, i.e., $v(t) = w(t)/N$, $t = 0 : (N-1)$.

   Subset II: $p = 1$ and $s \geq (N+1)/2$.

   For the generator $(1, s)$, the $(1, s - N)$-projection is considered, i.e., s is changed by $s - N$. The convolution equation $\mathbf{w} = \mathbf{A}\mathbf{v}$ is described by

$$
w(t) = N\frac{\sqrt{1+\bar{s}^2}}{\bar{s}}[v(t) + v(t-1) + v(t-2) + \dots + v(t-\bar{s}+1)], \tag{5.36}
$$
$$
t = (N-1) : -1 : -\bar{s}(N-1),
$$

where $\bar{s} = N - s$, and $v(-k) = 0$, when $k > \bar{s}(N-1)$.

   We denote by M the negative number $-\bar{s}(N-1)$. The inverse transform $\mathbf{b} = \mathbf{A}^{-1}\mathbf{w}$ can be calculated in the following recurrent form:

$$
\begin{aligned}
b_M &= w(M), \\
b_{M+1} &= w(M+1) - b_M, \\
b_{M+2} &= w(M+2) - (b_{M+1} + b_M), \\
b_{M+3} &= w(M+3) - (b_{M+2} + b_{M+1} + b_M), \\
&\dots \quad \dots \quad \dots \quad \dots \quad \dots \quad \dots \quad \dots \\
b_{M+\bar{s}-1} &= w(M+\bar{s}-1) - (b_{M+\bar{s}-2} + b_{M+\bar{s}-3} + \dots + b_{M+1} + b_M), \\
b_t &= w(t) - (b_{t-1} + b_{t-2} + \dots + b_{t-\bar{s}+2} + b_{t-\bar{s}+1)}), \\
&\quad t = M + \bar{s}, M + \bar{s} + 1, \dots, (N-1).
\end{aligned}
\tag{5.37}
$$

The required sums $v(t)$ of the discrete image are calculated by

$$
v(t) = b_t \frac{\bar{s}}{N\sqrt{1+\bar{s}^2}}, \quad t = (N-1) : -1 : -\bar{s}(N-1). \tag{5.38}
$$

   Subset III: $p = 0$ and $s = 1$. For the vertical projection, the line-integrals define the sums by $v(t) = w(t)/N$, $t = 0 : (N-1)$.

5.6.1 Splitting-signal composition

For (p, s)-projections, the corresponding splitting-signals $\{f_{p,s,t}; t = 0 : (N - 1)\}$ are calculated from the sums $v_t = v_{p,s}(t)$, by using only operations of addition.

Subset I: $p = 1$ and $0 < s \le (N - 1)/2$.
The components of the splitting-signal are calculated by

$$f_{p,s,t} = v_t + v_{t+N} + \cdots + v_{t+sN}, \quad t = 0 : (N - 1),$$

where $v_k = 0$, for $k > (1 + s)(N - 1)$.
Subset II: $p = 1$ and $s \ge (N + 1)/2$.
The components of the splitting-signal are calculated by

$$f_{p,s,t} = f_{p,-\bar{s},t} = v_t + v_{t-N} + \cdots + v_{t-\bar{s}N}, \quad t = (N - 1) : -1 : -\bar{s}(N - 1),$$

where $\bar{s} = N - s$, and $v_k = 0$, for $k < -\bar{s}(N - 1)$.
For the horizontal and vertical projections, when $(p, s) = (1, 0)$ and $(0, 1)$, respectively, the components of the splitting-signals are equal to the sums $v_t = v_{p,s}(t)$, i.e. $f_{p,s,t} = v_t$, $t = 0 : (N - 1)$.
For the image $f(x, y)$ shown in Figure 5.37, we consider a few line-integrals, sums, and splitting-signals. Figure 5.38 shows these characteristics for the $(1, 0)$-projection. The splitting-signal $\{f_{1,0,t}; t = 0 : 256\}$ and the vector of

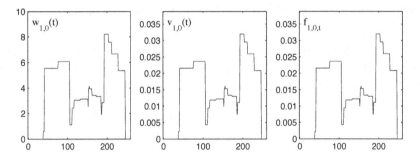

FIGURE 5.38
Line-integrals, sums, and the splitting-signal for the horizontal projection.

sums $v(t) = v_{1,0}(t)$ are the same, and equal the line-integrals, up to the constant 257, i.e., $v(t) = w(t)/257$. Figure 5.39 shows the data for the $(1, 2)$- and $(1, 254)$-projections. For the $(1, 2)$-projection, the line-sums $v(t) = v_{1,2}(t)$, where $t = 0 : 768$, are calculated from the line-integrals $w(t) = w_{1,2}(t)$ by

$$v(t) = b_t/(257\sqrt{5}/2), \quad t = 0 : 768,$$
$$b_0 = w(0), \quad b_t = w(t) - b_{t-1}, \quad t = 1 : 768.$$

The components of the splitting-signal $\{f_{1,2,t}; t = 0 : 256\}$ are calculated by $f_{1,2,t} = v_t + v_{t+257} + v_{t+514}$, $t = 0 : 256$, where $v_{769} = v_{770} = 0$.

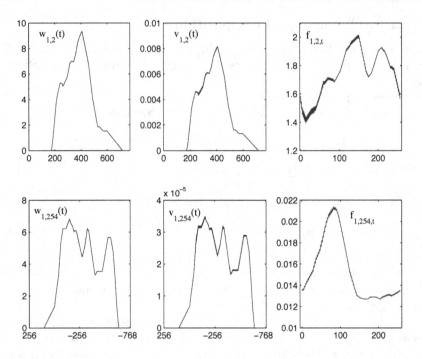

FIGURE 5.39
Line-integral, sums, and splitting-signal for the $(1, 2)$- and $(1, 254)$-projections.

For the $(1, 254)$-projection, the $(1, -3)$-projection is calculated instead. A total of 1025 line-sums $v(t) = v_{1,-3}(t)$, where $t = 256 : -1 : -768$, are calculated from the line-integrals $w(t) = w_{1,-3}(t)$ by

$$v(t) = b_t/(257\sqrt{10}/3), \quad t = 256 : -1 : -768,$$
$$\begin{cases} b_{-768} = w(-768), \\ b_{-767} = w(-767) - b_{-768}, \\ b_t = w(t) - (b_{t-1} + b_{t-2}) \quad t = -766 : 256. \end{cases}$$

The splitting-signal $\{f_{1,254,t}; \; t = 0 : 256\}$ is calculated by

$$f_{1,254,t} = v_t + v_{t-257} + v_{t-514} + v_{t-771}, \quad t = 0 : 256,$$

where $v_{-769} = v_{-770} = v_{-771} = 0$.

5.6.2 Inverse 2-D tensor transform

The 2-D discrete tensor transform of the image $f_{n,m}$ is considered in the form of the matrix $N \times (N+1)$. This matrix is composed from $N+1$ splitting-signals

as shown

$$\begin{bmatrix} f_{0,1,0} & f_{0,1,1} & f_{0,1,2} & \cdots & f_{0,1,N-1} \\ f_{1,0,0} & f_{1,0,1} & f_{1,0,2} & \cdots & f_{1,0,N-1} \\ f_{1,1,0} & f_{1,1,1} & f_{1,1,2} & \cdots & f_{1,1,N-1} \\ f_{1,2,0} & f_{1,2,1} & f_{1,2,2} & \cdots & f_{1,2,N-1} \\ \cdots & \cdots & \cdots & & \cdots \\ f_{1,N-1,0} & f_{1,N-1,1} & f_{1,N-1,2} & \cdots & f_{1,N-1,N-1} \end{bmatrix}.$$

For the image $f(x,y)$ shown in Figure 5.37, the 2-D DTT with 258 splitting-signals on its rows is shown in the form of the gray-scale image in Figure 5.40 in part a. The graph describing the energies of all splitting-signals

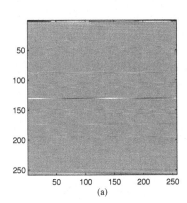
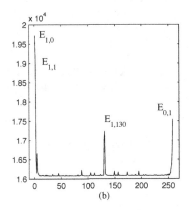

(a)   (b)

FIGURE 5.40
(a) Two-dimensional 258×257-point DTT of the image (in scaled format), and (b) the graph of the energy of splitting-signals.

is given in part b. Four splitting-signals have high energies

$$E_{p,s} = \sqrt{f_{p,s,0}^2 + f_{p,s,1}^2 + \ldots + f_{p,s,256}^2}$$

which are equal to $19718, 19060, 17235,$ and $17544,$ for the $(1,0),(1,1),(0,1),$ and $(1,130)$-projections, respectively.

The reconstructed discrete image $f_{n,m}$ is obtained by the inverse tensor transformation, which is calculated by (5.2). The calculation of the inverse 2-D DTT by this formula is used in the script ($\text{example7x7\_inverseDTT.m}$) in the considered 7×7 example. The following script (inverse_2DDTT.m) can be used for computing the inverse $N \times N$-point 2-D DTT in the general case, when N is prime.

```
% -----------------------------------------------------
%   Call: inverse_2DDTT.m
%   Jps={(1,s);s=0:N-1},{(0,1)}
%   A.M. Grigoryan and Nan Du,          10/27/2009
```

```
% ---------------------------------------------------
function image_fromTT=inverse_2DDTT(DTT_ofimage,N)
  % The inverse formula of the tensor transform:
  image_fromTT=zeros(N,N);
  N1=N-1;
  for n1=0:N1
      n=n1+1;
      for m1=0:N1
          m=m1+1;
          image_fromTT(n,m)=DTT_ofimage(1,n);
          col=m1-n1;
          for p=2:N+1
              col=col+n1;
              if col>=N
                 col=col-N;
              end
              image_fromTT(n,m)=image_fromTT(n,m)+...
                 DTT_ofimage(p,col+1);
          end
      end
  end
% Remove the redundancy of the tensor transform
  image_power=sum(DTT_ofimage(1,:));
  image_fromTT=(image_fromTT-image_power)/N;
% ---------------------------------------------------
```

5.7   MATLAB®-based code (N prime)

In this section, we give the script image_reconstructionA3A4m.m of the main program for image reconstruction. The beginning of this program is given in §5.4, and below is the remaining part of this program.

```
% ----------------------------------------------------
% call: image_reconstructionA3A4m.m
   ....
% ====================================================
%        START THE IMAGE RECONSTRUCTION
  % 1. The set J of all (N+1) generators (p,s):
  for p=0:N-1
      Jps(p+1,:)=[p,1];
  end
  Jps(p+2,:)=[1,0];
  Nps=N+1;
% Reconstruction of all rectangles
  % 1. Composition of the tensor from the integrals
  image_pst=reconstruct_AllrectanglesA34m(data_allrect,N,Jps);
```

```
% 2. Inverse 2-D tensor transform
  imageR=inverse_2DDTT(image_pst,N);
% -------------------------------------------
 figure(h_fig);
 subplot(1,2,2);
 imagesc(imageR');
 axis image;
 s_title=sprintf('reconstructed image %gx%g',N,N);
 title(s_title);
 display(' End of the code: image_reconstructionA3A4.m ');
% ========================================================
```

The script of the function `reconstruct_AllrectanglesA34m.m`, which accomplishes the calculation of all line-integrals $w(t)$ and the sums $v(t)$ is given below. This function uses the rectangle data (`data_allrect`) and set of generators (`Jps`) calculated in the main program (`image_reconstructionA3A4m.m`).

```
% ----------------------------------------------------------------------
% call: reconstruct_AllrectanglesA34m.m
% Processing: one projection for all rectangles.
% Set of generators: Jps={{(p,1);p=0:N-1},(1,0)}
% Each (p,s)-projection is processing and the 2-D tensor
% is composed in row # (p+s)+1 (and 1, when (p,s)=(0,1))
% Artyom and Meruzhan Grigoryan, June 30 / July 12, 2011
% ----------------------------------------------------------------------
 function image_pst=reconstruct_AllrectanglesA34m(data_allrect,N,Jps)
 %I   1. Determine the data of rectangles
    many_rectangles=size(data_allrect,1);
    d_xin1=1/N;        % the length of 1 cell (IE)
    d_yin1=d_xin1;     % the high of the cell (IE)
    % 2. The set of values of np=p,N-p in Jps
    Nps=N+1;  N_half=Nps/2-1;
    np_formatrixT=zeros(1,Nps);
    np_formatrixT(1:N_half+1)=0:N_half;   % case 1
    tnp=N-N_half-1:-1:1;
    np_formatrixT(N_half+2:N)=tnp;        % case 2
    np_formatrixT(Nps)=0;                 % case 3
 %II  Process each (p,s)-projection
    image_pst=zeros(Nps,N);   % matrix for the 2-D DTT
    number_ofallintegrals=0;
 for kps=1:Nps
    p=Jps(kps,1);  s=Jps(kps,2);
    if (s==0) % & (p==1)       % for (1,0)
         nrow=1;               % the row in 2-D DTT
    else nrow=p+2;             % for (p,1), p=0:(N-1)
    end
    np=np_formatrixT(kps);
    if np==0                   % cases (p,s)=(1,0) and (0,1)
       delta_r=d_xin1;         % 1/N;
```

```
     else
          delta_r=sqrt(1+np^2)/d/N;
     end
     % delta_r = length of the path of the ray in IE
  for k=1:many_rectangles
          % 1. Read the rectangle data
          data_1rect=data_allrect(k,:);
          n0=data_1rect(1);  m0=data_1rect(2);
          dx=data_1rect(3);  dy=data_1rect(4);  k_in=data_1rect(5);
          % The coordinates of the k-th rectangle:
          x0=n0;  y0=m0;  x1=(n0+dx);  y1=(m0+dy);
          p_rect=[x0 y0; x1 y1];
          % 2. Calculation of the (p,s)-projection, w(t):
          [W,number_ofw]=projection_inrectangle2N(p,s,N,p_rect,N_half);
          number_ofallintegrals=number_ofallintegrals+number_ofw;
          W=W/N;               % to go back to the grid [0,1]x[0,1]
          W=W*k_in;            % consider the rectangle intensity
          % 3. Translate geometry: G-rays into A-rays
          if np==0             % cases (p,s)=(1,0) and (0,1)
              B=W;
          else
              B=solution_of_linearsystemA(np-1,W);
          end
          % 4. Composition of the N-dim. vector a(n) from B(t)
          A=Avector_composition(N,B);
          % 5. Normalization by the path
          A=A/delta_r;
          % 6. Correction in case 2: reordering of data in A(n)
          A1=A;
          if (s==1) && (p>N_half)
              A1=vectorA_indexcorection(p-N,N,A);
          end
          % 7. Matrix composition for the Nx(N+1)-point DTT
          for i1=1:N
              image_pst(nrow,i1)=image_pst(nrow,i1)+A1(i1);
          end
          clear W B A A1;
  end
end
  fprintf('\n # of all integrals w(t) is %g \n',number_ofallintegrals);
% -----------------------------------------------------------------------
```

The time characteristics of calculation, when the program was implemented in MATLAB are given in Table 5.6. The time for calculation of the 2-D tensor transform and its inverse transform are also given.

Table 5.7 shows the total time of image processing when the program was implemented in Ci+. The time includes the calculation of all line-integrals $w(t)$, line-sums $v(t)$, and the calculation of the 2-D direct and inverse tensor transforms. The data in these tables were obtained from running programs on

TABLE 5.6
Time for MATLAB-based program [A3-A4]

$N \times N$	scan & reconstruct	2-D DTT	2-D iDTT
61×61	0.88s/rec	0.s	0s
101×101	4.11s/rec	0.2340s	0.5304s
127×127	8.45s/rec	0.2340s	0.5304s
257×257	1.86m/rec	1.9188s	4.5084s
401×401	24.01m/rec	0.2340s	0.5304s
509×509	24.01m/rec	0.2340s	0.5304s

TABLE 5.7
Time for the Ci-based program (the 1st version)

$N \times N$	projection data processing & reconstruction
37×37	$00 : 00 : 00.03$
67×67	$00 : 00 : 00.08$
101×101	$00 : 00 : 00.27$
127×127	$00 : 00 : 00.50$
257×257	$00 : 00 : 04.21$
401×401	$00 : 00 : 16.67$
509×509	$00 : 00 : xx.xx$
521×521	$00 : 00 : 37.24$
797×797	$00 : 02 : 15.82$
1031×1031	$00 : 04 : 55.69$
1777×1777	$00 : 25 : 29.33$
2053×2053	$00 : 39 : 40.44$

a Dell computer with Intel Dual CPU processor at 3.20-GHz speed.

Problems

Problem 5.1 Consider case A for all generators $(1, s)$, $s = 0 : (N - 1)$, when all G-rays are described by equation (5.29). Consider also that the $(1, s)$-projections are not substituted by $(1, s - N)$-projections, when $s > (N - 1)/2$. Write the code for image reconstruction using these rays. Discuss how fast or slow your program is, when compared with the program image_reconstructionA3A4m.m. For that, you may consider images on the square $[0, 1] \times [0, 1]$, which are composed of random rectangles, each with constant intensity, and the Cartesian lattices $N \times N$, for different prime N.

Problem 5.2 Suppose that you are given an image in the square $[0, 1] \times [0, 1]$ with image elements in the lattice 257×257. Generate such an image from the random rectangles.

A. For the $(1, 4)$-projection, calculate the line-integrals $w(t)$, sums $v(t)$, and the tensor splitting-signal $\{f_{1,4,t}; t = 0 : 256\}$. Sketch your results.

B. In most cases, the splitting-signals for the vertical and horizontal projections have high energy and represent big contributions in the reconstruction of the image. Discuss the reconstruction without these projections. In other words, calculate the image which is reconstructed from the $(1, s)$-projections, where $s = 1 : 256$.

Problem 5.3 Suppose that you are given an image in the square $[0, 1] \times [0, 1]$, which is composed of a few rectangles with constant intensity each. Each image element has the size $1/10 \times 1/10$. Discuss the reconstruction of the image on the lattice 10×10. Write your code and show results of reconstruction on examples.

Problem 5.4 The method of G-rays and tensor transform for image reconstruction, which is described in this chapter, can be applied for any size lattice $N \times N$, where $N > 1$ is an integer. Suppose that you are given an image in the square $[0, 1] \times [0, 1]$, which is composed of a few rectangles with constant intensity each. Each image element has the size $1/N \times 1/N$. Discuss the reconstruction of the image on the lattice $N \times N$, when $N = 100$.

Problem 5.5 Suppose that you are given an image in the square $[0, 1] \times [0, 1]$. As an example, generate such an image from the random rectangles on the Cartesian lattice 128×128. Sample this image on the lattice 127×127 and calculate the reconstruction of the image on this lattice. Discuss the result and calculate the minimum mean-square-root error of reconstruction.

Problem 5.6* Our program `image_reconstructionA3A4m.m` works 1.80 minutes (with graphics), when reconstructing the image on the lattice 257×257. The image is composed of ten random rectangles and it is shown in Figure 5.41. The realization of this program in Ci+ is much faster (by about 10 s). Develop

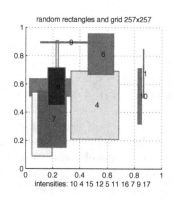
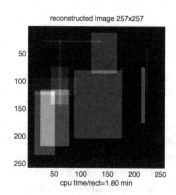

FIGURE 5.41 (See color insert)
Image with ten rectangles and its reconstruction 257×257.

the Ci version of this code, or write your own code in Ci+ to reconstruct the image from its projections.

6

Method of Particles

In this chapter, we describe a new method of summation of projection data by G-rays, which is based on the concept of *the point map of projections*. For this concept, the map of each point on the Cartesian lattice $N \times N$ is considered in the form of a matrix which describes all A-rays passing through this point. Each point is considered as a particle on the lattice, which is described by the field which we call *the field of the particle*. This concept is then generalized for the G-rays, and the field functions are defined for the G-particles, that represent the image elements as particles, or the smallest squares on the lattice. The field functions of such particles define all G-rays passing through this particle. The consideration of field functions for both A- and G-particles leads to a representation of the image by the field functions, and this representation allows us to reconstruct the image from its projections. We do consider that the image has a corpuscular structure, i.e. is composed of quantum-s, or small image elements which are described by their fields, and study of these fields is important for solving the problem of image reconstruction.

6.1 Point-map of projections

For the image $f(x, y)$ in the square $[0, 1] \times [0, 1]$, which is represented by image elements of size $\Delta x \Delta y = 1/N^2$ each, we derived the formula for the integral along each ray $l_{p,s}(t)$ of the (p, s)-projection. This integral is the linear sum of intensities of the image on the image elements, which are intersected by the ray. The coefficients of this sum are equal to the lengths of intersection of the ray with the image elements.

Thus, the ray is described by a set of image elements $\{[n_k, m_k]; k = 1 : K\}$, where K is the discrete length of the ray $l_{p,s}(t)$, or the number of image elements intersected by the ray. We denote the lengths of the intersections by $\Delta l(n_k, m_k)$. It is assumed that $\Delta l(n_k, m_k) \neq 0$ and intersection in only one point is not considered. The integral of the image $f(x, y)$ along the ray equals

$$w_{p,s}(t) = \sum_{k=1}^{K} f_{n_k, m_k} \Delta l(n_k, m_k). \tag{6.1}$$

For the scanning scheme that is used in the tensor and paired methods of image reconstructions, the lengths $\Delta l(n_k, m_k)$ are equal for all rays of the projection, i.e.. $\Delta l(n_k, m_k) = \Delta l(p, s)$. Therefore, the above equation can be written as

$$w_{p,s}(t) = \Delta l(p, s) \sum_{k=1}^{K} f_{n_k, m_k}. \tag{6.2}$$

Let $r_{p,s}(t)$ be the real value of the $f(x, y)$ image integral along the ray $l_{p,s}(t)$. We consider the mean square deviation of the sum of real rays from their discrete representation

$$\varepsilon = \sum_{(p,s) \in J_{N,N}} \left(w_{p,s}(t) - r_{p,s}(t) \right)^2, \tag{6.3}$$

where $J_{N,N}$ is the set of all generators (p, s). By minimizing this error by derivatives, we consider the system of equations

$$\frac{\partial \varepsilon}{\partial f_{n,m}} = \sum_{(p,s) \in J_{N,N}} 2 \left(w_{p,s}(t) - r_{p,s}(t) \right) \frac{\partial w_{p,s}(t)}{\partial f_{n,m}} = 0, \quad n, m = 0 : (N-1),$$

or

$$\sum_{(p,s) \in J_{N,N}} w_{p,s}(t) \frac{\partial w_{p,s}(t)}{\partial f_{n,m}} = \sum_{(p,s) \in J_{N,N}} r_{p,s}(t) \frac{\partial w_{p,s}(t)}{\partial f_{n,m}}.$$

When solving this system, we obtain the following:

$$\sum_{(p,s) \in J_{N,N}} w_{p,s}(t) l(p, s) \delta_{n,n_k} \delta_{m,m_k} = \sum_{(p,s) \in J_{N,N}} r_{p,s}(t) l(p, s) \delta_{n,n_k} \delta_{m,m_k},$$

$$\sum_{l_{p,s}(t) \cap [n,m] \neq \emptyset} w_{p,s}(t) l(p, s) = \sum_{l_{p,s}(t) \cap [n,m] \neq \emptyset} r_{p,s}(t) l(p, s).$$

One can notice that for each image element, or square $[n, m]$, the summations of integrals are accomplished only by the rays $l_{p,s}(t)$ passing through $[n, m]$. Therefore, we are faced with the concept of the "particle" as the sum of rays passing through the given image element. It should also be noted that if we consider equations (6.2) in normalized form,

$$\tilde{w}_{p,s}(t) = \frac{w_{p,s}(t)}{\Delta l(p, s)} = \sum_{k=1}^{K} f_{n_k, m_k},$$

the minimization of the mean-square error results in the following condition:

$$\sum_{l_{p,s}(t) \cap [n,m] \neq \emptyset} \tilde{w}_{p,s}(t) = \sum_{l_{p,s}(t) \cap [n,m] \neq \emptyset} r_{p,s}(t).$$

This means the equations and the concept of the particle, which we want to introduce, can be described in integer form, such as integer matrices.

This concept will be described first for the A-rays. For that, we create a picture of all rays passing through the center of a given (n_0, m_0)-th square, where $n_0, m_0 \in \{0, 1, ..., (N-1)\}$, in the discrete model of the image. First we consider the example. For the case when $N = 8$ and $(n_0, m_0) = (3, 3)$, we introduce the following table:

$$A(3,3) = A(8; 3, 3) = \begin{bmatrix} 1 & 0 & 1 & 1 & 1 & 0 & 1 & 0 \\ 0 & 1 & 1 & 1 & 1 & 1 & 0 & 1 \\ 0 & 1 & 1 & 1 & 1 & 1 & 0 & 1 \\ 1 & 1 & 1 & \mathbf{12} & 1 & 1 & 1 & 1 \\ 0 & 1 & 1 & 1 & 1 & 1 & 0 & 0 \\ 0 & 1 & 1 & 1 & 1 & 1 & 0 & 1 \\ 1 & 0 & 1 & 1 & 1 & 0 & 1 & 0 \\ 0 & 1 & 1 & 1 & 0 & 1 & 0 & 1 \end{bmatrix}. \tag{6.4}$$

We call this table *the map of projections for point* $(3, 3)$, or *the point map of projections for* $(3, 3)$. For $(n, m) \neq (3, 3)$, the coefficient $a(n, m) = a(3, 3, n, m)$ of this matrix equals 1, if there is an A-ray passing through the points $(3, 3)$ and (n, m) at the same time. This coefficient equals 0, if there is no such A-ray. In the center, the coefficient $a(3, 3)$ equals 12, or the number of projections. There are 12 rays passing through the point $(3, 3)$, which cover 46 points (n, m) on the Cartesian grid 8×8.

The descriptions of all 12 rays passing through the point $(3, 3)$ are given in Table 6.1. The coordinates of points (n, m), or squares $[n, m]$ that lie on

TABLE 6.1
Coordinates of the projection map for point $(3, 3)$

(p, s)	points $[n, m]$								card	$g(p, s)$
$(1, 0)$	$[4, 1]$	$[4, 2]$	$[4, 3]$	$[4, 4]$	$[4, 5]$	$[4, 6]$	$[4, 7]$	$[4, 8]$	8	1
$(1, 1)$	$[7, 1]$	$[6, 2]$	$[5, 3]$	$[4, 4]$	$[3, 5]$	$[2, 6]$	$[1, 7]$		7	1
$(1, 2)$	$[8, 2]$	$[6, 3]$	$[4, 4]$	$[2, 5]$					4	2
$(1, 3)$	$[7, 3]$	$[4, 4]$	$[1, 5]$						3	3
$(1, 4)$	$[8, 3]$	$[4, 4]$							2	4
$(1, 5)$	$[7, 5]$	$[4, 4]$	$[1, 3]$						3	3
$(1, 6)$	$[8, 6]$	$[6, 5]$	$[4, 4]$	$[2, 3]$					4	2
$(1, 7)$	$[8, 8]$	$[7, 7]$	$[6, 6]$	$[5, 5]$	$[4, 4]$	$[3, 3]$	$[2, 2]$	$[1, 1]$	8	1
$(0, 1)$	$[1, 4]$	$[2, 4]$	$[3, 4]$	$[4, 4]$	$[5, 4]$	$[6, 4]$	$[7, 4]$	$[8, 4]$	8	1
$(2, 1)$	$[2, 8]$	$[3, 6]$	$[4, 4]$	$[5, 2]$					4	2
$(4, 1)$	$[3, 8]$	$[4, 4]$							2	4
$(6, 1)$	$[3, 2]$	$[4, 4]$	$[5, 6]$	$[6, 8]$					4	2

the rays passing through the point $(3, 3)$ are given. On the 3-rd column of the table, the cardinalities $c(p, s)$, i.e., the numbers of the points (n, m) on the rays are shown.

To calculate tables $A(N; n_0, m_0)$ in the general case of N and (n_0, m_0), we can use the commands and scripts of the functions `Arays_nm.m` and `Aray_throu gh_nm.m`, which are given below. The variable `Anm` stands for the point map of projections $A(N; n_0, m_0)$. The numbering of squares is performing from $1 : N$, not $0 : (N - 1)$. The function Aray_through_nm.m calculates the point map A only for the single (p, s)-projection. It also calculates the ray number t, that passes through the point (n_0, m_0), as well as, it saves coordinates (n, m) of this ray in the 2-D array Ray_squares. We also include the command of printing these coordinates, to compose Table 6.1. In the code Arays_nm.m, all single maps are added together, to obtain the point map $A(N; m_0, n_0)$.

```
% from: run_pointmap.m
  N=8; n0=4; m0=4;
  Anm=Arays_nm(N,m,n);
% ------------------------------------------------------------
% call: Arays_nm.m                  / A.-M. Grigoryan, June 2011
function Anm=Arays_nm(N,m,n)
  Anm=zeros(N);
  ps=ps_generators(N); L=size(ps,1);
  for n_ps=1:L
       p=ps(n_ps,1); s=ps(n_ps,2);
       A1=Aray_through_nm(p,s,N,m,n);
       Anm=Anm+A1;
  end
% ------------------------------------------------------------
% call: Aray_through_nm.m  / A.-M. Grigoryan, June 2011
function [A,t,Ray_squares]=Aray_through_nm(p,s,N,m0,n0)
     S=Arays2grid(p,s,N); L=size(S,1);
     [t,ny]=find( S(:,:,1)==m0 & S(:,:,2)==n0 );
     card_of_IE=max(find(S(t,:,2),N,'first'));
     A=zeros(N);
     Ray_squares=zeros(card_of_IE,2);
     for k=1:card_of_IE
          m1=S(t,k,1); n1=S(t,k,2);
          A(m1,n1)=1;  fprintf('[%g,%g] ',m1,n1);
          Ray_squares(k,1)=m1; Ray_squares(k,2)=n1;
     end
     fprintf('\n');
% ------------------------------------------------------------
```

6.1.1 *A*-particle and the field

In this section, we introduce the concept of the particle and its field function. For that, we first consider a point (n_0, m_0) in the Cartesian lattice $N \times N$ and the set of A-rays passing through this point. The point map of projections is a key function for reconstructing the discrete image where each point is referred to as a particle on the lattice.

Definition 6.1 *The arithmetical particle, or A-particle of dimension N with the center in (n_0, m_0) is called the point (n_0, m_0) on the Cartesian lattice $N \times N$, which is described by the matrix $\|a(n_0, m_0, n, m)\|$ defining the number* *of A-rays passing through the points* (n_0, m_0) *and* (n, m) *at the same time.* *We call the two-dimensional function* $\psi_{n_0, m_0}(n, m) = a(n_0, m_0, n, m)$, *where* $n, m = 0 : (N - 1)$, *the field function of this particle.*

Example 6.1 In matrix form, the field functions of the A-particles $(0, 1)$ and $(3, 2)$ are

$$
\psi_{0,1} = \begin{bmatrix} 1 & 1 & 1 & 0 \\ \mathbf{6} & 1 & 1 & 1 \\ 1 & 1 & 0 & 0 \\ 1 & 0 & 1 & 0 \end{bmatrix} \quad \text{and} \quad \psi_{3,2} = \begin{bmatrix} 0 & 1 & 0 & 1 \\ 0 & 0 & 1 & 1 \\ 1 & 1 & 1 & \mathbf{6} \\ 0 & 1 & 1 & 1 \end{bmatrix}. \tag{6.5}
$$

Figure 6.1 illustrates two A-particles $(0, 1)$ and $(3, 2)$ and their field functions. Points $(0, 1), (1, 1), (2, 1)$, and $(3, 1)$ are on the first A-ray of the

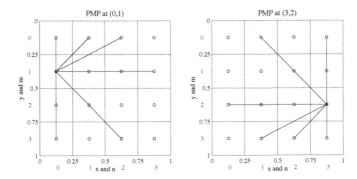

FIGURE 6.1
The field functions of the particles $(0, 1)$ and $(3, 2)$.

$(0, 1)$-projection. Points $(0, 1)$, and $(1, 0)$ are on the first A-ray of the $(1, 1)$-projection. Points $(2, 3), (1, 2)$, and $(0, 1)$ are on the first A-ray of the $(-1, 1)$-projection (or $(3, 1)$-projection). Points $(0, 0), (0, 1), (0, 2)$, and $(0, 3)$ are on A-ray number 0 of the $(1, 0)$-projection, and $(2, 0), (0, 1)$ are on the second A-ray of the $(1, 2)$-projection. The only point on the first A-ray of the $(2, 1)$-projection is $(0, 1)$. This single-point ray is shown by the bullet; therefore the number of rays seen on the figure is 5, but it is actually 6. The description of the particle $(0, 1)$ with its field function is also given in Table 6.2. The description of the A-particle $(3, 2)$ with its field function is given in Table 6.3. A-ray number 8 of the $(2, 1)$-projection passes through only the point $(3, 2)$ in the lattice 4×4, and this single-point ray is shown by the bullet in the figure.

Figure 6.2 illustrates two A-particles $(2, 2)$ and $(2, 1)$ and their field functions. One can notice six rays for both A-particles, because each of six A-rays passes through at least two points at the same time.

We can write the field functions of all 16 A-particles on the lattice 4×4 in the form of a block-table $[\psi_{n,m}]_{n,m=0:3}$, as shown in Table 6.4.

This table, or block-table 4×4, with all field functions can be packed into

TABLE 6.2
Description of the A-particle $(0, 1)$

(p, s)	points (n, m)	A-ray
$(0, 1)$	$(0, 1)$, $(1, 1)$, $(2, 1)$, $(3, 1)$	#1
$(1, 1)$	$(0, 1)$, $(1, 0)$	#1
$(2, 1)$	$(0, 1)$	#1
$(-1, 1)$	$(2, 3)$, $(1, 2)$, $(0, 1)$	#1
$(1, 0)$	$(0, 0)$, $(0, 1)$, $(0, 2)$, $(0, 3)$	#0
$(1, 2)$	$(2, 0)$, $(0, 1)$	#2

TABLE 6.3
Description of the A-particle $(3, 2)$

(p, s)	points (n, m)	A-ray
$(0, 1)$	$(0, 2)$, $(1, 2)$, $(2, 2)$, $(3, 2)$	#2
$(1, 1)$	$(2, 3)$, $(3, 2)$	#5
$(2, 1)$	$(3, 2)$	#8
$(-1, 1)$	$(3, 2)$, $(2, 1)$, $(1, 0)$	#-1
$(1, 0)$	$(3, 0)$, $(3, 1)$, $(3, 2)$, $(3, 3)$	#3
$(1, 2)$	$(3, 2)$, $(1, 3)$	#7

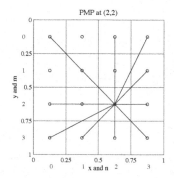 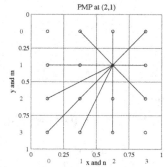

FIGURE 6.2
The field functions of the particles $(2, 2)$ and $(2, 1)$.

the following matrix of size 7×7, which we call *the base matrix of fields of A-particles*:

$$\Psi = \begin{bmatrix} 1 & 0 & 0 & 1 & 0 & 0 & 1 \\ 0 & 1 & 0 & 1 & 1 & 1 & 0 \\ 0 & 0 & 1 & 1 & 1 & 1 & 0 \\ 1 & 1 & 1 & \underline{\mathbf{6}} & 1 & 1 & 1 \\ 0 & 1 & 1 & 1 & 1 & 0 & 0 \\ 0 & 1 & 1 & 1 & 0 & 1 & 0 \\ 1 & 0 & 0 & 1 & 0 & 0 & 1 \end{bmatrix} . \tag{6.6}$$

TABLE 6.4
Basis of the field

6	1	1	1	1	6	1	1	1	1	6	1	1	1	1	6
1	1	0	0	1	1	1	0	1	1	1	1	0	1	1	1
1	0	1	0	1	1	0	1	1	1	1	0	0	1	1	1
1	0	0	1	0	1	0	0	0	0	1	0	1	0	0	1
1	1	1	0	1	1	1	1	0	1	1	1	0	0	1	1
6	1	1	1	1	6	1	1	1	1	6	1	1	1	1	6
1	1	0	0	1	1	1	0	1	1	1	1	0	1	1	1
1	0	1	0	1	1	0	1	1	1	1	0	0	1	1	1
1	1	1	0	0	1	1	1	1	0	1	1	0	1	0	1
1	1	1	0	1	1	1	1	0	1	1	1	0	0	1	1
6	1	1	1	1	6	1	1	1	1	6	1	1	1	1	6
1	1	0	0	1	1	1	0	1	1	1	1	0	1	1	1
1	0	0	1	0	1	0	0	0	0	1	0	1	0	0	1
1	1	1	0	0	1	1	1	1	0	1	1	0	1	0	1
1	1	1	0	1	1	1	1	0	1	1	1	0	0	1	1
6	1	1	1	1	6	1	1	1	1	6	1	1	1	1	6

Indeed, it is not difficult to notice that the following equality holds:

$$\psi_{n,m}(i,j) = \Psi(3 - n + i, 3 - m + j), \quad i,j = 0:3, \tag{6.7}$$

for any A-particle (n, m). The base field matrix is symmetric, i.e., $\Psi(i, j) = \Psi(j, i)$, for $i, j = 0 : 6$.

We here state that for other values of $N = 2^r$, $r > 2$, the similar base matrix of fields of A-particles can also be defined. As an example, we consider the $N = 8$ case, and two A-particles $(3, 3)$ and $(5, 4)$, whose field functions are illustrated in Figure 6.3. There are twelve A-rays passing through the A-

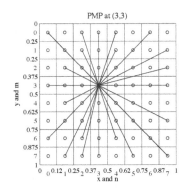
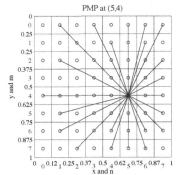

FIGURE 6.3
The field functions of the particles $(3, 3)$ and $(5, 4)$.

particle $(3, 3)$ in this example. The base matrix of fields of all 64 A-particles is the following matrix 15×15 :

$$\Psi = \begin{bmatrix}
1 & 0 & 0 & 0 & 0 & 0 & 0 & 1 & 0 & 0 & 0 & 0 & 0 & 0 & 1 \\
0 & 1 & 0 & 0 & 1 & 1 & 0 & 1 & 0 & 1 & 1 & 0 & 0 & 1 & 0 \\
0 & 0 & 1 & 0 & 0 & 0 & 0 & 1 & 0 & 0 & 0 & 0 & 1 & 0 & 0 \\
0 & 0 & 0 & 1 & 0 & 1 & 0 & 1 & 1 & 1 & 0 & 1 & 0 & 0 & 0 \\
0 & 1 & 0 & 0 & 1 & 0 & 1 & 1 & 1 & 0 & 1 & 0 & 0 & 1 & 0 \\
0 & 0 & 0 & 1 & 0 & 1 & 1 & 1 & 1 & 1 & 0 & 1 & 0 & 0 & 0 \\
0 & 0 & 0 & 0 & 0 & 1 & 1 & 1 & 1 & 1 & 0 & 1 & 0 & 0 & 0 \\
1 & 1 & 1 & 1 & 1 & 1 & 1 & \underline{12} & 1 & 1 & 1 & 1 & 1 & 1 & 1 \\
0 & 0 & 0 & 1 & 0 & 1 & 1 & 1 & 1 & 1 & 0 & 0 & 0 & 0 & 0 \\
0 & 0 & 0 & 1 & 0 & 1 & 1 & 1 & 1 & 1 & 0 & 1 & 0 & 0 & 0 \\
0 & 1 & 0 & 0 & 1 & 0 & 1 & 1 & 1 & 0 & 1 & 0 & 0 & 1 & 0 \\
0 & 0 & 0 & 1 & 0 & 1 & 1 & 1 & 0 & 1 & 0 & 1 & 0 & 0 & 0 \\
0 & 0 & 1 & 0 & 0 & 0 & 0 & 1 & 0 & 0 & 0 & 0 & 1 & 0 & 0 \\
0 & 1 & 0 & 0 & 1 & 1 & 0 & 1 & 0 & 1 & 1 & 0 & 0 & 1 & 0 \\
1 & 0 & 0 & 0 & 0 & 0 & 0 & 1 & 0 & 0 & 0 & 0 & 0 & 0 & 1
\end{bmatrix} . \qquad (6.8)$$

The field function of each A-particle (n, m), where $n, m \in \{0, 1, ..., 7\}$, can be calculated by

$$\psi_{n,m}(i, j) = \Psi(7 - n + i, 7 - m + j), \quad i, j = 0 : 7. \qquad (6.9)$$

One can notice that the matrix given in (6.4) represents the field function $\psi_{3,3}(i, j)$ of the A-particle $(3, 3)$. In general, this equation can be written as

$$\psi_{n,m}(i, j) = \Psi(N - 1 - n + i, N - 1 - m + j), \quad i, j = 0 : N - 1, \qquad (6.10)$$

where $n, m \in \{0, 1, ..., N-1\}$. The size of the base matrix is $(2N-1) \times (2N-1)$.

To calculate the base matrix Ψ in the general case of $N = 2^r$, when $r > 1$, we use the function bmf_Amatrix.m with the following script.

```
% call: bmf_Amatrix.m,    M. Grigoryan,   01/04/2012
% Calculate the base matrix of field of A-particles
function P=bmf_Amatrix(N)
 ps=ps_generatorsM(N);
 P=zeros(2*N-1);
 L=N+N/2;
 for n_ps=1:L
     p=ps(n_ps,1);
     s=ps(n_ps,2);
     for m=-(N-1):(N-1)
         sm=s*m; m1=N-m;
         for n=-(N-1):(N-1)
             if (p*n+sm)==0
                 P(N-n,m1)=R(N-n,m1)+1;
             end
         end
     end
 end
end
```

The matrix Ψ in (6.8) is calculated by calling the command P=bmf_Amatrix(8). The function ps_generatorsM is the modification of the function ps_generators, where generators $(1, s)$ and $(p, 1)$ are substituted by $(1, s - N)$ and $(p - N, 1)$, respectively, when s and $p > N/2$.

```
% call: ps_generatorsM.m
function Jps=ps_generatorsM(N)
   L=N+N/2;
   Jps=ones(2,L);
   Jps(1,1:N/2+1)=0:N/2; Jps(1,N/2+2:N)=-N/2+1:1:-1;
   Jps(2,N+1:N+N/4+1)=0:2:N/2; Jps(2,N+N/4+2:L)=-N/2+2:2:-2;
```

Figure 6.4 illustrates the base matrix of fields of A-particles for the $N = 4$ and 8 cases in parts a and b, respectively. These matrices are shown in the

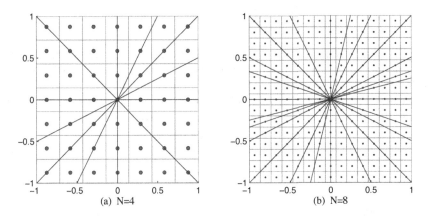

(a) N=4   (b) N=8

FIGURE 6.4
Base matrices of A-particles for (a) $N = 4$ and for (b) $N = 8$.

square $[-1, 1] \times [-1, 1]$ with the lattice $(2N - 1) \times (2N - 1)$. All A-rays passing through the center $(0, 0)$ of the base matrices are shown as well. Six A-rays pass through the original in the $N = 4$ case, and twelve when $N = 8$. If the A-ray passes through the centers of other squares $[n, m]$, then the coefficients $\Psi(n, m) = 1$.

6.1.2 Representation by field functions

It is clear that by introducing the field functions of A-particles, we define a new basis in the space of discrete images $f_{n,m}$, which is different from the basis of the 2-D paired transformation. In paired representation, the image $\{f_{n,m}\}$ is defined as the 3-D array $\{f'_{p,s,t}\}$,

$$f'_{p,s,t} = \sum_{m=0}^{N-1} \sum_{n=0}^{N-1} f_{n,m} \chi'_{p,s,t}(n, m), \quad (p, s, t) \in J'_{N,N}.$$

The paired representation is a transformation of the image into a set of 1-D splitting-signals generated by different frequency-points, which we call generators (p, s). Therefore the paired representation is referred to as the 2-D frequency and 1-D time transformation. All N^2 triplet-numbers (p, s, t) of the 2-D paired functions can be written in the matrix $N \times N$, as shown in §4.7.3. N^2 components $f'_{p,s,t}$ can be packed into the 2-D array $\{f'_{v,w}\}$ and the paired representation can be written as

$$f'_{v,w} = \sum_{m=0}^{N-1} \sum_{n=0}^{N-1} f_{n,m} \chi'_{v,w}(n, m), \quad v, w = 0 : (N-1).$$

Now we can represent the image by the field functions

$$a(i, j) = \sum_{m=0}^{N-1} \sum_{n=0}^{N-1} f_{n,m} \psi_{n,m}(i, j), \quad i, j = 0 : (N-1). \tag{6.11}$$

The physical meaning of this representation is clear. The image is transformed into the matrix, or field, the value of which at point (i, j), or A-particle (i, j), is determined by the field functions of A-particles which can be seen by A-particle (i, j) along the A-rays. In other words, $a(i, j)$ is the sum of the image along all rays passing through the point (i, j). By considering the base matrix of fields of A-particles, we can write the following:

$$a(i, j) = \sum_{m=0}^{N-1} \sum_{n=0}^{N-1} f_{n,m} \Psi(N-1-n+i, N-1-m+j). \tag{6.12}$$

We should notice that the obtained new basis of field functions $\psi_{n,m}$ is not orthogonal. That may be considered as the disadvantage of this basis. On the contrary, the paired transform is orthogonal, because of the complex way of combining parallel rays in each projection. This transform was originated from the tensor transform being non-orthogonal. However, in the paired transform the orthogonality was achieved by assigning the plus and minus signs to different parts of rays for each projection. In other words, each particle (p, s, t) in the 3-D space was considered with its co-particle $(p, s, t + N/2)$, and this pair of particles was described by the function $f'_{p,s,t} = (f_{p,s,t} - f_{p,s,t+N/2})$, and as a result, the principle of energy conservation was preserved. The same method may be considered for the basis field functions, to define the corresponding bi-orthogonal basis. For that, we first describe the $N = 4$ case.

All matrices of the field functions can be presented in the form of 1-D vectors, and then written into the matrix 16×16. To accomplish these calculations, we can run the code as R=A_rays2R(4), with the following script of the function.

```
function R=A_rays2R(N)
    NN=N*N; R=zeros(NN); k=1;
    for n=1:N
```

```
        for m=1:N
            Anm=Arays_nm(N,m,n);
            y=reshape(Anm,1,NN); R(k,:)=y; k=k+1;
        end
    end
```

As a result, we obtain the following matrix:

$$
R = \begin{bmatrix}
6 & 1 & 1 & 1 & 1 & 1 & 0 & 0 & 1 & 0 & 1 & 0 & 1 & 0 & 0 & 1 \\
1 & 6 & 1 & 1 & 1 & 1 & 1 & 0 & 1 & 1 & 0 & 1 & 0 & 1 & 0 & 0 \\
1 & 1 & 6 & 1 & 1 & 1 & 1 & 1 & 1 & 1 & 1 & 0 & 0 & 0 & 1 & 0 \\
1 & 1 & 1 & 6 & 0 & 1 & 1 & 1 & 0 & 1 & 1 & 1 & 1 & 0 & 0 & 1 \\
1 & 1 & 1 & 0 & 6 & 1 & 1 & 1 & 1 & 1 & 0 & 0 & 1 & 0 & 1 & 0 \\
1 & 1 & 1 & 1 & 1 & 6 & 1 & 1 & 1 & 1 & 0 & 1 & 1 & 0 & 1 \\
0 & 1 & 1 & 1 & 1 & 1 & 6 & 1 & 1 & 1 & 1 & 1 & 1 & 1 & 1 & 0 \\
0 & 0 & 1 & 1 & 1 & 1 & 1 & 6 & 0 & 1 & 1 & 1 & 0 & 1 & 1 & 1 \\
1 & 1 & 1 & 0 & 1 & 1 & 1 & 0 & 6 & 1 & 1 & 1 & 1 & 1 & 0 & 0 \\
0 & 1 & 1 & 1 & 1 & 1 & 1 & 1 & 1 & 6 & 1 & 1 & 1 & 1 & 1 & 0 \\
1 & 0 & 1 & 1 & 0 & 1 & 1 & 1 & 1 & 1 & 6 & 1 & 1 & 1 & 1 & 1 \\
0 & 1 & 0 & 1 & 0 & 0 & 1 & 1 & 1 & 1 & 1 & 6 & 0 & 1 & 1 & 1 \\
1 & 0 & 0 & 1 & 1 & 1 & 1 & 0 & 1 & 1 & 1 & 0 & 6 & 1 & 1 & 1 \\
0 & 1 & 0 & 0 & 0 & 1 & 1 & 1 & 1 & 1 & 1 & 1 & 1 & 6 & 1 & 1 \\
0 & 0 & 1 & 0 & 1 & 0 & 1 & 1 & 0 & 1 & 1 & 1 & 1 & 1 & 6 & 1 \\
1 & 0 & 0 & 1 & 0 & 1 & 0 & 1 & 0 & 0 & 1 & 1 & 1 & 1 & 1 & 6
\end{bmatrix}.
\qquad (6.13)
$$

This matrix is symmetric, $R = R'$, and its inverse matrix exists, although the determinant is the large number, $\det(R) = 4939060313245$. To calculate the inverse matrix R^{-1} and its 16 basis functions, we can use the code with script show_matrixIR.m, which is given below.

```
% call: show_matrixIR.m
    N=4;
    R=A_rays2R(N);
    fprintf('%16.0f \n',det(R));
    % inverse matrix 16x16
    Ri=inv(R);
    % represent this matrix as the block-matrix 4x4
    NN=N*N;
    iR=zeros(N,N,NN);
    for k=1:NN
        y=reshape(Ri(:,k),N,N); iR(:,:,k)=y;
    end
    % the 1st basis functions of transforms R and Ri
    k=1;
    field_function=reshape(R(:,k),N,N);    % at (0,0)
    cof_function=iR(:,:,k)
    y=field_function.*cof_function
```

```
      sum(sum(y))
% ========= calculated data ==========================
   % det(R)) = 4939060313245
   % field_function
      6      1      1      1
      1      1      0      0
      1      0      1      0
      1      0      0      1
   % cof_function
   0.2001     -0.0262     -0.0210     -0.0209
  -0.0262     -0.0120      0.0193      0.0110
  -0.0210      0.0193     -0.0259      0.0086
  -0.0209      0.0110      0.0086     -0.0266
   % y
   1.2007     -0.0262     -0.0210     -0.0209
  -0.0262     -0.0120           0           0
  -0.0210           0     -0.0259           0
  -0.0209           0           0     -0.0266
   % sum(sum(y))=1
% ========================================================
```

The following three matrices are printed when running this code:

$$\psi_{0,0} = \begin{bmatrix} 6 & 1 & 1 & 1 \\ 1 & 1 & 0 & 0 \\ 1 & 0 & 1 & 0 \\ 1 & 0 & 0 & 1 \end{bmatrix}, \quad \phi_{0,0} = \begin{bmatrix} 0.2001 & -0.0262 & -0.0210 & -0.0209 \\ -0.0262 & -0.0120 & 0.0193 & 0.0110 \\ -0.0210 & 0.0193 & -0.0259 & 0.0086 \\ -0.0209 & 0.0110 & 0.0086 & -0.0266 \end{bmatrix},$$

and

$$\psi_{0,0} .\times \phi_{0,0} = \begin{bmatrix} 1.2007 & -0.0262 & -0.0210 & -0.0209 \\ -0.0262 & -0.0120 & 0 & 0 \\ -0.0210 & 0 & -0.0259 & 0 \\ -0.0209 & 0 & 0 & -0.0266 \end{bmatrix}.$$

The first matrix describes the field function of the A-particle $(0,0)$; the second matrix is the first basis function of the inverse transform R^{-1}. This function is shown in matrix form. One can notice that in this matrix the maximum value has the coefficient at the same point $(0,0)$. This matrix probably describes a particle $(0,0)$ in another field. The third matrix is the point-wise multiplication of these two functions. The sum of all coefficients of the last matrix equals 1. The similar properties are valid for all other particles (n,m).

For instance, by using the same code for $k = 11$, we obtain the following matrices for the A-particle $(2,2)$:

$$\psi_{2,2} = \begin{bmatrix} 1 & 0 & 1 & 1 \\ 0 & 1 & 1 & 1 \\ 1 & 1 & 6 & 1 \\ 1 & 1 & 1 & 1 \end{bmatrix}, \quad \phi_{2,2} = \begin{bmatrix} -0.0259 & 0.0217 & -0.0186 & -0.0130 \\ 0.0217 & -0.0142 & -0.0148 & -0.0144 \\ -0.0186 & -0.0148 & 0.2014 & -0.0173 \\ -0.0130 & -0.0144 & -0.0173 & -0.0120 \end{bmatrix},$$

$$\psi_{2,2} \times \phi_{2,2} = \begin{bmatrix} -0.0259 & 0 & -0.0186 & -0.0130 \\ 0 & -0.0142 & -0.0148 & -0.0144 \\ -0.0186 & -0.0148 & \mathbf{1.2084} & -0.0173 \\ -0.0130 & -0.0144 & -0.0173 & -0.0120 \end{bmatrix},$$

and

$$\sum_{n=0}^{3} \sum_{m=0}^{3} \psi_{2,2}(n,m)\phi_{2,2}(n,m) = 1.$$

Thus, the discrete model of image reconstruction by A-rays can be solved by means of field functions of particles. It is assumed that each A-particle (n, m) which represents the corresponding square $[n, m]$ on the lattice, contributes equally to all rays passing through the center of this square. Here, we also assume that the normalization of the sums by the lengths of the rays inside the square has been considered.

We now consider the case when the particle does not contribute equally to different projections.

Example 6.2 For the $N = 4$ case, we consider the A-particle $(0, 1)$, its field function

$$\psi_{0,1} = \begin{bmatrix} 1 & 1 & 1 & 0 \\ \mathbf{6} & 1 & 1 & 1 \\ 1 & 1 & 0 & 0 \\ 1 & 0 & 1 & 0 \end{bmatrix}, \quad \text{and} \quad NP = \begin{bmatrix} 1 & 2 & 3 & 0 \\ \bullet & 5 & 5 & 5 \\ 1 & 6 & 0 & 0 \\ 1 & 0 & 6 & 0 \end{bmatrix}.$$

Here, the second matrix shows the way we number A-rays passing through the point $(0, 1)$. It is clock-wise numbering, starting from the horizontal $(0, 1)$-projection and considering the last projection to be the $(-1, 1)$-projection, which is the $(3, 1)$-projection. The $(2, 1)$-projection is the single-point $(0, 1)$ and has number 4, but is not shown. We denote the linear sums along these rays by $v_1, v_2, ..., v_6$. Therefore, the sums of all six projections can be written as

$$Y = \begin{bmatrix} v_1 & v_2 & v_3 & 0 \\ \displaystyle\sum_{k=1}^{6} v_k & v_5 & v_5 & v_5 \\ v_1 & v_6 & 0 & 0 \\ v_1 & 0 & v_6 & 0 \end{bmatrix}.$$

This matrix can be written column-wise into the raw-vector

$$Y \to y = \left(v_1, \sum_{k=1}^{6} v_k, v_1, v_1, v_2, v_5, v_6, 0, v_3, v_5, 0, v_6, 0, v_5, 0, 0 \right),$$

and then the matrix equation $y' = Rx'$ can be solved with respect to the original image-vector $x' = (x_{0,0}, x_{1,0}, x_{2,0}, x_{3,0}, x_{1,0}, ..., x_{3,2}, x_{3,3})'$.

It is expected, that by using this direct method, the reconstructed image will be smooth. To analyze this reconstruction, we first write the program to accomplish the reconstruction by the given equations. Below is the script `projection4.m` of this program. Here, `V` stands for the vector $V = (v_1, v_2, ..., v_6)$.

```
% --------------------------------------------------------
% call: projection4.m
function X=projection4(V)
    if (length(V)~=6) return; end
    y=zeros(1,16);
    y(1)=V(1); y(3)=V(1);  y(4)=V(1);  y(2)=sum(V);
    y(5)=V(2); y(6)=V(5);  y(10)=V(5); y(14)=V(5);
    y(7)=V(6); y(12)=V(6); y(9)=V(3);
    R=A_rays2R(4); x=inv(R)*y'; X=reshape(x',4,4);
% --------------------------------------------------------
```

If we consider the vector y with equal components, which corresponds to the case when all sums along A-rays passing through the point $(0, 1)$ are equal (the case of the discrete model), then we obtain the exact reconstruction. Indeed, the commands `V=ones(1,6); X=projection4(V)` result in the following image:

$$X = \begin{bmatrix} 0 & 0 & 0 & 0 \\ \underline{1} & 0 & 0 & 0 \\ 0 & 0 & 0 & 0 \\ 0 & 0 & 0 & 0 \end{bmatrix}.$$

We now consider the modified projection data

$$Y = \begin{bmatrix} 1 & 1 & 1 & 0 \\ \underline{6.2} & 1 & 1 & 1 \\ 1 & 1.2 & 0 & 0 \\ 1 & 0 & 1.2 & 0 \end{bmatrix}.$$

Here we changed the $(-1, 1)$-projection along the ray passing through the squares $[0, 1]$, $[1, 2]$, and $[2, 3]$. For this case, the vector $V = (1, 1, 1, 1, 1, 1.2)$ and, therefore, the vector y equals $(1, 6.2, 1, 1, 1, 1, 1.2, 0, 1, 1, 0, 1.2, 0, 1, 0, 0)$. The commands `V1=[1 1 1 1 1 1.2]; X1=projection4(V1)` result in the following image:

$$X_1 = \begin{bmatrix} 0.0003 & -0.0064 & -0.0117 & 0.0047 \\ \underline{1.0317} & -0.0021 & -0.0070 & -0.0117 \\ -0.0021 & 0.0330 & -0.0021 & -0.0064 \\ -0.0144 & -0.0021 & 0.0317 & 0.0003 \end{bmatrix}$$

$$= \begin{bmatrix} 0 & 0 & 0 & 0 \\ \underline{1} & 0 & 0 & 0 \\ 0 & 0 & 0 & 0 \\ 0 & 0 & 0 & 0 \end{bmatrix} + \begin{bmatrix} 0.0003 & -0.0064 & -0.0117 & 0.0047 \\ 0.0317 & -0.0021 & -0.0070 & -0.0117 \\ -0.0021 & 0.0330 & -0.0021 & -0.0064 \\ -0.0144 & -0.0021 & 0.0317 & 0.0003 \end{bmatrix}.$$

The particle $(0, 1)$ is not reconstructed exactly, as are not the remaining particles on the lattice, although they have small values (and can be thresholded).

The square error of this change equals 0.044. To obtain the full picture of image reconstruction, it is necessary to analyze other particles as well, and we will leave details of such analysis to the reader.

In general, the small squares, or image elements, contribute to the line-integrals of projections with different weights, which are defined by different characteristics, such as the lengths of intersections of squares with rays. Therefore we define the concept of a particle with the geometrical field function, which generalizes the concept of the A-particle.

6.2 Method of G-rays

In this section, we describe the process of summation and distribution of projection data between the points of the image that lie on the same rays of the projection. This method will be applied for reconstructing the image $f(x, y)$ which is represented by image elements in the Cartesian lattice $N \times N$ on the square $[0, 1] \times [0, 1]$.

For that, we consider the representation of each G-ray of the (p, s)-projection by the corresponding set of coordinates (n, m) of image elements, or small squares through which the ray passes. Our goal is to determine, for each small square with coordinate (n, m), the corresponding set of G-rays $l_{p,s}(t)$ that pass through this square. This set of rays will be used in the method of averaging the projection data. Given the (p, s)-projection, we denote by $ta(p, s; n, m)$ the number t, which shows the A-ray $l_{p,s}(t)$ passing through the center of the (n, m)-th square. To calculate the coordinates of these points on the lattice, we will use the function with script ps_Arays_points.m which is given in Chapter 4. It should be noted that in this function, the numbering of coordinates of points is performed from $(0, 0)$ to $(N - 1, N - 1)$. These sets of points will be used to determine the set of G-rays $l_{p,s}(t)$ passing through the (n, m)-th image element in the square $[0, 1] \times [0, 1]$, and determine the corresponding ray-sets, or set of image elements passed through by these rays.

6.2.1 G-rays for the first set of generators

We start with the $N = 8$ case and the generator $(p, s) = (1, 1)$. A-rays of the $(1, 1)$-projection are represented by the following pairs of coordinates (n, m)

on the lattice 8×8:

$$
\left\{
\begin{array}{l}
l_{1,1}(0) \\
l_{1,1}(1) \\
l_{1,1}(2) \\
l_{1,1}(3) \\
l_{1,1}(4) \\
l_{1,1}(5) \\
\overline{l_{1,1}(6)} \\
l_{1,1}(7) \\
l_{1,1}(8) \\
l_{1,1}(9) \\
\underline{l_{1,1}(10)} \\
\overline{l_{1,1}(11)} \\
l_{1,1}(12) \\
l_{1,1}(13) \\
l_{1,1}(14)
\end{array}
\right\}
\rightarrow
\left\{
\begin{array}{l}
(0,0) \\
(1,0)\ (0,1) \\
(2,0)\ (1,1)\ (0,2) \\
(3,0)\ (2,1)\ (1,2)\ (0,3) \\
(4,0)\ (3,1)\ (2,2)\ (1,3)\ (0,4) \\
(5,0)\ (4,1)\ (3,2)\ (2,3)\ (1,4)\ \underline{(0,5)} \\
(6,0)\ (5,1)\ (4,2)\ (3,3)\ (2,4)\ \overline{(1,5)}\ (0,6) \\
(7,0)\ (6,1)\ (5,2)\ (4,3)\ (3,4)\ (2,5)\ (1,6)\ (0,7) \\
(7,1)\ (6,2)\ (5,3)\ (4,4)\ (3,5)\ (2,6)\ (1,7) \\
(7,2)\ (6,3)\ (5,4)\ (4,5)\ (3,6)\ (2,7) \\
(7,3)\ (6,4)\ (5,5)\ \underline{(4,6)}\ (3,7) \\
(7,4)\ (6,5)\ (5,6)\ \overline{(4,7)} \\
(7,5)\ (6,6)\ (5,7) \\
(7,6)\ (6,7) \\
(7,7)
\end{array}
\right\}
$$

It is not difficult to see that $ta(1,1;n,m) = n+m$, $n,m = 0:7$. The rays are numbered from 0 to $L(p,s) = (p+s)(N-1) = 14$. For instance, the point $(0,5)$ lies on the A-ray with number $ta(1,1;0,5) = 5$, and the $(4,6)$-square lies on the A-ray with number $ta(1,1;4,6) = 10$. For this diagonal projection, each G-ray $l_{1,1}(t)$ passes through the squares whose coordinates coincide with the coordinates of the points that determine the A-ray $l_{1,1}(t)$. Thus, the G-ray passing through the (n,m)-th square has the number $t = n+m$ and it is described by the set of squares with the following coordinates:

$$l_{1,1}(t) = \{(n_0, m_0), (n_0 - 1, m_0 + 1), ..., (n,m), ..., (m_0 - 1, n_0 + 1), (m_0, n_0)\}$$

where $n_0 = t$ and $m_0 = 0$, if $t \leq 7$, and $n_0 = 7$ and $m_0 = t - 7$, if $t > 7$. Each square is intersected by a single G-ray. For example, if $(n,m) = (4,6)$, then $t = 4 + 6 = 10 = 7 + 3$, and the first square (from the left), which is intersected by the G-ray $l_{1,1}(10)$ is the $(7,3)$-th square.

We now consider the $(1,2)$-projection, i.e., the $(p,s) = (1,2)$ case. From the full set of A-rays of the $(1,2)$-projection, which is given in §4.8, we consider a few A-rays and their representation by the coordinates of the points on the lattice,

$$
\left\{
\begin{array}{l}
l_{1,2}(8) \\
l_{1,2}(9) \\
\overline{l_{1,2}(10)} \\
l_{1,2}(11)
\end{array}
\right\}
\rightarrow
\left\{
\begin{array}{l}
(6,1)\ (4,2)\ (2,3)\ (0,4) \\
(7,1)\ (5,2)\ \underline{(3,3)}\ (1,4) \\
(6,2)\ (4,3)\ \overline{(2,4)}\ (0,5) \\
(7,2)\ (5,3)\ (3,4)\ (1,5)
\end{array}
\right\}
$$

The A-ray passing through the point (n,m) on the lattice has the number calculated by

$$ta(1,2;n,m) = n + 2m, \quad n,m = 0:7.$$

For instance, the point $(3,3)$ lies on the A-ray $l_{1,2}(t)$ with number $t =$

$ta(1, 2; 3, 3) = 3 + 2(3) = 9$, and the point $(4, 3)$ lies on the A-ray $l_{1,2}(10)$, because $ta(1, 2; 4, 3) = 4 + 2(3) = 10$. Figure 6.5 shows the G-rays with numbers 9 and 10, as well as the filled $(3, 3)$-th square. These two rays are determined

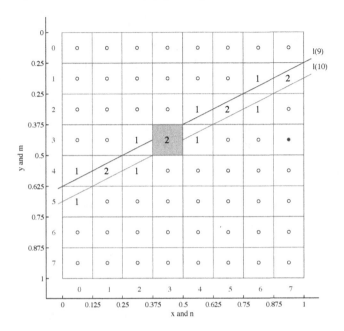

FIGURE 6.5
G-rays passing through the $(3, 3)$-th square in the $(1, 2)$-projection (case A).

by the squares with the following numbers (written from left to right):

$$l_{1,2}(9) = \{[0, 4], \underline{[1, 4]}, [2, 3], \underline{[3, 3]}, [4, 2], \underline{[5, 2]}, [6, 1], \underline{[7, 1]}\},$$
$$l_{1,2}(10) = \{[0, 5], [1, 4], [2, 4], [3, 3], [4, 3], [5, 2], [6, 2], [7, 1]\}.$$

Here, and later on, we use the square brackets in equations to distinguish the difference between the numbering of points and squares; the (n, m)-th square is denoted by $[n, m]$. We can write these two ray-sets as

$$l_{1,2}(9) = \{[n, m]; \ n + 2m \in \{8, 9\}\},$$
$$l_{1,2}(10) = \{[n, m]; \ n + 2m \in \{9, 10\}\}.$$

These rays pass together through the squares $[1, 4], [3, 3], [5, 2]$, and $[7, 1]$. In other words, these four squares are intersected twice by these G-rays. The remaining squares from these ray-sets are intersected only once, as shown in the figure. These coefficients of the intersections $c(n, m)$ can be calculated by

$$c(n, m) = 2 - |(n + 2m) - 9| = 2 - |(n - 3) + 2(m - 3)|,$$

considering only the squares of rays $l_{1,2}(9)$ and $l_{1,2}(10)$. Here 2 stands for s, and 9 is the number of the first G-rays passing through the square $[3, 3]$.

In general, the G-ray $l_{1,2}(t)$ can be defined by the following set of squares:

$$l_{1,2}(t) = \{[n, m]; \ n + 2m \in \{t - 1, t\}\}, \quad t = 0 : 21.$$

We consider the $(p, s) = (1, 3)$ case. The set of A-rays of the $(1, 3)$-projection contains 29 rays. We consider only the following three A-rays and their representations by points (n, m) on the lattice:

$$\begin{cases} l_{1,3}(12) \\ l_{1,3}(13) \\ l_{1,3}(14) \end{cases} \rightarrow \begin{cases} (6, 2)\,(3, 3)\,(0, 4) \\ (7, 2)\,(4, 3)\,(1, 4) \\ (5, 3)\,(2, 4) \end{cases}$$

The A-ray passing through the point (n, m) has the number which is calculated by $ta(1, 3; n, m) = n + 3m$, $n, m = 0 : 7$. To verify this formula, we consider the points $(3, 3)$ and $(7, 2)$. The first point is on the A-ray $l_{1,3}(t)$ with number

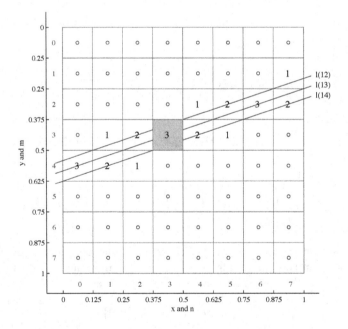

FIGURE 6.6
G-rays passing through the square $[3, 3]$ in the $(1, 3)$-projection (case A).

$ta(1, 3; 3, 3) = 3 + 3(3) = 12$. This is the number of the first G-ray passing through the $(3, 3)$-th square. Figure 6.6 shows the G-rays with numbers 12, 13, and 14, which pass through the $(3, 3)$-th square. The point $(7, 2)$ is on the A-ray $l_{1,3}(13)$, and $ta(1, 3; 7, 2) = 7 + 3(2) = 13$. This is the number of the first

G-ray passing through the $(7, 2)$-th square. This square is also on the G-ray $l_{1,3}(14)$. We now consider the square-representation of these three G-rays,

$$l_{1,3}(12) = \{[0, 4], [1, 3], [2, 3], [3, 3], [4, 2], [5, 2], [6, 2], [7, 1]\}$$
$$l_{1,3}(13) = \{[0, 4], [1, 4], [2, 3], [3, 3], [4, 3], [5, 2], [6, 2], [7, 2]\}$$
$$l_{1,3}(14) = \{[0, 4], [1, 4], [2, 4], [3, 3], [4, 3], [5, 3], [6, 2], [7, 2]\}$$

and for each of these rays, we have the following description:

$$l_{1,3}(t) = \{[n, m]; \ n + 3m \in \{t - 2, t - 1, t\}\}, \quad t = 12, 13, 14.$$

These three rays together pass through the squares $[0, 4], [3, 3]$, and $[6, 2]$. Other squares are intersected once or twice by the G-rays, as shown in the figure. These coefficients of the intersections $c(n, m)$ can be calculated by

$$c(n, m) = 3 - |(n + 3m) - 12| = 3 - |(n - 3) + 3(m - 3)|,$$

when considering only the three rays $l_{1,3}(12), l_{1,3}(13)$, and $l_{1,3}(14)$. Here, 3 stands for s, and 12 is the number of the first G-rays passing through the square $[3, 3]$. The square-set representation of all G-rays of this projection is described by

$$l_{1,3}(t) = \{[n, m]; \ n + 3m \in \{t - 2, t - 1, t\}\}, \quad t = 0 : 21.$$

We consider the next (p, s)-projection, when $(p, s) = (1, 4)$. G-rays of this projection can be described by the square-set similar to the cases $(p, s) = (1, 2)$ and $(1, 3)$. The unique A-ray $l_{1,4}(t)$ of this projection, which intersects the point (n, m) on the lattice, has the number $t = n + 4m$. Let us consider as an example, the point $(3, 3)$, which is on the A-ray with number 15. This A-ray passes through the point $(7, 2)$ on the lattice. Figure 6.7 shows four G-rays, starting with the ray $l_{1,4}(15)$, which is the first ray passing through this point.

It is not difficult to see from this figure that these rays can be described by the following squares:

$$l_{1,4}(15) = \{[0, 3], [1, 3], [2, 3], [3, 3], [4, 2], [5, 2], [6, 2], [7, 2]\}$$
$$l_{1,4}(16) = \{[0, 4], [1, 3], [2, 3], [3, 3], [4, 3], [5, 2], [6, 2], [7, 2]\}$$
$$l_{1,4}(17) = \{[0, 4], [1, 2], [2, 3], [3, 3], [4, 3], [5, 3], [6, 2], [7, 2]\}$$
$$l_{1,4}(18) = \{[0, 4], [1, 2], [2, 4], [3, 3], [4, 3], [5, 3], [6, 3], [7, 2]\}.$$

We also can write that

$$l_{1,4}(15) = \{[n, m]; \ n + 4m \in \{12, 13, 14, 15\}\}$$

and, in general,

$$l_{1,4}(t) = \{[n, m]; \ n + 4m \in \{t - 3, t - 2, t - 1, t\}\}.$$

These four rays together pass through the squares $[3, 3]$ and $[7, 2]$. Other

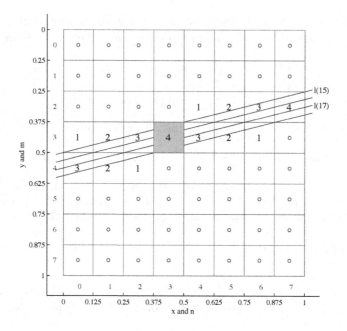

FIGURE 6.7
G-rays passing through the square $[3,3]$ in $(1,4)$-projection (case A).

squares are intersected once, twice, or three times by the G-rays, as shown in the figure. The coefficients of the intersections $c(n,m)$ for other squares can be calculated by

$$c(n,m) = 4 - |(n + 4m) - 15| = 4 - |(n - 3) + 4(m - 3)|,$$

when considering only these four rays. Here, 4 stands for s, and 15 is the number of the first G-rays passing through the square $[3,3]$.

Thus, for the G-rays of the $(1,s)$-projection, when $1 \leq s \leq 4$, the following holds:

$$
\begin{aligned}
l_{1,s}(t) &= \{[n,m]; \, n + sm \in \{t - s - 1, ..., t - 1, t\}\}, \quad t = 0 : 7(1 + s), \\
c(n,m) &= s - |(n - 3) + s(m - 3)|, \quad n, m = 0 : 7,
\end{aligned}
\tag{6.14}
$$

and $ta(1, s; n, m) = n + sm$.

6.2.2 G-rays for the second set of generators

We now consider the $(p, s) = (1, 5)$ case. In this case, $s = 5 > N/2 = 4$ and the $(1, -3)$-projection is used. We consider the A-ray of the projection that intersects the point $(3, 3)$ of the lattice. As shown in Chapter 4, this A-ray $l_{1,-3}(t)$ has the number calculated by $t = n + 3(7 - m) = 3 + 3(4) = 15$, when using the positive numbering, and the number $15 - 21 = -6$ in the original numbering of the rays. This A-ray also passes through the points $(0, 2)$ and $(6, 4)$ in the lattice, and we can write $l_{1,-3}(-6) = \{(0, 2), (3, 3), (6, 4)\}$.

Figure 6.8 shows three G-rays which pass through the $(3, 3)$-th square. The ray $l_{1,4}(-6)$ is considered the first ray that passes through this square. The other two G-rays are $l_{1,4}(-5)$ and $l_{1,4}(-4)$. In positive numbering, the ray $l_{1,4}(-5)$ is $l_{1,4}(16)$, and the ray $l_{1,4}(-4)$ is $l_{1,4}(17)$.

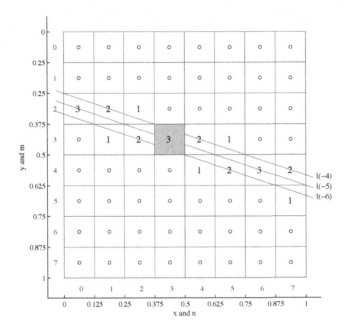

FIGURE 6.8
G-rays passing through the square $[3, 3]$ for the $(1, -3)$-projection (case A).

The following square-set representation holds for these rays:

$$l_{1,-3}(-4) = \{[0, 2], [1, 2], [2, 2], [3, 3], [4, 3], [5, 3], [6, 4], [7, 4]\}$$
$$l_{1,-3}(-5) = \{[0, 2], [1, 2], [2, 3], [3, 3], [4, 3], [5, 4], [6, 4], [7, 4]\}$$
$$l_{1,-3}(-6) = \{[0, 2], [1, 3], [2, 3], [3, 3], [4, 4], [5, 4], [6, 4], [7, 3]\}.$$

These rays can also be written as

$$l_{1,-3}(-4) = \{[n, m]; \ n + 4m \in \{15, 16, 17\}\}$$
$$l_{1,-3}(-5) = \{[n, m]; \ n + 4m \in \{14, 15, 16\}\}$$
$$l_{1,-3}(-6) = \{[n, m]; \ n + 4m \in \{13, 14, 15\}\}.$$

For positive numbering, when $t \to \bar{t} = t + 21$, we can use the following equation for these rays:

$$l_{1,-3}(t) = l_{1,-3}(\bar{t}) = \{[n, m]; \ n + 3(7 - m) \in \{\bar{t} - 2, \bar{t} - 1, \bar{t}\}\}.$$

These three rays together pass through the squares $[0,2],[3,3]$, and $[6,4]$. Other squares are intersected once or twice by these G-rays, as shown in the figure. The coefficients of the intersections $c(n,m)$ for other squares can be calculated by

$$c(n,m) = 3 - |(n+3(7-m))-15| = 3 - |(n-3)+3(7-(m-3))-21|,$$

when considering only these three rays. Here 3 stands for $-s$, and 15 is the number of the first G-rays passing through the square $[3,3]$.

We consider the next projection, the $(1,-2)$-projection, when $(p,s) = (1,6)$. Two G-rays of this projection, which pass through the square $[3,3]$, are shown in Figure 6.9. These rays have numbers -2 and -3, respectively, or 12

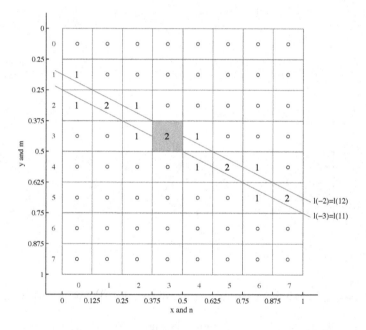

FIGURE 6.9
G-rays pass the square $[3,3]$ in the $(1,-2)$-projection (case A).

and 11, when numbering after the transform $t \to \bar{t} = t + 14$. The number 11 is calculated by $3 + 2(7-3)$, and the 11-th G-ray is considered to be the first ray intersecting the square $[3,3]$. The square set representation of these two rays is described by

$$l_{1,-2}(-2)=\{[0,1],[1,2],[2,2],[3,3],[4,3],[5,4],[6,4],[7,5]\}$$
$$l_{1,-2}(-3)=\{[0,2],[1,2],[2,3],[3,3],[4,4],[5,4],[6,5],[7,5]\}.$$

We can also write the following:

$$l_{1,-2}(-2)=\{[n,m];\ n+2(7-m)\in\{11,12\}\}$$
$$l_{1,-2}(-3)=\{[n,m];\ n+2(7-m)\in\{10,11\}\}.$$

In general, for positive numbering of the rays, we write these two rays as

$$l_{1,-2}(t) = l_{1,-2}(\bar{t}) = \{[n, m]; \; n + 2(7 - m) \in \{\bar{t} - 1, \bar{t}\}\}.$$

The rays together pass through the squares $[1, 2], [3, 3], [5, 4]$, and $[7, 5]$. Other squares are intersected once by these G-rays, as shown in the figure. The coefficients of the intersections $c(n, m)$ for other squares can be calculated by

$$c(n, m) = 2 - |(n + 2(7 - m) - 11| = 2 - |(n - 3) + 2(7 - (m - 3)) - 14|,$$

when considering only these two rays. Here, 2 stands for $-s$, and 11 is the number of the first G-rays passing through the square $[3, 3]$.

For the $(-1, 1)$-projection, the G-rays are described by the following squares:

$$\{l_{-1,1}(t)\} = \begin{cases} l(0) \\ l(1) \\ l(2) \\ l(3) \\ l(4) \\ l(5) \\ l(6) \\ l(7) \\ l(8) \\ l(9) \\ l(10) \\ l(11) \\ l(12) \\ l(13) \\ l(14) \end{cases} \rightarrow \begin{cases} [7, 0] \\ [6, 0] [7, 1] \\ [5, 0] [6, 1] [7, 2] \\ [4, 0] [5, 1] [6, 2] [7, 3] \\ [3, 0] [4, 1] [5, 2] [6, 3] [7, 4] \\ [2, 0] [3, 1] [4, 2] [5, 3] [6, 4] [7, 5] \\ [1, 0] [2, 1] [3, 2] [4, 3] [5, 4] [6, 5] [7, 6] \\ [0, 0] [1, 1] [2, 2] [3, 3] [4, 4] [5, 5] [6, 6] [7, 7] \\ [0, 1] [1, 2] [2, 3] [3, 4] [4, 5] [5, 6] [6, 7] \\ [0, 2] [1, 3] [2, 4] [3, 5] [4, 6] [5, 7] \\ [0, 3] [1, 4] [2, 5] [3, 6] [4, 7] \\ [0, 4] [1, 5] [2, 6] [3, 7] \\ [0, 5] [1, 6] [2, 7] \\ [0, 6] [1, 7] \\ [0, 7] \end{cases}$$

The number of the G-ray passing through the square $[n, m]$ is defined by the same formula as for the A-ray passing through the point (n, m) on the grid,

$$ta(7, 1; n, m) = ta(-1, 1; n, m) = (7 - n) + m, \quad n, m = 0 : 7.$$

For example, the square $[3, 3]$ is in the same direction as the G-ray number $(7 - 3) + 3 = 7$. In this diagonal projection, each square is intersected by only one G-ray.

6.2.3 G-rays for the third set of generators

For the $(2, 1)$-projection, two G-rays passing through the $(3, 3)$-th square are shown in Figure 6.10. The first G-ray has the number $3(2) + 3(1) = 9$. These rays are determined by the following squares:

$$l_{2,1}(9) = \{[1, 7], [1, 6], [2, 5], [2, 4], [3, 3], [3, 2], [4, 1], [4, 0]\}$$
$$l_{2,1}(10) = \{[1, 7], [2, 6], [2, 5], [3, 4], [3, 3], [4, 2], [4, 1], [5, 0]\}.$$

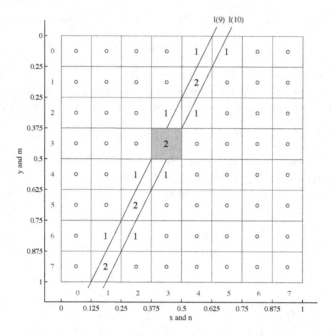

FIGURE 6.10
G-rays passing through the square $[3, 3]$ in the $(2, 1)$-projection (case A).

We can also write $l_{2,1}(9) = \{[n, m]; \; 2n + m \in \{8, 9\}\}$, and in general

$$l_{2,1}(t) = \{[n, m]; \; 2n + m \in \{t - 1, t\}\}.$$

The coefficients of the intersections $c(n, m)$ of squares along these two G-rays with number 8 and 9 can be calculated by

$$c(n, m) = 2 - |(n + 2m - 9)| = 2 - |2(n - 3) + (m - 3))|.$$

Here 2 stands for p, and 9 is the number of the first G-rays passing through the square $[3, 3]$.

The above given description of the G-rays corresponds to case A, when the A-rays are shifted by $t \to t - 0.5$. In our code, for the $(p, 1)$-th projections, when $p \leq N/2$, we use case B, when the A-rays are shifted by $t \to t + 0.5$. Therefore, the first ray passing through the $(3, 3)$-th square has number 8, and therefore, $l_{2,1}(8) = \{[n, m]; \; 2n + m \in \{8, 9\}\}$. In general for case B, the G-ray number t can be described by

$$l_{2,1}(t) = \{[n, m]; \; 2n + m \in \{t, t + 1\}\}.$$

For the $(4, 1)$-projection, the first of four G-rays intersecting the square $[3, 3]$ has the number $4(3) + 1(3) = 15$. In Figure 6.11, these four rays are

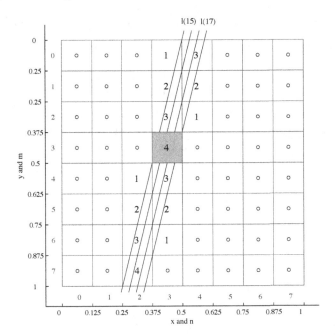

FIGURE 6.11
G-rays passing through $(3, 3)$ for the $(4, 1)$-projection (case A).

shown. The first G-ray can be written by squares as

$$l_{4,1}(15) = \{[2, 7], [2, 6], [2, 5], [2, 4], [3, 3], [3, 2], [3, 1], [3, 0]\}$$

or $l_{4,1}(15) = \{[n, m]; 4n + m \in \{12, 13, 14, 15\}\}$. In general, the G-rays are described similarly,

$$l_{4,1}(t) = \{[n, m]; 4n + m \in \{t - 3, t - 2, t - 1, t\}\}, \quad t = 15 : 18.$$

The coefficients of the intersections $c(n, m)$ of squares along these four G-rays can be calculated by

$$c(n, m) = 4 - |(4n + m - 15| = 4 - |4(n - 3) + (m - 3))|.$$

Here 4 stands for p, and 15 is the number of the first G-rays passing through the square $[3, 3]$.

The numbering of the G-rays in the figure corresponds to case A, when A-rays are shifted by $t \to t - 3/2$. In case B, the shift of A-rays is by $t \to t + 3/2$, and ray number 15 has number 12, and ray number 17 is 14. Therefore, we can write that

$$l_{4,1}(12) = \{[n, m]; 4n + m \in \{12, 13, 14, 15\}\}.$$

In general case B, the representation of the G-rays by squares is described by

$$l_{4,1}(t) = \{[n,m]; \; 4n + m \in \{t, t+1, t+2, t+3\}\}, \quad t = 0:35.$$

6.2.4 G-rays for the fourth set of generators

The last projection to be described is the $(6,1)$-projection, which is conside-red as the $(-2,1)$-projection. Two G-rays intersecting the square $[3,3]$ have numbers -3 and -2 and, are shown in Figure 6.12. The first G-ray passing

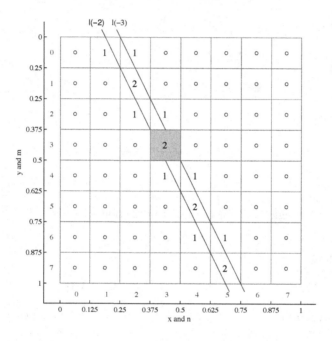

FIGURE 6.12
G-rays passing through the square $[3,3]$ in the $(-2,1)$-projection (case A).

through the square is the ray with number $-2(3) + 1(3) = -3$. The square representations of this ray and the next ray $l(-2)$ are the following:

$$l_{-2,1}(-3) = \{[2,0], [2,1], [3,2], [3,3], [4,4], [4,5], [5,6], [5,7]\}$$
$$l_{-2,1}(-2) = \{[1,0], [2,1], [2,2], [3,3], [3,4], [4,5], [4,6], [5,7]\}.$$

We also can write that

$$l_{-2,1}(-3) = \{[n,m]; \; 2(7-n) + m \in \{10,11\}\}$$
$$l_{-2,1}(-2) = \{[n,m]; \; 2(7-n) + m \in \{11,12\}\}.$$

For positive numbering, when $t \to \bar{t} + 14$, the rays with numbers -3 and -2 have numbers 11 and 12, respectively, and the above representations can be written as

$$l_{1,-2}(t) = l_{1,-2}(\bar{t}) = \{[n,m]; \ 2(7-n) + m \in \{\bar{t} - 1, \bar{t}\}\},$$

where $\bar{t} = 11$ and 12. In general, this representation holds for all G-rays of the $(-2, 1)$-projection. It is clear, that for the horizontal and vertical projections, i.e., for the $(1, 0)$- and $(0, 1)$-projections, each square $[n, m]$, including $[3, 3]$, is intersected by only one G-ray.

6.2.5 Map of projections for one square

Considering all 12 projections for the case $N = 8$, and the number of G-rays in each of these projections that pass through the square $[3, 3]$, we obtain the following matrix:

$$G(3,3) = \begin{bmatrix} 1 & 1 & 1 & 2 & 4 & 1 & 1 & 0 \\ 1 & 1 & 2 & 3 & 4 & 1 & 1 & 3 \\ 4 & 4 & 4 & 6 & 6 & 6 & 7 & 6 \\ 2 & 5 & 10 & \mathbf{26} & 10 & 5 & 2 & 1 \\ 7 & 6 & 6 & 6 & 4 & 4 & 4 & 2 \\ 1 & 1 & 4 & 3 & 2 & 1 & 1 & 3 \\ 1 & 1 & 4 & 2 & 1 & 1 & 1 & 0 \\ 0 & 2 & 4 & 1 & 0 & 2 & 0 & 1 \end{bmatrix}. \tag{6.15}$$

We call this matrix *the map of projections for square* $[3, 3]$. The coefficient $c_{n,m}$ of this matrix equals the number of G-rays passing through the squares $[n, m]$ and $[3, 3]$ at the same time. If there is no G-ray passing through the squares $[n, m]$ and $[3, 3]$ at the same time, then $c_{n,m} = 0$. There are 26 rays passing through the square $[3, 3]$, which all together pass through 46 squares.

The set descriptions of all 12 rays passing through the square $[3, 3]$ are given in Table 6.5. The coordinates of squares $[n, m]$, which together with the square $[3, 3]$ pass through by G-rays, are shown in the table.

It can be seen from the matrix in (6.15), that the coefficient $c(3, 3) = 26$ is in the center for the sub-matrix of size 7×7, and this sub-matrix is symmetric with respect to this center. It is interesting to note the following property of the frame-sets of this matrix. The k-th frame-set at center $(3, 3)$ is defined as the following set of points:

$$\Upsilon^k = \Upsilon^k(n_0, m_0) = \{(n, m); \ \max(|n - n_0|, |m - m_0|) = k\}, \quad k > 1.$$

The values of the map of projections $C(3, 3)$ on the first three frame-sets have

$$\begin{bmatrix} 4 & 6 & 6 \\ 10 & & 10 \\ 6 & 6 & 4 \end{bmatrix}, \quad \begin{bmatrix} 1 & 2 & 3 & 4 & 1 \\ 4 & & & & 6 \\ 5 & & & & 5 \\ 6 & & & & 4 \\ 1 & 4 & 3 & 2 & 1 \end{bmatrix}, \quad \begin{bmatrix} 1 & 1 & 1 & 2 & 4 & 1 & 1 \\ 1 & & & & & & 1 \\ 4 & & & & & & 7 \\ 2 & & & \cdot & & & 2 \\ 7 & & & & & & 3 \\ 1 & & & & & & 1 \\ 1 & 1 & 4 & 2 & 1 & 1 & 1 \end{bmatrix},$$

TABLE 6.5

Coordinates of the map of projection for square $[3,3]$

(p,s)	squares $[n,m]$								card	$g(p,s)$
$(1,0)$	$3,0]$	$[3,1]$	$[3,2]$	$[3,3]$	$[3,4]$	$[3,5]$	$[3,6]$	$[3,7]$	8	1
$(1,1)$	$6,0]$	$[5,1]$	$[4,2]$	$[3,3]$	$[2,4]$	$[1,5]$	$[0,6]$		7	1
$(1,2)$	$0,4]$	$[1,4]$	$[2,3]$	$[3,3]$	$[4,2]$	$[5,2]$	$[6,1]$	$[7,1]$	8	1
	$0,5]$	$[1,4]$	$[2,4]$	$[3,3]$	$[4,3]$	$[5,2]$	$[6,2]$	$[7,1]$	8	1
$(1,3)$	$0,4]$	$[1,3]$	$[2,3]$	$[3,3]$	$[4,2]$	$[5,2]$	$[6,2]$	$[7,1]$	8	1
	$0,4]$	$[1,4]$	$[2,3]$	$[3,3]$	$[4,3]$	$[5,2]$	$[6,2]$	$[7,2]$	8	1
	$0,4]$	$[1,4]$	$[2,4]$	$[3,3]$	$[4,3]$	$[5,3]$	$[6,2]$	$[7,2]$	8	1
$(1,4)$	$0,3]$	$[1,3]$	$[2,3]$	$[3,3]$	$[4,2]$	$[5,2]$	$[6,2]$	$[7,2]$	8	1
	$0,4]$	$[1,3]$	$[2,3]$	$[3,3]$	$[4,3]$	$[5,2]$	$[6,2]$	$[7,2]$	8	1
	$0,4]$	$[1,2]$	$[2,3]$	$[3,3]$	$[4,3]$	$[5,3]$	$[6,2]$	$[7,2]$	8	1
	$0,4]$	$[1,2]$	$[2,4]$	$[3,3]$	$[4,3]$	$[5,3]$	$[6,3]$	$[7,2]$	8	1
$(1,-3)$	$0,2]$	$[1,2]$	$[2,2]$	$[3,3]$	$[4,3]$	$[5,3]$	$[6,4]$	$[7,4]$	8	1
	$0,2]$	$[1,2]$	$[2,3]$	$[3,3]$	$[4,3]$	$[5,4]$	$[6,4]$	$[7,4]$	8	1
	$0,2]$	$[1,3]$	$[2,3]$	$[3,3]$	$[4,4]$	$[5,4]$	$[6,4]$	$[7,3]$	8	1
$(1,-2)$	$0,1]$	$[1,2]$	$[2,2]$	$[3,3]$	$[4,3]$	$[5,4]$	$[6,4]$	$[7,5]$	8	1
	$0,2]$	$[1,2]$	$[2,3]$	$[3,3]$	$[4,4]$	$[5,4]$	$[6,5]$	$[7,5]$	8	1
$(1,-1)$	$7,7]$	$[6,6]$	$[5,5]$	$[4,4]$	$[3,3]$	$[2,2]$	$[1,1]$	$[0,0]$	8	1
$(0,1)$	$0,3]$	$[1,3]$	$[2,3]$	$[3,3]$	$[4,3]$	$[5,3]$	$[6,3]$	$[7,3]$	8	1
$(2,1)$	$1,7]$	$[1,6]$	$[2,5]$	$[2,4]$	$[3,3]$	$[3,2]$	$[4,1]$	$[4,0]$	8	1
	$(1,7]$	$[2,6]$	$[2,5]$	$[3,4]$	$[3,3]$	$[4,2]$	$[4,1]$	$[5,0]$	8	1
$(4,1)$	$2,7]$	$[2,6]$	$[2,5]$	$[2,4]$	$[3,3]$	$[3,2]$	$[3,1]$	$[3,0]$	8	1
	$2,7]$	$[2,6]$	$[2,5]$	$[3,4]$	$[3,3]$	$[3,2]$	$[3,1]$	$[4,0]$	8	1
	$2,7]$	$[2,6]$	$[3,5]$	$[3,4]$	$[3,3]$	$[3,2]$	$[4,1]$	$[4,0]$	8	1
	$2,7]$	$[3,6]$	$[3,5]$	$[3,4]$	$[3,3]$	$[4,2]$	$[4,1]$	$[4,0]$	8	1
$(-2,1)$	$2,0]$	$[2,1]$	$[3,2]$	$[3,3]$	$[4,4]$	$[4,5]$	$[5,6]$	$[5,7]$	8	1
	$[1,0]$	$[2,1]$	$[2,2]$	$[3,3]$	$[3,4]$	$[4,5]$	$[4,6]$	$[5,7]$	8	1

the same sum of values on the frame-sets: $2 \cdot 26 = 52$.

The total number of intersections of G-rays with the $(3,3)$-th square is 26, and each of these G-rays intersects this square twice. The next frame-set has

$$
\Upsilon^4 = \begin{bmatrix}
1 & 0 & 2 & 0 & 1 & 4 & 2 & 0 & 1 \\
0 & & & & & & & & 0 \\
3 & & & & & & & & 3 \\
2 & & & & & & & & 6 \\
1 & & & & & & & & 1 \\
6 & & & & & & & & 2 \\
3 & & & & & & & & 3 \\
0 & & & & & & & & 0 \\
1 & 0 & 2 & 4 & 1 & 0 & 2 & 0 & 1
\end{bmatrix}
$$

the same sum of values $2 \cdot 26 = 52$. One also can notice that all frame-sets are the same for two maps of projections $C(3,3)$ and $C(4,4)$.

We now extend the map $C(3,3)$ to the square matrix 9×9 with the 4-th frame,

$$U_{9,9}(26) = \begin{bmatrix} 1 & 0 & 2 & 0 & 1 & 4 & 2 & 0 & 1 \\ 0 & 1 & 1 & 1 & 2 & 4 & 1 & 1 & 0 \\ 3 & 1 & 1 & 2 & 3 & 4 & 1 & 1 & 3 \\ 2 & 4 & 4 & 4 & 6 & 6 & 6 & 7 & 6 \\ 1 & 2 & 5 & 10 & \mathbf{26} & 10 & 5 & 2 & 1 \\ 6 & 7 & 6 & 6 & 6 & 4 & 4 & 4 & 2 \\ 3 & 1 & 1 & 4 & 3 & 2 & 1 & 1 & 3 \\ 0 & 1 & 1 & 4 & 2 & 1 & 1 & 1 & 0 \\ 1 & 0 & 2 & 4 & 1 & 0 & 2 & 0 & 1 \end{bmatrix}. \tag{6.16}$$

We call this matrix *the UH-square* 9×9 of the number 26. In our study, we also use the UH-square 15×15 of the number 26, which equals

$$U_{15,15}(26) = \begin{bmatrix} 1 & 0 & 0 & 1 & 1 & 0 & 0 & 1 & 1 & 3 & 1 & 1 & 0 & 0 & 1 \\ 0 & 1 & 0 & 0 & 2 & 0 & 0 & 1 & 2 & 2 & 2 & 0 & 0 & 1 & 0 \\ 0 & 0 & 1 & 0 & 1 & 1 & 0 & 1 & 3 & 2 & 1 & 0 & 1 & 0 & 0 \\ 1 & 0 & 0 & 1 & 0 & 2 & 0 & 1 & 4 & 2 & 0 & 1 & 0 & 0 & 1 \\ 2 & 2 & 1 & 0 & 1 & 1 & 1 & 2 & 4 & 1 & 1 & 0 & 1 & 2 & 2 \\ 2 & 3 & 3 & 3 & 1 & 1 & 2 & 3 & 4 & 1 & 1 & 3 & 4 & 5 & 5 \\ 0 & 0 & 1 & 2 & 4 & 4 & 4 & 6 & 6 & 6 & 7 & 6 & 4 & 2 & 1 \\ 1 & 1 & 1 & 1 & 2 & 5 & 10 & \mathbf{26} & 10 & 5 & 2 & 1 & 1 & 1 & 1 \\ 1 & 2 & 4 & 6 & 7 & 6 & 6 & 6 & 4 & 4 & 4 & 2 & 1 & 0 & 0 \\ 5 & 5 & 4 & 3 & 1 & 1 & 4 & 3 & 2 & 1 & 1 & 3 & 3 & 3 & 2 \\ 2 & 2 & 1 & 0 & 1 & 1 & 4 & 2 & 1 & 1 & 1 & 0 & 1 & 2 & 2 \\ 1 & 0 & 0 & 1 & 0 & 2 & 4 & 1 & 0 & 2 & 0 & 1 & 0 & 0 & 1 \\ 0 & 0 & 1 & 0 & 1 & 2 & 3 & 1 & 0 & 1 & 1 & 0 & 1 & 0 & 0 \\ 0 & 1 & 0 & 0 & 2 & 2 & 2 & 1 & 0 & 0 & 2 & 0 & 0 & 1 & 0 \\ 1 & 0 & 0 & 1 & 1 & 3 & 1 & 1 & 0 & 0 & 1 & 1 & 0 & 0 & 1 \end{bmatrix}. \tag{6.17}$$

In general, *the UH-square* of the number M is such a square matrix, that all frame-sets of this matrix satisfy the following property: the sum of coefficients on the frame-set equals $2M$.

Thus, for the square $[3,3]$ we constructed the matrix $C(3,3)$ with coefficients $c(n,m) = c(3,3,n,m)$ defining the number of G-rays passing through the squares $[3,3]$ and $[n,m]$ at the same time. The coefficient $c(3,3)$ is considered to be the center of this matrix. In general, for a given square $[n_0, m_0]$), we can construct a similar matrix with the center in (n_0, m_0). For instance, for the square $[4,4]$, we obtain the following map of projections, or the matrix:

$$C(4,4) = \begin{bmatrix} 1 & 0 & 2 & 0 & 1 & 4 & 2 & 0 \\ 0 & 1 & 1 & 1 & 2 & 4 & 1 & 1 \\ 3 & 1 & 1 & 2 & 3 & 4 & 1 & 1 \\ 2 & 4 & 4 & 4 & 6 & 6 & 6 & 7 \\ 1 & 2 & 5 & 10 & \underline{26} & 10 & 5 & 2 \\ 6 & 7 & 6 & 6 & 6 & 4 & 4 & 4 \\ 3 & 1 & 1 & 4 & 3 & 2 & 1 & 1 \\ 0 & 1 & 1 & 4 & 2 & 1 & 1 & 1 \end{bmatrix} \tag{6.18}$$

with the coefficient $c(4, 4, 4, 4) = 26$ at the center. The sum of all coefficients in this matrix equals 207 as for the matrix $C(3, 3)$, and it is a sub-matrix of the UH-square 15×15.

Definition 6.2 *The geometrical particle, or G-particle (n_0, m_0) of dimension N is called the square $[n_0, m_0]$, which is described by the matrix $\|c(n_0, m_0, n, m)\|$ defining the number of G-rays passing through the squares $[n_0, m_0]$ and $[n, m]$ at the same time. We call the two-dimensional function $\phi_{n_0, m_0}(n, m) = c(n_0, m_0, n, m)$, where $n, m = 0 : (N-1)$, the field function of this particle.*

Hereinafter we consider the set of projections which are defined by the tensor representation of the discrete image. Therefore, G-particles are referred to as the tensor G-particles. For the same reason, the A-particles defined earlier are referred to as the tensor A-particles.

Example 6.3 In the $N = 4$ case, the field functions of the G-particles $(1, 1)$ and $(2, 1)$ are

$$\phi_{1,1} = \begin{bmatrix} 3 & 2 & 1 & 0 \\ 2 & \mathbf{8} & 2 & 1 \\ 1 & 2 & 3 & 2 \\ 0 & 1 & 2 & 1 \end{bmatrix} \quad \text{and} \quad \phi_{2,1} = \begin{bmatrix} 2 & 1 & 0 & 1 \\ 3 & 2 & 1 & 0 \\ 2 & \mathbf{8} & 2 & 1 \\ 1 & 2 & 3 & 2 \end{bmatrix}. \tag{6.19}$$

Figure 6.13 illustrates two G-particles $(1, 1)$ and $(2, 1)$ and their field functions. The number of G-rays passing through each G-particle equals 8. One

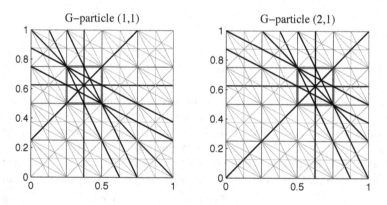

FIGURE 6.13 (See color insert)
The field functions of two G-particles $(1, 1)$ and $(2, 1)$.

can notice that the picture of the intersection of G-rays with each square $[n, m]$ is the same. It is the geometric representation, or symbol, of the G-particle. To have such a "uniform" set of rays, we add a few rays for each (p, s)-projection, when $(p, s) \neq (0, 1), (1, 0), (1, 1)$ and $(-1, 1)$.

TABLE 6.6

Field functions of *G*-particles

8	2	1	1	2	8	2	1	1	2	8	2	1	1	2	8
2	3	2	1	1	2	3	2	0	1	2	3	0	0	1	2
1	2	1	1	0	1	2	1	1	0	1	2	0	1	0	1
1	1	1	1	0	1	1	1	0	0	1	1	1	0	0	1
2	1	0	0	3	2	1	0	2	3	2	1	1	2	3	2
8	2	1	1	2	8	2	1	1	2	8	2	1	1	2	8
2	3	3	1	1	2	3	2	0	1	2	3	0	0	1	2
1	2	1	1	0	1	2	1	1	1	1	2	0	1	0	1
1	0	1	0	2	1	0	1	1	2	1	0	1	1	2	1
2	1	0	0	3	2	1	0	2	3	2	1	1	2	3	2
8	2	1	1	2	8	2	1	1	2	8	2	1	1	2	8
2	3	2	1	1	2	3	2	0	1	2	3	0	0	1	2
1	0	0	1	1	1	0	0	1	1	1	0	1	1	1	1
1	0	1	0	2	1	0	1	1	2	1	0	1	1	2	1
2	1	0	0	3	2	1	0	2	3	2	1	1	2	3	2
8	2	1	1	2	8	2	1	1	2	8	2	1	1	2	8

We can write the field functions of all 16 *G*-particles in the lattice 4×4 in the form of the block-table $[\psi_{n,m}]_{n,m=0:3}$, as shown in Table 6.6.

This table, or block-table 4×4 with all field functions can be packed into the following matrix of size 7×7, which we call *the base matrix of fields of G-particles*:

$$\Psi = \begin{bmatrix} 1 & 1 & 1 & 1 & 0 & 0 & 1 \\ 1 & 1 & 2 & 1 & 0 & 1 & 0 \\ 1 & 2 & 3 & 2 & 1 & 0 & 0 \\ 1 & 1 & 2 & \mathbf{8} & 2 & 1 & 1 \\ 0 & 0 & 1 & 2 & 3 & 2 & 1 \\ 0 & 1 & 0 & 1 & 2 & 1 & 1 \\ 1 & 0 & 0 & 1 & 1 & 1 & 1 \end{bmatrix}. \qquad (6.20)$$

Indeed, it is not difficult to notice that the following equality holds:

$$\psi_{n,m}(i,j) = \Psi(3-n+i, 3-m+j), \quad i,j = 0:3, \qquad (6.21)$$

for any *A*-particle (n,m). The base matrix is symmetric, i.e. $\Psi(i,j) = \Psi(j,i)$, for $i,j = 0:6$. One can notice that the base matrix is the UH-square 7×7 of number 8.

All 16 field functions of *G*-particles can be represented in the form of 1-D vectors of length 16 each, and then written into the matrix 16×16. The result of this operation is the following matrix:

$$R = \begin{bmatrix}
8 & 2 & 1 & 1 & 2 & 3 & 2 & 1 & 1 & 2 & 1 & 1 & 1 & 1 & 1 \\
2 & 8 & 2 & 1 & 1 & 2 & 3 & 2 & 0 & 1 & 3 & 1 & 0 & 1 & 1 \\
1 & 2 & 8 & 2 & 0 & 1 & 2 & 3 & 1 & 0 & 1 & 2 & 0 & 0 & 1 \\
1 & 1 & 2 & 8 & 0 & 0 & 1 & 2 & 0 & 1 & 0 & 1 & 1 & 0 & 0 \\
2 & 1 & 0 & 0 & 8 & 2 & 1 & 1 & 2 & 3 & 2 & 1 & 1 & 2 & 1 \\
3 & 2 & 1 & 0 & 2 & 8 & 2 & 1 & 1 & 2 & 3 & 2 & 0 & 1 & 2 \\
2 & 3 & 2 & 1 & 1 & 2 & 8 & 2 & 0 & 1 & 2 & 3 & 1 & 0 & 1 \\
1 & 2 & 3 & 2 & 1 & 1 & 2 & 8 & 0 & 0 & 1 & 2 & 0 & 1 & 0 \\
1 & 0 & 1 & 0 & 2 & 1 & 0 & 0 & 8 & 2 & 1 & 1 & 2 & 3 & 2 \\
2 & 1 & 0 & 1 & 3 & 2 & 1 & 1 & 2 & 8 & 2 & 1 & 1 & 2 & 3 \\
1 & 2 & 1 & 0 & 2 & 3 & 2 & 1 & 1 & 2 & 8 & 2 & 0 & 1 & 2 \\
1 & 1 & 2 & 1 & 1 & 2 & 3 & 2 & 1 & 1 & 2 & 8 & 0 & 0 & 1 \\
1 & 0 & 0 & 1 & 1 & 0 & 1 & 0 & 2 & 1 & 0 & 0 & 8 & 2 & 1 \\
1 & 1 & 0 & 0 & 2 & 1 & 0 & 1 & 3 & 2 & 1 & 0 & 2 & 8 & 2 \\
1 & 1 & 1 & 0 & 1 & 2 & 1 & 0 & 2 & 3 & 2 & 1 & 1 & 2 & 8
\end{bmatrix}. \tag{6.22}$$

This matrix is symmetric, $R = R'$, its inverse matrix exists, and the determinant is $\det(R) = 17023383606024$.

In general, the field functions of the G-particle can be calculated by means of the base field function as

$$\psi_{n,m}(i,j) = \Psi(N - 1 - n + i, N - 1 - m + j), \quad i,j = 0 : N - 1, \tag{6.23}$$

where $n, m \in \{0, 1, ..., N-1\}$. The size of the base matrix is $(2N-1) \times (2N-1)$ and the matrix is symmetric. The base field function for the $N = 8$ case is given in (6.17), which is the UH-square 15×15 of number 26. This number, as the central coefficient of the base field function, can be calculated by

$$\Psi(N - 1, N - 1) = 2 + \frac{3}{8}N^2, \quad N = 4, 8, 16, 32, ...$$

6.2.5.1   Codes for particles

We present a few codes that were used to calculate all matrices that were described above for the $N = 4$ and 8 examples. First, we consider the script UH_square.m of the program for calculating the UH-squares, for the cases when $N = 2^r$, $r > 1$. Here, the UH-squares are defined for the set of generators (p, s), which is used in the tensor and paired transforms. Therefore, we call such UH-squares *the tensor UH-squares*.

The UH-square has size $(2N - 1) \times (2N - 1)$; therefore, we put this square into the square $[-1, 1] \times [-1, 1]$ with the lattice $(2N - 1) \times (2N - 1)$. The central point of the UH-square is in the central square $[0, 0]$. The UH-square defines all G-rays passing through the square $[0, 0]$.

```
% call: UH_square.m          /          M. Grigoryan,   2011
% The tensor UH-square (2N-1)x(2N-1), for N=2,4,8,16, ...
function Q=UH_square(N)
  N2=2*N; N12=N/2;
  delta=2/(N2-1); delta2=delta/2;
  ps=ps_generators(N); L=size(ps,1);
  Q=zeros(N2-1);
  for i1=1:L
      p=ps(i1,1);   s=ps(i1,2);
      if p==1
            if (s>N12) s=-N+s; end
            t_min=-(abs(s)+1)/N2; k=1;
            if (s~=0) k=1:abs(s); end
      else
            if (p>N12) p=-N+p; end
            t_min=-(abs(p)+1)/N2;
            if (p==0) k=1; else k=1:abs(p); end
      end
      t=t_min+k/N;
      if (s==0)
          Q(:,N)=Q(:,N)+ones(N2-1,1);
      elseif (p==0)
          Q(N,:)=Q(N,:)+ones(1,N2-1);
      else
          if (s==1)
              for n1=1:length(k)
                  for xi=-1+delta2:delta:1
                      n=ceil(xi/delta-0.5)+N;
                      m=N-ceil((t(n1)-xi)/delta/p+0.5)+1;
                      Q(m,n)=Q(m,n)+1;
                  end
              end
          else
              for n1=1:length(k)
                  for yi=-1+delta2:delta:1
                      m=N-ceil(yi/delta+0.5)+1;
                      n=ceil((t(n1)-yi)/delta/s-0.5)+N;
                      Q(m,n)=Q(m,n)+1;
                  end
              end
          end
      end
  end
```

Figure 6.14 illustrates the UH-square 15×15 of number 26 and all G-rays passing through the central component of the square, which is the square $1/8 \times 1/8$. This central component is shown separately on the right. The UH-square is in the square $[-1,1] \times [-1,1]$. All field functions are defined from this UH-square by sliding the window of size 1×1.

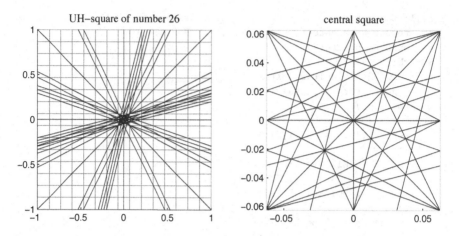

FIGURE 6.14 (See color insert)
Map of G-rays in the UH-square of number 26 and its central component.

Figure 6.15 illustrates the UH-square 7×7 of number 8 and its central component with all eight G-rays passing through this component.

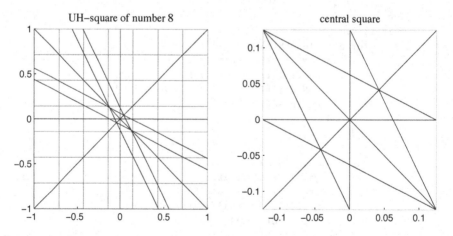

FIGURE 6.15
Map of G-rays in the UH-square of number 8 and its central component.

For reconstruction of the image $f(x, y)$ with high resolution, i.e. as the discrete image $f_{n,m}$, where $n, m = 0 : N_1 - 1, M_1 - 1$ and $N_1, M_1 > N$, it will be desired to reconstruct the values of the image at all or a few points of intersections of G-rays inside the central square, as well as other squares.

To calculate the block-table $[\psi_{n,m}]_{n,m=0:(N-1)}$ of the field functions of the G-particles in the lattice 4×4, we use the code `tableR_FFofGrays.m`. The script is given below for the $N = 4$ example. Here, the 2-D array UH8 stands for the base matrix.

```
% call: tableR_FFofGrays.m        /       A.-M. Grigoryan, 2011
  N=4;
  UH8=UH_square(N);    % Base matrix, UH-square (2N-1)x(2N-1)
  TR=zeros(N*N);       % For the table of all field functions
  R2=zeros(N);
  m=1;
  for i1=1:N
     n=1;
     for j1=1:N
        R2=UH8(N-j1+1:N-j1+N,N-i1+1:N-i1+N);
        TR(n:n+N-1,m:m+N-1)=R2;   n=n+4;
     end
     m=m+4;
  end
```

To calculate the matrix R in (6.22) from the base matrix Ψ, we use the code matrixR_FFofGrays.m with the following script.

```
% call: matrixR_FFofGrays.m       /       A.-M. Grigoryan, 2011
  N=4;
  UH8=UH_square(N);    % Base matrix, or the UH-square in (6.20)
  R=zeros(N*N);        % Array for the matrix R in (6.22)
  R2=zeros(N); k=1; NN=N*N;
  for i1=1:N
     for j1=1:N
        R2=UH8(N-j1+1:N-j1+N,N-i1+1:N-i1+N);
        R(k,:)=reshape(R2,1,NN);   k=k+1;
     end
  end
```

To calculate the inverse matrix R^{-1} and print all of its 16 basic functions, we use the code matrixR_FFofGrays.m, which is given below.

```
% ----------------------------------------------------------------
% call: matrixRin_FFofGrays        /       A.-M. Grigoryan, 2011
% ----------------------------------------------------------------
  matrixR_FFofGrays; % It calculates matrix R for the N=4 case
  fprintf('%16.0f \n',det(R)); % 17128697013225
% The inverse matrix NxN (or 16x16)
  Ri=inv(R);
% Represent this matrix as the block-matrix 4x4
  NN=N*N;
  iR=zeros(N,N,NN);
  for k=1:NN
     y=reshape(Ri(:,k),N,N); iR(:,:,k)=y;
  end
% The 1st basis functions of R and Ri
```

```
k=1;   % for G-particle (0,0)
field_function=reshape(R(:,k),N,N)
cof_function=iR(:,:,k)
sum(sum(field_function.*cof_function)) %1
% ========= calculated data ===============
% field_function #1
%  8  2  1  1
%  2  3  2  1
%  1  2  1  1
%  1  1  1  1
% cof_function
%  0.1636   -0.0192   -0.0049   -0.0107
% -0.0192   -0.0500   -0.0183   -0.0008
% -0.0049   -0.0183    0.0177    0.0056
% -0.0107   -0.0008    0.0056   -0.0056
% ==========================================
k=6;   % for G-particle (1,2)
field_function=reshape(R(:,k),N,N)
cof_function=iR(:,:,k)
sum(sum(field_function.*cof_function)) %1
% ========= calculated data ===============
% field_function #6
%  3  2  1  0
%  2  8  2  1
%  1  2  3  2
%  0  1  2  1
% cof_function
% -0.0500   -0.0120   -0.0008    0.0102
% -0.0120    0.1812   -0.0093   -0.0025
% -0.0008   -0.0093   -0.0459   -0.0219
%  0.0102   -0.0025   -0.0219    0.0177
% ==========================================
```

For the first G-particle $(0,0)$, we obtain the following matrices of the first basic functions of the transform, R and R^{-1}, respectively:

$$\psi_{0,0} = \begin{bmatrix} 8 & 2 & 1 & 1 \\ 2 & 3 & 2 & 1 \\ 1 & 2 & 1 & 1 \\ 1 & 1 & 1 & 1 \end{bmatrix}, \quad \phi_{0,0} = \begin{bmatrix} \mathbf{0.1636} & -0.0192 & -0.0049 & -0.0107 \\ -0.0192 & -0.0500 & -0.0183 & -0.0008 \\ -0.0049 & -0.0183 & 0.0177 & 0.0056 \\ -0.0107 & -0.0008 & 0.0056 & -0.0056 \end{bmatrix}.$$

The basic functions of the G-particle $(1,2)$ are also calculated separately in this code:

$$\psi_{1,2} = \begin{bmatrix} 3 & 2 & 1 & 0 \\ 2 & 8 & 2 & 1 \\ 1 & 2 & 3 & 2 \\ 0 & 1 & 2 & 1 \end{bmatrix}, \quad \phi_{1,2} = \begin{bmatrix} -0.0500 & -0.0120 & -0.0008 & 0.0102 \\ -0.0120 & \mathbf{0.1812} & -0.0093 & -0.0025 \\ -0.0008 & -0.0093 & -0.0459 & -0.0219 \\ 0.0102 & -0.0025 & -0.0219 & 0.0177 \end{bmatrix}.$$

One can notice that in matrices $\phi_{0,0}$ and $\phi_{1,2}$, the maximum values are at the points $(0, 0)$ and $(1, 2)$, respectively, i.e. at the exact locations of the considered G-particles.

6.3 Reconstruction by field transform

As shown above for A-rays and G-rays, we can define the systems of field functions that describe the image in terms of A- and G-particles, respectively. In both cases, the unique form of these fields allows for calculating the sums of fields of such particles

$$a(i, j) = \sum_{m=0}^{N-1} \sum_{n=0}^{N-1} \psi_{n,m}(i, j) f_{n,m}, \quad i, j = 0 : (N - 1). \tag{6.24}$$

Here, $\{f_{n,m}\}$ stands for the image to be reconstructed, and $\psi_{n,m}(i, j)$ are coefficients $c(i, j, n, m)$ defining the number of rays passing through the particle (i, j). The 4-D matrix $\|c(i, j, n, m)\|$ of size $(N \times N \times N \times N)$ can be represented in the form of a 2-D matrix of size $(N^2 \times N^2)$. We denote this 2-D matrix by $R = \|r_{k,l}\|$, as shown in (6.13) and (6.22) for the example when $N = 4$. We here remind the reader that the matrix R is composed of all N^2 field functions $(N \times N)$ written row-wise in the form of 1-D vectors,

$$r_{(i-1)N+j,(n-1)N+m} = c(i, j, n, m), \quad i, j, n, m = 0 : (N - 1).$$

The matrices of data $a(i, j)$ and $f_{n,m}$ can also be written row-wise

$$a_{i,j} \rightarrow a_{(i-1)N+j} = a_{i,j}, \quad f_{n,m} \rightarrow f_{(n-1)N+m} = f_{n,m},$$

in the form of 1-D vectors. Equation 6.24 can therefore be written as

$$a_k = \sum_{l=0}^{N^2-1} r_{k,l} f_l, \quad k = 0 : (N^2 - 1). \tag{6.25}$$

In matrix form, we obtain the equation $\mathbf{a} = R\mathbf{f}$, where $\mathbf{a} = (a_0, a_1, ..., a_{N^2-1})'$ and $\mathbf{f} = (f_0, f_1, ..., f_{N^2-1})'$. The existence of the inverse matrix, R^{-1}, allows for calculating the vector $\mathbf{f}$,

$$\mathbf{f} = R^{-1}\mathbf{a}. \tag{6.26}$$

Thus, the reconstruction of the image from its projections is achieved by using the sums of fields of particles. The reconstruction is by the G-rays, i.e. the matrix R and the solution in the above equation are considered with respect the systems of field functions of G-particles. The matrix R is referred to as the matrix of the field transformation of G-particles. Therefore, we call the described method *the method of field transformation of G-particles* (FTGP). We here consider the example of such reconstruction when the image is on the Cartesian lattice 64×64.

Example 6.4 In the square $[0,1] \times [0,1]$ we consider the lattice 64×64 and the following image composed randomly of ten rectangles. Each of the rectangles is given by coordinates $(x_{k,1}, y_{k,1})$ and $(x_{k,2}, y_{k,2})$, $k = 1:10$, as shown in Table 6.7. The coordinates are written in integer form, i.e. they should be normalized by the factor of $N = 64$. The intensities of these rectangles are given in the last column.

TABLE 6.7
Data of 10 rectangles

k	$x_{k,1}$	$y_{k,1}$	$x_{k,2}$	$y_{k,2}$	R_k
1	10	15	20	30	1
2	12	40	30	50	2
3	40	40	50	60	6
4	45	10	50	30	3
5	15	20	30	35	4
6	24	20	40	35	2
7	32	40	38	50	5
8	5	15	10	20	7
9	50	20	55	25	2
10	10	30	25	40	3

These ten rectangles are shown in Figure 6.16 in part a, along with reconstruction of the image from $3N/2 = 96$ projections in part b.

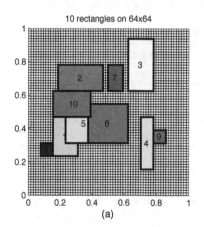
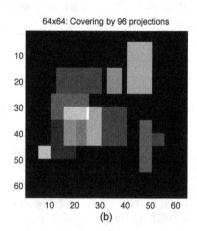

FIGURE 6.16
(a) Ten rectangles on the lattice 64×64 and (b) the reconstructed image.

We also describe another example, where a few rectangles are shifted and do not lie on the lattice. For that, we modify the coordinates of five rectangles, as shown in Table 6.8.

Figure 6.17 shows ten rectangles, five of which are not on the lattice, in part a. The result of image reconstruction from 96 projections is shown

TABLE 6.8

Data of five shifted rectangles

k	$x_{k,1}$	$y_{k,1}$	$x_{k,2}$	$y_{k,2}$	R_k
2	12	40.5	30	50	2
3	40	40.25	50	60	6
4	45.35	10	50	30	3
6	24	20	40	35.2	2
9	50	20	55.45	25	2

in part b. One can notice that the rectangles are well reconstructed with

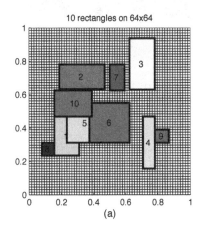 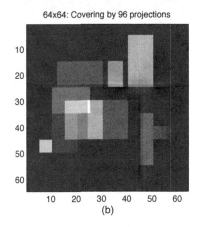

FIGURE 6.17

(a) Ten rectangles and the lattice 64×64 and (b) the reconstructed image.

their exact edges, and additional work can be done to remove the effect of shifting the rectangles. The result of reconstruction will look better when reconstructing with high resolution. Figure 6.18 shows the same ten rectangles and the lattice 128×128 on the square $[0,1] \times [0,1]$ in part a. The result of image reconstruction from 192 projections is shown in part b. The 2nd rectangle is now on the lattice and does not cause the error of reconstruction.

To accomplish the image reconstruction by FTGP, which is described in Example 6.4, we use the code `reconstruction_byR.m` with the following script.

```
% ------------------------------------------------------------
% call: reconstruction_byR        / A.-M. Grigoryan, 2012
% ------------------------------------------------------------
 N=64;
 R=zeros(N,N,N,N);    % 4-D matrix R
 N_rect=10;
 MNrects=[10 15 20 30
         12 40 30 50
```

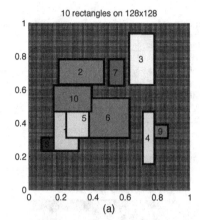
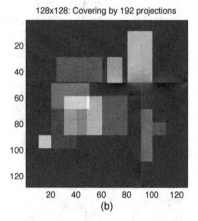

FIGURE 6.18 (See color insert)
(a) Ten rectangles and the lattice 128×128 and (b) the reconstructed image.

```
                40 40 50 60
                45 10 50 30
                15 20 30 35
                24 20 40 35
                32 40 38 50
                 5 15 10 20
                50 20 55 25
                10 30 25 40];
Matr_rect=MNrects/N;
E=[1 2 6 3 4 2 5 7 2 3];    % intensities of rectangles
% Plot the lattice NxN and rectangles -----------------
figure;
colormap(gray);
subplot(1,2,1); hold on
draw_grid; axis image;
h_x1=xlabel('(a)'); set(h_x1,'FontSize',12);
S_title=sprintf('%g rectangles on %gx%g',N_rect,N,N);
h_title=title(S_title);
% -----------------------------------------------------
ps=ps_generators(N);
L=N+N/2;        % number of projections
A=zeros(N,N);   % for sums a((i,j) of fields of particles
for k=1:L       % loop for (p,s)-projections
    p=ps(k,1); s=ps(k,2);
    [p1,s1,t]=tensor_scheme_pst(p,s,N);
    Len_t=length(t);
    for it=1:Len_t  % loop for (p,s,t)-G-rays
        sum_A=fieldpst_tensor_scheme(p1,s1,t(it),N,N_rect,Matr_rect,E);
        f=general_scheme_pst2(p1,s1,t(it),N);
```

```
        f_sz=size(f);
        for nf=1:f_sz(2)
            m1=f(1,nf); n1=f(2,nf); L1=f(3,nf);
            A(m1,n1)=A(m1,n1)+sum_A/L1;        % Eq. 6.24
        end
        %========= calculate the 4-D matrix R ===========
        for nf=1:f_sz(2)
            for mf=1:f_sz(2)
                R(f(1,nf),f(2,nf),f(1,mf),f(2,mf))=...
                    R(f(1,nf),f(2,nf),f(1,mf),f(2,mf))+1;
            end
        end
        %=================================================
    end   % for (p,s,t)-rays
end   % for (p,s)-projection
fprintf(' 4-D Matrix R has been calculated \n');
NN=N^2;
R2=reshape(R,NN,NN);
a=reshape(A,NN,1);
x=R2\a;        % instead of x=inv(R2)*a;                % Eq. 6.26
X=reshape(x,N,N);   % reconstructed image
% Plot the reconstructed image
subplot(1,2,2);
imagesc(X); axis image;
S_title=sprintf('%gx%g: Covering by %g projections ',N,N,L);
h_title=title(S_title);
h_x2=xlabel('(b)'); set(h_x2,'FontSize',12);
% print -dps Figure6_16.ps
% -------------------------------------------------------------------
```

This main program uses the following functions:

1. draw_grid.m (to draw the rectangles and grid as in Figure 6.17(a))
2. ps_generators.m (to calculate the generators (p,s) of projections)
3. tensor_scheme_pst.m (to calculate all values of 't' in equations pn+ms=t of G-rays that pass through the square [0,1]x[0,1], for a given (p,s). It also calculates p=p-N and s=s-N, when p,s>N/2).
4. general_scheme_pst2.m (to represent the G-ray as a set of triplets [x,y,d] of image-elements [x,y] and lengths d of intersections of this ray with [x,y])
5. fieldpst_tensor_scheme.m (to calculate the field function along the G-ray l(p1,s1,t1) for the sum of rectangles).
5a. raylength_inie.m (to calculate the ray-rectangle intersections)
6. sort_data.m (to reorder the coordinates of the ray-image-elements)

These programs are written not only for the tensor scheme of scanning, when the set of 3N/2 generators (p,s) is defined as {(1,s)},{(2p,1)}, and sets of G-rays and A-rays correspond to these (p,s)-projections. Arbitrary scheme of scanning can be used, including the fan scheme, and not only integer values of generators (p,s) can be considered.

However in this section, these codes are used for the tensor scheme.

```
% =====================================================================
% call: tensor_scheme_pst.m
function [p,s,t]=tensor_scheme_pst(p,s,N)
 N2=N/2;
 delta=1/N; delta2=delta/2;
 if p*s==0
    t_min=delta2; t_max=1-delta2;
 else
     if p>N2
         p=p-N; t_min=p+delta; t_max=1-delta;
     elseif s>N2
         s=s-N; t_min=s+delta; t_max=1-delta;
     else
         ps=max(p,s); t_min=delta; t_max=1+ps-delta;
     end
 end
 t=t_min:delta:t_max;
% =====================================================================
% call: general_scheme_pst2.m
% Brief description:
% G-ray is represented as a set of triplets [x,y,d] of image-elements
% and lengths d of intersections of this ray with [x,y]. The triplets
% are written in the 3-D array F={f(1,2,3,:)=(x,y,d,:)}.
% Algorithm is simple:
%   1. Calculate all intersections with horizontal lines of the grid.
%   2. Calculate all intersections with vertical lines of the grid.
%   3. Unite these coordinates in array xy=[x y].
%   4. Sort xy by the first coordinate.
%   5. Find the coordinates of the image-elements (IE) along the G-ray.
%   6. Calculate the lengths of intersection of the ray with these IEs.
%      If the intersection is zero or close to zero (as eps*100), then
%      it is considered that the G-ray does not pass through the IE.
%      The small value (eps*100) should be chosen more accurately.
% ---------------------------------------------------------------------
function F=general_scheme_pst2(p,s,t,N)
 t=-t;
 delta=1/N;
 eps100=eps*100;
 ones1N=ones(1,N)*delta;
% 1,2
 if abs(p)<eps100;                        % (p==0)  in the tensor scheme
     f=zeros(3,N);
     x=-t/s;
     if x>0 & x<=1
         x_n=ceil(x/delta);
     end
     f(2,:)=ones(1,N)*x_n;
     f(1,:)=1:N; f(3,:)=ones1N;
```

```
      F=f;  return
elseif abs(s)<eps100;                % (s==0)  in the tensor scheme
    f=zeros(3,N);
    y=-t/p;
    if y>0 & y<=1
        y_n=ceil(y/delta);
    end
    f(1,:)=1+N-ones(1,N)*y_n;
    f(2,:)=1:N;  f(3,:)=ones1N;
    F=f;  return
else                              % (p,s)~=(0,0)  in the tensor scheme
   a(1,:)=0:N;  a=a*delta;
   in_x=1;  in_y=1;  x=[];y=[];
   for i1=1:length(a)
        ai1=a(i1);
        xn=(-t-p*ai1)/s;
        if(xn >=0 & xn<=1)
            x(:,in_x)=[xn;ai1];  in_x=in_x+1;
        end
        yn=(-t-s*ai1)/p;
        if(yn >=0 & yn<=1)
            y(:,in_y)=[ai1;yn];  in_y=in_y+1;
        end
   end
% 3.
   xy=[x y];
% 4.
   t=sort_data(xy);  sz=size(t);
% 5.
   t=t/delta;
   for i1=1:sz(2)-1
        i2=i1+1;
        ff=(t(1,i2)+t(1,i1))/2;  f(2,i1)= ceil(ff);
        ff=(t(2,i2)+t(2,i1))/2;  f(1,i1)= N-ceil(ff)+1;
   end
   xs(1,:)=t(1,:);  ys(1,:)=t(2,:);
% 6.
   m=1;
   for i1=1:sz(2)-1
        i2=i1+1;
        sd=(xs(i2)-xs(i1))^2+(ys(i2)-ys(i1))^2;
        f(3,i1)=sqrt(sd)*delta;
        % consider only image elements with length > eps*100
        if f(3,i1)>eps100
            F(:,m)=f(:,i1);   m=m+1;
        end
   end
end
```

It should be noted, that after the command "t=sort_data(xy);"
we probably should remove the equal pairs in 'xy', in a case
if such repetition takes place. For the tensor scheme, such
situation will not happen, but not in the general case.
Therefore in the script of the above function, we could add
the following command: "t=delete_equalpairs(t);" and use the
following script for the function 'delete_equalpairs.m'.

```
% -------------------------------------------------------------------
function y=delete_equalpairs(x)
  sz=size(x);
  y(:,1)=x(:,1);
  z=x(:,1); m=2;
  for i1=2:sz(2)
      if x(1,i1)~=z(1) || x(2,i1)~=z(2)
          z=x(:,i1); y(:,m)=z; m=m+1;
      end
  end
% ===================================================================
function sum_E=fieldpst_tensor_scheme(p,s,t,N,N_rect,Matr_rect,E)
  sum_E=0;
  for k=1:N_rect
      [Length_inter,F]=raylength_inie(p,s,t,N,Matr_rect(k,:));
      if isempty(Lengt_inter)
          Length_inter=0;
      end
      sum_E=sum_E+E(k)*Length_inter;
  end
% ===================================================================
% call: raylength_inie.m
% Brief description:
% Coordinates (x1,y1) and (x2,y2) and length of intersection of the
% given G-ray l(p,s,t) with the rectangle ('rec') are calculated.
% These coordinates are written into the 2-D array F=[x1 y1; x2 y2]
% and the length of intersection is written in 'Length_inter'.
% Algorithm is the same as in the code "general_scheme_pst2.m", the
% only difference is that instead of image elements we consider the
% one given rectangle. Calculations for the horizontal and vertical
% projections are separated from all other (p,s)-projections.
% -------------------------------------------------------------------
function [Length_inter,F]=raylength_inie(p,s,t,N,rec)
  rx1=rec(1); ry1=rec(2); rx2=rec(3); ry2=rec(4);
  x=[]; y=[]; eps100=eps*100;
  if abs(p)<eps100
      x=t/s;
      if x>rx2 | x<rx1
          Length_inter=[]; F=[];
      else
          Length_inter=ry2-ry1; F=[x,ry1;x,ry2];
      end
```

```
elseif abs(s)<eps100
    y=t/p;
    if y>ry2 | y<ry1
        Length_inter=[]; F=[];
    else
        Length_inter=rx2-rx1; F=[rx1,y;rx2,y];
    end
else
    x=[]; y=[];
    for i1=1:2
        i12=2*i1; ri12=rec(i12);
        xn=(t-p*ri12)/s;
        if(xn >=rx1 & xn<=rx2)
            x=[x [xn; ri12]];
        end
        ri12=rec(i12-1);
        yn=(t-s*ri12)/p;
        if(yn >=ry1 & yn<=ry2)
            y=[y [ri12;yn]];
        end
    end
    xy=[x y];
    if isempty(xy)
        Length_inter=[]; F=[]; return
    end
    xy=sort_data(xy);
end
% calculate the length of the intersection
if (exist('xy')==1)
    x1=xy(1,1); x2=xy(1,end);
    y1=xy(2,1); y2=xy(2,end);
    Length_inter=sqrt((x2-x1)^2+(y2-y1)^2);
    F=[x1 y1;x2 y2];
    % do not consider such intersection if length<eps*100
    if Length_inter<eps100
        Length_inter=[]; F=[];
    end
end
% ====================================================================
function f=sort_data(x)
 sz=size(x);
 [y,I]=sort(x(1,:));
 for i1=1:sz(2)
     f(:,i1)=x(:,I(i1));
 end
% ====================================================================
```

The scripts of these programs can be found at http://fasttransforms.com in the folder Lectures/Tensor and Tomography.

In the examples considered above for the G- and A-particles on the lattice 4×4, the reconstruction of the image by FTGP has been obtained by using the inverse matrix R^{-1}. The described program `reconstruction_byR.m` also accomplishes the calculation of this inverse matrix. The calculation of this matrix can be separated and calculated and stored in advance. It is clear that the calculation of such matrix $(N^2 \times N^2)$ is difficult for large values of $N = 2^r$, when $r > 2$. For instance, our personal computer can perform the reconstruction of the image with maximal size 128×128, when using this program. The $N = 256$ case faces a lack of the required memory when running the program, and therefore new methods of calculation should be applied when programming in MATLAB®. In the main program `reconstruction_byR.m`, the command `x=R2\a` is used instead of calculation of the inverse matrix, when using the command `x=inv(R2)*a`. We also mention the minimization techniques of nonlinear programming [36], which may be applied for calculating the vector x.

The direct method of FTGP may not be very effective for high values of N. We therefore consider another, more effective solution of the problem, without calculation of the inverse matrix R^{-1}.

6.4   Method of circular convolution

We now consider the transformation of the image, which is determined by the system of field functions in (6.24)

$$a(i,j) = \sum_{m=0}^{N-1} \sum_{n=0}^{N-1} f_{n,m} \psi_{n,m}(i,j), \quad i,j = 0 : (N-1). \tag{6.27}$$

As mentioned above, each field function can be calculated by the base field matrix, $\psi_{n,m}(i,j) = \Psi(N-1-n+i, N-1-m+j)$, $i,j = 0 : N-1$. Therefore, we have the following property for the field function of the G-particle (n,m):

$$\psi_{n,m}(i,j) = \psi(i-n, j-m), \quad i,j = 0 : N-1. \tag{6.28}$$

The transform of the image in (6.27) can be written as the linear convolution

$$a(i,j) = (f * \psi)(n,m) = \sum_{m=0}^{N-1} \sum_{n=0}^{N-1} f_{n,m} \psi(i-n, j-m), \tag{6.29}$$

where $i,j = 0 : (N-1)$. The simple way to solve this 2-D convolution equation is based on the 2-D discrete Fourier transform. Because the base field matrix Ψ is of size $(2N-1) \times (2N-1)$, we will use the zero padding to extend the image $f_{n,m}$ to this size. We denote the extended image by $\tilde{f}_{n,m}$. Then, the following is valid:

$$
\begin{aligned}
a(i,j) &= \sum_{m=0}^{2N-1} \sum_{n=0}^{2N-1} \tilde{f}_{n,m}\psi(i-n,j-m) \\
&= \sum_{m=0}^{2N-1} \sum_{n=0}^{2N-1} \tilde{f}_{n,m}\Psi(N-1+i-n, N-1+j-m) \\
i,j &= 0 : (N-1).
\end{aligned}
\tag{6.30}
$$

Thus, the value of the 2-D linear convolution $a(i,j)$ can be calculated from the incomplete 2-D circular convolution of the extended image $\tilde{f}_{n,m}$ with the base field matrix Ψ. To illustrate this property of zero padding, we describe the following example.

Example 6.5 For the $N = 4$ case, we consider the script of the program run_ccN4.m.

```
% call: run_ccN4.m
  % f=round(4*rand(4));    % random input 4x4
    f=[ 1  3  2  1
        0  1  2  2
        0  4  3  2
        3  0  3  3];
  UH8=[ 1  1  1  1  0  0  1
        1  1  2  1  0  1  0
        1  2  3  2  1  0  0
        1  1  2  8  2  1  1
        0  0  1  2  3  2  1
        0  1  0  1  2  1  1
        1  0  0  1  1  1  1];
  % The direct method of 2-D convolution:
  R=zeros(4);
  for i1=1:4
      for j1=1:4
          R=UH8(4-1+j1-3:4+j1-1,4-1+i1-3:4+i1-1);
          A1(j1,i1)=sum(sum(R.*f));
      end
  end;  A1
      %  48  61  49  34
      %  40  55  66  46
      %  32  68  69  56
      %  38  33  63  68
      % The method of 2-D circular convolution:
  F=zeros(7);
  F(1:4,1:4)=f(:,:);
  for i1=1:4
      for j1=1:4
          s=0;
          for m1=1:7
```

```
            m2=4-1+j1-3+m1-1;
            if m2<1 m2=m2+7; end
            if m2>7 m2=m2-7; end
            for n1=1:7
                n2=4-1+i1-3+n1-1;
                if n2<1 n2=n2+7; end
                if n2>7 n2=n2-7; end
                s=s+UH8(m2,n2)*F(m1,n1);
            end
        end
        A2(j1,i1)=s;
    end
end;  A2
```

The input image 4×4 as a result of command `f=round(4*rand(4))` is the matrix $f = [1\,3\,2\,1;\ 0\,1\,2\,2;\ 0\,4\,3\,2;\ 3\,0\,3\,3]$. The linear convolution calculated by the direct method and the method of circular convolution results in the same image $A1 = A2 = [48\,61\,49\,34;\ 40\,55\,66\,46;\ 32\,68\,69\,56;\ 38\,33\,63\,68]$.

Because the linear convolution $(f * \psi)$ can be enclosed in the circular convolution $(\tilde{f} \otimes \psi)$, one can use the discrete Fourier transform to solve equation (6.30) and obtain the reconstruction $f = \{f_{n,m}\}$. At the same time, this method allows for processing the reconstructed image in the frequency domain for filtration or enhancement, if such processing is desired or required.

Let W be the twiddle coefficient $W = W_{2N-1} = e^{-2\pi i/(2N-1)}$. The 2-D DFT of the circular convolution $a(n,m)$ can be calculated as follows:

$$
\begin{aligned}
A_{p,s} &= \sum_{i=0}^{2N-2}\sum_{i=0}^{2N-2} a(i,j)W^{pi+sj} \\
&= \sum_{m=0}^{2N-2}\sum_{n=0}^{2N-2} \tilde{f}_{n,m} \sum_{i=0}^{2N-2}\sum_{i=0}^{2N-2} \Psi(N-1+i-n, N-1+j-m)W^{pi+sj} \\
&= \sum_{m=0}^{2N-2}\sum_{n=0}^{2N-2} \tilde{f}_{n,m} \sum_{i=0}^{2N-2}\sum_{i=0}^{2N-2} \Psi(i,j)W^{p(i-N+1+n)+s(j-N+1+m)} \\
&= \sum_{m=0}^{2N-2}\sum_{n=0}^{2N-1} \tilde{f}_{n,m}W^{pn+sm} \sum_{i=0}^{2N-1}\sum_{i=0}^{2N-2} \Psi(i,j)W^{pi+sj} \cdot W^{-(N-1)(p+s)} \\
&= \tilde{F}_{p,s}\Psi_{p,s} \cdot W^{-(N-1)(p+s)}
\end{aligned}
$$

where we denote the $(2N-1)\times(2N-1)$-point DFTs of the convolution $a(n,m)$, extended image $\tilde{f}$, and base matrix Ψ by $A_{p,s}$, $\tilde{F}_{p,s}$, and $\Psi_{p,s}$, respectively. Thus, we obtain the following formulas for image reconstruction:

$$
\tilde{F}_{p,s} = \frac{A_{p,s}}{\Psi_{p,s}}W^{(N-1)(p+s)}, \quad p, s = 0 : (2N-1), \tag{6.31}
$$

and

$$f_{n,m} = \sum_{p=0}^{2N-1} \sum_{s=0}^{2N-1} \tilde{F}_{p,s} W^{-(np+ms)}, \quad n, m = 0 : (N-1). \tag{6.32}$$

This formula can be used if the 2-D DFT of the base field matrix does not have zero values. This matrix and its 2-D DFT can be calculated for any values of $N = 2^r$, and the simple program can be written to verify the fact that $\Psi_{p,s} \neq 0$, for all (p,s).

As an example, we consider the DFT of the UH-square for the $N = 4$ case. The required calculations are accomplished by using the program with the following script.

```
% call: checkzeros_inDFTofUHsquare.m
  UH8=[ 1 1 1 1 0 0 1
        1 1 2 1 0 1 0
        1 2 3 2 1 0 0
        1 1 2 8 2 1 1
        0 0 1 2 3 2 1
        0 1 0 1 2 1 1
        1 0 0 1 1 1 1];
  FUH8=fft2(UH8);
  mn=min(min(abs(FUH8)));   % 0.1491
  R_ofFUH8=real(FUH8)
   56.0000  -14.0036    4.6250   -1.1213   -1.1213    4.6250  -14.0036
  -14.0036    0.5625   -0.3351   -0.0332    4.7409   -7.1528   24.0858
    4.6250   -0.3351   -2.3107    3.6860   -4.4058    7.8358   -7.1528
   -1.1213   -0.0332    3.6860   -8.7518    3.0785   -4.4058    4.7409
   -1.1213    4.7409   -4.4058    3.0785   -8.7518    3.6860   -0.0332
    4.6250   -7.1528    7.8358   -4.4058    3.6860   -2.3107   -0.3351
  -14.0036   24.0858   -7.1528    4.7409   -0.0332   -0.3351    0.5625
  I_ofFUH8=imag(FUH8)
        0   -6.7438    5.7995   -4.9129    4.9129   -5.7995    6.7438
  -6.7438    0.7053   -1.4683    0.1454   -5.9449    3.4446   -0.0000
   5.7995   -1.4683   10.1237   -4.6221    2.1217    0.0000   -3.4446
  -4.9129    0.1454   -4.6221    4.2147        0   -2.1217    5.9449
   4.9129   -5.9449    2.1217        0   -4.2147    4.6221   -0.1454
  -5.7995    3.4446   -0.0000   -2.1217    4.6221  -10.1237    1.4683
   6.7438    0.0000   -3.4446    5.9449   -0.1454    1.4683   -0.7053
```

The real and imaginary parts of the DFT of the UH-square are printed separately, and one can see that there is no zero value in this DFT. In absolute scale, the minimum number of the DFT is 0.1491.

Below is the remaining part of the script, which calculates the DFT of the UH-square of number 26 for the $N = 8$ case. The minimum number of the DFT is 0.6347 in absolute scale.

```
  UH26=[ 1 0 0 1 1 0 0 1 1 3 1 1 0 0 1
```

```
    0  1  0  0  2  0  0  1  2  2  2  0  0  1  0
    0  0  1  0  1  1  0  1  3  2  1  0  1  0  0
    1  0  0  1  0  2  0  1  4  2  0  1  0  0  1
    2  2  1  0  1  1  1  2  4  1  1  0  1  2  2
    2  3  3  3  1  1  2  3  4  1  1  3  4  5  5
    0  0  1  2  4  4  4  6  6  6  7  6  4  2  1
    1  1  1  1  2  5 10 26 10  5  2  1  1  1  1
    1  2  4  6  7  6  6  6  4  4  4  2  1  0  0
    5  5  4  3  1  1  4  3  2  1  1  3  3  3  2
    2  2  1  0  1  1  4  2  1  1  1  0  1  2  2
    1  0  0  1  0  2  4  1  0  2  0  1  0  0  1
    0  0  1  0  1  2  3  1  0  1  1  0  1  0  0
    0  1  0  0  2  2  2  1  0  0  2  0  0  1  0
    1  0  0  1  1  3  1  1  0  0  1  1  0  0  1 ];
FUH26=fft2(UH26);
mn=min(min(abs(FUH26)))   % 0.6347
```

It is not difficult to verify that the 2-D DFT of the UH-squares does not have zeros (see also Table 6.9, for other values of $N = 2^r$, $r > 3$. The number, k, of these squares UH(k) is also shown in the table. We here omit the analytical proof of this fact and leave it to the reader.

TABLE 6.9
Tensor UH(k)-squares

N	k	$\min\{|\Psi_{p,s}|\}$
16	98	0.1585
32	386	0.1204
64	1538	0.0283
128	6146	0.0385
256	24578	0.0346
512	98306	0.0086
1024	393218	0.0004

We consider the base matrix Ψ as the function $\phi(n, m)$ with the center at $(0, 0)$, i.e.

$$\phi(n, m) = \Psi(N + n - 1, N + m - 1), \quad n, m = -(N - 1) : (N - 1).$$

We now calculate the 2-D DFT of the base function,

$$\Psi_{p,s} = \sum_{n,m=0}^{2N-2} \Psi(n, m) W^{np+ms}$$

$$= \sum_{n,m=-(N-1)}^{N-1} \Psi(N + n - 1, N + m - 1) W^{(N+n-1)p+(N+m-1)s}$$

$$= \sum_{n,m=-(N-1)}^{N-1} \phi(n, m) W^{np+ms} \cdot W^{(N-1)(p+s)} = \phi_{p,s} W^{(N-1)(p+s)},$$

where $W = W_{2N-1} = \exp(-2\pi i/(2N-1))$. Now we consider the transform of the function $\phi(n,m)$ shifted as $(n,m) \to (n+N, m+N)$,

$$\Psi_{p,s} W^{N(p+s)} = \phi_{p,s}, \quad p,s = 0 : (2N-2).$$

This function is real and, therefore, the phase of the base matrix equals

$$\text{phase}(\Psi_{p,s}) = \frac{2\pi N}{2N-1}(p+s) = \pi(p+s) + \frac{\pi}{2N-1}(p+s).$$

Below is the script of the program DFTofUHsquare.m to calculate the real function $\phi_{p,s}$ and show that it is real, namely, that the imaginary part of this spectral function is zero. The minimum absolute value 0.038486 of this function is also calculated. The $N = 128$ case is written, and it can also be used for other values of $N = 2^r$, when integer $r > 2$.

```
% ---------------------------------------------------------------
% call: DFTofUHsquare.m   / A.-M. Grigoryan 2011
% ---------------------------------------------------------------
  N=input(' Set the value of N=4,8,16,32, ...  : ');
  N=128;
  UHN=UH_square(N);     % UH-square
  FUH=fft2(UHN);
  mn=min(min(abs(FUH)));     % 0.0385
  fprintf('\n    min(abs(FUH)=%8.6f \n',mn);
  M=2*N-1;
  wN=exp(-1j*2*N*pi/M);
  AUH=zeros(M);
  for p=1:M
      for s=1:M
          ww=wN^(p+s-2);
          AUH(p,s)=FUH(p,s)*ww;
      end
  end
  i1=min(min(imag(AUH))); i2=max(max(imag(AUH)));
  absmin=min(min(abs(AUH)));
  fprintf('   imag is in [%f,%f], min(abs(FUH))=%8.6f \n',...
     i1,i2,absmin');
  % imag is in [-0.000000,0.000000], min(abs(FUH))=0.038486
% ---------------------------------------------------------------
```

6.4.1 Uniform frames

We consider equation (6.29) of the linear convolution, where the values of basic functions $\psi(n,m,i,j) = \psi(i-n, j-m)$ are written with consideration of the lengths of the intersection of rays with the image elements, or squares

$$a(i,j) = \sum_{m=0}^{N-1} \sum_{n=0}^{N-1} f_{n,m}\psi(i-n, j-m), \quad i,j = 0 : (N-1). \tag{6.33}$$

This convolution equation can be written in terms of frames, and it is easy to do that for the following "ideal scheme of scanning." This scanning scheme is explained as follows.

The frames Υ^k of the tensor UT-square for number 26 have different coefficients, the number of which equals $c(k) = (2k+1)^2 - (2k-1)^2 = 8k$, when $k \geq 1$, and $c(0) = 1$ for the center. We now assume that the numbers on the frames are distributed uniformly. Let r_k be the coefficients on frames Υ^k, respectively. Then the linear convolution in (6.33) can be written as

$$a(i,j) = \sum_{k=0}^{N-1} \sum_{(n,m)\in\Upsilon^k} f_{n,m}\psi(i-n, j-m)$$

$$= \sum_{k=0}^{N-1} \sum_{(n,m)\in\Upsilon^k} r_k f_{i-n,j-m} = \sum_{k=0}^{N-1} r_k \sum_{(n,m)\in\Upsilon^k} f_{i-n,j-m}.$$

The projection data in point (i,j) is calculated by integrating values of the image on the frames, or frames shifted to the center with coordinates (i,j). It should be noted for comparison, that in the method of backprojections, the each projection is amplified by the filter $|w_1|$ (see (7.3)), which in the frequency domain $(w_1 \cos\theta, w_1 \sin\theta)$ describes the circle of radius $|w_1|$.

Problems

Problem 6.1 (Model 1) Consider the symmetric tensor scheme of scanning, when $2N-1$ generators (p,s) are used from the following set:

$$J^1_{N,N} = \{(1,s), s = 0 : (N-1)\} \cup \{(p,1), p = 0, 2 : (N-1)\}.$$

A. Determine the base matrix Ψ of G-particles in the $N=4$ case. Show that this matrix Ψ is the UH-square.

B. Determine the field functions of the G-particles $(1,1)$ and $(3,2)$ in the $N=4$ case.

C. All N^2 field functions of G-particles can be written row-wise into the matrix $N^2 \times N^2$. Determine such matrix for $N=4$.

Problem 6.2 Consider the symmetric tensor scheme of scanning with the set of generators $J^1_{N,N}$. Determine the base matrix Ψ of G-particles in the $N=8$ case. Show that this matrix Ψ is the UH-square.

Problem 6.3 Consider the symmetric tensor scheme of scanning with the set of generators $J^1_{N,N}$. Show that, similar to the tensor scanning scheme, the 4-dimensional matrix $R = ||c(i,j,n,m)||$ provides the representation in the form of the linear convolution. Here the coefficients $c(i,j,n,m)$ equal the number of rays passing through the particles (i,j) and (n,m) at the same time.

Problem 6.4 Write the code and show that reconstruction of the image $N \times N$, where N is a power of two, is possible when using the scanning scheme with the set $J_{N,N}^1$.

Problem 6.5 (Model 2) Consider the following scheme of scanning, when $N - 1$ generators (p, s) are used from the following set:

$$J_{N,N}^2 = \{(1, s), s = 0 : (N/2 - 1)\} \cup \{(p, 1), p = 0, 2 : (N/2 - 1)\}.$$

For the $N = 4$ case, determine the following:

    A. The base matrix Ψ of G-particles and show that the matrix Ψ is the UH-square. Calculate $\det(\Psi)$.

    B. The field functions of the G-particles $(1, 0)$ and $(1, 1)$.

    C. All N^2 field functions of G-particles can be written row-wise into the matrix $N^2 \times N^2$. Determine this matrix for $N = 4$. Determine the rank of this matrix and determine if the inverse matrix exists.

    D. Show that, similar to the tensor scanning scheme, the 4-dimensional matrix $R = ||c(i, j, n, m)||$ provides the representation in the form of the linear convolution.

Problem 6.6 Repeat Problem 6.4 for the $N = 8$ case.

Problem 6.7 Write the code and show that reconstruction of the image $N \times N$, where N is a power of two, is possible when using the scanning scheme with the set $J_{N,N}^2$.

The solutions of the above problems show that different scanning schemes can be used with a number of projections greater or less than $3N/2$.

Problem 6.8 (Model 3) Given integer $M > 0$, consider the following scheme of scanning, when $2M - 1$ generators (p, s) are used from the set

$$J_{N,N}^3 = \{(1, s), s = 0 : (M - 1)\} \cup \{(p, 1), p = 0, 2 : (M - 1)\}.$$

Write the code and find the minimum value of $M = M(N)$ for which reconstruction of the image $N \times N$ is possible. Here, N is a power of two.

Problem 6.9 (Model 4) Given integers $M > 0$ and $K > 0$, consider the following scheme of scanning, when $M + K - 1$ generators (p, s) are used from the following set:

$$J_{N,N}^4 = \{(1, s), s = 0 : (M - 1)\} \cup \{(p, 1), p = 0, 2 : (K - 1)\}.$$

Write the code and find the minimum values of $M = M(N)$ and $K = K(N)$ for which reconstruction of the image $N \times N$ is possible. Here N is a power of two.

Problem 6.10 (Model X) Given integer $M > 0$, consider the scheme of scanning, when M generators (p, s) are selected randomly with p and s uniformly distributed in the interval $[0, N - 1]$.

$$J_{N,N}^{x} = \{(p, s), s, p \in [0, N - 1]\}.$$

Write the code and find out if the exact reconstruction of the image $N \times N$ is possible with this random scanning scheme. Here, N is a power of two. Determine for which values of M the reconstruction can be achieved. Provide examples and explain the results. *(Note that p and s may have non-integer values.)*

7

Methods of Averaging Projections

In this chapter we describe a few methods of averaging for image reconstruction. The main result with a statistical model of the image is described in §7.4. It allows us to achieve a high-quality image reconstruction through the process of iterations and renewal of statistics after each iteration. This approach does not require any transformation and its effective realization can be used for reconstructing the image in discrete form with any size.

In Chapters 3, 4, and 5, the method of image reconstruction, which is based on transferring the geometry from A-rays to G-rays, was described. The reconstruction of the image is exact, when considering the image $f(x, y)$ as a set of small cells, or image elements inside the Cartesian lattice $N \times N$. We started the description of this method with the images composed of random rectangles, and then the proposed method was applied to other images as well. The exact reconstruction was achieved under the assumption that the image elements are exactly on the lattice. This is a real model of the image for reconstruction, which is used in the known methods of finite series expansion [29, 31, 47].

We now consider an example where the image consists of a few rectangles $r_k(x, y)$ with intensities R_k and positions $(x_k, y_k, \Delta x_k, \Delta y_k)$,

$$f(x, y) = \frac{1}{(\Delta x)^2} \sum_{k=1}^{K} R_k \text{rect}\left(\frac{x - x_k}{\Delta x_k} - \frac{1}{2}, \frac{y - y_k}{\Delta y_k} - \frac{1}{2} \right). \tag{7.1}$$

When all data of random rectangles, $x_k, y_k, \Delta x_k$, and Δy_k, are integer multiples of $\Delta x = 1/N$, the image can be reconstructed exactly in the form of the corresponding discrete analog $\{f_{n,m}\}$ of this image. Given (n, m), the value of $f_{n,m}$ equals the average image intensity in the square $[n, m]$. We now consider the case when data of such rectangles may not be integer multiple to $1/N$, i.e., when one or a few rectangles of the image are not divided by the Cartesian lattice $N \times N$ by small squares of size $1/N \times 1/N$ each. In this case, we say that the image $f(x, y)$ is not on the lattice, and it is shifted.

As a example, Figure 7.1 shows the image $f(x, y)$ with seven random rectangles, which is not on the Cartesian lattice 256×256. The reconstructed image of the size 256×256 is also shown. One can observe that although the rectangles with their straight edges have almost been reconstructed and can be observed well, there are many "noisy" points in the image. This image therefore needs to be processed in order to filter such a noise, as is done, for

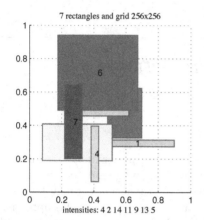
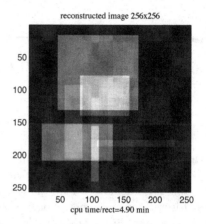

FIGURE 7.1 (See color insert)
Image with seven random rectangles and the reconstructed discrete image
256×256.

instance, in the known method of filtered backprojection (BP). This method
is also called the method of summation of projection data by rays, and it is
used widely in computed tomography [28, 29, 30].

7.1   Filtered backprojection

First we describe the method of summation of projection data by the rays for
the set of projections, which is described in the tensor representation of the
image. In other words, we consider the method of summation with the tensor
scheme of scanning the image. The concept of back projecting the measured
data obtained on the receiver side of X-ray tomography can be described
briefly as follows. Let $\mathcal{L}$ be the set of all G-rays l which are used for scanning
the object or image $f(x, y)$, and let Δl be the length of the ray l in the image.
The value w_l of the projection along the ray l is divided equally among all
points (x, y) lying on this ray. Then, for every point (x, y) on the image plane,
all rays passing through this point are considered. The sum of averaging line-
integrals calculated along these rays determines the value of the reconstruction
at the point (x, y),

$$f_{bp}(x, y) = \sum_{l \in \mathcal{L}} \sum_{l \ni (x,y)} \frac{w_l}{\Delta l}. \tag{7.2}$$

We first apply this idea on the discrete images, and then on images modeled
by random rectangles in the square $[0, 1] \times [0, 1]$. To all points in the rectangles,

as well as outside them, some values of intensity are assigned. The resulting image is smooth, because of averaging.

This method can be described by considering the 2-D Fourier transform of the image $f(x, y)$ in the polar system of coordinates in the frequency domain [30]. Indeed, let us consider the line-integral of the image

$$w(t) = w_\theta(t) = \int f(x,y)dl_\theta(t) = \int_0^1 \int_0^1 f(x,y)\delta(x\cos\theta + y\sin\theta - t)dxdy$$

along the line $l_\theta(t) = \{x\cos\theta + y\sin\theta = t\}$. The Fourier transform of $w(t)$ coincides with the 2-D Fourier transform of the image along the line $\omega_2 = \omega_1\tan(\theta)$ in the frequency domain (ω_1, ω_2). In other words,

$$F(w_1\cos\theta, w_1\sin\theta) = \int_0^1 \int_0^1 f(x,y)e^{-j(x\omega_1\cos\theta + y\omega_1\sin\theta)}dxdy =$$

$$\int_{-\infty}^{\infty}\int_0^1\int_0^1 f(x,y)e^{-j\omega_1 t}\delta(x\cos\theta + y\sin\theta - t)dxdydt = \int_{-\infty}^{\infty} w_\theta(t)e^{-j\omega_1 t}dt = W_\theta(\omega_1).$$

This is the statement of *the Fourier Slice Theorem*. Therefore, by using the inverse Fourier transform, the image can be written as

$$f(x,y) = \frac{1}{2\pi}\int_0^\pi \left[\frac{1}{2\pi}\int_{-\infty}^{\infty} W_\theta(\omega_1)|\omega_1|e^{j\omega_1 t}d\omega_1 \right] d\theta, \tag{7.3}$$

where $t = x\cos\theta + y\sin\theta$. For a given angle θ, the same value of the integral

$$b_\theta(t) = \frac{1}{2\pi}\int_{-\infty}^{\infty} W_\theta(\omega_1)|\omega_1|e^{j\omega_1 t}d\omega_1$$

will be added to the image $f(x, y)$ at all points (x, y) on the line $l_\theta(t) = \{x\cos\theta + y\sin\theta = t\}$. This value is the inverse Fourier transform of the projection amplified by the filter $|\omega_1|$. Therefore, Equation (7.3) describes the process of *filtered backprojection*,

$$f(x,y) = \frac{1}{2\pi}\int_0^\pi b_\theta(x\cos\theta + y\sin\theta)d\theta.$$

The realization by this formula requires the full, or infinite number of projections $\{w_\theta(t); \theta \in [0, \pi]\}$ and convolution with the impulse response $h(t)$, which is defined by the frequency characteristics $H(\omega_1) = |\omega_1|$, where $\omega_1 \in (-\infty, \infty)$. These two complex operations cannot be accomplished exactly in practice, but only approximated under some assumptions. The number of projections is reduced to a certain number by considering, for instance, only 180 angles distributed uniformly in the interval $[0, \pi]$. The filter $H(\omega_1)$ is not bounded, but it will be cut by some frequency ω_0,

$$\tilde{H}(\omega_1) = \text{rect}\,(\omega_1/(2\omega_0)) - \text{tr}\,(\omega_1/(2\omega_0)) = \begin{cases} |\omega_1|, & \text{when } \omega_1 \in (-\omega_0, \omega_0), \\ 0, & \text{otherwise}, \end{cases}$$

when assuming that there is no aliasing for the projection data when sampling with a period $T < \pi/\omega_0$. To smooth the sharp cutoffs at $\pm\omega_0$ and reduce the ringing artifacts in the reconstruction, different smoothing window functions can be used to modify the Ran-Lak filter $\tilde{H}(\omega_1)$. For instance, we consider the generalized Hamming window [29, 30]

$$H(\omega_1) = |w_1| \cdot \begin{cases} \alpha + (1 - \alpha)\cos(\pi\omega_1/\omega_0), & \text{when } \omega_1 \in (-\omega_0, \omega_0), \\ 0, & \text{otherwise}, \end{cases}$$

where the value of α is selected from the interval $(0, 1)$. As an example, Figure 7.2 shows the Ran-Lak filter with cutoff frequency $\omega_0 = 1.45\text{rad/s}$, the

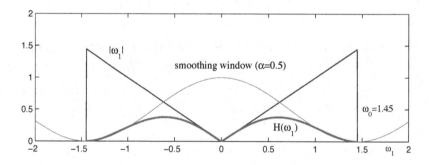

FIGURE 7.2
Modification of the Ran-Lak filter by Humming smoothing window.

Humming window with the parameter $\alpha = 0.5$, and the function $H(\omega_1)$.

7.2   BP and method of splitting-signals

Each splitting-signal of the discrete image $f_{n,m}$ can be defined as a sum of the image along parallel lines, i.e., by projection data. At the same time, the splitting-signal defines the direction image as being a component of the image, as stated in (2.49), for the image of size $N \times N$, where N is a power of 2. This direction image has constant values along the parallel lines. The number of such constants equals $N/2$ for the first series of generators (p, s), and $N/2^k$, for the kth series of generators, where $k = 2, 3, ..., r$. This property explains the main difference between the paired transform-based method of image reconstruction and the method of backprojection. In the method of backprojection, direction images also are used, and for many projections, these images are incomplete. This means they are not calculated for all points on the Cartesian grid $N \times N$. The number of such images equals the number of projections, and the value of each of these images at the point (n, m) is set to be equal to the average (and filtered) value of the projection along the ray passing through this point. This rough assumption does not allow for accurate reconstruction of the image, even for a large number of projections.

For example, consider the case when the full set of $2(128) - 1$ measurements are used for calculating the projection by the angle arctan(2) of the image on the lattice 128×128. The corresponding direction image in BP will contain these 255 values along the parallel lines. In the paired method, from these 255 measurements, 128 values of the splitting-signal in tensor representation, and then 64 values in paired representation are calculated and used in the direction image. The number of projections, or direction images, for image reconstruction by BP is considered to be 180, assuming the uniform change of the angle of projections by 1 degree. The projections which are used in the tensor and paired methods are calculated by the angles

$$\Phi = \Phi(256) = \{\pi/2 - \arctan(s/p); \; (p, s) \in J_{256,256}\},$$

which differ from the angles used in BP. It follows from the principle of super-position, that the number of required projections equals 192, from which 382 direction images are calculated for image reconstruction by the paired method. All direction images in BP are non-negative, and in paired representation the direction images have both positive and negative values. The BP method results in a smooth image, and the paired method allows reconstruction of the discrete image (noise is not considered).

As an example, Figure 7.3 shows the discrete tomo image of size 256×256 in part a, together with the results of image reconstruction by the filtered backprojection in b and c, when respectively 384 and 360 projections are used at angles equally spaced between 0 and 180 degrees. The mean-square-root error of image reconstruction equals 0.010308 in (c) and 0.011086 in (d). One can notice that many details on these images are not seen well. For comparison, the reconstruction of the image by the paired transform with the same number, 360, of projections is shown in d. The error of reconstruction equals 0.002065 and 360 angles are defined by the frequency-points (p, s) of the set of generators $J_{256,256}$, whose splitting-signals have high energies,

$$E_{p,s} = \frac{1}{256^2} \sum_{t=0}^{255} f^2_{p,s,t} > 8773.85.$$

The graph of the energy function $E_{p,s}$ is given in Figure 7.4, where the generators (p, s) are ordered, as written in the set $J_{256,256}$, i.e., the first is

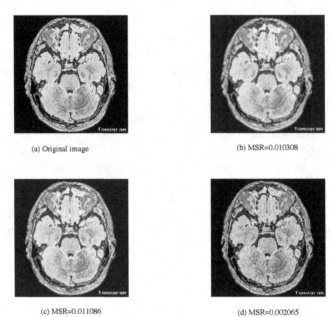

(a) Original image

(b) MSR=0.010308

(c) MSR=0.011086

(d) MSR=0.002065

FIGURE 7.3
(a) The image 256 × 256. The image reconstruction by the filtered backprojection algorithm by (b) 384 and (c) 360 projections (by MATLAB® function `iradon.m`), and (d) reconstruction by the 2-D DPT with 360 projections.

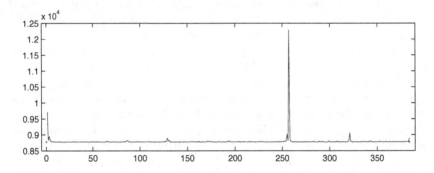

FIGURE 7.4
The energy curve of all 384 splitting-signals of the tomo image.

$(1,0)$, the 257-th is $(0,1)$, and the last is $(254,1)$. The splitting-signal generated by $(p,s) = (0,1)$ has the highest energy 12290.08, and the splitting-signal with generator $(1,0)$ is the second with high energy signal, $E(1,0) = 9719.93$. These two signals are defined by the vertical and horizontal projections, respectively. It is interesting to note that the next high-energy splitting-signal with energy 8896.03 is the signal generated by $(1, 128)$, not the signal defined by the diagonal projection with $E(1, 1) = 8827.80$. The value of the minimum energy of the splitting-signals equals 8771.83.

In the case when N is prime, the image $f_{n,m}$ can be reconstructed from $f(x, y)$ by direction images of number $N + 1$,

$$f_{n,m} = \frac{1}{N} \left[\sum_{s=0}^{N-1} f_{1,s,(n+sm) \bmod N} + f_{0,1,m} \right] - NE[f], \quad n, m = 0 : (N - 1).$$

The image is the average of direction images $d^{(p,s)}(n, m) = f_{p,s,(pn+sm) \bmod N}$ generated by frequencies $(1, s)$, $s = 0 : N - 1$, and $(0, 1)$. We consider, for example, the image of size 257×257 and the $(1, 5)$-projection with $L = (1 + 5)(256 - 1) + 1 = 1531$ parallel rays. In the traditional backprojection, all values of L line-integrals $w(t)$, $t = 0 : L - 1$, along these rays will be put back and distributed equally at points of these rays. The splitting-signal $\{f_{1,5,t}; t = 0 : 256\}$ has 257 different values. Thus, in the tensor method, from these 1531 integrals, 257 values of the splitting-signal are calculated, and these 257 values (not 1531) will be distributed back along 1531 rays. This is the main difference: along the same rays, these two methods put back different values. BP adds average values of the integrals. In the tensor method, these integrals $w(t)$ are transformed into 257 new combinations $f_{1,5,t}$ of integrals, and then put back along the rays, to get the exact reconstruction. We here remind the reader that the tensor method works on the Cartesian grid, and the BP uses the polar grid, which does not cover the Cartesian grid.

The following should also be noted when applying the Fourier Slice Theorem for image reconstruction [29, 30, 54]. The calculation of the 2-D Fourier transform of the unknown image $f(x, y)$ is incomplete. Such an incomplete 2-D DFT is defined on the polar grid along a finite number of radial lines, or better said, is filled by calculating 1-D DFTs over the projections. Then, these spectral components on the radial points are transferred to the Cartesian grid, by using methods of interpolation. Even in the case when such transformation of spectral data from the polar grid to the Cartesian grid is performed with a small error, this transformation does not preserve, but breaks the mathematical structure of the 2-D DFT of the image. Consider for example, any projection which is different from the projections by $0°, 45°$, and $90°$, such as the projection calculated by the angle $\text{arctg}(2)$. Our example is for the discrete image 16×16. When using the Fourier Slice Theorem, the 2-D DFT of the discrete image will be calculated at 8 frequency-points on the line l_1 which is considered in the square grid $Y_{16,16} = \{(p, s); p, s = -8 : 7\}$ with the original $(0, 0)$ in the center, as shown in Figure 7.5 part a. Since the tensor and paired representations are described on the square grid $X_{16,16} = \{(p, s); p, s = 0 : 15\}$, and the 2-D DFT is periodic, we transfer the four parts of $Y_{16,16}$ into $X_{16,16}$ by the following simple rule:

$$Y = \begin{bmatrix} A & B \\ C & D \end{bmatrix} \rightarrow \begin{bmatrix} D & C \\ B & A \end{bmatrix} = X.$$

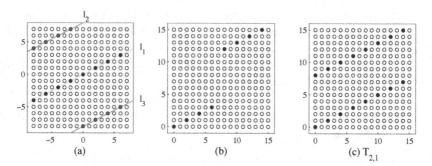

FIGURE 7.5
(a) The grid $Y_{16,16}$ with 16 points on the lines l_1, l_2, and l_3, and the grid $X_{16,16}$ with (b) 8 points on the line l_1, and (c) 16 points of the group $T_{2,1}$.

As a result, eight frequency-points on the line l_1 will be located on the grid $X_{16,16}$ as shown in part b. According to the tensor representation, the projection data at the angle arct(2) define the 2-D DFT of the image $f_{n,m}$ at 16 frequency-points of the cyclic group $T_{2,1}$. This complete group of frequency-points on two parallel lines with eight points each is shown in part c. The same group on the grid $Y_{16,16}$ is shown in part a, as the set of frequency-points on the line l_1 and two additional parallel lines l_2 and l_3, which are missing when the Fourier Slice Theorem is used to calculate the 2-D DFT of the discrete image by the given projection. The 2-D DFT of the image does not have a smooth form. The missing frequency-points on these two additional lines may only be substituted by interpolating data from other projections (which are assumed to be "nearly independent" [30]), or filled by zeros, which leads to an error of reconstruction in both cases. Similar reasoning can be applied for many other projections as well.

7.2.1 Tensor method of summation of projections

As mentioned above, we consider that the set of uniformly distributed angles in the interval $[0°, 180°]$, for collecting the projection data, does not provide the correct calculation of the reconstruction image in the frequency and spatial Cartesian grids. The discrete images of size $N \times N$, when N is a power of 2 and prime, are defined by very different sets of angles for projections and in the frequency domain they are defined by different sets $T_{p,s}$ of frequency-points, and this fact is not taken into consideration in the method of summation.

As examples, we consider the tomo image of size 256×256, as well as the same image in 257×257, by adding a zero column and row to the image. Following the tensor representation, we take 384 and 258 projections defined by the angles of sets $\Phi(256)$ and $\Phi(257)$, respectively. We use the following simple formula for reconstructing the discrete image $\hat{f}_{n,m}$ from projection data (without any filtration):

$$\hat{f}_{n,m} = \frac{1}{M} \sum_{\vartheta = \vartheta(p,s) \in \Phi(N)} \frac{p_\vartheta(np + ms)}{\text{card}(l_\vartheta(np + ms))},$$

$$n, m = 0 : (N - 1), \tag{7.4}$$

where M is the number of projections, i.e., $M = 394$ and $M = 258$, for $N = 256$ and $N = 257$, respectively. Here, $p_\vartheta(t) = v(t)$ denotes the sums of

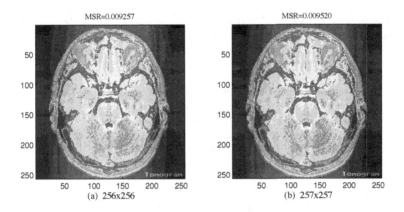

(a) 256x256 (b) 257x257

FIGURE 7.6
The reconstruction of the tomo image of size (a) 256×256, by 384 projections, and (b) 257×257, by 258 projections.

the image $f_{n,m}$ at the points on the lines

$$l_\vartheta(t) = \{(n_1, m_1); \, n_1 p + m_1 s = t\}.$$

The values of $v(t)$ can be calculated from the line-integrals $w(t)$ of the image $f(x, y)$ along the G-rays of the (p, s)-projection, as described in Chapter 3. Function $\text{card}(l_\vartheta(t))$ is used for the cardinality, or number of grid points on these lines. The weight $\text{card}(l_\vartheta(t))$ of the projection $p_\vartheta(t)$ takes values from the interval $[1, N]$, and the higher the frequency (p, s) generating the angle ϑ, the smaller this weight. Figure 7.6 shows the result of image reconstruction in part a, for the 256×256 case, and in b, for the 257×257 case. The mean-square-root error of image reconstruction equals 0.009257 in (a) and 0.009520 in (b). The obtained reconstructions are the high-quality images with all fine details of the original image. The time data from running MATLAB-based codes on a personal computer with Intel Core Duo processor at 2.99 GHz speed are equal to 4.2 min and 3 min, for calculating the reconstruction for images 256×256 and 257×257, respectively.

Below is the script of the program `bptt_tomo256book.m`, which is used to accomplish the above described tensor method of backprojection of the tomo image 256×256, which is shown in Figure 7.6 in part a. The image seen when backprojecting only one (p, s)-projection is calculated by the function projection_call.m. The concept of tensor representation is considered with direct calculation of the coordinates of points (x, y) on the grid $N \times N$, which lie on A-rays of the projection. Namely, for a given (p, s)-projection and each integer t from $[0, N-1]$, the line-integrals, or sums of the image along the coordinates of the following set of A-rays

$$l_{p,s}(t), \; l_{p,s}(t+N), \; l_{p,s}(t+2N), \; \cdots, \; l_{p,s}(t+(p+s-1)N)$$

are calculated separately. Each such set of rays determines the corresponding component of the tensor transform, $f_{p,s,t}$.

```
%========================================================
% call: bptt_tomo256book.m
% Backprojection method with 3N/2 specified angles.
% Function "projection_call.m" is used to calculate
% the image back-projected by one (p,s)-projection.
%
% Art Grigoryan,  07/21/2010,   EE UTSA
%========================================================
   N=256;
   fid=fopen('tomo256.img','rb');
   X=fread(fid,[N,N]); fclose(fid); clear fid; X=X';
   % The set of generators (p,s)
   L=N+N/2;
   Jps=ones(2,L);
   Jps(1,1:N)=0:N-1; Jps(2,N+1:end)=0:2:N-2;
   % Reconstruction of the image
   image_out=zeros(N);
   for k=1:L
      p=Jps(1,k); s=Jps(2,k);
      image_out=image_out+projection_call(X,p,s,N);
   end
   image_out=image_out/L;          % MSR=0.009257
   % Display the images
   figure
   subplot(1,2,1); imshow(X,[]);
   subplot(1,2,2); imshow(image_out,[]);
% -----------------------------------------------------
   function image_back=projection_call(X,p,s,N)
      image_back=zeros(N);
      TN=0:N:max(p,s)*N;
      ps1=p+s;
      for t=0:N-1
          value=zeros(1,ps1);   % for line-integrals
          L=zeros(1,ps1);       % for number of IE
          save_xy=zeros(2,ps1,ps1); % for coordinates of IE
          np=0;
```

```
for n=1:N
    ms=0;
    for m=1:N
        t_see=np+ms;
        k=floor(t_see/N); k1=k+1;
        t1=t_see-TN(k1);
        if t1==t
            Lk1=L(k1)+1; L(k1)=Lk1;
            save_xy(1,k1,Lk1)=m;
            save_xy(2,k1,Lk1)=n;
            value(k1)=value(k1)+X(m,n);
        end
        ms=ms+s;
    end
    np=np+p;
end
value_projected=value(1:k1)./L(1:k1);
for k2=1:k1
    for n=1:L(k2)
        x_onray=save_xy(1,k2,n);y_onray=save_xy(2,k2,n);
        image_back(x_onray,y_onray)=value_projected(k2);
    end
end
end
% ================================================================
```

This code can be written in an effective way, if we determine the rays that intersect all image elements, or squares $1/N \times 1/N$, on which the square $[0,1] \times [0,1]$ is divided by the Cartesian grid $N \times N$, as in the method of fast projection integrals by squares (FPIS).

Example 7.1 Consider the $N = 64$ case, where the gray-scale image $f(x,y)$ of size 64×64 is composed of seven rectangles with the data given in Table 7.1.

TABLE 7.1
Integer data of rectangles

k	x'_k	y'_k	$\Delta x'_k$	$\Delta y'_k$	R_k
1	18	19	6	15	9
2	29	29	14	25	6
3	39	39	18	18	13
4	41	14	17	17	3
5	14	38	2	10	10
6	36	11	10	8	15
7	24	9	7	25	4

Figure 7.7 shows seven rectangles that compose the image $f(x,y)$ (on the left), together with the gray-scale discrete image $f_{n,m}$ (on the right). The reconstruction $\hat{f}_{n,m}$ of the image $f(x,y)$ by the described tensor method, when

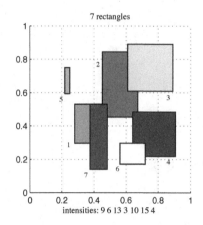
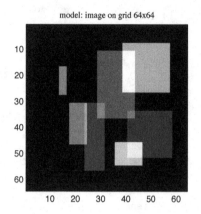

FIGURE 7.7

The image with seven rectangles and gray-scale image of size 64×64.

96 projections are used, is shown Figure 7.8 in part a. The mean-square-root error (MSR) of reconstruction equals 0.035878. All rectangles can be seen in their exact form with clear contours. For comparison, the image reconstructed by the filtered BP with 100 projections is shown in b. BP is accomplished by using the MATLAB function `iradon.m` with the standard Ram-Lak filter. These projections are defined by 100 angles distributed uniformly in the interval $[0, 180°]$. The reconstruction is an unfocused image with the error equal to 0.037464.

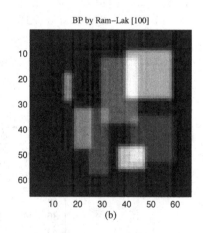

FIGURE 7.8

The reconstruction of the image (a) by 96 projections (without any filtering), and (b) by 100 projections in BP method with the cropped Ram-Lak filter.

Below is the main part of the program bptt_rectangles.m, which accomplishes the reconstruction of the image with the tensor method described by
(7.4), as well as the reconstruction by the filtered BP method with the ramp,
or Ram-Lak filter, which is shown in Figure 7.8.

```
%======================================================================
% call: bptt_rectangles.m
% Backprojection methods with 3N/2 specified angles.
% Art Grigoryan,  07/21/2010,  EE UTSA
%================================================================
N=64; many_rectangles=7;
data_allrect=zeros(many_rectangles,5);
data_allrect=[ 18      19      6      15      9
               29      29     14      25      6
               39      39     18      18     13
               41      14     17      17      3
               14      38      2      10     10
               36      11     10       8     15
               24       9      7      25      4];
% composition of the gray-scale image fd(x,y) from rectangles
imageSQ_model=zeros(N,N);
for k=1:many_rectangles
    data_1rect=data_allrect(k,:);
    x=data_1rect(1); dx=data_1rect(3);
    y=data_1rect(2); dy=data_1rect(4);
    d_intensity=data_1rect(5);
    x1=x+dx; y1=y+dy;
    imageSQ_model(x:x1,y:y1)=imageSQ_model(x:x1,y:y1)+d_intensity;
end
imageSQ_model=imageSQ_model(:,N:-1:1);
% Art Method    [96 projections]  ------------------
  X=imageSQ_model';
  % The set of generators (p,s)
  L=N+N/2;
  Jps=ones(2,L);
  Jps(1,1:N)=0:N-1; Jps(2,N+1:end)=0:2:N-2;
  % Reconstruction of the image
  image_out=zeros(N);
  for k=1:L
      p=Jps(1,k); s=Jps(2,k);
      image_out=image_out+projection_cal1(X,p,s,N);
  end
  image_out=image_out/N;
  % BP method [100 projections] -------------------
  f=linspace(0,180,100);
  Radon_ofX = radon(X,f);                      % 95x100
  X_iradon = iradon(Radon_ofX,f,'Ram-Lak');    % 66x66
  % Display the images
  figure;  colormap(gray);
```

```
subplot(1,2,1); imshow(image_out,[]); title('BP by Art [96]');
subplot(1,2,2); imshow(X_iradon,[]);  title('BP by Ram-Lak [100]');
%==================================================================
```

Example 7.2 Consider the image $f(x, y)$ on the square $[0, 1] \times [0, 1]$, which is composed of eight circles and two ellipses,

$$f(x, y) = \sum_{k=1}^{10} f_k(x, y) = \sum_{k=1}^{10} \begin{cases} I_k; \text{ if } \dfrac{(x - x_k)^2}{a_k^2} + \dfrac{(y - y_k)^2}{b_k^2} \leq r_k^2 \\ 0; \text{ otherwise} \end{cases}$$

The spatial data for these circles and ellipses, as well as the intensities, are given below.

k	1	2	3	4	5	6	7	8	9	10
xk	24	48	88	80	64	20	112	48	24	116
yk	40	52	60	24	92	104	112	16	76	24
ak	1	1	1	5	5	1	1	1	1	1
bk	1	1	1	4	8	1	1	1	1	1
rk	12	20	28	4	4	8	4	4	4	4
Ik	20	8	12	16	20	20	24	20	24	24

The geometry of these figures is in the integer form, but it will be normalized by factor of 127. The number 127 is prime and the reconstruction of the image is considered on the lattice 127×127. Figure 7.9 shows the sampled discrete image in part a. The method of backprojection by the tensor scanning scheme uses 128 projections. The reconstruction of the image is shown in b. The image reconstructed by the method of BP with 128 projections is shown in c.

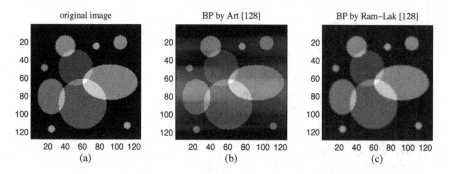

FIGURE 7.9
(a) The image 127×127, the reconstruction with 128 projections by the (b) tensor method and (c) BP method with the cropped Ram-Lak filter.

In the tensor method of summation, we can consider different sets of generators $J_{N,N}$. For example, the result of reconstruction in Figure 7.6(b) for the tomo image 257×257 is calculated when the set of generators is $J_1 = \{\{(1,s);\, s = 0 : 256\}, (0,1)\}$. Figure 7.10 shows the reconstruction by another set of 258 generators in part a, which equals $J_2 = \{\{(p,1);\, p = 0 : 256\}, (1,0)\}$.

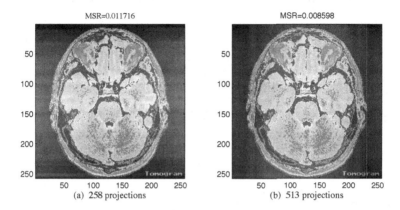

| MSR=0.011716 | MSR=0.008598 |

(a) 258 projections   (b) 513 projections

FIGURE 7.10
The reconstruction of the tomo image 257×257 by (a) the set of 384 projections J_2 and (b) the set J_3 of 513 projections.

MSR error of the reconstruction is 0.011716. One can also consider the symmetric scheme of scanning, when the set of generators is combined by these two sets, namely when

$$J_3 = J_1 \cup J_2 = \{(1,s);\, s = 0 : 256\} \cup \{(p,1);\, p = 0, 2 : 256\}\}.$$

The reconstruction of the image by 513 projections with this set is shown in part b. MSR error of this reconstruction is 0.008598.

7.3 Method of summation of line-integrals

Let P be the set of projections, i.e., the set of such triplets (p,s,t) for which integrals $w = w_{p,s}(t)$ of the image along the G-rays $px + ys = t$ are calculated. The image $f(x,y)$ is considered on the square $S = [0,1] \times [0,1]$, which is divided by the set of N^2 disjoint squares, or image elements, by the Cartesian lattice $N \times N$. If a G-ray passes through the square $[i,j]$, we denote the component of the integral in this square by $w_t^{(i,j)}$. Each G-ray is determined by the corresponding set of squares along which the ray passes. By introducing the field functions of G-particles, we calculate the 4-dimensional matrix $||c(n,m,i,j)|| = ||\psi_{n,m}(i,j)||$, coefficients of which define the number of G-rays passing through the squares $[n,m]$ and $[i,j]$ at the same time.

The backprojection method of averaging the values of all integrals along the rays passing through the squares on S can be described briefly as follows. Let $P(n, m, i, j)$ be the set of G-rays passing the squares $[n, m]$ and $[i, j]$ at the same time. Then, we consider the following system of equations:

$$q(n, m) = \sum_{[i,j] \in S, \, P(n,m,i,j) \neq \emptyset} \sum_{t=1}^{c(n,m,i,j)} w_t^{(i,j)}, \quad n, m = 0 : (N-1).$$

Here, $q(n, m)$ is the sum of all components of the line-integrals in the intersection with the square $[n, m]$.

If we propose the value of the reconstructed image at each point (n, m) in the Cartesian lattice by collecting all rays passing through the square $[n, m]$ and averaging their contribution, then this value will be calculated by

$$\bar{f}(n, m) = <q(n, m)> = \sum_{[i,j] \in S, \, P(n,m,i,j) \neq \emptyset} \sum_{t=1}^{c(n,m,i,j)} <w_t^{(i,j)}> .$$

In the model of image reconstruction under consideration with the tensor scanning scheme, the lengths of each G-ray intersecting the squares are equal, for each (p, s)-projection. This means the values $<w_t^{(i,j)}>$ are equal, i.e., $<w_t^{(i,j)}> = w^{(i,j)}$, and we can write that

$$\bar{f}(n, m) = \sum_{[i,j] \in S, \, P(n,m,i,j) \neq \emptyset} \sum_{t=1}^{c(n,m,i,j)} w^{(i,j)} = \sum_{[i,j] \in S} \psi_{n,m}(i, j) w^{(i,j)}.$$

7.4   Models with averaging

In this section, we analyze models of image reconstruction and their solutions, without considering orthogonal basis functions. The G-ray passes through a certain number of image elements that have, in general, a non-uniform distribution of intensities. The contribution of each image element in the line-integral is different. One can consider that each G-ray represents a sum of random variables, ξ. With no preliminary information about such variables, we first can consider that the contribution of each variable is uniform with the remaining elements of this sum. In other words, if the G-ray passes through m image elements $[i_k, j_k]$, $k = 1 : m$, then the line-integral along this ray is $w = \xi_{[i_1,j_1]} + \xi_{[i_2,j_2]} + \cdots + \xi_{[i_m,j_m]}$, where we assume $\xi_{[i_k,j_k]} = w/m$. By collecting the information about all G-rays passing through image elements, we can obtain some information about the mean or other statistics of random variables. We will describe simple models of image reconstruction and analyze approximations of the image, by considering first the mean values. The study of such models will allows us to approach the next steps of processing, to achieve a better approximation of the reconstruction. What is important to us is to keep the scanning schemes proposed in the tensor and paired transform-based methods of image reconstruction. In such scanning schemes, there is no coefficients in the geometry $w = \xi_{[i_1,j_1]} + \xi_{[i_2,j_2]} + \cdots + \xi_{[i_m,j_m]}$, because the length of intersection of the G-rays with any image element is the same. This is the main principle of scanning in our algorithms, which is valid for all projections.

We now consider the example of the nonuniform image and the lattice 2×2 on it. Let $f(x, y)$ be the following image:

$$f(x,y) = \begin{array}{|c|c|} \hline a(x,y) & b(x,y) \\ \hline c(x,y) & d(x,y) \\ \hline \end{array} = \begin{array}{|c|c|} \hline 1 & 2 \\ \hline 3 & 4 \\ \hline 5 & 6 \\ \hline 7 & 8 \\ \hline \end{array}. \tag{7.5}$$

This image has two different values on each square, as shown in Figure 7.11. For instance, in the first image element, or square $[0,0]$, the image has two values of $c(x,y)$, 5 and 7. The image has value 5 on the area which is two times smaller than the area with intensity 7. For simplicity of calculation, this image is considered on the square $[0,2] \times [0,2]$.

Our goal is to obtain a reconstruction, which is a discrete image $f_{n,m}$ of size 2×2. If the lattice 2×6 or 4×6 is considered on the image, then the reconstruction of the image as the discrete image of size 2×6 or 4×6 is exact. We here stand on the case 2×2. What 2×2-point discrete representation for this image should be considered as the discrete analog or reconstruction of the image? Can we consider it as one of the following images

$$\begin{bmatrix} 1 & 2 \\ 5 & 6 \end{bmatrix}, \quad \begin{bmatrix} 3 & 4 \\ 7 & 8 \end{bmatrix}, \quad \begin{bmatrix} 2 & 3 \\ 6 & 7 \end{bmatrix}, \quad \text{and} \quad \begin{bmatrix} 4 & 6 \\ 12 & 14 \end{bmatrix}?$$

It is difficult to select one of these or other images, without describing the mathematical model of image reconstruction and its solution.

7.4.1 Method of proportion

Because the tensor transform-based scanning scheme is used, we consider four projections, which are generated by $(p, s) = (1,0), (0,1), (1,1)$, and $(-1,1)$. For the given image, ten integrals of four projections can be written as

$w_{1,0}(0) = a_1 + b_1$	$w_{0,1}(0) = a_2 + c_2$
$w_{1,0}(1) = c_1 + d_1$	$w_{0,1}(1) = b_2 + d_2$
$w_{-1,1}(0) = a_3$	$w_{1,1}(0) = c_4$
$w_{-1,1}(1) = c_3 + b_3$	$w_{1,1}(1) = a_4 + d_4$
$w_{-1,1}(2) = d_3$	$w_{1,1}(2) = b_4$

The numbers a_k, b_k, c_k, and d_k define the contributions of the image elements $a(x,y), b(x,y), c(x,y)$, and $d(x,y)$ in the integrals.

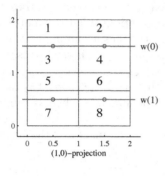

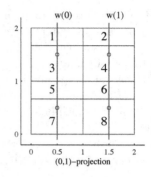

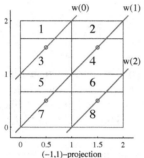

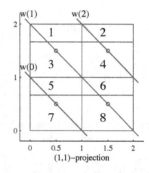

FIGURE 7.11
The lattice 2×2 on the image and four projections.

For the image $f(x, y)$ given above, these integrals are calculated as follows:

$w_{1,0}(0) = 3 + 4 = 7$	$w_{0,1}(0) = (1 + 3) + (5 + 7) = 16$
$w_{1,0}(1) = 7 + 8 = 15$	$w_{0,1}(1) = (2 + 4) + (6 + 8) = 20$
$w_{-1,1}(0) = \dfrac{2 \cdot 3 + 1}{3} = \dfrac{7}{3}$	$w_{1,1}(0) = \dfrac{2 \cdot 7 + 5}{3} = \dfrac{19}{3}$
$w_{-1,1}(1) = \dfrac{(2 \cdot 7 + 5) + (2 \cdot 4 + 2)}{3} = \dfrac{29}{3}$	$w_{1,1}(1) = \dfrac{(2 \cdot 3 + 1) + (2 \cdot 8 + 6)}{3} = \dfrac{29}{3}$
$w_{-1,1}(2) = \dfrac{2 \cdot 8 + 6}{3} = \dfrac{22}{3}$	$w_{1,1}(2) = \dfrac{2 \cdot 4 + 2}{3} = \dfrac{10}{3}$

The mean values of the image in image elements equal

$$\bar{a} = \frac{2 \cdot 3 + 1}{3} = \frac{7}{3}, \bar{b} = \frac{2 \cdot 4 + 2}{3} = \frac{10}{3}, \bar{c} = \frac{2 \cdot 7 + 5}{3} = \frac{19}{3}, \bar{d} = \frac{2 \cdot 8 + 6}{3} = \frac{22}{3},$$

and therefore, the discrete block-mean image is considered to be

$$\bar{f} = \{\bar{f}_{n,m}; \ n, m = 0, 1\} = \boxed{\begin{array}{|c|c|} 7/3 & 10/3 \\ \hline 19/3 & 22/3 \end{array}} = \boxed{\begin{array}{|c|c|} 2.33 & 3.33 \\ \hline 6.33 & 7.33 \end{array}}. \qquad (7.6)$$

We now consider the simple method of summation, when all line-integrals of rays passing through the certain image element are added. The values of $a, b, c,$ and d are estimated as follows:

$$\hat{a} = \frac{w_{1,0}(0) + w_{0,1}(0) + w_{-1,1}(0) + w_{1,1}(1)}{4} = \bar{a} + \frac{b_1 + c_2 + d_4}{4},$$

$$\hat{b} = \frac{w_{1,0}(0) + w_{0,1}(1) + w_{-1,1}(1) + w_{1,1}(2)}{4} = \bar{b} + \frac{a_1 + d_2 + c_3}{4},$$

$$\hat{c} = \frac{w_{1,0}(1) + w_{0,1}(0) + w_{-1,1}(1) + w_{1,1}(0)}{4} = \bar{c} + \frac{d_1 + a_2 + b_3}{4},$$

$$\hat{d} = \frac{w_{1,0}(1) + w_{0,1}(1) + w_{-1,1}(2) + w_{1,1}(1)}{4} = \bar{d} + \frac{c_1 + b_2 + a_4}{4}.$$

For the image given above, the estimated values differ from their means as follows: $\hat{a} - \bar{a} = 5.83$, $\hat{b} - \bar{b} = 5.83$, $\hat{c} - \bar{c} = 3.83$, and $\hat{d} - \bar{d} = 3.83$. We can conduct the evaluation of this method with respect to the exact model for image $f = [a, b; c, d]$. The estimations of image values can be written as

$$\hat{a} = \bar{a} + \frac{b+c+d}{4} = \frac{3}{4}\bar{a} + \frac{a+b+c+d}{4} = \frac{3}{4}\left(\bar{a} + \frac{e}{3}\right),$$

$$\hat{b} = \bar{b} + \frac{a+d+c}{4} = \frac{3}{4}\left(\bar{b} + \frac{e}{3}\right),$$

$$\hat{c} = \bar{c} + \frac{d+a+b}{4} = \frac{3}{4}\left(\bar{c} + \frac{e}{3}\right),$$

$$\hat{d} = \bar{d} + \frac{c+b+a}{4} = \frac{3}{4}\left(\bar{d} + \frac{e}{3}\right),$$

where $e = a + b + c + d$. The error of estimation is thus defined as $e/3$.

Another algorithm with the following estimations can also be considered:

$$\hat{a} = \frac{\frac{1}{2}w_{1,0}(0) + \frac{1}{2}w_{0,1}(0) + w_{-1,1}(0) + \frac{1}{2}w_{1,1}(1)}{4},$$

$$\hat{b} = \frac{\frac{1}{2}w_{1,0}(0) + \frac{1}{2}w_{0,1}(1) + \frac{1}{2}w_{-1,1}(1) + w_{1,1}(2)}{4},$$

$$\hat{c} = \frac{\frac{1}{2}w_{1,0}(1) + \frac{1}{2}w_{0,1}(0) + \frac{1}{2}w_{-1,1}(1) + w_{1,1}(0)}{4},$$

$$\hat{d} = \frac{\frac{1}{2}w_{1,0}(1) + \frac{1}{2}w_{0,1}(1) + w_{-1,1}(2) + \frac{1}{2}w_{1,1}(1)}{4}.$$

For the exact model, these equations can be written as

$$\hat{a} = \frac{\frac{1}{2}(a+b) + \frac{1}{2}(a+c) + a + \frac{1}{2}(a+d)}{4} = \frac{1}{2}a + \frac{e}{8} = \frac{1}{2}\left(a + \frac{e}{4}\right),$$

$$\hat{b} = \frac{\frac{1}{2}(a+b) + \frac{1}{2}(b+d) + \frac{1}{2}(c+b) + b}{4} = \frac{1}{2}b + \frac{e}{8} = \frac{1}{2}\left(b + \frac{e}{4}\right),$$

$$\hat{c} = \frac{\frac{1}{2}(c+d) + \frac{1}{2}(c+a) + \frac{1}{2}(c+b) + c}{4} = \frac{1}{2}c + \frac{e}{8} = \frac{1}{2}\left(c + \frac{e}{4}\right),$$

$$\hat{d} = \frac{\frac{1}{2}(c+d) + \frac{1}{2}(b+d) + d + \frac{1}{2}(a+d)}{4} = \frac{1}{2}d + \frac{e}{8} = \frac{1}{2}\left(d + \frac{e}{4}\right).$$

The error of estimation for this method is defined as $e/4$, which is less than $e/3$ in the first method. This example shows that the first averaging method results in a big error, in comparison with the second method. It also gave us the idea of using probability characteristics of the image along each ray, and defining a recursive method of image reconstruction. For that, we need to know only which image elements are intersected by each G-ray.

In the method described above, we use the information of the proportionality of the areas of rectangles into which the square $[0,2] \times [0,2]$ is divided. In other words, we use the proportional approach of distribution of random variables in the line-integrals. We therefore can call this method *the method of proportion*. In general, such information is unknown.

7.4.2   Method with probability model

We now describe a method of image reconstruction which is based on probability characteristics of the image. In the first stage, we consider the initial states; the probabilities of random variables along the rays equal $1/2$. The values $\hat{a}, \hat{b}, \hat{c}$, and $\hat{d}$ are calculated as shown above. The second stage of iteration will lead to non-uniform distribution of the contributions of each image element in line-integrals. For example, for the integral $w_{-1,1}(1) = c_3 + b_3$, the contributions of image elements c and b are considered to be $\hat{c}/(\hat{c}+\hat{b}) \cdot w_{-1,1}(1)$ and $\hat{b}/(\hat{c}+\hat{b}) \cdot w_{1,1}(1)$, respectively. By continuing such iteration, we obtain the image reconstruction if the convergence holds for such iteration.

Thus, the initial values of $\hat{a}, \hat{b}, \hat{c}$, and $\hat{d}$ are considered to be 1. Then, the marginal probabilities along the rays equal

$$
\begin{aligned}
&P_{ab}(\xi = a) = \frac{\hat{a}}{\hat{a} + \hat{b}}, \quad P_{ab}(\xi = b) = \frac{\hat{b}}{\hat{a} + \hat{b}}, \quad P_{ac}(\xi = a) = \frac{\hat{a}}{\hat{a} + \hat{c}}, \\
&P_{ac}(\xi = c) = \frac{\hat{c}}{\hat{a} + \hat{b}}, \quad P_{ad}(\xi = a) = \frac{\hat{a}}{\hat{a} + \hat{d}}, \quad P_{ad}(\xi = d) = \frac{\hat{d}}{\hat{a} + \hat{d}}, \\
&P_{bc}(\xi = b) = \frac{\hat{b}}{\hat{b} + \hat{c}}, \quad P_{bc}(\xi = c) = \frac{\hat{c}}{\hat{b} + \hat{c}}, \quad P_{bd}(\xi = b) = \frac{\hat{b}}{\hat{b} + \hat{d}}, \\
&P_{bd}(\xi = d) = \frac{\hat{d}}{\hat{b} + \hat{d}}, \quad P_{cd}(\xi = c) = \frac{\hat{c}}{\hat{c} + \hat{d}}, \quad P_{cd}(\xi = d) = \frac{\hat{d}}{\hat{c} + \hat{d}}.
\end{aligned}
\tag{7.7}
$$

In the next stage of the iteration, new values of variables $\hat{a}, \hat{b}, \hat{c}$, and $\hat{d}$ are calculated as

$$
\hat{a} = \frac{P_{ab}(\xi = a)w_{1,0}(0) + P_{ac}(\xi = a)w_{0,1}(0) + w_{-1,1}(0) + P_{ad}(\xi = a)w_{1,1}(1)}{4},
$$

$$
\hat{b} = \frac{P_{ab}(\xi = b)w_{1,0}(0) + P_{bd}(\xi = b)w_{0,1}(1) + P_{bc}(\xi = b)w_{-1,1}(1) + w_{1,1}(2)}{4},
$$

$$
\hat{c} = \frac{P_{cd}(\xi = c)w_{1,0}(1) + P_{ac}(\xi = c)w_{0,1}(0) + P_{bc}(\xi = c)w_{-1,1}(1) + w_{1,1}(0)}{4},
$$

$$
\hat{d} = \frac{P_{cd}(\xi = d)w_{1,0}(1) + P_{bd}(\xi = d)w_{0,1}(1) + w_{-1,1}(2) + P_{ad}(\xi = d)w_{1,1}(1)}{4}.
$$

Consider first the implementation of this algorithm on the exact model, when the image is $f = [1, 2; 3, 4]$. Below is the script run_iterations.m of the program, which can be used to accomplish the reconstruction of this image.

```
% ========================================================
% run_iterations.m         M.M. Grigoryan, 11/17/2012
x=[1 2
   3 4];
N=4;
y=mit_2by2Model2(N)
%    0.8873      2.0452
%    2.9937      3.9738
e1=norm(x-y);
%  N =     1       2       3       4       10
%  e1= 0.9045  0.3723  0.1326  0.1215  0.2198
% ----------------------------------------------------
function y=mit_2by2Model2(N)
%   image is x=[1 2; 3 4];
w101=3;    w102=7;
w011=4;    w012=6;
w1m11=1;   w1m12=5;   w1m13=4;
w111=3;    w112=5;    w113=2;
% initial values of states:
ap=1; bp=1; cp=1; dp=1;
for i1=1:N
    Pab_a=ap/(ap+bp); Pab_b=1-Pab_a;
    Pac_a=ap/(ap+cp); Pac_c=1-Pac_a;
    Pad_a=ap/(ap+dp); Pad_d=1-Pad_a;
    Pbc_b=bp/(bp+cp); Pbc_c=1-Pbc_b;
    Pbd_b=bp/(bp+dp); Pbd_d=1-Pbd_b;
    Pcd_c=cp/(cp+dp); Pcd_d=1-Pcd_c;
    ap=(Pab_a*w101+Pac_a*w011+w1m11+Pad_a*w111)/4/2;
    bp=(Pab_b*w101+Pbd_b*w012+Pbc_b*w1m12+w113)/4/2;
    cp=(Pcd_c*w102+Pac_c*w011+Pbc_c*w1m12+w111)/4/2;
    dp=(Pcd_d*w102+Pbd_d*w012+w1m13+Pad_d*w112)/4/2;
    y=[ap, bp; cp, dp];
end
y=2*y;
% ========================================================
```

With $N = 1, 2, 3,$ and 4 steps of iteration, we obtain, respectively, the following approximations of the image:

$$\begin{bmatrix} 1.5000 & 2.2500 \\ 2.7500 & 3.2500 \end{bmatrix}, \quad \begin{bmatrix} 1.1398 & 2.1261 \\ 2.8866 & 3.6895 \end{bmatrix}, \quad \begin{bmatrix} 0.9718 & 2.0668 \\ 2.9549 & 3.8884 \end{bmatrix}, \quad \begin{bmatrix} 0.8873 & 2.0452 \\ 2.9937 & 3.9738 \end{bmatrix}$$

The curve of the mean-square-error of approximation of the image on the first 30 iterations is shown in Figure 7.12, along with the approximation of the image after ten iterations. The convergence takes place in this algorithm, and the minimum error (0.1215) of approximation is achieved on the 4-th iteration.

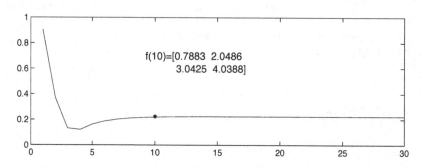

FIGURE 7.12
Errors of reconstruction of the image.

To apply the described method to the image $f(x,y)$ with a non-uniform distribution of intensities on image elements, we use the code with script `mit_2by2Model1.m`, which is similar to the script `mit_2by2Model2.m`, where only the values of integrals have been changed, as shown below.

```
% =========================================================
function y=mit_2by2Model1(N)
% integrals wps(t) for 4 projections
 w101=7;        w102=15;
 w011=16;       w012=20;
 w1m11=7/3;     w1m12=29/3;    w1m13=24/3;
 w111=19/3;     w112=29/3;     w113=10/3;
 . . .
% =========================================================
```

With $N = 2, 3, 4$, and 10 steps of iteration, we obtain, respectively, the following approximations of the image:

$$\begin{bmatrix} 3.4786 & 4.9806 \\ 7.1138 & 8.4610 \end{bmatrix}, \quad \begin{bmatrix} 3.0779 & 4.7116 \\ 7.4040 & 8.8971 \end{bmatrix}, \quad \begin{bmatrix} 2.8563 & 4.5628 \\ 7.5889 & 9.1112 \end{bmatrix}, \quad \begin{bmatrix} 2.5510 & 4.4206 \\ 7.8601 & 9.3220 \end{bmatrix}.$$

7.4.3   Reconstruction of the shifted image

In this section, we evaluate the method of proportion for an image which is not on the lattice 2×2. To simplify our calculations, we modify the image given in (7.5) in the following way. In the square $[0,2] \times [0,2]$, we consider the image

$$f(x,y) = \begin{array}{|c|c|} \hline a(x,y) & b(x,y) \\ \hline c(x,y) & d(x,y) \\ \hline \end{array} = \begin{array}{|c|c|} \hline 0 & 0 \\ \hline 3 & 4 \\ \hline 5 & 6 \\ \hline 7 & 8 \\ \hline \end{array}. \tag{7.8}$$

This image and rays of four projections are shown in Figure 7.13.

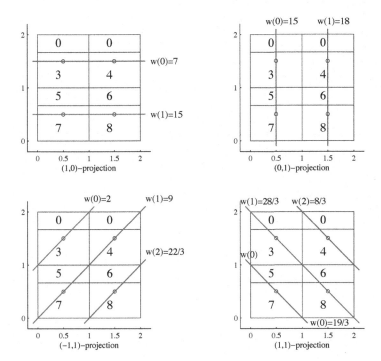

FIGURE 7.13
The lattice 2×2 on the image and four projections.

It is not difficult to calculate all ten line-integrals for this image,

$$
\begin{array}{|l|l|}
\hline
w_{1,0}(0) = a_1 + b_1 = 7 & w_{0,1}(0) = a_2 + c_2 = 15 \\
w_{1,0}(1) = c_1 + d_1 = 15 & w_{0,1}(1) = b_2 + d_2 = 18 \\
\hline
w_{-1,1}(0) = a_3 = 2 & w_{1,1}(0) = c_4 = \frac{19}{3} \\
w_{-1,1}(1) = c_3 + b_3 = 9 & w_{1,1}(1) = a_4 + d_4 = \frac{28}{3} \\
w_{-1,1}(2) = d_3 = \frac{22}{3} & w_{1,1}(2) = b_4 = \frac{8}{3} \\
\hline
\end{array}
\tag{7.9}
$$

The block-mean discrete image is considered to be

$$
\bar{f} = \begin{array}{|c|c|}
\hline
2 & 8/3 \\
\hline
19/3 & 22/3 \\
\hline
\end{array}
= \begin{array}{|c|c|}
\hline
2 & 2.67 \\
\hline
6.33 & 7.33 \\
\hline
\end{array}.
$$

To accomplish the calculation of image reconstruction, we use the code with script `run_iterations3.m`, which is given below. The function `mit_2by2Model3.m` is similar to `mit_2by2Model2.m`, where only the values of line-integrals are taken from (7.9). This main program calculates the approximation of the image on the 4-th stage of iteration and the mean-square distance. The distance is relative to the block-mean image $\bar{f}$.

```
% =======================================================
% run_iterations3.m      A.-M. Grigoryan, 01/11/2012
N=3;
y3=mit_2by2Model3(N)
%     2.9209     4.1578
%     7.2468     8.3708
% the block-mean image
xm =[2     8/3
     19/3 22/3];
d1=norm(xm-y3)      % 2.2247
% ----------------------------------
function y=mit_2by2Model3(N)
% integrals wps(t) for 4 projections
  w101=7;        w102=15;
  w011=15;       w012=18;
  w1m11=2;       w1m12=9;       w1m13=22/3;
  w111=19/3;     w112=28/3;     w113=8/3;
  ...
% =======================================================
```

With $N = 3, 4$, and 5 steps of iteration, we obtain, respectively, the following approximations $\hat{f}(N)$ of the image:

$$\hat{f}(3) = \begin{bmatrix} 2.9209 & 4.1578 \\ 7.2468 & 8.3708 \end{bmatrix}, \quad \hat{f}(4) = \begin{bmatrix} 2.7090 & 4.0082 \\ 7.4258 & 8.5796 \end{bmatrix}, \quad \hat{f}(5) = \begin{bmatrix} 2.5881 & 3.9326 \\ 7.5321 & 8.6840 \end{bmatrix}.$$

The curve of the mean-square distance of approximation of the image in
the first 30 iterations is shown in Figure 7.14, along with two approximations
of the image after three and ten iterations. The minimum distance (2.2247)
of approximation is achieved on the 3-rd iteration.

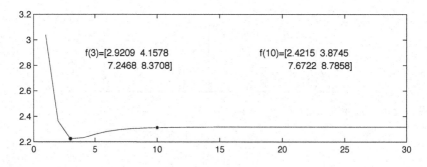

FIGURE 7.14
The distance of the approximation to the block-mean image.

7.4.4 Method of minimization of error

We now consider another approach and find the approximation of the re-constructed image by minimizing the mean-square error. For that, we again consider the example 2×2, with the known integrals, which we denote by y_k as follows:

$w_{1,0}(0) = a_1 + b_1 = y_0$	$w_{0,1}(0) = a_2 + c_2 = y_2$
$w_{1,0}(1) = c_1 + d_1 = y_1$	$w_{0,1}(1) = b_2 + d_2 = y_3$
$w_{1,1}(0) = c_4 = y_4$	$w_{-1,1}(0) = a_3 = y_7$
$w_{1,1}(1) = a_4 + d_4 = y_5$	$w_{-1,1}(1) = c_3 + b_3 = y_8$
$w_{1,1}(2) = b_4 = y_6$	$w_{-1,1}(2) = d_3 = y_9$

To find the approximation of values for a, b, c, and d, we demand the minimum of the mean-square-root error

$$\mathcal{L} = (a+b-y_0)^2 + (c+d-y_1)^2 + (a+c-y_2)^2 + (b+d-y_3)^2 + (c-y_4)^2$$
$$+ (a+d-y_5)^2 + (b-y_6)^2 + (a-y_7)^2 + (c+b-y_8)^2 + (d-y_9)^2.$$

By differentiating this expression by a, b, c, and d, we obtain the following system of linear equations:

$$(a+b-y_0) + (a+c-y_2) + (a+d-y_5) + (a-y_7) = 0,$$
$$(a+b-y_0) + (b+d-y_3) + (b-y_6) + (c+b-y_8) = 0,$$
$$(c+d-y_1) + (a+c-y_2) + (c-y_4) + (c+b-y_8) = 0,$$
$$(c+d-y_1) + (b+d-y_3) + (a+d-y_5) + (d-y_9) = 0.$$

These equations can be written in a new form and new variables Ω_x, which we introduce for $x = a, b, c, d$, as follows:

$$(a+b) + (a+c) + (a+d) + a = \Omega_a = y_0 + y_2 + y_5 + y_7,$$
$$(a+b) + (b+d) + b + (c+b) = \Omega_b = y_0 + y_3 + y_6 + y_8,$$
$$(c+d) + (a+c) + c + (c+b) = \Omega_c = y_1 + y_2 + y_4 + y_8,$$
$$(c+d) + (b+d) + (a+d) + d = \Omega_d = y_1 + y_3 + y_5 + y_9.$$

We write this system of linear equations in matrix form

$$\begin{pmatrix} 4 & 1 & 1 & 1 \\ 1 & 4 & 1 & 1 \\ 1 & 1 & 4 & 1 \\ 1 & 1 & 1 & 4 \end{pmatrix} \begin{pmatrix} a \\ b \\ c \\ d \end{pmatrix} = \begin{pmatrix} \Omega_a \\ \Omega_b \\ \Omega_c \\ \Omega_d \end{pmatrix} \tag{7.10}$$

as well as its solution

$$\begin{pmatrix} a \\ b \\ c \\ d \end{pmatrix} = \frac{1}{21} \begin{pmatrix} 6 & -1 & -1 & -1 \\ -1 & 6 & -1 & -1 \\ -1 & -1 & 6 & -1 \\ -1 & -1 & -1 & 6 \end{pmatrix} \begin{pmatrix} \Omega_a \\ \Omega_b \\ \Omega_c \\ \Omega_d \end{pmatrix}. \tag{7.11}$$

The matrix 4×4 can be written as

$$\begin{pmatrix} 6 & -1 & -1 & -1 \\ -1 & 6 & -1 & -1 \\ -1 & -1 & 6 & -1 \\ -1 & -1 & -1 & 6 \end{pmatrix} = 7 \begin{pmatrix} 1 & 0 & 0 & 0 \\ 0 & 1 & 0 & 0 \\ 0 & 0 & 1 & 0 \\ 0 & 0 & 0 & 1 \end{pmatrix} + \begin{pmatrix} -1 & -1 & -1 & -1 \\ -1 & -1 & -1 & -1 \\ -1 & -1 & -1 & -1 \\ -1 & -1 & -1 & -1 \end{pmatrix}.$$

By denoting the sum of projections by $S_w = \Omega_a + \Omega_b + \Omega_c + \Omega_d$, the solution can be written as

$$\begin{pmatrix} a \\ b \\ c \\ d \end{pmatrix} = \frac{1}{3} \begin{pmatrix} \Omega_a \\ \Omega_b \\ \Omega_c \\ \Omega_d \end{pmatrix} - \frac{1}{21} S_w. \tag{7.12}$$

We now describe this solution for the case when the image $f(x, y)$ is determined by (7.5). To accomplish the calculations, we consider the program with the following script mse_2by2Model1.m.

```
% call: mse_2by2Model1.m      A.-M. Grigoryan, Jan 11, 2012
% integrals wps(t) for 4 projections
w101=7;        w102=15;
w011=16;       w012=20;
w111=19/3;     w112=29/3;     w113=10/3;
w1m11=7/3;     w1m12=29/3;    w1m13=24/3;
% ========================================
y0=w101;       y1=w102;
y2=w011;       y3=w012;
y4=w111;       y5=w112;       y6=w113;
y7=w1m11;      y8=w1m12;      y9=w1m13;
Qa=y0+y2+y5+y7;
Qb=y0+y3+y6+y8;
Qc=y1+y2+y4+y8;
Qd=y1+y3+y5+y9;
Q=[Qa Qb Qc Qd];
A=Q/3-sum(Q)/21;
y=reshape(A,2,2)'
%    3.3492      5.0159
%    7.3492      9.2381
% ========================================
```

The result of calculations is the approximation

$$\bar{f} = \begin{bmatrix} 3.3492 & 5.0159 \\ 7.3492 & 9.2381 \end{bmatrix}.$$

We now change the values of all ten integrals, y_k, for the image $f = [1, 2; 3, 4]$, as shown below in the script mse_2by2Model2.m.

```
% call: mse_2by2Model2.m      A.-M. Grigoryan, Jan 11, 2012
```

```
% integrals wps(t) for 4 projections
w101=3;    w102=7;
w011=4;    w012=6;
w1m11=1;   w1m12=5;   w1m13=4;
w111=3;    w112=5;    w113=2;
% =======================================

...    ...    ...
A=Q/3-sum(Q)/21;
y=reshape(A,2,2)'
%    1.0000    2.0000
%    3.0000    4.0000
% =======================================
```

The calculation by this program results in the exact reconstruction of the image.

Thus we obtain the simple solution in (7.12) of the image reconstruction, when stating the problem in terms of minimization of the mean-square error. The only question is whatever this solution may have negative values. We leave the answer to this question to the reader as an exercise.

7.4.5 Corpuscular approach

In the considered tensor scanning scheme, four rays pass through each image element in the lattice 2×2. The geometrical presentation of each image element as a particle has the form of $\bigstar$. In matrix form, the following matrices correspond to four particles:

$$\phi_{0,0} = \begin{bmatrix} 4 & 1 \\ 1 & 1 \end{bmatrix}, \quad \phi_{0,1} = \begin{bmatrix} 1 & 4 \\ 1 & 1 \end{bmatrix}, \quad \phi_{1,0} = \begin{bmatrix} 1 & 1 \\ 4 & 1 \end{bmatrix}, \quad \phi_{1,1} = \begin{bmatrix} 1 & 1 \\ 1 & 4 \end{bmatrix}.$$

Presentation of the field by the particles is

$$A = a\phi_{0,0} + b\phi_{0,1} + c\phi_{1,0} + d\phi_{1,1} = \begin{bmatrix} 4a + b + c + d & a + 4b + c + d \\ a + b + 4c + d & a + b + c + 4d \end{bmatrix}.$$

Considering this 2-D field in the form of a 1-D vector

$$\begin{pmatrix} 4a + b + c + d \\ a + 4b + c + d \\ a + b + 4c + d \\ a + b + c + 4d \end{pmatrix} = \begin{pmatrix} 4 & 1 & 1 & 1 \\ 1 & 4 & 1 & 1 \\ 1 & 1 & 4 & 1 \\ 1 & 1 & 1 & 4 \end{pmatrix} \begin{pmatrix} a \\ b \\ c \\ d \end{pmatrix},$$

we obtain the same equation as in (7.10). If we consider the inverse matrix given in (7.11) as a field of particles, which we call *anti-particles*, these anti-particles in matrix form will be written as

$$\psi_{0,0} = \begin{bmatrix} 6 & -1 \\ -1 & -1 \end{bmatrix}, \quad \psi_{0,1} = \begin{bmatrix} -1 & 6 \\ -1 & -1 \end{bmatrix}, \quad \psi_{1,0} = \begin{bmatrix} -1 & -1 \\ 6 & -1 \end{bmatrix}, \quad \psi_{1,1} = \begin{bmatrix} -1 & -1 \\ -1 & 6 \end{bmatrix}.$$

Normalizing the particles as

$$\phi_{n,m} = \frac{1}{\sqrt{21}}\phi_{n,m}, \quad \psi_{n,m} = \frac{1}{\sqrt{21}}\psi_{n,m}, \quad n,m = 0,1,$$

we obtain the orthogonality between particles and anti-particles

$$(\phi_{n,m}, \psi_{n_1,m_1}) = \delta_{n,n_1}\delta_{m,m_1}, \quad n,m,n_1,m_1 = 0,1.$$

It should be noted that the anti-particles $\psi_{n,m}$ do not have representation in the form of four directions as the particles $\phi_{n,m}$. One may think about simplification of the orthogonal basis for effective realization. For instance, let us remove one projection from the scanning scheme, for instance, the $(-1,1)$-projection. In this case, the geometrical presentation of the particle is ⊢, and four particles are defined as

$$\phi_{0,0} = \begin{bmatrix} 3 & 1 \\ 1 & 1 \end{bmatrix}, \quad \phi_{0,1} = \begin{bmatrix} 1 & 3 \\ 0 & 1 \end{bmatrix}, \quad \phi_{1,0} = \begin{bmatrix} 1 & 0 \\ 3 & 1 \end{bmatrix}, \quad \phi_{1,1} = \begin{bmatrix} 1 & 1 \\ 1 & 3 \end{bmatrix}.$$

The field defined by these particles is described by the following matrix:

$$R = \begin{pmatrix} 3 & 1 & 1 & 1 \\ 1 & 3 & 0 & 1 \\ 1 & 0 & 3 & 1 \\ 1 & 1 & 1 & 3 \end{pmatrix}, \quad \text{and} \quad R^{-1} = \frac{1}{48}\begin{pmatrix} 21 & -6 & -6 & -3 \\ -6 & 20 & 4 & -6 \\ -6 & 4 & 20 & -6 \\ -3 & -6 & -6 & 21 \end{pmatrix}.$$

Then, the set of anti-particles becomes "heavy," in the sense that the coefficients of the matrices of these particles contain large numbers, when compared with the original case. Indeed, the following matrices determine these anti-particles:

$$\psi_{0,0} = \begin{bmatrix} 21 & -6 \\ -6 & -3 \end{bmatrix}, \quad \psi_{0,1} = \begin{bmatrix} -6 & 20 \\ 4 & -6 \end{bmatrix}, \quad \psi_{1,0} = \begin{bmatrix} -6 & 4 \\ 20 & -6 \end{bmatrix}, \quad \psi_{1,1} = \begin{bmatrix} -3 & -6 \\ -6 & 21 \end{bmatrix}.$$

When using four projections, one can observe the movement of the particles from one to another. The same movement holds for their anti-particles. The movement is periodic, and four consecutive movements correspond to one full circle. What we have just seen is the fact that the removal of one projection breaks this property.

If in addition we remove another diagonal projection, the $(1,1)$-projection, and consider the geometry of the particle as ⊤, the particles will be defined as

$$\phi_{0,0} = \begin{bmatrix} 2 & 1 \\ 1 & 0 \end{bmatrix}, \quad \phi_{0,1} = \begin{bmatrix} 1 & 2 \\ 0 & 1 \end{bmatrix}, \quad \phi_{1,0} = \begin{bmatrix} 1 & 0 \\ 2 & 1 \end{bmatrix}, \quad \phi_{1,1} = \begin{bmatrix} 0 & 1 \\ 1 & 2 \end{bmatrix},$$

and we obtain a singular field

$$R = \begin{pmatrix} 2 & 1 & 1 & 0 \\ 1 & 2 & 0 & 1 \\ 1 & 0 & 2 & 1 \\ 0 & 1 & 1 & 2 \end{pmatrix}, \quad \det(R) = 0.$$

These examples show that for a large size $N \times N$ of image in reconstruction, an extra number of projections may significantly simplify the structure of the anti-particles, which in turn will result in an effective reconstruction of the image. We may assume that the tensor scanning scheme is one of the "optimal" schemes in image reconstruction.

7.5 General case: Probability model

In this section, we describe the probability model method for reconstructing the image of size $N \times N$, where $N > 2$. Each G-rays $l_{p,s}(t)$ of the (p, s)-projection is described by the following set of triplets:

$$l_{p,s}(t) = \{(n_k, m_k, \Delta l_k); \; k = 1 : \operatorname{card} l_{p,s}(t)\} \tag{7.13}$$

where the pairs (n_k, m_k) define the image elements (IEs) trough which the G-ray passes. We denote by $I(p, s)$ the set of these IEs, and by card its cardinality. It is the discrete length of the G-ray. Δl_k are the lengths of intersections of the ray with the corresponding k-th IEs in the set $I(p, s)$.

On the first stage of the algorithm, the line-integral $w_{p,s}(t)$ along the G-ray is distributed between image elements $[n_k, m_k]$ of the set $I(p, s)$ in proportion to the lengths Δl_k,

$$w_{p,s,t}(n_k, m_k) = w_{p,s,t}^{(n_k, m_k)} = w_{p,s}(t) \frac{\Delta l_k}{\sum \Delta l_i}. \tag{7.14}$$

Here, the summation of Δl_i is by all image elements of the set $I(p, s)$. The intersection of the G-ray with each image element has the same length. Therefore, the above equation can be written as

$$w_{p,s,t}(n_k, m_k) = \frac{w_{p,s}(t)}{\operatorname{card} l_{p,s}(t)}, \quad k = 1 : \operatorname{card} l_{p,s}(t). \tag{7.15}$$

The values of the discrete image are therefore defined as

$$a_{n,m} = \frac{1}{c(n, m)} \sum_{(p,s,t)} w_{p,s,t}(n, m), \quad n, m = 0 : (N - 1). \tag{7.16}$$

Here $c(n, m)$ is the number of G-rays $l_{p,s}(t)$ passing through the image element, or square $[n, m]$. The summation of $w_{p,s,t}(n, m)$ is by all triplets (p, s, t) defining such G-rays. In the considered tensor scheme of scanning the image by G-rays, the matrix $\|c(n, m)\|$ is known, and there is no need to calculate these coefficients. All coefficients $c(n, m)$ are equal in this matrix.

To demonstrate the result of approximation of the image, $a_{n,m}$, on this first stage, we consider the following example when $N = 64$. The gray-scale

TABLE 7.2
Integer data of rectangles

k	x'_k	y'_k	$\Delta x'_k$	$\Delta y'_k$	R_k
1	10	15	10	15	1
2	12	40	18	10	2
3	40	40	10	20	1
4	45	10	5	20	3

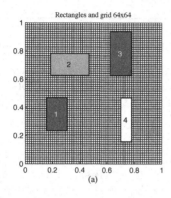
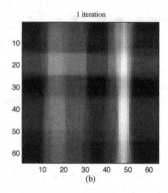

FIGURE 7.15
(a) Image with 4 rectangles on the square 64×64 and (b) the first iteration.

image on the square $[0,1] \times [0,1]$ is composed of four rectangles with the data given in Table 7.2. The data is normalized with respect to the lattice 64×64. Figure 7.15 shows these four rectangles in part a. The result of the calculation of the image $a_{n,m}$ after the first iteration is shown in part b. The appearance of each rectangle is spread in two major directions, vertical and horizontal.

On the first stage of the algorithm, which we also call *the zero stage*, we receive the image which is considered to be a rough estimation of the original image; it gives us some information about the probability of the image.

We now consider the next stage of the reconstruction. The line-integrals are considered along the new image $a_{n,m}$. The values of the new image on the squares $[n_k, m_k]$, through which the ray $l_{p,s}(t)$ passes, will be denoted by $s^1_{p,s,t}(n_k, m_k)$. Along this ray, the probability density function at point (n_k, m_k), which is referred to as the square $[n_k, m_k]$, is calculated by

$$p^{(1)}_{p,s,t}(n_k, m_k) = \frac{s^{(1)}_{p,s,t}(n_k, m_k)}{\sum\limits_i s^{(1)}_{p,s,t}(n_i, m_i)}, \quad k, (i) = 1 : \mathsf{card}\, l_{p,s}(t). \qquad (7.17)$$

Therefore, the calculation in (7.15) will be changed to

$$w^{(1)}_{p,s,t}(n_k, m_k) = w_{p,s}(t) p^{(1)}_{p,s,t}(n_k, m_k), \quad k = 1 : \mathsf{card}\, l_{p,s}(t). \qquad (7.18)$$

The image at this stage is calculated by

$$a_{n,m}^{(1)} = \frac{1}{c(n,m)} \sum_{(p,s,t)} w_{p,s,t}^{(1)}(n,m), \quad n,m = 0 : (N-1). \tag{7.19}$$

This process will be repeated a few times, and the probability density functions will be recalculated on each new image, until the given number of iterations is accomplished. At each new stage of the iteration, the obtained image $a_{n,m}^{(n)}$, $n \geq 1$, is used to get probability information for the next image, by calculating the probability density function by

$$p_{p,s,t}^{(n+1)}(n_k, m_k) = \frac{s_{p,s,t}^{(n+1)}(n_k, m_k)}{\sum_i s_{p,s,t}^{(n+1)}(n_i, m_i)}, \quad k, (i) = 1 : \mathbf{card}\, l_{p,s}(t). \tag{7.20}$$

Here, $s_{p,s,t}^{n+1}(n_k, m_k)$ denotes the values of the new image on the squares $[n_k, m_k]$ through which the ray $l_{p,s}(t)$ passes. The image at this stage is calculated by

$$a_{n,m}^{(n+1)} = \frac{1}{c(n,m)} \sum_{(p,s,t)} w_{p,s,t}^{(n+1)}(n,m), \quad n,m = 0 : (N-1), \tag{7.21}$$

and the distribution of the line-integral among the image elements $[n_k, m_k]$ is defined as

$$w_{p,s,t}^{(n+1)}(n_k, m_k) = w_{p,s}(t)p_{p,s,t}^{(n+1)}(n_k, m_k), \quad k = 1 : \mathbf{card}\, l_{p,s}(t). \tag{7.22}$$

As an example, we consider the same image 64×64 with four rectangles. Figure 7.16 illustrates the process of reconstruction of this image on different stages of the calculation. In parts a, b, c, and d, the results of image reconstruction after $5, 10, 20$, and 50 iterations are shown, respectively.

The described process of iterations converges and results in a good-quality reconstruction. Figure 7.17 shows the curve of differences between the approximations after k and $k-1$ iterations, where $k = 2 : 100$,

$$\varepsilon_k = \varepsilon(a^{(k)}, a^{(k-1)}) = \frac{1}{N^2}\sqrt{\sum_{n=0}^{N-1}\sum_{n=0}^{N-1}[a_{n,m}^{(k)} - a_{n,m}^{(k-1)}]^2}, \quad (N = 64).$$

The difference ε_k decreases exponentially to zero, and $\varepsilon_{100} = 8.3545 \cdot 10^{-6}$. The approximation process results in a high-quality image when N increases. Figure 7.18 shows the results of image reconstruction after 100 and 200 iterations in parts a and b, respectively. The following is important to mention. Given the number of iterations, the final image should be normalized by the factor M_1, which is approximately equal to $1/N$, for large N. This factor is used to display the image, and calculate the differences ε_k and the errors of reconstruction. This factor is not used in the process of iterations described above. Below are a few values of the factor M_1:

N	8	16	32	64	128	256
M_1	0.1362667	0.06465907	0.03160831	0.0156805	0.0078208	0.0039045

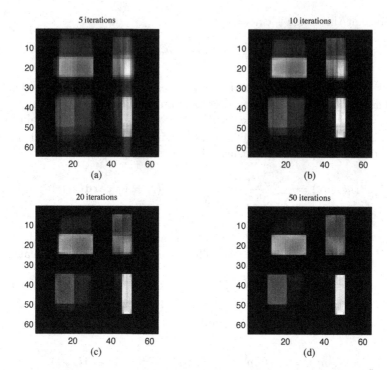

FIGURE 7.16
Reconstruction of the image with 5, 10, 20, and 50 iterations.

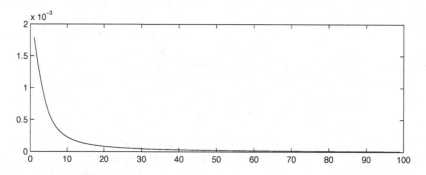

FIGURE 7.17
Difference between the first 100 iterations.

7.5.0.1   Code of the reconstruction

To implement the proposed algorithm with the probability model, and print all figures described above for this method, we use the program with the script `mreconstruction_bypm1.m`, which is given below. The image is composed of four

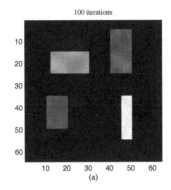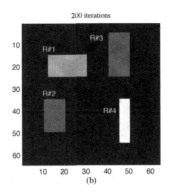

FIGURE 7.18
Reconstruction of the image with (a) 100 and (b) 200 iterations.

rectangles with data given in Table 7.2. The number of iterations is 200. This program uses the same functions `tensor_scheme_pst`, `general_scheme_pst2`, and `fieldpst_tensor_scheme`, which are described in Section 6.3 in the program `reconstruction_byR.m`.

```
% --------------------------------------------------------
% call: mreconstruction_bypm1.m   / A.-M. Grigoryan, 2011
N=64;
M=3*N*N/8+2; M1=1/N;
delta=1/N;
N_rect=4;
MNrects=[10 15 20 30
         12 40 30 50
         40 40 50 60
         45 10 50 30];
Matr_rect=delta*MNrects;
E=[1 2 1 3];       % intensities of 4 rectangles
% Plot the grid NxN
figure;  colormap(gray);
subplot(1,2,1);
draw_grid; axis image; pause(1);
% ==================== Processing ======================
ps=ps_generators(N); L=size(ps,1);
eps100=eps*100;
N_iterations=200; % number of iteration
subplot(1,2,2);
for iter=1:N_iterations  % loop for iterations
    Y=zeros(N,N);
    for k=1:L   % loop for (p,s)-projections
        p=ps(k,1); s=ps(k,2);
        %-------------------------------
        [p1,s1,t]=tensor_scheme_pst(p,s,N);
```

```
%-----------------------------------
Length_t=length(t);
for it=1:Length_t % loop for (p,s,it) G-rays
    sum_E=fieldpst_tensor_scheme(p1,s1,t(it), ...
        N,N_rect,Matr_rect,E);
    if (sum_E>eps)
        f=general_scheme_pst2(p1,s1,t(it),N);
        f_sz2=size(f,2);
        L0=0;
        for nf=1:f_sz2
            L0=L0+f(3,nf);
        end
        if iter==1
            for nf=1:f_sz2
                m1=f(1,nf); n1=f(2,nf);
                Y(m1,n1)=Y(m1,n1)+sum_E/L0;
            end
        else
        % =========  STATISTICS =============
        %      Integrate for statistical rays
        S_po=0;
        for nf=1:f_sz2
            m1=f(1,nf); n1=f(2,nf);
            S_po=S_po+R(m1,n1);
        end
        if (S_po>eps100)
            % Calculate statistical rays
            for nf=1:f_sz2
                m1=f(1,nf); n1=f(2,nf);
                Y(m1,n1)=Y(m1,n1)+ss*R(m1,n1);
            end
        end
        % ===================================
        end
    end
    end % for loop: (p,s,t)-rays
  end  % for loop: (p,s)-projections
  R=Y/M ;
  % display the normalized reconstruction on this stage
  subplot(1,2,2);
  if iter==1
    XR=R;
  else
      XR=R/M1;
  end
  imagesc(XR); axis image;
  title(sprintf(' %g Iterations',iter));  pause(0.1);
end % for loop: iterations
% -------------------------------------------------------------
```

In conclusion, we consider the example when image $f(x, y)$ is composed of rectangles that are not on the grid. Let $f(x, y)$ be the gray-scale image on the square $[0, 1] \times [0, 1]$, which is composed of four rectangles with the data given in Table 7.3. The data is normalized with respect to the lattice 64×64.

TABLE 7.3
Integer data of rectangles

k	x'_k	y'_k	$\Delta x'_k$	$\Delta y'_k$	R_k
1	10	15	10	$15 + 0.25$	1
2	$12 + 0.1$	40	18	10	2
3	40	$40 + 0.2$	10	20	1
4	45	10	$5 + 0.5$	20	3

Figure 7.19 shows four shifted rectangles on the square $[0, 1] \times [0, 1]$ in part a, along with the results of image reconstruction after 10 iterations in part b.

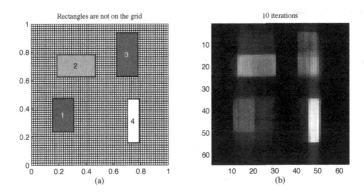

FIGURE 7.19
(a) Image with 4 rectangles and (b) image approximation after 10 iterations.

Figure 7.20 shows the results of image reconstruction after 50 and 100 iterations in parts a and b, respectively.

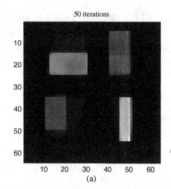
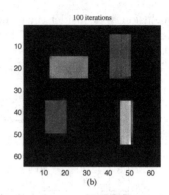

FIGURE 7.20
Reconstruction of the image after (a) 50 and (b) 100 iterations.

Problems

Problem 7.1 (Discrete Model) Consider the following two images on the square $[0,1] \times [0,1]$:

$$f_1(x,y) = \begin{cases} 2; & \text{if } x \in [0.6875, 0.8125], \ y \in [0.750, 0.875], \\ 3; & \text{if } x \in [0.1875, 0.3125], \ y \in [0.125, 0.250], \\ 0; & \text{otherwise,} \end{cases}$$

$$f_2(x,y) = \begin{cases} 1; & \text{if } \dfrac{(x-0.5)^2}{a^2} + \dfrac{(y-0.5)^2}{b^2} \le \dfrac{1}{4}, \\ 0; & \text{otherwise.} \end{cases}$$

where $a = 0.25$ and $b = 0.5$. Figure 7.21 shows rectangles and the ellipse together in part a, along with the discrete representation of the image composed of these figures in b. The discrete image is of size 61×61, and 61 is prime. Therefore, only 62 projections are required for the reconstruction of the image in b.

    A. Use the tensor method of summation by (7.4) and calculate the reconstruction of the image $f_1(x,y)$ on the Cartesian lattice 61×61.

    B. Repeat A for the image $f_2(x,y)$, as well as for the sum of images $f(x,y) = f_1(x,y) + f_2(x,y)$.

    Display the results and calculate errors of reconstruction in A, B, and C.

Problem 7.2 (Continuous Model)* Consider the following three-rectangle image on the square $[0,1] \times [0,1]$:

$$f(x,y) = \begin{cases} 1; & \text{if } x,y \in [3/8 - 1/16, 3/8 + 1/16], \\ 2; & \text{if } x,y \in [1/2 - 1/16, 1/2 + 1/16], \\ 1; & \text{if } x,y \in [5/8 - 1/16, 5/8 + 1/16], \\ 0; & \text{otherwise.} \end{cases}$$

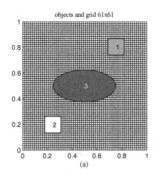
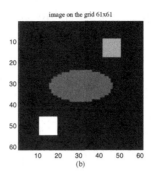

FIGURE 7.21
(a) Rectangles and ellipse on $[0,1] \times [0,1]$ and (b) the discrete image 61×61.

A. Use the tensor method of summation by (7.4) and calculate the reconstruction of this image on the Cartesian lattices 128×128 and 256×256. Calculate all required line-sums $p_\vartheta(np + ms)$ through the line-integrals, by transferring the geometry of G-rays to A-rays. Display the results and calculate the MRS error of reconstruction in both cases.

B. Find a way to improve the results of the image reconstruction by removing the shadows created by these rectangles. Show your results.

Problem 7.3 Consider the image on the square $[0,1] \times [0,1]$, which is composed of the sum of seven circles with the following coordinates (x_k, y_k), radii r_k, and intensities I_k:

k	1	2	3	4	5	6	7
x_k	6	12	22	20	16	5	28
y_k	10	13	15	6	23	26	28
r_k	3	5	7	4	7	2	1
I_k	5	2	3	4	5	5	6

The data of coordinates and radii are given in integer form, and should be normalized by the factor of N, when reconstructing on the lattice $N \times N$.

A. Use the tensor method of summation and calculate the reconstruction of this image on the Cartesian lattices 128×128 and 256×256. Display the results and calculate the MRS error of reconstruction in each case.

B. Remove the vertical and horizontal projections in the tensor scheme of scanning and reconstruct the image on the lattices 128×128 and 256×256.

C. Compare MRS errors of reconstruction in *A* and *B*. If errors of reconstruction in *B* are smaller than in *A*, explain this fact.

Problem 7.4 (Probability Model) Consider the following image on the square $[0,1] \times [0,1]$:

$$[f(x,y);\ 0 \le x, y \le 1] =$$

0	0	0	0	0	0	0	0
0	0	1	1	1	1	1	0
0	0	1	1	1	1	1	0
0	0	1	1	1	1	1	0
0	0	1	1	3	1	1	0
0	0	1	1	1	1	1	0
0	0	0	0	0	0	0	0
0	0	0	0	0	0	0	0

Each square is referred to as an image element of size $1/8 \times 1/8$. Use the method with statistics and calculate the 8×8 reconstruction of this image after $1, 2, 20$, and 30 iterations. Show the result and calculate the MSR error of reconstruction for each of these iterations.

Problem 7.5 (Probability Model) Consider the following image on the square $[0,1] \times [0,1]$:

$$[f(x,y);\ 0 \le x, y \le 1] =$$

1	0	0	0	0	0	1
0	0	0	1	0	0	0
0	0	0	1	0	0	0
0	1	1	7	1	1	0
0	0	0	1	0	0	0
0	0	0	1	0	0	0
1	0	0	0	0	0	1

Each square is referred to as an image element of size $1/7 \times 1/7$. Use the method with statistics and calculate the 7×7 reconstruction of this image after $10, 20$, and 30 iterations. Show the result and calculate the MRS error of reconstruction for each of these iterations.

Problem 7.6 (Probability Model) Consider the following image on the square $[0,1] \times [0,1]$:

$$[f(x,y);\ 0 \le x, y \le 1] =$$

0	0	2	2	0	2	0	0
1	1	0	0	4	6	4	0
1	1	2	0	0	2	0	1
0	0	2	0	0	2	0	1
2	0	2	0	0	2	0	0
0	4	0	1	1	0	0	0
3	3	3	0	0	2	2	2
0	0	0	2	2	0	0	0

Each square is referred to as an image element of size $1/8 \times 1/8$.

A. Use the method with statistics and calculate the 8×8 approximation of this image after $10, 50$, and 100 iterations.

B. Calculate the errors of approximation in $10, 50$, and 100 iterations.

C. Repeat *A* and *B* for the 16×16 approximation of the image.

Problem 7.7 In the program `mreconstruction_bypm1.m`, which is given in §7.5.0.1, the calculation of normalized integrals $w_{p,s}(t)/\sum_i \Delta l_i$ in (7.18) is accomplished for each projection inside the loop for iterations.

A. Improve this program and write a new program which calculates all these values separately, saves them, and then uses these integrals during the next process of iterations.

B. Determine if your program performs all calculations in MATLAB faster than the program `mreconstruction_bypm1.m` for the same four rectangles shown in Figure 7.15 in part a. Show your results for the $N = 64$ and 128 cases.

Problem 7.8 Consider the image on the square $[0, 1] \times [0, 1]$, which is composed of the sum of seven rectangles with the following coordinates $[x_{k,1}, y_{k,1}, x_{k,2}, y_{k,2}]$ and intensities I_k :

k	1	2	3	4	5	6	7
$x_{k,1}$	10	12	40	45	29	20	30
$y_{k,2}$	15	40	40	10	29	7	2
$x_{k,1}$	20	30	50	50	35	40	33
$y_{k,2}$	30	50	60	30	35	10	16
I_k	1	2	1	3	2	1	1

The data of coordinates are given in integer form, and should be normalized by the factor of 64.

A. Use the method with statistics and the set of 96 generators (p, s) $J^1 = \{(1, s), s = 0 : 63\} \cup \{(2p, 1), p = 0 : 31\}$, and calculate the 64×64 approximation of this image after $10, 50$, and 100 iterations.

B. Repeat *A* for the following set of 127 generators: $J^3 = \{(1, s), s = 0 : 63\} \cup \{(p, 1), p = 0, 2 : 63\}$.

C. Show the results in *A* and *B*, calculate the MSR errors of the obtained approximations, and compare the results.

Problem 7.9 Consider the image composed from the ten circles and ellipses,

$$f(x, y) = \sum_{k=1}^{10} \begin{cases} I_k; \text{ if } \dfrac{(x - x_k)^2}{a_k^2} + \dfrac{(y - y_k)^2}{b_k^2} \leq r_k^2 \\ 0; \text{ otherwise} \end{cases}$$

whose data (in integer form) and intensities I_k are the following:

k	1	2	3	4	5	6	7	8	9	10
x_k	24	48	88	80	64	20	112	48	24	116
y_k	40	52	60	24	92	104	112	16	76	24
a_k	1	1	1	5	5	1	1	1	1	1
b_k	1	1	1	4	8	1	1	1	1	1
r_k	12	20	28	4	4	8	4	4	4	4
I_k	20	8	12	16	20	20	24	20	24	24

The data of coordinates are given in integer form, and should be normalized by the factor of $N = 128$. Use the method with statistics and the set of $3N/2$ generators (p, s) $J^1 = \{(1, s), s = 0 : N - 1\} \cup \{(2p, 1), p = 0 : N/2 - 1\}$, and calculate the $N \times N$ approximation of this image after 50 and 100 iterations. Show the results and calculate the MRS errors of the obtained approximations.

Problem 7.10 Consider the image composed of one ellipse,

$$f(x, y) = \begin{cases} 1; \text{ if } \dfrac{[(x - 0.5)\cos\alpha - (y - 0.5)\sin\alpha]^2}{0.2^2} + \\ \quad + \dfrac{[(x - 0.5)\sin\alpha + (y - 0.5)\cos\alpha]^2}{0.1^2} \leq 1 \\ 0; \text{ otherwise} \end{cases}$$

where $x, y \in [0, 1]$, and the angle $\alpha = 30°$.

A. Calculate the results of reconstruction of this image on the lattice 128×128 after performing 20 and 50 iterations. Round the results of reconstruction and print them.

B. Sample the given image $f(x, y)$ on the lattice 128×128 and calculate the error of reconstructions in A.

Bibliography

[1] J.W. Cooley and J.W. Tukey, "An algorithm the machine computation of complex Fourier series," *Math. Comput,* vol. 9, no. 2, pp. 297-301, 1965.

[2] D.F. Elliot and K.R. Rao, *Fast Transforms: Algorithms, Analyzes, and Applications,* New York, Academic, 1982.

[3] L. Rabiner and R.W. Schaefer, *Speech Signal Processing,* Prentice Hall, Englewood Cliffs, New Jersey, 1983.

[4] R.E. Blahut, *Fast Algorithms for Digital Signal Processing,* Addison-Wesley, Reading, Massachusetts, 1985.

[5] P. Duhamel and M. Vetterli, "Fast Fourier transforms: A tutorial review and state of the art," *Signal Processing,* vol. 19, pp. 259-299, 1990.

[6] *Handbook for Digital Signal Processing,* edited by J.F. Kaiser and S.K. Mitra, Wiley-Interscience, New York, 1993.

[7] W.K. Pratt *Digital Image Processing,* Wiley-Interscience, New York, 1978.

[8] A. Rosenfeld and C. Kak, *Digital Image Processing,* vols. I and II. Academic Press, New York, 1982.

[9] Anil K. Jain, *Fundamentals of Digital Image Processing,* Prentice Hall, New Jersey, 1989.

[10] R.C. Gonzalez and R.E. Woods, *Digital Image Processing,* 2nd edition, Prentice Hall, New-Jersey, 2002.

[11] R.V. L. Hartley, A more symmetrical Fourier analysis applied to transmission problems," *Proc. IRE,* 30, pp. 144-150, March 1942.

[12] R.N. Bracewell, *The Hartley Transform,* Oxford University Press, 1986.

[13] R.N. Bracewell, O. Buneman, H. Hao, and J. Villasenor, "Fast two-dimensional Hartley transform," *Proc. IEEE,* 1986, 74, pp. 1282-1283.

[14] J.L. Wu and Pei Soo-Chang, "The vector split-radix algorithm for 2-D FHT," *IEEE Transactions on Signal Processing,* 1993, vol. 41, no. 2, pp. 960-965.

[15] G. Bi, "New split-radix algorithm for the discrete Hartley transform," *IEEE Transactions on Signal Processing,* February 1997, vol. 45, no. 2, pp. 297-302.

[16] N. Ahmed and R. Rao, *Orthogonal Transforms for Digital Signal Processing,* Springer-Verlag, Berlin, 1975.

[17] K.R. Rao and P. Yip, *Discrete Cosine Transform—Algorithms, Advantages, Applications,* Academic Press, Boston, 1990.

[18] W.B. Pennebaker and J.L. Mitchell, *JPEG Still Image Compression Standard,* Van Nostrand Reinhold, New York, 1993.

[19] G. Mandyam, N. Ahmed, and N. Magotra, "Lossless image compression using the discrete cosine transform," *JVCIR,* vol. 8, no. 1, pp. 21-26, Mar. 1997.

[20] M. Iwahashi, N. Kambayashi, and H. Kiya, "Bit reduction of DCT basis for transform coding," *ECJ,* vol. 80, no. 8, pp. 81-91, Aug. 1997.

[21] W. Philips, "Lossless DCT for combined lossy/lossless image coding," *Proc. IEEE Int. Conf. Image Process.,* vol. 3, pp. 871-875, 1998.

[22] M.S. Moellenhoff and M.W. Maier, "DCT transform coding of stereo images for multimedia applications," *IndEle,* vol. 45, no. 1, pp. 38-43, Feb. 1998.

[23] L.R. Welch, "Walsh functions and Hadamard matrices. Application of Walsh functions," *Proc. 1970, Symp.* Washington, D.C., pp. 163-165, 1970.

[24] S.S. Agaian, *Hadamard Matrices and Their Applications,* Lecture Notes in Mathematics 1168. Springer Verlag, Berlin, Heitelbery, New York, Tokyo, 1985.

[25] R. Gordon, "A tutorial on ART (algebraic reconstruction techniques)," *IEEE Trans. Nuclear Science,* vol. 21, pp. 78-93, 1974.

[26] R. A. Brooks and G. Di Chiro, "Principles of computer assisted tomography (CAT) in radiographic and radioisotopic imaging," *Phys. Med. Biol.,* vol. 21, no. 5, pp. 689-732, 1976.

[27] G.T. Herman and A. Lent, "Iterative reconstruction algorithms," *Comput. Biol. Med.,* vol. 6, pp. 273-294, 1976.

[28] S. Helgason, *The Radon Transform.* Progress in Mathematics 5, Birkhäuser Boston, 1980.

[29] G.T. Herman, *Image Reconstruction From Projections. The Fundamentals of Computerized Tomography,* Academic Press, New York, 1980.

[30] A. Rosenfeld and A.C. Kak, *Digital Picture Processing: Vol. 1,* Academic Press, Orlando, 1982.

[31] Y. Censor, "Finite series-expansion reconstruction methods," *Proc. of IEEE,* vol. 71, no. 3, pp. 409-419, 1983.

[32] A.H. Andersen and A.C. Kak, "Simultaneous algebraic reconstruction technique (SART): A superior implementation of the ART algorithm," *Ultrasonic Imaging,* vol. 6, pp. 81-94, 1984.

[33] R.C. Gonzalez and Wintz, *Digital Image Processing,* 2nd edition, Prentice Hall, Englewood Cliffs, New Jersey, 2001.

[34] G.N. Ramachandran and A.V. Lakshminaryanan, "Three-dimensional reconstruction from radiographs and electron micrographs," *Proc. Nat. Acad. Sci. USA,* vol. 68, pp. 2236-2240, 1971.

[35] A.N. Kotelnikov, *Theory of Potential Noise Stability,* Moscow, Nauka, 1956.

[36] A.V. Fiacco and G.P. McCormick, *Nonlinear Programming: Sequential Unconstrained Minimization Techniques,* John Wiley and Sons, Inc., New York, 1968.

[37] Lankaster P., *Theory of Matrices.* Academic Press, New York, London, 1969.

[38] H.J. Nussbaumer, *Fast Fourier Transform and Convolution Algorithms,* 2nd ed., Springer Verlag, Berlin, 1982.

[39] A.M. Grigoryan, "An algorithm of the two-dimensional Fourier transform," *Izvestiya VUZov SSSR, Radioelectronica,* vol. 27, no. 10, pp. 52-57, USSR, Kiev 1984 (translated at http://fasttransforms.com/Art-USSR-papers/Art_IzvuzovSSSR1984.pdf).

[40] A.M. Grigoryan, "An optimal algorithm for computing the two-dimensional discrete Fourier transform," *Izvestiya VUZ SSSR, Radioelectronica,* vol. 29, no. 12, pp. 20-25, USSR, Kiev 1986 (translated at http://fasttransforms.com/Art-USSR-papers/Art_IzvuzovSSSR1986.pdf).

[41] A.M. Grigoryan, "New algorithms for calculating discrete Fourier transforms," *Journal Vichislitelnoi Matematiki i Matematicheskoi Fiziki,* vol. 26, no. 9, pp. 1407-1412, AS USSR, Moscow, 1986 (translated at http://fasttransforms.com/Art-USSR-papers/Art_JVMATofASUSSR1986.pdf).

[42] A.M. Grigoryan and M.M. Grigoryan, "Two-dimensional Fourier transform in the tensor presentation and new orthogonal functions," *Avtometria, AS USSR Siberian section,* no. 1, pp. 21-27, Novosibirsk, 1986.

[43] A.M. Grigoryan, "An algorithm for computing the two-dimensional Fourier transform of equal orders," Proceedings of the IX All-Union Conference on Coding and Transmission of Information, Part 2, pp. 49-52, USSR, Odessa, 1988 (translated at http://fasttransforms.com/Art-USSR-papers/Art_AllUnConf1988.pdf).

[44] A.M. Grigoryan, "An algorithm for computing the discrete Fourier transform with arbitrary orders," *Journal Vichislitelnoi Matematiki i Matematicheskoi Fiziki,* AS USSR, vol. 30, no. 10, pp. 1576-1581, Moscow, 1991 (translated at http://fasttransforms.com/Art-USSR-papers/Art_JVMATofASUSSR1991.pdf).

[45] A.M. Grigoryan, and S.S. Agaian, "Split manageable efficient algorithm for Fourier and Hadamard transforms," *IEEE Trans. on Signal Processing,* vol. 48, no. 1, pp. 172-183, Jan. 2000.

[46] A.M. Grigoryan, "2-D and 1-D multi-paired transforms: Frequency-time type wavelets," *IEEE Trans. on Signal Processing,* vol. 49, no. 2, pp. 344-353, Feb. 2001.

[47] A.M. Grigoryan, "Method of paired transforms for reconstruction of images from projections: Discrete model," *IEEE Trans. on Image Processing,* vol. 12, no. 9, pp. 985-994, Sep. 2003.

[48] A.M. Grigoryan and S.S. Agaian, *Multidimensional Discrete Unitary Transforms: Representation, Partitioning and Algorithms,"* Marcel Dekker Inc., New York, 2003.

[49] A.M. Grigoryan and S.S. Agaian, "Transform-Based Image Enhancement Algorithms with Performance Measure," *Advances in Imaging and Electron Physics, Academic Press,* vol. 130, pp. 165-242, May 2004.

[50] F.T. Arslan and A.M. Grigoryan, "Fast splitting α-rooting method of image enhancement: Tensor representation," *IEEE Trans. on Image Processing,* vol. 15, no. 11, pp. 3375-3384, Nov. 2006.

[51] A.M. Grigoryan and M.M. Grigoryan, *Brief Notes in Advanced DSP: Fourier Analysis with MATLAB*, CRC Press, Taylor and Francis Group, Boca Raton, Florida, 2009.

[52] A.M. Grigoryan and K. Naghdali, "On a method of paired representation: Enhancement and decomposition by series direction images," *Journal of Mathematical Imaging and Vision*, vol. 34, no. 2, pp. 185-199, June 2009.

[53] A.M. Grigoryan and Nan Du, "2-D images in frequency-time representation: Direction images and resolution map," *Journal of Electronic Imaging*, vol. 19, no. 3, (033012) pp. 1-14, July-September 2010.

[54] Nan Du and A.M. Grigoryan, "Paired transform method of image reconstruction from projections," in *Object Modeling, Algorithms and Applications*, edited by R.P. Barneva, V. Brimkov, et al., Research Publishing, June 18, 2010.

[55] A.M. Grigoryan, *Multidimensional Discrete Unitary Transforms*, book-chapter 19 (69 pages) in Transforms and Applications Handbook, 3rd ed., in the Electrical Engineering Handbook series (Editor in Chief, Prof. Alexander Poularikas) CRC Press, Taylor and Francis Group, Boca Raton, Florida, 2010.

[56] A.M. Grigoryan and Nan Du, "Principle of superposition by direction images," *IEEE Trans. on Image Processing*, vol. 20, no. 9, pp. 2531-2541, Sept. 2011.

[57] A.M. Grigoryan, "Comments on 'The discrete periodic Radon transform,'" *IEEE Trans. on Signal Processing*, vol. 58, no. 11, pp. 5962-5963, Nov 2010.

[58] T. Hsung, D.P.K. Lun, and W. Siu, "Reply to 'Comments on The Discrete Periodic Radon Transform,'" *IEEE Trans. on Signal Processing*, vol. 58, no. 11, pp. 5963-5964, Nov 2010.

[59] A.M. Grigoryan and Nan Du, "Comments on 'Generalised finite Radon transform for $N \times N$ images,'" *Image and Vision Computing*, vol. 29, pp. 797-801, November 2011.

Appendix A

AN ALGORITHM OF THE TWO-DIMENSIONAL
FOURIER TRANSFORM[1]

GRIGORYAN A. M.

A new algorithm for calculating the two-dimensional Fourier transform is proposed, by constructing the tensor of the third order. It is shown, that such an algorithm requires operations of multiplication by 1.6-2 times fewer, than known algorithms and does not require an additional temporary storage for saving intermediate results of calculations.

Consider an arbitrary array $\{f_{n,k}\}$ of the discrete signal, whose size for simplicity of exposition will be considered equal, i.e., $n,k = 1 : N$, for an integer number N.

At each point (p,s) of the spectrum, the Fourier transform of the signal up to the normalized factor equals

$$F_{p,s} = \sum_{n=1}^{N} \sum_{k=1}^{N} f_{n,k} W^{np+ks}, \tag{A.1}$$

where $W = W_N = \exp(2\pi i/N)$.

For arbitrary $p,s,t = 1 : N$, define the sets

$$V_{p,s,t} = \{(n,k); \, n,k = 1 : N, \, np + ks = t(\mathrm{mod}\, N)\} \tag{A.2}$$

and let

$$f_{p,s,t} = \sum_{V_{p,s,t}} f_{n,k} \,. \tag{A.3}$$

Then, due to (A.2) and (A.3), at arbitrary point (p,s) of the spectrum we obtain

$$F_{p,s} = \sum_{t=1}^{N} f_{p,s,t} W^t. \tag{A.4}$$

Thus, by constructing the tensor of the 3rd order $\{f_{p,s,t}; \, p,s,t = 1 : N\}$ by (A.3), we represent the spectrum of the signal at each point (p,s) in the form of $F_{p,s} = (f_{p,s,1}, f_{p,s,2}, ..., f_{p,s,N})$.

Such tensor representation is one-to-one and is characterized by the following property. If for any integer l we denote $\bar{l} = l(\mathrm{mod}\, N)$, then for any p,s and k

$$F_{\overline{kp},\overline{ks}} = \sum_{t=1}^{N} f_{p,s,t} W^{tk}. \tag{A.5}$$

Indeed, due to definitions (A.2) and (A.3), as well the property of periodicity of the discrete Fourier transform, we have

$$\sum_{t=1}^{N} f_{p,s,t} W^{tk} = \sum_{t=1}^{N} \left(\sum_{V_{p,s,t}} f_{n,m} \right) W^{tk} = \sum_{n=1}^{N} \sum_{m=1}^{N} f_{n,m} W^{(np+ms)k} = F_{kp,ks} = F_{\overline{kp},\overline{ks}},$$

[1] *Vol. 27, no. 10, Izv. vuzov MV and SSO SSSR. Radioelectronica, 1984. UDK 517.443*

because, for arbitrary p, s, the sets $V_{p,s,t}$ divide the domain of definition of the spectrum $X_N = \{(n, m); n, m = 1 : N\}$ in such a way, that $X_N = \bigcup_{t=1}^{N} V_{p,s,t}$ and $V_{p,s,t_1} \cap V_{p,s,t_2} = \emptyset$ when $t_1 \neq t_2$.

Formula (A.5) means for fixed values of (p, s), from components $f_{p,s,1}, \ldots, f_{p,s,N}$ of the tensor, by means of the Fourier transform, one can obtain the spectrum at points $(\overline{kp}, \overline{ks})$ which compose the cyclic group $T_{p,s}^N$ with the generator (p, s), i.e.,

$$T_{p,s}^N = \{(\overline{kp}, \overline{ks}); \ k = 1 : N\}. \tag{A.6}$$

For instance, for $(p, s) = (1, 1)$, from components of the tensor $f_{1,1,t}$, we obtain the values of the spectrum of the signal at samples $(1, 1), (2, 2), \ldots, (N, N)$.

As an example, let N be a power of two, then, by performing the fast Fourier transform (FFT) over the sequence $f_{p,s,1}, f_{p,s,2}, \ldots, f_{p,s,N}$, we define due to (A.5) the values of the spectrum in the group $T_{p,s}^N$, by performing about $0.5N \log_2 N$ (the exact number $0.5N(\log_2 N - 2) + 1)^2$ operations of complex multiplications by non-trivial rotation factors (powers of the exponent W), i.e., an average for each sample of the spectrum

$$v_{p,s} = 0.5N \log_2 N / \operatorname{card} T_{p,s}^N \tag{A.7}$$

operations of multiplication; card means the cardinality (order) of the group.

If p, s and N do not have common dividers, then card $T_{p,s}^N = N$, therefore in each point of form $(\overline{kp}, \overline{ks})$, for such p, s, the spectrum of the two-dimensional signal is calculated by means of $0.5 \log_2 N$ operations of complex multiplication, i.e., the same number as is necessary in the case of the one-dimensional Fourier transform. It should be noted, that the existing algorithms of the two-dimensional FFT require $1.5 - 2$ times more similar operations [1,2].

For N prime, the similar assertion is valid and the corresponding estimate $v_{p,s}$ can be considered equal to (A.7), where N in the numerator of the formula is substituted by the nearest number equal to the power of two. Indeed, for such N, the algorithm of FFT does not exist and, therefore, as done often in most practical problems, the artificial lengthening of the processed sequence is performed by adding zeros to the minimum size which is equal to the power of two. Let us show how to use formula (A.5) for effective, in the sense of high speed, calculation of the complete spectrum of the two-dimensional signal.

It is obvious, that to do this it is necessary to select the values (p, s) of the samples in an optimal way, such that the corresponding groups $T_{p,s}^N$ cover the domain X_N with a minimum number of intersections. Indeed, for N equal to powers of 2, one can show that there is no such covering of the spectral domain

$$X_N = \bigcup_J T_{p,s}^N, \tag{A.8}$$

where J is a set in X_N, that the conditions of independence are fulfilled, i.e.,

$$T_{p,s}^N \cap T_{p_1,s_1}^N = (N, N) \tag{A.9}$$

for arbitrary (p, s) and (p_1, s_1) from the set J.

---

[2]It should be 0.5$N(\log_2 N - 3) + 2$, as in the paired FFT.   A.M. Grigoryan, Dec. 2010

For prime N such covering of the domain of the spectrum exists; for instance, as such a set one can take

$$J_p = \{(1,p),\, p = 1 : N\} \cup (N,1)$$

or

$$J'_p = \{(p,1),\, p = 1 : N\} \cup (1,N). \tag{A.10}$$

Indeed, the following lemma is valid.

L e m m a. If the number N is prime, then for any $l \neq N$

$$X_N = \sum_{k=1}^{N} {}^{\bullet} \left(T_{k,l}^{N} \setminus (N,N) \right) \stackrel{\bullet}{+} T_{l,N}^{N} = \sum_{k=1}^{N} {}^{\bullet} \left(T_{l,k}^{N} \setminus (N,N) \right) \stackrel{\bullet}{+} T_{N,l}^{N}, \tag{A.11}$$

where the signs $\sum^{\bullet}$ and $\stackrel{\bullet}{+}$ denote the union of disjoint sets.

P r o o f. For simplicity, we take $l = 1$, and let k_1 and k_2, $k_1 \neq k_2$, be such that $T_{k_1,1}^{N} \cap T_{k_2,1}^{N} \neq (N,N)$, i.e. there exists $(p_0, s_0) = n_1(k_1, 1) = n_2(k_2, 1)$, which however takes place only with fulfillment of the equalities $n_1 = n_2$ and $n_1 k_1 = n_2 k_2 (\bmod N)$, from which it follows that $n_1(k_1 - k_2) = 0 (\bmod N)$. And, because by the assumption, $n_1 \neq N$ and N is prime, the last equality when $k_1 \neq k_2$ could not be fulfilled. Similarly, one can prove that for any $k \in [1, N]$, we have $T_{k_1,1}^{N} \cap T_{1,N}^{N} = (N,N)$.

Then, because all sets $T_{k,1}^{N} \setminus (N,N)$ and $T_{1,N}^{N}$ are contained in X_N and together have the cardinality equal $N(N-1) + N = N^2 = \operatorname{card} X_N$, from here the validity of the decomposition (A.11) follows. The lemma has been proved.

Thus, the existence of the optimal covering of the domain X_N of the spectrum is proved for N prime. By similar reasoning, it can be proved that for N equal powers of two, the optimal covering of the domain X_N of the spectrum does not exist, in the sense of (A.8) and (A.9).

Therefore, as indicated above, for N, being a power of 2, in the covering (A.8) it is necessary to pick up such generators (p,s), whose corresponding cyclic groups intersect in minimum in the aggregate. One can prove that for such N it is enough to take, in (A.8), the set equal

$$J_r = \{(l,p),\, p = 1 - N\} \cup \{(2k,l),\, k = 1 : N/2\}$$

or

$$J'_r = \{(p,l),\, p = 1 - N\} \cup \{(l,2k),\, k = 1 : N/2\}, \tag{A.12}$$

whose power card $J_r = \operatorname{card} J'_r = 3N/2$ (l is prime).

Thus, due to (A.5), (A.8), (A.12), and (A.10), we find that the spectrum of the original signal can be defined completely by means of $3N/2$ one-dimensional Fourier transforms, for N equal to a power of two, and for prime N, it is enough to perform $(N+1)$ times such procedures.

The existing methods of calculation of the two-dimensional Fourier transform, which are based on the separability of the transform, perform $2N$ such procedures [3]. Therefore, from the point of view of computation, the proposed tensor method of calculation of the DFT gains in the number of the used one-dimensional FTs, when compared with the existing two-dimensional FFT, and the gain equals, for N being a power of two and prime number, respectively:

$$\gamma_N^r = 2N/(3N/2) = 4/3 \quad \text{and} \quad \gamma_N^p = 2N/(N+1), \tag{A.13}$$

from which one can see that for N prime, the number of non-trivial complex multiplications required for calculating the complete spectrum in the proposed method is almost twice fewer than in the existent methods of the DFT, and 4/3 times fewer for N equal the power of two.

At that, if for prime N the estimate of the relation in (A.13) is exact, for N equals a power of two, the number of complex multiplications can be reduced in the following way. Consider the groups covering the domain of the spectrum

$$T_{l,1}^N, \ T_{l,2}^N, \ldots, T_{l,N}^N, \ T_{2,l}^N, \ T_{4,l}^N, \ldots, T_{N,l}^N, \tag{A.14}$$

when $N = 2^r$, where r is an integer number, and l is any prime. It is not difficult to show that for all $p = 1 : N$, the following is valid:

$$T_{l,p}^N \cap T_{l,p+N/2^k}^N = \{(n,m); \ \exists k_1, \ n = 2^k k_1 l, \ m = 2^k k_1 p\}, \tag{A.15}$$

i.e., intersections of these groups is a subgroup which consists of all elements with coordinates multiple to 2^k. Indeed, for arbitrary integer k_1, we have $(\overline{2^k k_1 l}, \overline{2^k k_1 p}) = (\overline{2^k k_1 l}, \overline{2^k k_1 (p + N/2^k)})$ and on the contrary, if for some k_1 and k_2 we have $(\overline{k_2 l}, \overline{k_2 p}) = (\overline{k_1 l}, \overline{k_1 (p + N/2^k)})$, then the following conditions should be fulfilled: $(k_2 - k_1)l = 0 (\mathrm{mod}\, N)$, $(k_2 - k_1)p = k_1 N/2^k (\mathrm{mod}\, N)$, and because L is prime, N is the power of two, it follows from the first equality that $k_1 = k_2$, and from the second one that k_1 is the number multiple to 2^k.

Thus, the equality in (A.15) has been proved. For any k and $m = 1 : r$, we denote by $T_{l,k}^{N,m}$ the set

$$T_{l,k}^{N,m} = T_{l,k}^N \setminus \left(T_{l,k}^N \cap T_{l,k+N/2^m}^N \right). \tag{A.16}$$

Consider first the sets $T_{l,k}^{N,1}$, i.e., the $m = 1$ case. Due to (A.12) and (A.15), we obtain

$$X_N = \bigcup_{k=1}^{N/2} \left(T_{l,k}^N \cup T_{l,k+N/2}^{N,1} \right) \bigcup_{m=1}^{N/4} \left(T_{2m,l}^N \cup T_{2m+N/2,l}^{N,1} \right), \tag{A.17}$$

and it is obvious that card $T_{l,k}^{N,m} = (1 - 1/2^m)$card $T_{l,k}^N$.

The sets $T_{l,k}^{N,1}$, for $k = 1 : N$, consist of all odd elements of groups $T_{l,k}^N$ and have cardinality card $T_{l,k}^{N,1} = 0,5N$.

In [3], the algorithms of the FFT with decimation in frequency are described, which after the first stage of iteration, when the N-point transform is divided by two $N/2$-point Fourier transforms, all values of the spectrum with odd and even coordinates are defined separately. Therefore, using such algorithms of the FFT for calculating the two-dimensional spectrum by formula (A.5), one can define the values of the spectrum at the sets $T_{l,k}^{N,1}$, by performing twice fewer operations of complex multiplication than for the corresponding group $T_{l,k}^N$.

Thus, because of the selected covering (A.17), we find that the number of operations of complex multiplication sufficient for calculating the two-dimensional Fourier transform by the tensor method is no greater than the number

$$v_1 = N/2 v_0 + N/2\, 1/2 v_0 + N/4 v_0 + N/4\, 1/2 v_0 = 9/8 N v_0, \tag{A.18}$$

where v_0 is the number of operations of multiplication by non-trivial rotated factors, which are necessary for the one-dimensional FFT with decimation by frequency, equals, as was already mentioned above,

$$v_0 = N/2(\log_2 N - 2) + 1. \tag{A.19}$$

Therefore, substituting (A.19) in (A.18), we have the following: $v_1 < 9/16N^2(\log_2 N - 2)$.

Continuing with similar reasoning, by means of (A.15) one can also exclude intersections in the groups $T_{l,k}^N$, where $k = 1 : N/2$, and $T_{m,l}^N$, where $m = 1 : N/2 - 1$. For instance, in the next stage, when $m = 2$, we obtain the decomposition

$$X_N = \bigcup_{k=1}^{N/4} \left(T_{l,k}^N \cup T_{l,k+N/4}^{N,2} \right) \bigcup_{k=1}^{N/2} T_{l,k+N/2}^{N,1} \cup,$$

$$\bigcup_{m=1}^{N/4} T_{2m+N/2,l}^{N,1} \bigcup_{m=1}^{N/8} \left(T_{2m,l}^N \cup T_{2m+N/4,l}^{N,2} \right), \tag{A.20}$$

in addition, due to (A.15), card $T_{l,k}^{N,2} = 3N/4$.

As in the $m = 1$ case, if the spectrum of the signal was defined in the group $T_{l,k}^{N,1}$, then for calculating the spectrum in the set $T_{l,k}^{N,2}$, it is enough on the second stage of iteration in the algorithm of the FFT with decimation by frequency, after when two $N/2$-point Fourier transforms are divided by four $N/4$-point transforms, to use only the first three $N/4$-point Fourier transforms. Therefore, to define the spectrum in $T_{l,k}^{N,2}$, it is enough to perform $4/3$ times fewer operations of the complex multiplication than for the corresponding $T_{l,k}^{N,1}$.

Thus due to (A.20), on the second stage of improvement, when $m = 2$, the number of operations of multiplication for calculating the DFT by the tensor method will not be greater than the number

$$v_2 = N/4v_0 + N/4 \cdot 3/4v_0 + N/2 \cdot 0.5v_0 + N/8v_0 + N/8 \cdot 3/4v_0 + N/4 \cdot 0.5v_0 = 33/32Nv_0.$$

Similarly, on the third stage, when $m = 3$, we obtain the corresponding number $v_3 = 129/128Nv_0$ and so on; if continue with $m = 4, 5, ..., r$, for large N, we find that the number of necessary operations of multiplication by the rotation factors in the algorithm is approximated from above by the number $v = (N + 1)/N\, Nv_0 = (N + 1)v_0$.

Because the known methods of two-dimensional FFT, which use the property of separability of the two-dimensional transform, perform $2N$ one-dimensional FFTs, the number of necessary operations of multiplications for them equals $v = 2Nv_0$.

It follows that the proposed algorithm of calculation of the DFT, by the number of operations of multiplication, is more than χ_N times effective, where $\chi_N = 2N/(N + 1)$, than the mentioned algorithms; therefore, for large N, we have $\chi_N \approx 2$.

If we compare the proposed algorithm of the DFT with the known algorithm of the two-dimensional FFT, which uses the two-dimensional "butterfly" (which is similar to the one-dimensional "butterfly"; see, for instance [1]), the number of multiplications for which, as is not difficult to show, equals $v' = 0.25\,N^2(3\log_2 N - 4) + 1$, then the corresponding coefficient of effectiveness by the number of the specified operations will be:

$$\chi_N' = \frac{v'}{v} = \frac{1}{4} \frac{N^2(3\log_2 N - 4) + 1}{(N + 1)(N/2(\log_2 N - 2) + 1)}.$$

Hence, we obtain $\chi_N' > 1.6$ for large N; for instance, for values of N equal 128 and 1014, we have $\chi_{128}' > 1.68$ and $\chi_{1024}' > 1.62$.

Thus, the proposed tensor algorithm of calculation of the two-dimensional Fourier transform is more than $1.6 - 2$ times as effective as all known algorithms of the FFT, by the number of necessary operations of complex multiplication. It should also be noted that, if in the one-dimensional FFTs $\omega_0 = v_0/N$ operations of multiplication on average are used for each sample of the spectrum, in the tensor algorithm of the DFT in average per sample $\omega_1 = v/N^2$ operations of multiplication are used, i.e., $\omega_1 = (N+1)/N^2 v_0 = (N+1)/N\omega_0$; hence, $\omega_1 > \omega_0$ and, for large values of N, we have $\omega_1 \approx \omega_0$.

Thus, in the proposed algorithm of calculation of the DFT, the number of operations that are used for each sample of the spectrum is the same as in the algorithm of the one-dimensional Fourier transform, for any large size of the signal, $N \times N$, where N is a power of 2.

Since for primes N in the tensor method of the DFT, it is enough to perform $(N+1)$ one-dimensional Fourier transforms, in this case the number of multiplications on average per sample also equals

$$\tilde{\omega}_1 = (N+1)\tilde{v}_0/N^2 = (N+1)/N\tilde{\omega}_0 \qquad (A.21)$$

where $\tilde{v}_0$ denotes the number of multiplications required for calculating the N-point FT, and $\tilde{\omega}_0 = \tilde{v}_0/N$. As already mentioned, one can consider $\tilde{v}_0$ to be equal to the estimate in (A.19), for the corresponding nearest to the prime N size equal to a power of two. According to the Lemma on covering of the spectrum domain, the estimate in (A.21) cannot be improved, and therefore is the minimum.

In comparison with the existing algorithms of the two-dimensional FFT, the proposed algorithm is also more effective by the volume of required temporary storage. Indeed, in the known algorithms of the two-dimensional FFT, it is necessary to have additional temporary storage for constantly storing the intermediate results of computing during the entire process of calculation. The volume of such storage exceeds a few times the volume of the original signal, which essentially reduces the possibility of using the algorithms of the two-dimensional FFT for the large size of the signal. In the proposed tensor algorithm, such additional storage is not required, because the values of the spectrum in each group of samples are calculated right after defining the corresponding components of the tensor.

Thus, the existing methods of calculation of the two-dimensional Fourier transform yield the tensor method by the main characteristics. In spite of the fact that the sizes of the two-dimensional signal were assumed equal, one can easily extend the tensor method of calculation of the spectra in the general case. Moreover, similar reasoning can be transferred to the case of the n-dimensional Fourier transform, where $n > 2$, as well.

References

[1] Sereda L. A. An algorithm of fast calculation of the two-dimensional discrete Fourier transform. *Izv. vuzov SSSR, Radioelectronica,* 1983, vol. 26, no. 7, pp. 18–22.
[2] Gold B., Raider C. Digital processing of signals / Tran. from English, M.: Sov. radio, 1973, 367 p.
[3] Yaroslavsky L. P. Introduction to digital image processing. M.: Sov. radio, 1979, 312 p.

Appendix B

NEW ALGORITHMS FOR COMPUTING DISCRETE FOURIER TRANSFORMS[1]

GRIGORYAN A. M.

(Yerevan)

Effective methods of calculation of the multi-dimensional discrete Fourier transform are proposed, which are based on its new representation.

In this work, a new general approach to consideration of the multi-dimensional discrete Fourier transform (MDFT) is described, the main idea of which is a unique possibility of presentation of each component of the transform by the corresponding one-dimensional vector. This approach allows for performing separate calculations of the MDFT in each disjoint with other groups of samples, on which the whole domain of definition of the spectrum is divided, which allows for constructing the effective algorithms of calculation of the MDFT by means of the minimum number of one-dimensional DFTs.

The case of the two-dimensional Fourier transform is described in detail, and the corresponding algorithms are compared with the most current algorithms of calculation of the two-dimensional DFT, which are based on the method of decimation by Cooley-Tukey [1],[2], the polynomial transforms [3], and the operation of the two-dimensional "butterfly." As the particular case, the corresponding algorithm for the one-dimensional Fourier transform is considered, too.

1. Vector representation of the spectrum of the MDFT

Consider an arbitrary array $\{f_{k_1,...,k_n}\}$ for the n-dimensional discrete signal, whose size for simplicity of exposition will be considered equal, i.e., $1 \leq k_i \leq N$, $i = 1, 2, ..., n$, for an integer number N. Each spectral component at sample $(p_1, ..., p_n)$, where $p_i \in Z_N^1 = 1, 2, ..., N$, $i = 1, 2, ..., n$, which up to the normalized factor equals

$$F_{p_1,...,p_n} = \sum_{k_1=1}^{N} \cdots \sum_{k_n=1}^{N} f_{k_1,...,k_n} W^{k_1 p_1 + ... + k_n p_n}, \tag{B.1}$$

where $W = W_N = \exp(2\pi i/N)$, can be represented in the form of the N-dimensional vector

$$\bar{F}_{p_1,...,p_n} = (f_{p_1,...,p_n,1}, ..., f_{p_1,...,p_n,N}), \tag{B.2}$$

for which

$$F_{p_1,...,p_n} = \sum_{t=1}^{N} f_{p_1,...,p_n,t} W^t.$$

For that, as follows from (B.1), each component of the vector (B.2) must be

---

[1] Journal of Computation Mathematics and Mathematical Physics, AS USSR. Volume 26, No. 9, 1986. *SCIENTIFIC REPORTS*, UDK 517.97:537.812

calculated by summing values of the original signal at points of the corresponding set:

$$V_{p_1,\dots,p_n,t} = \left\{ (k_1, \dots, k_n);\ 1 \le k_i \le N,\ i = 1, 2, \dots, n,\ \sum_{i=1}^{n} k_i p_i = t \bmod N \right\}, \quad (B.3)$$

i.e.

$$f_{p_1,\dots,p_n,t} = \sum_{V_{p_1,\dots,p_n,t}} f_{k_1,\dots,k_n}. \quad (B.4)$$

It is obvious that for a fixed value of point $(p_1, \dots, p_n)$, the sets (??) for different $t = 1, 2, \dots, N$ together cover the whole domain of definition Z_N^n of the original signal without intersecting, therefore we have

$$F_{\overline{kp_1},\dots,\overline{kp_n}} = \sum_{t=1}^{N} f_{p_1,\dots,p_n,t} W^{kt} \quad (B.5)$$

for any integer k; here $\overline{kp_i} = (kp_i) \bmod N$. It follows from (B.5), that the Fourier transform over the vector (B.2), which corresponds to the spectral component at point $(p_1, \dots, p_n)$, defines the values of the n-dimensional spectrum at all points of the group with generator $(p_1, \dots, p_n)$:

$$T_{p_1,\dots,p_n} = \{ (\overline{kp_1}, \dots, \overline{kp_n});\ k = 1, 2, \dots, N \}. \quad (B.6)$$

By selecting the minimum number of points of the spectrum, whose corresponding groups (B.6) together would cover the whole domain Z_N^n of definition of the spectrum, one can construct an effective algorithm of calculation of the n-dimensional MDFT. It should be noted that in the particular case, when $n = 1$, formula (B.5) degenerates into the original formula (B.1) for odd points p ($T_p = Z_N^1$), and therefore it is not very interesting, but for any $n > 1$ it is useful when calculating the spectrum. Below, as an example, the $n = 2$ case is described.

2. Algorithm of computation of the two-dimensional DFT

For the spectrum of the two-dimensional discrete signal $\{f_{k_1,k_2}\}$, due to (B.5), at any point (p_1, p_2) with an integer k, we have

$$F_{\overline{kp_1},\overline{kp_2}} = \sum_{t=1}^{N} f_{p_1,p_2,t} W^{kt}. \quad (B.7)$$

Consider the cases when the size of the signal N is an arbitrary prime number or power of two. It is not difficult to show that for N prime, the totality of groups (B.6), which corresponds to the points of the set $J_p = \{(1,1), (1,2), \dots, (1,N), (N,1)\}$, cover the whole domain Z_N^2 in optimal way, in the sense that all covering groups in such totality intersect between themselves only at the point (N, N). In the case when N equals a power of two, such optimal covering does not exist. Indeed, one can be easily convinced that in any totality of groups (B.6) that entirely cover the domain of definition of the spectrum, there will, without fail, be intersections in the even points as well. Moreover, each such point belongs to at least two different groups of that totality, besides each covering totality contains no less than $3N/2$ groups (B.6), because for powers of two N one can fulfill the covering of the domain of definition of the spectrum, for instance, by the totality of groups corresponding to points of the set $J_r = \{(1,1), (1,2), ..., (1,N), (2,1), (4,1), (6,1), ..., (N,1)\}$.

Thus, the two-dimensional $(N \times N)$-point DFT can be calculated completely by formula (B.7), by means of $N+1$ and $3N/2$ one-dimensional N-point DFTs, for N equal to arbitrary prime numbers and powers of two, respectively.

Such algorithms are more effective than the known algorithms from [1],[2], because the latter perform the Fourier transform first by all rows, and then by the columns of the obtained signal, i.e., perform $2N$ indicated transforms. From here it follows that the proposed algorithms perform 2 or 4/3 times fewer operations of multiplication, if N is prime or a power of two, respectively. Further, the calculation at each group of points (B.6) is performed by formula (B.7) independently of calculation at other groups; therefore, the large additional storage for saving intermediate results is not required in the proposed algorithms. It also is obvious that when computer-aided programming such algorithms, all accesses to memory, where the original signal is stored, can be fulfilled only by rows and columns.

It should be noted that the analogous result for prime N has been obtained in the reducible algorithm [3] of calculation of the two-dimensional DFT, by means of more complicated polynomial transformations.

In the case when N is a power of 2, in such algorithm the calculation of the $(N \times N)$-point DFT is fulfilled by means of the polynomial transforms and not only $3N/2$ reducible N-point DFTs, but also $(N/2 \times N/2)$-point DFT, which in its turn can be calculated in a similar way by $3N/4$ reducible $N/2$-point DFTs and one $(N/4 \times N/4)$-point DFT, and so on.

The proposed algorithm of the two-dimensional DFT for N being powers of two is also more effective, by the number of operations of multiplication, than the known algorithms which use the operation of the two-dimensional butterfly (see, for instance, [4]), where $\bar{v}_{N,N} = (N^2/4)(3\log_2 N - 4) + 1$ such operations are required. Indeed, because the N-point DFT, when fulfilled by the algorithm of the FFT, requires $(N/2)(3\log_2 N - 3) + 2$ non-trivial complex operations of multiplication, then the proposed algorithm of calculation of the two-dimensional DFT requires $v_{N,N} = 3/4N^2(3\log_2 N - 3) + 3N$ such operations. Hence, for $N \geq 8$, we have $\bar{v}_{N,N} - v_{N,N} = 5/4N^2 - 3N > 0$.

Similar conclusions are valid for the algorithms of calculation of the n-dimensional DFT, which are based on (B.5) for the $n > 2$ case, too. As already mentioned, for N equal to a power of two, groups (B.6), which together cover the domain of definition of the spectrum Z_N^n, define in their intersections all even points of the spectrum, the number of which equals $(N/2)^n$. In this connection, the essential reduction of operations of multiplication at such points can be achieved by removing the repeated calculations of the MDFT by formula (B.5).

For instance, in the two-dimensional case, such improvement of the described algorithm is reduced to the fact that the $(N \times N)$-point DFT is calculated by means of $3N/2$ incomplete N-point DFT, which decreases the number of operations of multiplication almost by 1.5 times.

Below, a more effective algorithm of calculation of the n-dimensional DFT is proposed for the case when N is a power of two.

3. Paired representation of the spectrum of the MDFT

Consider the vector representation of the arbitrary component of the spectrum, which was defined in §1. Because for all $t \in \{1, 2, ..., N/2\}$, the following takes place $W^t = -W^{t+N/2}$, one can represent uniquely the component (B.1) at each point $(p_1, ..., p_n)$ by the $N/2$-dimensional vector

$$\bar{F}'_{p_1,...,p_n} = (f'_{p_1,...,p_n,1}, ..., f'_{p_1,...,p_n,N/2}), \tag{B.8}$$

where each component is calculated from (B.4):

$$f'_{p_1,...,p_n,t} = f_{p_1,...,p_n,t} - f_{p_1,...,p_n,t+N/2}, \tag{B.9}$$

where $t = 1, 2, ..., N/2$, and for which

$$F_{p_1,...,p_n} = \sum_{t=1}^{N/2} f'_{p_1,...,p_n,t} W^t.$$

Moreover, as follows from (B.5), for an arbitrary odd value of k the following takes place

$$F_{\overline{kp_1},...,\overline{kp_n}} = \sum_{t=1}^{N/2} f'_{p_1,...,p_n,t} W^{kt}.$$

By opening here the exponential factors for odd values of $k = 2m - 1$, where m is any integer, we obtain

$$F_{\overline{kp_1},...,\overline{kp_n}} = \sum_{t=1}^{N/2} \left(f'_{p_1,...,p_n,t} W_N^{-t}\right) W_{N/2}^{mt}. \tag{B.10}$$

Thus, by fulfilling the $N/2$-point Fourier transform over the vector

$$\overline{G(F)}_{p_1,...,p_n} = (f'_{p_1,...,p_n,1} W_N^{-1}, ..., f'_{p_1,...,p_n,N/2} W_N^{-N/2}), \tag{B.11}$$

we obtain all values of the spectrum of the signal at points composing the odd half of the corresponding group (B.6):

$$T'_{p_1,...,p_n} = \{(\overline{(2m-1)p_1}, ..., \overline{(2m-1)p_n}); \ m = 1, 2, ..., N/2\}. \tag{B.12}$$

For an example, for the point $(1, ..., 1)$ by the Fourier transform over the components of the vector $\overline{G(F)}_{1,...,1}$ the spectral components at the points $(1, ..., 1)$, $(3, ..., 3)$, ..., $(N-1, ..., N-1)$ are defined.

By using formula (B.10), we construct the algorithm for computing the complete n-dimensional DFT. As in the algorithms based on the method of computation of the MDFT by (B.5), it is obvious one need select a minimum number of points whose corresponding sets (B.12) together cover the entire domain Z_N^n of definition of the spectrum.

Unlike the method described in §1 for calculation of the MDFT, this method of calculation, which is based on the representation of each spectral component in form of (B.8), which we call the paired representation, makes it possible to cover the domain of definition of the spectrum, for N equal powers of two, optimally, i.e., such that the following conditions will be fulfilled

$$Z_N^n = \bigcup_J T'_{p_1,...,p_n}, \qquad T'_{p_1,...,p_n} \cap T'_{g_1,...,g_n} = \emptyset, \tag{B.13}$$

where the union is performed by points of some subset J in Z_N^n, and the condition of non-intersection is fulfilled for any different points of this set.

We show this with examples with the two-dimensional and one-dimensional DFTs, whose corresponding algorithms are described below. More general cases of the MDFT are considered in a similar way.

4. Paired algorithm of the two-dimensional DFT

Considering the two-dimensional case, we take the set J in (B.13) equal to

$$J^2 = J_N \cup 2J_{N/2} \cup 4J_{N/4} \cup \ldots \cup NJ_1, \tag{B.14}$$

where $kJ_{N/k} = \{(kp_1, kp_2); (p_1, p_2) \in J_{N/k}\}$, for all $k = 2^r$, $r = 0, 1, ..., \log_2 N$, and $J_{N/k}$ are the above defined sets $\{(1, p_2), p_2 = 1, 2, ..., N/k\} \cup \{(2p_1, 1), p_1 = 1, 2, ..., N/(2k)\}$ all powers of two k from 1 to $N/2$; for $k = N$ we have $J_1 = \{1, 1\}$. It is easy to show that in set (B.14) when $k_1 \neq k_2$ will be $J_{N/k_1} \cap J_{N/k_2} \neq \emptyset$, and thus J^2 satisfies the condition of optimality (B.13) of covering the domain of definition of the spectrum. Further, the cardinalities of sets (B.12) in (B.13) for such J equal $\operatorname{card} T'_{p_1, p_2} = 0.5N/2^r$, if $(p_1, p_2) \in 2^r J_{N/2^r}$, where $r = 0, 1, ..., \log_2 N$, and for the sets in (B.14) we have $\operatorname{card} J_{N/2^r} = 1.5N/2^r$. Therefore, it follows easily from (B.14), $\operatorname{card} J^2 = 3N - 2$.

Thus, in the optimal covering (B.13), which unites $3N - 2$ sets (B.12), $3N/2$ of them contain $N/2$ elements each, $3N/4$ sets contain $N/2$ elements each, and so on. Therefore, to define the complete spectrum of the two-dimensional signal of size $N \times N$, it is enough to fulfill $3N/2$ one-dimensional $N/2$-point DFTs, $3N/4$ one-dimensional $N/4$-point DFTs, and so on.

Indeed, let the arbitrary point (p_1, p_2) from J^2 belong to the set $2^r J_{N/2^r}$ for some $r \in [0, \log_2 N]$, i.e., g.c.d.$(p_1, p_2) = 2^r$. Due to definitions (B.3),(B.4) and (B.9), it is obvious, that in the corresponding vector $\overline{G(F)}_{p_1, p_2}$, all components with numbers not multiple to the number 2^r equal zero. Therefore, one can associate uniquely to such vector the $N/2^{r+1}$-dimensional vector $\overline{G(F)}_{p_1, p_2}^r = (g_{p_1, p_2, 1}, g_{p_1, p_2, 2}, ..., g_{p_1, p_2, N/2^{r+1}})$, wherein the components $g_{p_1, p_2, t} = f'_{p_1, p_2, 2^r t} W_{N/2^r}^{-t}$ for all $t = 1 - N/2^{r+1}$. At that, for any values of m, the following equality is valid

$$\sum_{t=1}^{N/2} \left(f'_{p_1, p_2, t} W_N^{-t} \right) W_{N/2}^{mt} = \sum_{t=1}^{N/2^{r+1}} g_{p_1, p_2, t} W_{N/2^{r+1}}^{mt}. \tag{B.15}$$

Therefore the $N/2$-point DFT of the vector $\bar{F}'_{p_1, p_2}$, for arbitrary point $(p_1, p_2) \in 2^r J_{N/2^r}$, where $r \in \{0, 1, ..., \log_2 N\}$, is equivalent to fulfillment of the $N/2^{r+1}$-point DFT over the corresponding vector $\overline{G(F)}_{p_1, p_2}^r$. It follows from here, that the results of $3N - 2$ one-dimensional DFTs of order $N/2$, which are sufficient for calculating the complete two-dimensional DFT by the proposed algorithm with formula (B.10), are equivalent in union to the results of fulfillment of the mentioned DFTs: $3N/2$ of order $N/2$ each, $3N/4$ of order $N/4$ each, and so on.

Calculate the volume $v'_{N,N}$ of non-trivial operations of multiplication by the rotated exponential factors, which are necessary for fulfillment of the proposed algorithm of calculation of the $(N \times N)$-point DFT. If we denote by v_M (with arbitrary power of two M) the volume of such operations for the M-point DFT, then it is obvious, that from (B.10) and (B.15) we obtain

$$v'_{N,N} = (3N/2)(v_{N/2} + N/2 - 2) + (3N/4)(v_{N/4} + N/4 - 2) + \dots$$

$$\dots + 3 \cdot 2(v_2 + 2 - 2) = 3N \sum_{r=1}^{\log_2 N - 2} 2^{-r} v_{N/2^r} + N^2 - 6N + 8. \tag{B.16}$$

This formula of estimation of the number of non-trivial operations of multiplication, required for calculation of the two-dimensional DFT by means of the one-dimensional transforms, due to optimality of the covering (B.13) is exact, i.e., can not be improved. Therefore if one has minimum estimates of such operations for the one-dimensional DFT, one can define by means of (B.16) a minimum number of complex operations for the two-dimensional DFT, too. If we estimate all $v_{N/2^r}$ by means of the one-dimensional algorithms from [1],[2], for which these estimates equal, as already mentioned, $(N/2^{r+1})[\log_2(N/2^r) - 3] + 2$, then it is not difficult to obtain

$$v'_{N,N} = \frac{N^2}{2}\left(\log_2 N - \sum_{r=1}^{\log_2 N - 2} \frac{r}{4^r} - 1\right) - 8(\log_2 N - 1),$$

from which it follows, that for $N > 8$

$$v'_{N,N} \approx (N^2/2)(\log_2 N - 5/2) - 8(\log_2 N - 1). \tag{B.17}$$

When compared with the algorithm described in §1 with formula of calculation (B.5), which performs $3N/2$ one-dimensional N-point DFTs by means of $v_{N,N} = (3N/2)v_N$ operations of multiplication, we find that $v_{N,N} > 1.5v'_{N,N}$. Thus, by the number of operations of multiplication the described here algorithm, which we call paired, is effective more than 1.5 times the algorithm in §1.

5. Algorithm of the one-dimensional DFT

Consider the $n = 1$ case, which corresponds to the one-dimensional Fourier transform. The set J in (B.13), by which one must construct the algorithm based on the paired representation of the spectrum, it obviously can be taken as equal to $J^1 = \{1, 2, 4, 8, ..., N\}$ with cardinality $\log_2 N + 1$. This set represents the projection of the set J^2 defined above on the one-dimensional domain Z_N^1. At that, $T'_1 = \{1, 3, ..., N-1\}$, ..., $T'_m = \{m, 3m, ..., N-m\}$, for $m = 2^r$, $r = 1, 2, ..., \log_2 N - 1$, and $T'_N = \{N\}$.

Similar to the $n = 2$ case, we find that for calculating the N-point DFT it is enough to fulfill one $N/2$-point DFT over the vector $\overline{G(F)}_1$, one $N/4$-point DFT over the vector $\overline{G(F)}_{2^1}$, and so on, one $N/2^r$-point DFT over the vector $\overline{G(F)}_{2^r}^{2^{r+1}}$, where $r = 1, 2, ..., \log_2 N - 1$; here,

$$\overline{G(F)}_{2^r}^{2^{r+1}} = \left(f'_{2^r,1} W_{N/2^r}^{-1}, \; f'_{2^r,2} W_{N/2^r}^{-2}, \; ..., f'_{2^r,N/2^r} W_{N/2^r}^{-N/2^{r+1}}\right).$$

At that their components are calculated easily; for example, for $r = 0, 1$, we have $f'_{1,t} = f_t - f_{t+N/2}$, $t = 1, 2, ..., N/2$, $f'_{2,t} = f_t + f_{t+N/2} - f_{t+3N/4} - f_{t+N/4}$, $t = 1, 2, ..., N/4$. Consequently, the number of non-trivial operations of multiplication in this algorithm is calculated by the following recurrent formula for $N > 8$:

$$v'_N = (v'_{N/2} + N/2 - 2) + (v'_{N/4} + N/4 - 2) + \dots + (v'_8 + 6) + v'_8;$$

from here the estimate for multiplications follows:

$$v'_N = 2v'_{N/2} + N/2 - 2 = (N/2)(\log_2 N - 3) + 2.$$

Thus, the same estimate of operations of multiplication has been obtained as in the algorithms from [1],[2], which was used above when deriving estimate (B.17) for the two-dimensional DFT.

It should be noted that the described method of calculation of the one-dimensional DFT, which is based on the paired representation of the latter, allows for constructing algorithms for calculation of the DFT with a minimum number of the same transforms but smaller orders for any orders of the transform, which are even numbers. Indeed, the existence of the necessary optimal covering of the domain of definition of the spectrum in form (B.12) for any even order N can be proved by the simple construction of sets J in (B.13). And such set J for any even N can be taken as equal to $J = \{1, 2, 4, 8, ..., 2^r\}$, where 2^r is the greatest power of two, which is the division of N. It is clear that in the general $n > 2$ case, when calculating the n-dimensional DFT by the method based on the paired representation, the algorithms using the minimal number of the one-dimensional DFTs can be constructed similar to the algorithms described above for the $n = 1, 2$ cases, and for any orders whose at least one dimension is the even number.

It should be noted in conclusion that with corresponding organization of the procedure of fulfillment of the paired representation of the necessary spectral components of the signal by the specialized processors, one can realize the process of calculation of the MDFT in a time period that is close to the time of fulfillment of the $N/2$-point DFT. Indeed, the process of calculation of the complete MDFT by the mentioned algorithm is divided into separate procedures of calculation of the spectrum for each disjoint group of samples, which all together define the whole domain of definition of the spectrum, and the maximum order of the one-dimensional DFT by means of which the values of the spectrum of the original signal are defined at each such group equals $N/2$.

References

[1] Rabiner L., Gold B. *Theory and application of the digital signal processing.* M.: Mir, 1978.
[2] Yaroslavsky L. P. *Introduction to digital image processing.* M.: Sov. radio, 1979.
[3] Nussbaumer, H. J. Calculation of two-dimensional convolutions and discrete Fourier transforms, in *Fast algorithms in digital image processing.* M.: Radio and communication, 1984, pp. 43-88.
[4] Sereda L. A. An algorithm of fast calculation of the two-dimensional discrete Fourier transform. *Izv. vuzov SSSR, Radioelectronica*, 1983, vol. 26, no. 7, pp. 18–22.

Index

Printed in the United States
by Baker & Taylor Publisher Services